Price $ 29.95

ISBN: 13: 978-1892530103

Printed in the United States of America

Book of Astrology by LIGIA BALU

CHINESE ASTROLOGY - HOW TO FIND YOUR SOUL-MATE, STARS, AND DESTINY

Visit: www.ligiabalu.com

FORWARD

Astrology means the study of the stars. It is an ancient course of prehistoric times which concentrates on the correlation between celestial even events. *HOW TO FIND YOUR SOUL-MATE, STARS AND DEST* on traditional research and offers a basic introduction to this age old science. It is know how to find one's soul mate and destiny, and one important aspect of that sear understanding of one's self. Understanding ourselves and our motivations sets us better understanding the other persons in our lives and those persons we come in cont you are very careful in your decision making, when you learn to match the signs by months, and years with your own sign, you are less likely to make mistakes in choosing soul mate, and destiny for life. This is a unique book designed for the reader to unnecessary mistakes in life: broken hearts, broken families, lost loves, lonely times, cr screaming over loved ones. Through the study of Astrology, you the reader, may come to understanding of the problems you have suffered in life. Perhaps you have found yourself why other people appear happy together and why other people appear to have so few pro Perhaps you have asked yourself, "Why not me? What is wrong with me? Why can't happy? I have many good qualities just like other people. Why am I attracted to the w person? Why do bad things happen to me?" By thoroughly understanding not only conscious self but your subconscious reasoning (often referred to in Astrology as unconscious), you may finally find the answers you have been searching for. Just as effective you can study and begin to understand persons born to other signs as well. Why does a person a or behave the way they do, either obstinate, or changeable, or impulsive? Perhaps a basic answe to that question can be found in the nature of the Sun sign.

The Zodiac refers to the "Circle of Animals" or the "Circle of Life" and is the name for the belt of constellations viewed in the celestial heavens. It is made up of twelve signs: Aries, Taurus, Gemini, Cancer, Leo, Virgo, Libra, Scorpio, Sagittarius, Capricorn, Aquarius, and Pisces. These are referred to as the Sun signs or the sign which the Sun was in on the date a person was born. The Sun, as the most important celestial body, is considered the greatest influence in one's personal horoscope.   A further understanding of Astrology, however, depends upon an understanding of one's Moon sign, Ascendant, Houses, Planets, and the Aspects these Planets make to one another. Each of these influences the person as well which is why two people with the same Sun sign can be so different. *STARS AND DESTINY* provides the reader with an explanation of each of these influences. And because each person is different, *STARS AND DESTINY* presents both the positive and the negative characteristics of each of these influences. And it is the combination of these positives and negatives that produce the individual. And it is left to each individual to determine how to react to these positives and negatives. In other words,

...erson may use assertiveness to develop leadership qualities while another becomes ...sive, obstinate, and even hostile. Life is a long road of self-improvement, but the basics ...n the same. How one person reacts and relates to another depends to a large degree on each ...n's security and self-esteem. That is what this book strives to explain: Why we are the way ...e.

Over the last twenty years, in my journey through life, many friends, co-workers, ...aintances, and others who chose the wrong soul mate for better or for worse have called me ...counseling. Because love can be blind, these situations often change into disaster. I have ...erved over time that these decisions lead to misunderstandings, disappointments, tragedies, ...rce, and loneliness. What suffers the most is the creation of love. The resulting lonely time is ...a life to live. Life is short, and this is no way to find happiness or to be content. A broken ...rt can lead a person to believe that all men and/or women are the same, and it is no use to ...ntinue looking for the right one. This attitude can result in the person choosing to be alone for a ...g time or to attempt to protect themselves and the creations of love: their children. Rather ...an trying again to find the right partner, they choose to remain alone when it, otherwise, would ...e possible to be happy and never lonely. It may be difficult to believe, but Astrology can help ...eople learn the lessons they need to know in order never to be lonely. The simple truth comes ...rom God above. God doesn't like to see us alone. He never intended people to go through life ...by themselves, but instead, He planned for each of us to discover our true soul mates and to go ...through life - and eternity - together. This is simple logic because the soul mate is the other half of ourselves.

In life, it is easy to make mistakes especially if one is attracted to the wrong person in the wrong place for the wrong reasons. We are all really looking for love rather than to be left lonely and by ourselves. Why have the men and women in your life been mistakes? When thinking about love, remember it is the most powerful force in life. It is nice to have someone to share your life with. Success is an empty bag when you have no one to share it with. You ask yourself, "What is wrong with me?" Nothing is wrong with you. You have just been attracted to the wrong person. It is difficult to find Mr. or Mrs. Right when you are attracted to Mr. or Mrs. Wrong. Once in this relationship, you struggle to make it work. Many times, when we are caught up in these wrong relationships, we miss the opportunity to meet Mr. or Mrs. Right. Why do we struggle so much when there is a way to make life easier and never to be alone again? Take time for yourself. Study not only yourself but the others around you. Of the people you meet, select one for your destiny with whom to share love and life. Astrology provides you with this necessary information. What is required is the time and effort to grasp a fuller understanding of yourself. When you truly understand yourself, you better understand what it is you want, what you are looking for, and what makes you happy. You are a unique individual with a specific combination of characteristics, personality traits, likes, and dislikes. What you must also accept is that others have particular tendencies which you must either accept and learn to live with or make the decision to find that special person with whom you are the most compatible. Love is the magic which propels us all to initiate this search throughout the universe for our true soul mate and love match. I like to see people happy and to see less sorrow and loneliness. Lift is short. Today we are here, and tomorrow we are not. Learn to live each day to its fullest.

House Rulerships

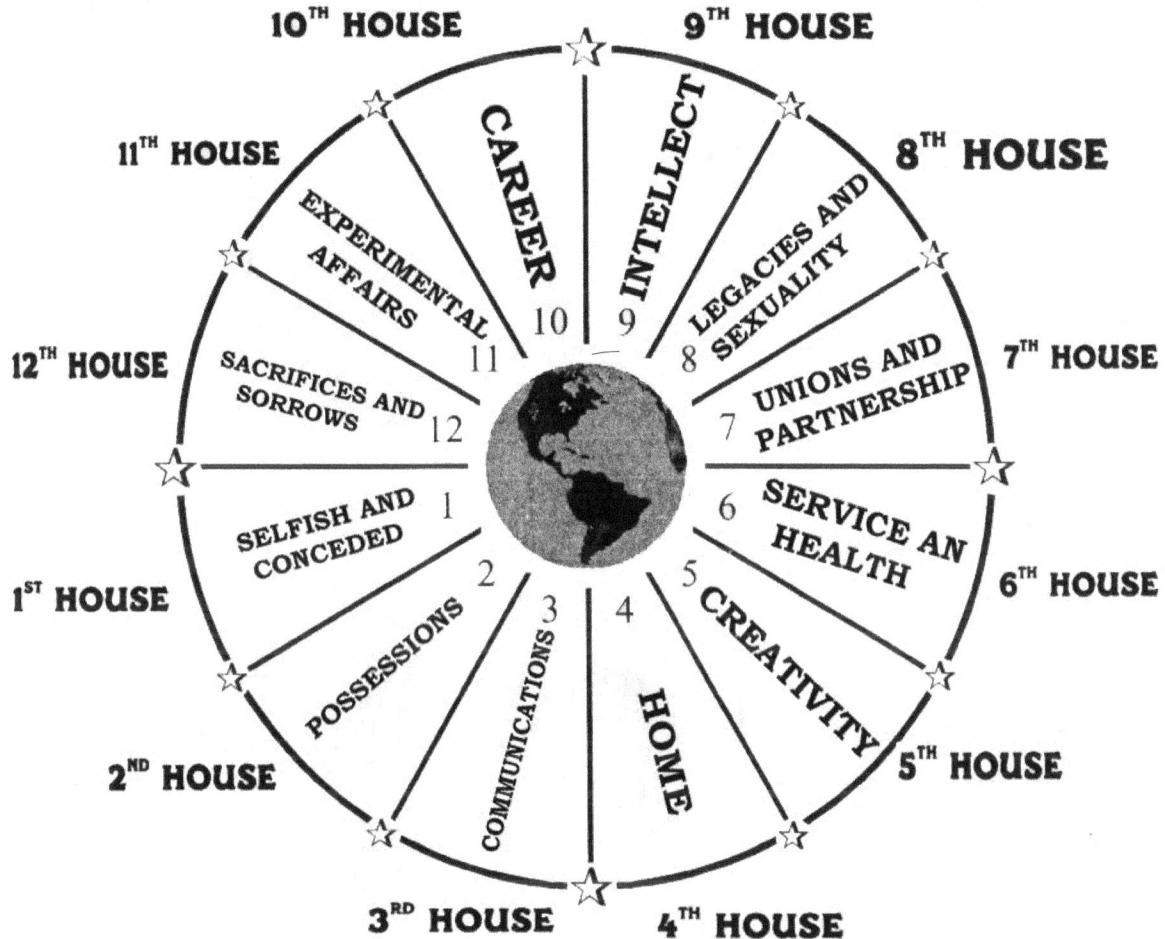

10TH HOUSE
9TH HOUSE
11TH HOUSE
8TH HOUSE
12TH HOUSE
7TH HOUSE
1ST HOUSE
6TH HOUSE
2ND HOUSE
5TH HOUSE
3RD HOUSE
4TH HOUSE

CAREER

INTELLECT

EXPERIMENTAL AFFAIRS

LEGACIES AND SEXUALITY

SACRIFICES AND SORROWS

UNIONS AND PARTNERSHIP

SELFISH AND CONCEDED

SERVICE AN HEALTH

POSSESSIONS

CREATIVITY

COMMUNICATIONS

HOME

10 9 8 7 6 5 4 3 2 1 11 12

Combinations of Elements - Our Mission on The Earth

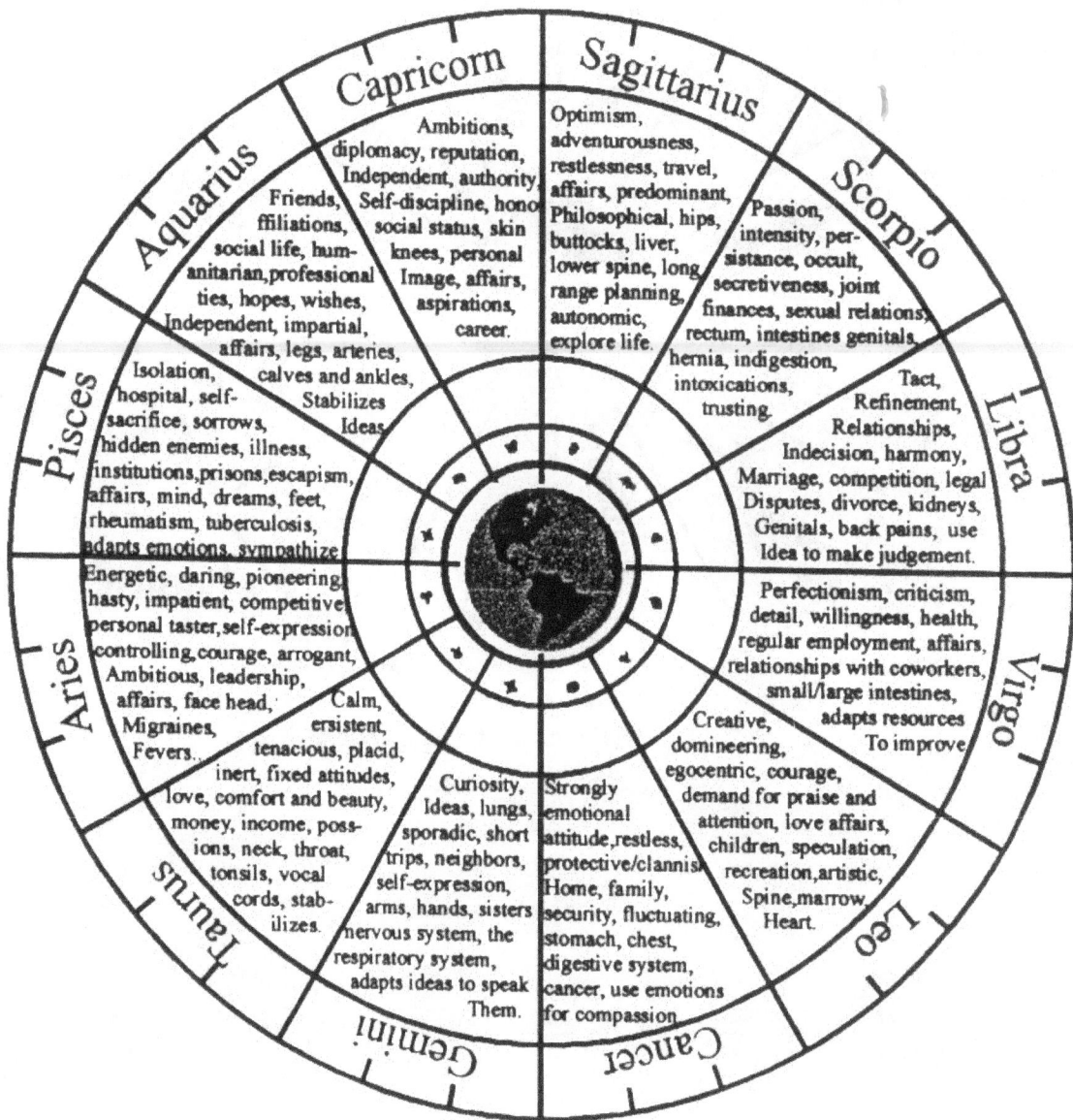

Capricorn — Ambitions, diplomacy, reputation, Independent, authority, Self-discipline, honor, social status, skin, knees, personal Image, affairs, aspirations, career.

Sagittarius — Optimism, adventurousness, restlessness, travel, affairs, predominant, Philosophical, hips, buttocks, liver, lower spine, long range planning, autonomic, explore life.

Scorpio — Passion, intensity, persistance, occult, secretiveness, joint finances, sexual relations, rectum, intestines genitals, hernia, indigestion, intoxications, trusting.

Aquarius — Friends, ffiliations, social life, humanitarian, professional ties, hopes, wishes, Independent, impartial, affairs, legs, arteries, calves and ankles, Stabilizes Ideas

Libra — Tact, Refinement, Relationships, Indecision, harmony, Marriage, competition, legal Disputes, divorce, kidneys, Genitals, back pains, use Idea to make judgement.

Pisces — Isolation, hospital, self-sacrifice, sorrows, hidden enemies, illness, institutions, prisons, escapism, affairs, mind, dreams, feet, rheumatism, tuberculosis, adapts emotions, sympathize

Virgo — Perfectionism, criticism, detail, willingness, health, regular employment, affairs, relationships with coworkers, small/large intestines, adapts resources To improve

Aries — Energetic, daring, pioneering, hasty, impatient, competitive, personal taster, self-expression controlling, courage, arrogant, Ambitious, leadership, affairs, face head, Migraines, Fevers.

Taurus — Calm, ersistent, tenacious, placid, inert, fixed attitudes, love, comfort and beauty, money, income, possions, neck, throat, tonsils, vocal cords, stabilizes.

Leo — Creative, domineering, egocentric, courage, demand for praise and attention, love affairs, children, speculation, recreation, artistic, Spine, marrow, Heart.

Gemini — Curiosity, Ideas, lungs, sporadic, short trips, neighbors, self-expression, arms, hands, sisters nervous system, the respiratory system, adapts ideas to speak Them.

Cancer — Strongly emotional attitude, restless, protective/clannish Home, family, security, fluctuating, stomach, chest, digestive system, cancer, use emotions for compassion

THE MEANING OF LIFE ON EARTH

THE METEMPSYCHOSIS OR TRANSMIGRATION OF EVERYONE'S SOULS, THEIR SOUL-MATES, STARS, AND DESTINIES

How to Find Your Soul-Mate, Stars, and Destiny

Ligia Balu, Author

LIGIA BALU blends the wisdom of the ancient astrologers together with her insight of discovering your *destiny*, fulfilling your *fantasies*, and finding your *soul-mate*. This book provides the stepping stones to unlocking the mysteries of A*strology*.   Through

LIGIA BALU'S sensitivities, the reader comes to understand the influence of the stars on our very lives.

LIGIA BALU explains not only how the stars lead us to our destinies but why we make the decisions we do, why we are attracted to certain people and certain situations, and why we face the challenge in our daily lives of overcoming our individual problems.

LIGIA BALU provides us with an understanding of ourselves and of other people. She explains how to recognize your *soul-mate*, how to become more compatible with others, and most importantly, how to be happier and more content through an understanding of your own destiny. The stars lead us and guide us, influencing us daily, and challenging us with obstacles that mold us into the people we become. Learn to read the future and how to make better decisions through the stars.

LIGIA BALU takes her readers on the grand adventure of inner exploration through an understanding of the vastness of the Universe. Sensitive by nature, she explores with her readers the dilemmas of everyday life that face us all. She brings to her readers the needed information for finding the *soul-mate* of our dreams. She explains why some Sun signs are compulsively drawn to others even when it is against all logical explanation while other Sun signs seldom waver from a logical course of action and rarely experience spur of the moment activities.

LIGIA BALU takes her readers on a guided tour of the mysteries of the Universe and explains why:

SAGITTARIUS are impulsively drawn to the stubborn ARIES.

LEO and CAPRICORN disagree about money.

CANCER women are patient with GEMINI men.

TAURUS develop either love or hate relationships with SCORPIO.

VIRGO women never approve of how SAGITTARIUS men spend money.

GEMINI may upset the boat for AQUARIUS.

LIBRA can be easily unbalanced by SCORPIO moods.

PISCES women find excitement with ARIES men.

SCORPIO has trouble understanding the flaky AQUARIUS.

ARIES men with an ARIES women spells competition.

SCORPIO can exact retribution from SAGITTARIUS.

CAPRICORN men delight in PISCES women if they can rule the magic.

There is a relationship between the causes and effects, a relationship born of fatalistic attractions in an order established by Ethereal reasoning. The Sun, Moon, Planets, and Stars are celestial bodies adding to and influencing our daily lives and spiritual beings. Through an intangible electromagnetic force this influence rides like waves of energy pulsating and dispensing impending causes and effects and leading us on a journey through life and understanding. This law can appear to be a terrible experience for some people because, quite often, a person is destined to suffer during either part or all of his life. No matter how hard this person attempts to avoid this suffering it seems that whatever is destined will be accomplished. While this law of cause and effect can seem like a terrible injustice, it is respectful and Holy in that it also serves a purpose. In its mystery, this law is superior to our ability to fully understand it or its reason for being. The law precedes our development and in no way fits comfortably into our way of thinking.

The aspects of the sky during our birth, our dreams, and even during our human misgivings, indicate a very strange but real connection between our perception of life and this Ethereal influence which surrounds us. Through the Universe designed by God comes these guides offering us warnings that we are destined for these experiences, and that the suffering serves a purpose in our lives.

Ancient scholars living in remote and desolate areas of the old world studied this phenomena. These scholars devoted their lives to studying God's purpose for man. Their lives were difficult, and they maintained minimum contact with the outside world. They turned their thoughts to studying the mysteries of the Universe and the secrets of the world. From time to time, they shared these secrets with a few chosen people who sought them out. They taught that all of our present lives are linked to our past lives.

This transmigration of the souls, as it iscalled, was studied by Pythagora, the Greek philosopher. Pythagora explained that the visible Universe, the sky with all its stars, are only passing stages of the soul of the whole world. Matter is concentrated then dissolved and seeded in all of the Cosmic and imponderable space. Every solar whirlpool contains a part of the Universal soul which is developing within itself during the millions of centuries and which contains an impressive, impulsive force and measure. When considering these Powers, the species of the live souls, which appear one by one on the stage of our little World, are given by God and descend from the Father. They are coming from an unalterable and superior spiritual order, conforming to a preceding material evolution, and belongs to a dead solar system. Some of these invisible, endless powers are guiding the existence of this World, and others are waiting, in a cosmic divine dream-like sleep. They blossom to re-enter into later generations according to the Divine law.

THE PLANETS ARE THE DAUGHTERS OF THE SUN, born of the

Sun, and each is in tune with the attractive forces with its inherent material rotating. Each possesses a semiconscious soul which rises from the solar heart, and each possesses a specific character relating to its special evolutionary role. As every Planet is a different expression of God's will, it has a specific function in the chain of the Planets and, therefore, in the chain of events. The ancient scholars identified the characteristics associated with each of these Planets and with those of the gods--characteristics which represent divine faculties of the action and reaction in the Universe.

The ancients identified the four elements as the fundamental indicators of the four graduated stages of the material world. The first element, the densest and roughest one, is the most unmanageable to the Spirit while the last element, the finest one, has the closest relationship with the Spirit. Earth represents the solid state, water the liquid state, air the gaseous state, and fire the imponderable state. There is another fifth element, the Ethereal one, which represents such a subtle force that it does not exist in the material state. This is the original Cosmic Fluid, the Astral Light or the Soul of the World.

What is the human soul but a part of the Great Soul of the World, a sparkle from the Divine Spirit--a coin for immortality. Everyone of us has within himself GOD, but to find Him, we need to develop ourselves. We must build a moral foundation upon which we can remain next to Him.

How to Find Your *Soul-Mate, Stars,* and *Destiny*

The Creator made man to have His face and to be like Him; however, Man does not accomplish this until after many incarnations. These successive lives are given to man so that he can improve himself and atone for previous sins thus helping him to differentiate the good from the bad and the light from the darkness.

Only these cycles can explain terrible injustices and unfairness which are suffered by some people, great happiness which is given to other people, sudden deaths, twin souls searching for each other all their lives, enemies and friendships, and unexplainable passions. Is there a Director behind the scenes whose existence we cannot explain? Are we not the same actors just performing in different plays? Is it not possible for one to have moments of lucid retrospective, which seems to be reminiscences of a previous life?

And what happens to the soul at the time of death of the body? When death approaches, the soul may have misgivings about separating from the body, and in some instances, pictures of the life flashes before it in rapid succession and frightful clearness. The approaching death disturbs the soul as it slowly loses consciousness.

In a saintly and pure soul, there is a Spiritual Awakening that occurs during this gradual detachment from the body. Through introspection this soul perceives the existence of another world before the body's last dying breath. It hears a remote call and responds to a pale, invisible beam of light. This soul feels happiness when it is at last released from the dying body. There is a feeling of escaping and being caught up in the middle of that great light that takes it to the spiritual world where it will belong from now on.

Most likely this does not happen with the many people whose lives were a fight between the material and any superior aspiration. In these cases, the soul may awake as if in a nightmare with no guiding hand to lead it. With no voice to cry out, this soul remembers the suffering and may exist in fear and darkness. It longs for its earthly body which it may still see and which holds an unbearable attraction to it. This soul was living only by its body and for its body, and at death, it searches through the cold body and the dead brain matter, but it cannot find itself. Whether dead or alive, it does not know, and while it wishes to see and understand, it does not. The darkness is all encompassing as is chaos and obscurity. This soul may cling to the phosphorescence of its mortal remains which is frightening but attracting it at the same time. Then the ugly dream and chaos begins again. This state may continue for several months or years depending upon the forces of the material instincts of the soul. Whether good or bad, this soul becomes, little by little, conscious of the new stage of existence. It leads itself, finally free from the body, to drift and fly between the hollows of the Earth's atmosphere as if it is carried upon electrical rivers and where it will see other lost souls. In this way a journey begins like a dizzy and fiery flight. It will climb higher and higher in an effort to escape the Earth's atmosphere and travel to a region of the solar system where it will find guides who, in some cases, are friends and relatives from the former life. The Earth slowly disappears like in a dream, and a new sleep which is like a delicious swoon wraps the soul like a sweet caress. The soul sees only its flying guide which carries it into the deepest infinity of space. It reawakens on a star where the mountains, flowers, and vegetation provide a sweet embrace. The soul is surrounded by lightening creatures, both men and women, who overlook and initiate it into the mystery of this new life. Here, the aspect of the body does not become the mask of the soul. The transparent soul appears in its true shape as if shining in the daylight. The soul's psyche, led by a sublime wisdom, finds the Divine Country in which it attempts to understand the Symphony of the

Universe. The soul rests on the golden beaches of this star paradise, and it rests under the transparent veil of a dream filled with sublime light, perfumes, and melodies.

This celestial life of the soul can last for hundreds of thousands of years depending upon its scale and impulsive force. Only the most perfect soul or the most sublime; however, can prolong this existence endlessly. Other souls are recalled by the law of reincarnation to suffer new trials in order to forget previous sins. Exactly like the human life, the spiritual life has a beginning, an apogee or culmination, and a decline. When the spiritual life is ending, the soul is trapped in a whirlpool of melancholy, but an undefinable force is attracting it again to the pain and suffering of the Earth. This desire is filled with terrible misgivings of leaving the Earthly life. But the day has arrived, and the law must be accomplished. A veil-like mist covers the face, and the soul can no longer see its companions through this veil which becomes thicker and thicker. The soul hears their sad farewells, and the tears of the people who loved him are penetrating him like a celestial dew, leaving it thirsty for a now unknown happiness. Then the soul solemnly swears and makes promises to remember the truth from a world of love or lies and pain from a world of hate.

When the soul awakens again, it is in the heavy atmosphere of the Earth, in the abyss of birth and death, not having yet lost its heavenly memories. It is here that its guide introduces the soul to its new mother who is carrying the child's seed. Then, the most impenetrable mystery of life on Earth, the mystery of reincarnation and maternity transpires. This mysterous fusion is carried out slowly, organ by organ, fiber by fiber. Step by step, the soul loses its Divine Self-Conscience, and the Light becomes dimmer and dimmer. With birth, that horrible pain pushes the soul, and a bloody convulsion uproots it from the Eternal Soul and places it in a newborn body. The new-baby arrives into this world and is yelling frightfully. The Celestial Memory has entered the deep recesses of the unconsciousness.

The Law of Reincarnation and Deincarnation is the true sense of live and death. This law represents the main mode of the soul's evolution and enables us to watch the past or the future and into the depth of Nature and Divinity. This law shows us the rhythm and the measure, the immortality and the goal. From the Spiritual point of view, it shows us that the correspondence of Devine life and death - as birth on Earth is like a Devine death - and the death is like a revival. The alternation of these two lives is necessary for the development of the soul, and each is the consequence and the explanation of the other. According to this Law, the facts from a specific life have an influence on the next life.

These lives follow one by one, but they do not seem to be alike, and yet they are linked by an undefeatable logic. Each of these lives has his or her own law and his or her different destiny. According to the Law of Repercussion, the events in a specific live have a punishing or rewarding influence on the next life. The individual will be reborn owning the instincts and the talents molded in his previous incarnation, and the quality of his next existence will be determined, in most cases, by the quality of his choices made in his previous life.

Pytagora taught that the apparent injustices of destiny, misery, suffering, and misfortunes, can be explained by the fact that each existence is the reward or punishment for the previous life.

There are No Words or Actions Without an Echo in Eternity.

When the soul finally wins over the material, then it finds itself at the beginning and ending of all things. By development of all its Spiritual Faculties, the soul then enters into a Divine Stage and in full agreement with the Holy Will. According to this, when the soul arrives at that progressed state, it does not go back but becomes immortal in a place where there is no pain and no sadness, but only endless love.

A sinful life will only allow a painful next life; an imperfect life will only allow a hardworking next life; and that is the way morality, while imperfect during a single life, can be perfectly achieved in successive lives.

What is the Ultimate Goal of man and of mankind, according to this doctrine? After so many lives, deaths, reincarnations, leisure, and painful awakens is there an end for the infinite soul?

Yes, undoubtedly yes! Only through the development of all his spiritual faculties, will the soul finally matter. He will find within himself the beginning and ending of all things. Only then will he enter into the Divine stage in full agreement with the Holy Will. We know that once the soul has arrived at this Superior Stage, it cannot go back. He will be immortal forever.

The above is an explanation of why some people have "lucky" stars, why some people suffer and why, after a sad or happy time on Earth, nobody will encounter a final death. Then, when our missions on Earth come to an end, and when our souls are lost in the land where:

"THERE IS NO PAIN, NO SADNESS, NO TEARS, BUT ONLY AN ENDLESS HAPPY, LOVING LIFE FOREVER AND EVER." Only then shall we encounter the GREAT GATE, written upon it:

"IDEALS, DREAMS, LOVE, FORGIVENESS, FORGETFULNESS, PEACE"

TABLE OF CONTENTS

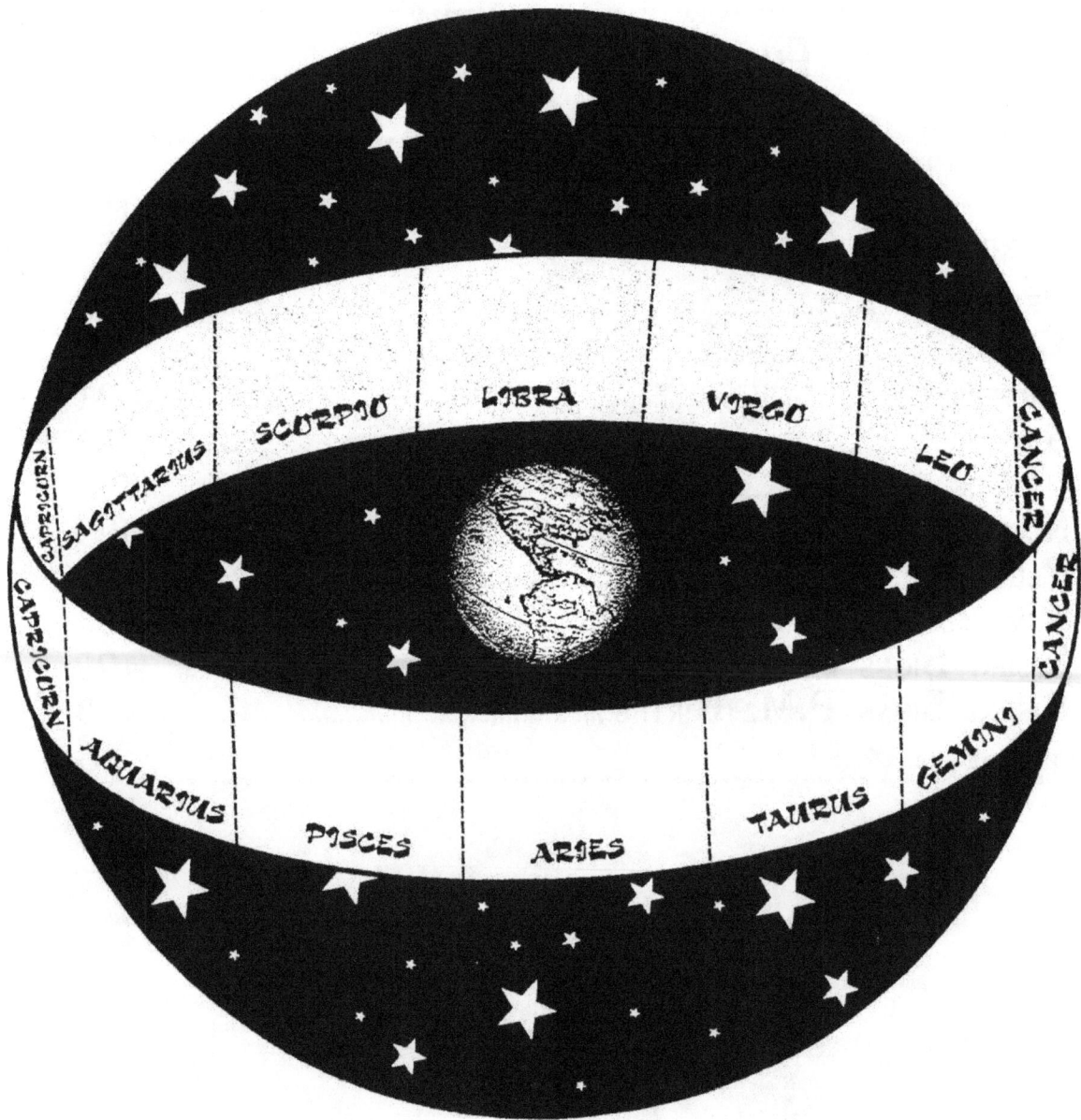

Chinese Astrology

The ancient art of astrology practiced in China is distinctively different from Western astrology. In fact, these two systems of astrology are so different that there is no comparison or correlation between them. Chinese astrology is based on the study of the stars, but it is a different group of stars altogether. of importance is the star Tzu Wei or the Pole Star and the constellations which are closely associated with it. Tzu Wei is the name of the god who rules the Pole Star which is the center of the astronomy system and the calendar which is based on astrology. Tzu Wei is considered the center of all life.

Created with this system is a cycle and the twelve animals which represent each cycle. Unlike Western astrology which is based on the apparent movement of the Sun through the Zodiac, Chinese astrology focuses on a Lunar calendar. In other words, the Chinese Zodiac is made up of a twelve year cycle, and each year of this cycle is named after a different animal. Each animal bestows distinct characteristics to its year. The Chinese system is based on the belief that the year of a person's birth is the primary factor in determining that individuals personality traits, physical and mental attributes and degree of success and happiness throughout life. This system of astrology is actually much more complicated than this description implies and is based on drawing up a chart much like a Birth Chart in Western astrology, and then determining which stars fall with each segment of that chart. However, a look at the animal signs can be instructive especially when an individual begins to compare compatibility between these signs. For this purpose, each animal sign will be described and discussed.

Once you have determined your animal sign and studied how it applies to you, the next step is to determine your lover's animal sign. Then read the pages which describe how those two signs get along. These recommendations are an important part of the Chinese system and considered important in determining the traits and make up of the persons in question. The more you understand about yourself and your own personal make up, the easier it is for you to decide just who it is that you can best get along with or who you might decide to share your life with. And it goes without saying that the more you know about the other person, the better off you will be in the long run.

Discovering your animal sign is as easy as finding your year of birth on the chart below. Once you know your animal sign, you are ready to learn more about yourself and to discover your perfect love match. This can be both enlightening and entertaining as you find yourself within the pages that follow. You will be amazed at the accuracy of Chinese astrology at describing your traits and your tendencies. Once you know yourself well, look up your best friends, family members, and, yes, that special person in whom you are most interested. Is he or she right for you? You are about to discover the answer to your questions. Then you can decide the accuracy of this system for yourself. Is your current affair sizzling with excitement, or do you prefer a relationship that offers security? Your animal sign can answer those questions about you and offer suitable matches for your sign.

Table of Years from Europe Which Corresponds to The Chinese Signs

YEAR	SIGN	DATES
1900	Rat	1/31/00--2/18/01
1901	Ox	2/19/01--2/07/02
1902	Tiger	2/08/02--1/28/03
1903	Cat	1/29/03--2/15/04
1904	Dragon	2/16/04--2/03/05
1905	Snake	2/04/05--1/24/06
1906	Horse	1/25/06--2/12/07
1907	Goat	2/13/07--2/01/08
1908	Monkey	2/03/08--1/21/09
1909	Rooster	1/22/09--2/09/10
1910	Dog	2/10/10--1/29/11
1911	Pig	1/30/11--2/17/12
1912	Rat	2/18/12--2/05/13
1913	Ox	2/06/13--1/25/14
1914	Tiger	1/26/14--2/13/15
1915	Cat	2/14/15--2/02/16
1916	Dragon	2/03/16--1/22/17
1917	Snake	1/23/17--2/10/18
1918	Horse	2/11/18--1/31/19
1919	Goat	2/01/19--2/19/20
1920	Monkey	2/20/20--2/07/21
1921	Rooster	2/08/21--1/27/22
YEAR	**SIGN**	**DATES**
1922	Dog	1/28/22--2/15/23
1923	Pig	2/16/23--2/04/24
1924	Rat	2/05/24--1/23/25
1925	Ox	1/24/25--2/12/26
1926	Tiger	2/13/26--2/01/27
1927	Cat	2/02/27--1/22/28
1928	Dragon	1/23/28--2/09/29
1929	Snake	2/10/29--1/21/30
1930	Horse	1/22/30--2/16/31

YEAR	SIGN	DATES
1931	Goat	2/17/31--2/05/32
1932	Monkey	2/06/32--1/25/33
1933	Rooster	1/26/33--2/13/34
1934	Dog	2/14/34--2/03/35
1935	Pig	2/04/35--1/23/36
1936	Rat	1/24/36--2/10/37
1937	Ox	2/11/37--1/30/38
1938	Tiger	1/31/38--2/18/39
1939	Cat	2/19/39--2/07/40
1940	Dragon	2/08/40--1/26/41
1941	Snake	1/27/41--2/14/42
1942	Horse	2/15/42--2/04/43
1943	Goat	2/05/43--1/24/44
1944	Monkey	1/25/44--2/12/45
1945	Rooster	2/13/45--2/01/46
1946	Dog	2/02/47--1/21/47
1947	Pig	1/22/47--2/09/48
1948	Rat	2/10/48--1/28/49
1949	Ox	1/29/49--2/16/50
1950	Tiger	2/17/50--2/05/51
1951	Rabbit	2/06/51--1/26/52
1952	Dragon	1/27/52--2/13/53
1953	Snake	2/14/54--2/02/54
1954	Horse	2/03/54--1/23/55
1955	Goat	1/24/55--2/01/56
1956	Monkey	2/12/56--1/30/57
1957	Rooster	1/31/57--2/17/58
1958	Dog	2/18/58--2/07/59
1959	Pig	2/08/59--1/27/60
1960	Rat	1/28/60--2/14/61
YEAR	**SIGN**	**DATES**
1961	Ox	2/15/61--2/04/62
1962	Tiger	2/05/62--1/24/63
1963	Cat	1/25/63--2/12/64
1964	Dragon	2/13/64--2/01/65
1965	Snake	2/02/65--1/20/66
1966	Horse	1/21/66--2/08/67
1967	Goat	2/09/67--1/29/68
1968	Monkey	1/30/68--2/16/69
1969	Rooster	2/17/69--2/05/70
1970	Dog	2/06/70--1/26/71

1971	Pig	1/27/71--2/14/71
1972	Rat	2/15/72--2/02/73
1973	Ox	2/03/73--1/22/74
1974	Tiger	1/23/74--2/10/75
1975	Rabbit	2/11/75--1/30/76
1976	Dragon	1/31/76--2/17/77
1977	Snake	2/18/77--2/06/78
1978	Horse	2/07/78--1/27/79
1979	Goat	1/28/79--2/15/80
1980	Monkey	2/16/80--2/04/81
1981	Rooster	2/05/81--I/24/82
1982	Dog	1/25/82--2/12/83
1983	Pig	2/13/84--2/19/85
1984	Rat	2/02/84--2/19/85
1985	Ox	2/20/85--2/08/86
1986	Tiger	2/09/86--1/28/87
1987	Rabbit	1/29/87--2/16/88
1988	Dragon	2/17/88--2/05/89
1989	Snake	2/06/89--1/26/90
1990	Horse	1/27/90--2/14/91
1991	Goat	2/15/91--2/03/92
1992	Monkey	2/04/92--1/22/93
1993	Rooster	1/23/93--2/09/94
1994	Dog	2/10/94--1/30/95
1995	Pig	1/31/95--2/18/96
1996	Rat	2/19/96--2/06/97
1997	Ox	2/07/97--1/27/98
1998	Tiger	1/28/98--2/15/99
1999	Rabbit	2/16/99-2/04/2000

Date With Destiny

RAT

Share With Me Your Fantasy

THE TWELVE ANIMAL SIGNS

THE RAT

The Rat is considered a favorable sign, and you are known for having a good heart even though there are times when you can be inconsiderate and selfish. You are attractive, you like to work hard, and you like to play hard. In fact, no matter what it is you are doing, you much prefer to excel. The Rat represents people who find luck most when they are generous and when they use their intelligence and ability to work hard to the best of their abilities. You may be quiet at times, but you can be gregarious and sociable and you enjoy an active social life and a wide variety of friends. In your personal life, however, you may have only a few very close friends. Your energetic drive combined with your adaptable and versatile nature produces financial security, however, a second business may be just as successful as your first. The Rat has to watch a tendency to spend money as fast as he or she makes it, and can be especially generous with a loved one. In fact, because of your spending habits, your list of assets may be shorter than your liabilities. Whatever your career, the Rat is usually successful as long as the intelligence and intellect is put to good use. You may make an excellent manager or discover that you have a talent for selling. You may find that you have a tendency to be secretive about some aspects of your life, and you can be critical and just a bit self-centered. Although you are open, friendly, and generous at the beginning of a relationship, you sooner or later become a little over critical of the other person. This tendency doesn't hamper your active social life, but many of your friendships may not last long. If not checked, this can also carry over to pettiness toward co-workers and colleagues which may put a damper on your big plans. In fact, you may find that you are much more successful with less ambitious or smaller businesses. Many times, the Rat is a person who finds success later in life or at least after forty. Your investments will, by that time, have had time to pay off but your returns may not be overly large. You also are not the luckiest sign when it comes to speculation mainly because others may distrust your intentions. On the other hand, you can be cunning and shrewd and usually come out on top in any law suits, deciding to settle out of court.

In love, you are highly emotional and lavish attention on the other person. But no matter how generous you can be, you don't like to be pushed and there will be times when you don't deliver on your promises. Sexually, your lover may complain that you can be selfish, and in this area you must learn that the secret to love is often satisfying the other person and not to just find your own pleasure. However, you do enjoy sexual diversity and a bit of experimentation. Being that you are attentive, however, and that you usually take the initiative in romance, you may find that you enjoy many lovers in your life time. But you are happiest when your lover is supportive, admiring, and attentive. One of your most admirable traits is that you aren't easily upset, and you find that this adds to the compatibility of your relationships. Whether or not your family approves of your final choice, when you make your decision on who to marry, you have a tendency to stick to your commitment. As a parent, the Rat is loving and encouraging.

Within your immediate family, you are a dutiful and respectful which brings the admiration of your parents, but you may find that one of your parents is more supportive and sympathetic toward you. You do have that little tendency to hold a grudge, though, and this can cause problems or prolong disagreements with your siblings.

You have a yearning to travel and to see the sights, preferring the natural beauty of the outdoors to cities and manmade structures. In fact, a Chinese saying describes the Rat as most enjoying the mountains and the seas.

The Rat is most fortunate in that this sign is known for a long life and good health with only minor illnesses. The Rat is a person who has many great ideas, high hopes, and aspiring ambitions, but there is a tendency to run ram shod over other people. Because of this, many of your plans are not realized. Whatever you circumstances are, though, you are adaptable and a born survivor. Your gregarious and sociable nature keeps you on the move with many friends and an active life, but you must watch a tendency to be revengeful when you don't get your own way.

You are imaginative, however, as well as discerning, and you usually have the ability to transform your dreams into solid ideas and reality. You have a preference for the eloquent but can be carefree and outgoing. But there are those times when you can be more than a bit miserly. on the other hand, you make every attempt to pay your own way, and you are not opposed to accepting the generosity of others. And you are quite adept at being sociable while making others feel comfortable and at ease, but this can be only superficial in many instances. Underneath all these traits, you can be impulsive and at time persistent. You are also well aware when opportunities present themselves, ready to assess them and to take advantage of them. Because of your critical nature, you don't always trust others.

In a permanent relationship or marriage, you make every attempt to be cooperative, loving and supportive of your partner, and you remain passionate throughout the relationship. You may find a long-lasting relationship with a Dragon in spite of numerous arguments. In a marriage to an Ox or a Rat you will feel secure and able to trust the other person. A person born in the year of the Monkey is well suited for you if the person isn't overly controlling. Less perfect are relationships with either a Snake, Dog, Pig or Tiger. You may find that a person born in the year of the Horse is too independent and overly honest in business affairs. You and the Rabbit get along well if a strong friendship and mutual understanding form the basis of the relationship otherwise be prepared for a difficult time. You will be tempted to cheat on a person born in the year of the Horse.

In Western culture, the Rat is hardly considered an appealing animal, but this doesn't hold true in the Chinese culture. This is a most favorable sign because the Rat is considered industrious and hard working. While you can be impatient with other people at times, you are a good planner and want to see your ideas through to completion. There are many careers for which you are well suited such as publishing, trading, accounting, business, or in the arts. You are not always lucky in speculation. Your most positive trait is your good heart.

January

Those persons born in the month of January during the year of the Rat are independent, proud and self-assured with a natural regal bearing that draws the attention of others. Their enthusiasm and overconfidence in some instances puts off less self-reliant people. They can be obstinate in their endurance and pursuit of what they want in life. But up until about the age of thirty to thirty-five they spend money as fast as they earn it, always believing they can simply make more. A romantic and powerful figure, these people rarely exhibit signs of anger or of being upset by situations. They can be patient and calm with a great deal of inner strength that sees them through difficult situations. This person is not prone to accepting advice from others, but prefers to seek their own course and to pursue financial security, fame, or prestige. major successes, however, generally come after the age of thirty-five, especially if the person is in a private business. This person may also be the recipient of unexpected money at some time in his or her life. The January Horse can be extremely sociable and extroverted, but they have a tendency to use a great deal of self-control against temptations to become too pleasure-seeking. This is a person who stays well balanced during difficult times, learning from the experience and developing a personal strength and security from any hardships they may have to endure. Many a January Horse must guard against becoming too self-centered and only thinking of their own success or desires.

All the same, this is a romantic person who pursues love and romance. The January Rat finds compatibility with anyone born in the sign of the Ox, the Rat, or the Dragon. The January Rat female discovers lasting love and romance with a man born in the sign of the January Ox or the December Rat. The January Rat man is most suited for a relationship with a woman born in the sign of the September Ox, an April Rat, or an October Dragon. Both men and women of this sign should avoid romantic relationships with others born in the year of the Horse. You are straightforward and direct in your relationships. Because this sign is on the cusp with the Boar, you may also share those qualities.

February

A person born in the month of February during the year of the Rat may discover that he or she is sensitive and intuitive with precognitive skills that lead to extrasensory perceptions. When this person learns to trust in their intuitions, they make excellent decisions based on this skill rather on their intellect or logical reasoning ability. But they must learn to be pragmatic about their abilities and not become complacent or overly self-confident. The February Rat can be thrown off balance during those times in his or her life when this perceptive or psychic ability wanes, and it is during these periods that they may experience the most difficulties in life. This person is extroverted and kind, and may find life a social whirlwind. However, they can become impatient with the logical and methodical approach that other people take when making plans. And other people can become upset with the impulsive nature of the February Rat. This person does well in a career which require a great deal of imagination or a flair for the creative.

The February Rat female is known for her stylish, neat appearance and an enchanting charm which is natural and unaffected. It is said that they women of this sign attract men easily and effortlessly, more so than any other sign. The men of this sign are particular and neat, and look for a companionable, feminine woman who manages her career and home well. Both men and women born in the sign of the February Rat are loving, passionate and romantic with a tendency to be faithful and loyal to their loved ones. The women of this sign do best with a romantic partner who is born in the sign of the June Dragon, the November Ox, or the a man born in the sign of the Monkey. The February Rat man finds love and wonderful companionship with a woman born in the sign of October Rabbit, September Dragon, or July Ox. Both men and women who are a February Rat do well with a person born in any month in the year of the Dragon. But men and women of this sign should more than likely make an attempt to avoid romantic relationships with people born in the year of the Horse or the year of the Snake. You seek a partner who is wise.

March

Those persons born in the month of March during the year of the Rat are well liked and accepted by other people. This person makes every effort to appear self-assured and can be assertive in social and business matters. But, in reality, they possess many hidden doubts and fears and may worry extensively about their personal affairs. They can also become upset and react in an unpredictable manner when surrounded by too much confusion. There are those March Rat persons who prefer quieter settings and may even become loners in an effort to avoid large crowds or too many distractions in their life or work. This is a person who is supportive and open to other people, but there are times when others take advantage of his or her caring and giving nature. Their generally positive outlook becomes unbalanced when this type of difficulty presents itself, but it is extremely hard for the March Rat to turn down a friend in need. They must guard against this inability to say no to others. In some situations, a friend or lover will take advantage of this person to the point of wrecking havoc with their finances. But there is also indications that it is a friend who recommends the March Rat to an influential person which may result in a good position or a promotion. The March Rat, in many cases, experiences a financial roller coaster ride with his or her finances fluctuating from very good to bad. This produces many changes in their lives and they are constantly seeking to be more financially stable and secure. It is wise for this person to curtail spending during the good times and to save as much as possible for eventual difficult periods.

The March Rat is no novice to romance, but may experience numerous emotional upsets in his or her love life. The March Rat female is best suited for a romantic relationship with a person born in the sign of the April Snake, the December Rat, or the June Rabbit. The March Rat man should look for a relationship with an April Dragon, March Ox, or December Monkey. Both men and women of the March Rat sign should avoid relationships with those people born in the year of the Horse. Those people born in the year of the Tiger are also companionable for the March Rat.

APRIL

Those persons born in the month of April during the year of the Rat appear mature for their age and exude a self-confidence that can even make others feel good about themselves. They can actually be less secure than they appear, however, and may privately be questioning their own actions and thoughts. But generally, this is a person who gives careful thought to his or her actions. They accept responsibilities seriously, never saying they'll do something unless they intend to do so, and they carry through to completion what they start. In this manner, they can be somewhat calculating but they are also logical in their thinking processes. This can be an especially intuitive person who is aware of psychic abilities which can be used with some degree of accuracy to predict future occurrences. For the most part, the April Rat achieves success one step at a time with careful planning and a lot of hard work. Rarely is anything given to this person that they haven't worked and strifed to accomplish. On occasion, an April Rat will feel the call to do something artistic rather it be in art, writing, or acting, and this person will put everything else to one side in an effort to achieve this goal. But it can be difficult for the April Rat to accomplish this task without much effort, determination and hard work. Those April Rat's that do become prosperous may also be extravagant in their spending habits. The April Rat must guard against making mistakes or misjudgments which can result in unnecessary misfortune or difficulties. The April Rat is a good communicator who does well in numerous career fields or in their own business.

The April Rat is a romantic person who much prefers to be in love, and who falls in love easily. But most often this is a person who has one great love in their life. The female April Rat may find that a man born in the sign of the December Dog answers her need for romance. Or she may do well with a man born in the year of the Dragon. A man April Rat finds true love with a woman born in the sign of the September Rabbit or the year of the Snake. Both men and women born in the sign of the April Rat should avoid relationships with those born in the year of the Horse.

MAY

Born in the month of May during the year of the Rat, this person is attractive, cheerful, open, and direct. Others find them appealing and approachable with a special grace and charming nature. This is a genuine and kind person with who isn't superficial with others which brings them many friends and admirers. The May Rat works hard at endeavors but also enjoys an active social life and prefers a little excitement in life. This person is also ambitious and aspires to a good position in either management, a leadership role, or in politics. They can also do well in their own business when they apply all their efforts in this direction. This person can also be idealistic and just which can develop into a philosophical approach to much that they undertake. They can aspire to great things, but find that they must work diligently and with great effort to reach this accomplishment.

The May Rat is active and energetic and socially likes the fun and excitement of going out and mingling with other people. Wherever there are groups of people hanging out just for the fun of being with other people, that is where the young May Rat can be found. The May Rat is also a romantic at heart who sexually is direct, open, and uninhibited. They must guard against romance becoming a preoccupation that interferes with other areas of their lives. once they find a romantic partner or marry, however, the May Rat is faithful, loyal, and honest in his or her relationship. They are also loving and supportive of their children. A May Rat female does well in a relationship with a man born in the sign of September Dragon or with a June Tiger. The May Rat man finds love and compatibility with a woman who is born in the sign of the August Ox or the September Monkey. Those persons born in the sign of the May Rat should make an attempt to avoid relationships with others born in the year of the Horse.

The May Rat must seek a balance between romance and work, making every effort to remain sensible in decisions and to maintain a positive approach and outlook by developing your analytical skills. This approach leads to both happy relationships and good luck in other endeavors.

JUNE

Those born in the month of June during the year of the Rat are people who are cheerful and optimistic with an outgoing and extroverted nature. They can be kind, compassionate, and sympathetic as well and are adaptable to new situations and open and accepting of other people. This can be somewhat of an extravagant person who is generous and spends money freely with little thought of the future. They are congenial and love to laugh and joke with friends and family.

This is an energetic person who moves quickly through his or her tasks and is ready to start on something else. Their original ideas and innovative plans can work to their benefit. They can also be a bit ostentatious with a tendency to show off their possessions, and they like to appear financially well off by wearing designer clothes or the latest fashions. This tendency may attract attention and help them to make a certain type of friend, but it can also cost them the friendship of more genuine people and if not controlled can become a problem in forming lasting relationships. A tendency to be somewhat argumentative can also develop into problems and difficulties at work and in relationships. This is a person who may find that he or she is discontent and unhappy at work or in their relationships until about the age of thirty when they become more settled and assured of their position. The June Rat must also guard against selfishness and greed which can add to their discontentment. Once they find a job they excel at and learn to control their spending habits, the June Rat becomes happier with life in general.

This person is sensitive and romantic and desires love and romance. In a relationship they seek compatibility but may be attracted to complete opposite types of people. The June Rat female finds compatibility with a man born in the sign of the January Tiger or the April Monkey. The June Rat man is successful in a relationship with a woman born in the sign of the February Monkey or the March Rabbit. Both men and women of the sign June Rat may want to avoid serious relationships with others born in the year of the Boar, the Snake, or the Horse. The June Rat person loves to travel and to share his or her experiences with others.

JULY

Born in the month of July during the year of the Rat, this person is gentle and kind and prefers to be considerate and attentive to other people. They are good listeners but may actually make few close friends. This is an independent person who doesn't seek the help or advice of others, and who doesn't very often get in a position of overextending himself for others. As a young adult, the July Rat is changeable and may strive to prove his or her independence by relocating or changing jobs until a situation is found where they are satisfied with their positions. However, once a career catches their attention, they become dedicated and devoted and are capable of working hard to accomplish their goals. While they may not attain great success, the July Rat is competent and succeeds gradually by taking things one step at a time. Generally, there are indications that a sensible nature prevents unnecessary difficulties and losses. This can be a strong willed person who has an inner strength that gets them through any difficulties or which helps them to overcome obstacles in life. The July Rat is drawn to family ties and will most generally stay close to his or her family. And this is especially true with their mother.

A happy marriage or lasting relationships is indicated for the July Rat, but they must learn to find contentment within the relationship and to overcome a tendency to be critical of their romantic partner. A July Rat man may find it difficult to be faithful to a romantic partner. Once they find the right partner, a happy relationship develops. The July Rat female is compatible, well liked, and sought after by men of numerous signs because she is easy to get along with. She may find the most suitable relationships with a man born in the sign of the May Dragon, a September Ox, or a February Dragon. The July Rat man can develop a lasting relationship with a woman born in the sign of June Monkey, June Tiger, or May Rabbit. Both men and women born in the sign of the July Rat should avoid serious relationships with persons born in the year of the Horse, the Boar, or the Sheep. You find that life becomes more settled and secure for you after the age of thirty.

August

Those persons born in the month of August during the year of the Rat are intelligent and knowledgeable. The August Rat is aware of their surroundings and is adept at sizing up situations. Their active minds make them alert and often ingenious with a natural curiosity and many times a yearning for adventure and excitement or an element of risk. And their cleverness and risk-taking tendency can lead them to make mistakes from time to time which can result in misfortune. However, for the most part, this is a person who appears blessed with good luck and even good fortune. And their alert attentiveness and astute intelligence helps them in financial matters. Their skills in observation leads them to understand other people well, and they are for the most part diplomatic and congenial while at the same time remaining assertive and even aggressive when pursuing their goals. They are accepting and tolerant of others, fitting in well in group situations, and they love to socialize. That combined with a romantic tendency results in many an August Rat enjoying numerous romantic affairs. This is an adaptable person who easily moves from one group to another, and in other areas of life, is comfortable with new ideas and new situations. They are cheerful and friendly and make friends easily. The August Rat is well suited for any number of career fields and is especially good in fields requiring a creative flair, innovative outlook, or a daring nature.

Often the August Rat will enjoy his or her own lifestyle and prefer a relationship with a person who shares similar interests, values, and tastes. The August Rat female finds the most suitable partner with a man born in the sign of the February Ox or the March Dragon. The August Rat man is well suited for a lasting relationship with a woman born in the sign of the December Ox, a March Dog, or a February Snake. Both men and women born in the sign of the August Rat should preferably avoid relationships with persons born in the year of Horse, the Cock, or the year of the Boar. After finding the right person, the August Rat develops a loving and lasting relationship and finds good fortune as well.

September

Those people born in September during the year of the Rat are intellectual, intelligent, and self-disciplined. They are sensitive and intuitive to the feelings of others and use this ability in their interpersonal relationships both socially and in business. They appreciate beauty in all its form and prefer lovely settings and surroundings. They also appreciate beautiful people and are drawn to others who are physically attractive and appealing. There are times when other people mistake this appreciation as well as the September Rat's sensitivities and forgiving and understanding nature as flattery. This person is kind and sociable and while they know many people and make friends easily, they form few close friendships. At the same time, the September Rat is most comfortable with security and stability in his or her career and makes every effort to work productively and diligently toward promotions or a better position. This is also a person who uses their intellectual ability toward self-improvement which results in efforts to better themselves. However, there are those September Rats who are perfectionists and are never quite happy with themselves. This type of person is just as critical of others as of themselves which can cause difficulties in relationships. Others make the effort to improve their minds and their inner selves through their self-discipline. Their intellectual abilities also provide them with excellent memories and recall of facts and figures, names and dates, and other important information which benefits them in their careers. These traits can lead to success in business or in careers.

The September Rat may feel unlucky in romantic relationships and feel misunderstood because of a difficulty communicating with members of the opposite sex. They can become impatient with their partners. The female September Rat while being honest prefers not to be assertive, but she seems to know intuitively how satisfy her partner. She finds the most success with a man born in the sign of May Snake or October Tiger. A September Rat man is well suited for a woman born in the sign of the August ox. They should avoid relationships with the Boar or the Horse.

OCTOBER

Those persons born in the month of October during the year of the Rat are perceptive, quick thinkers with a strong sense for the right move to make. Their valiant efforts help to prevent them from making a wrong decision or taking unwise action. This person is good at making plans and carrying through with them. Often, the October Rat does best in his or her own business rather than working for other people or for a large organization. This person is courageous and faces difficulties and obstacles with an undaunted perseverance. Success for the October Rat usually comes after the age of thirty and requires much effort and hard work which pays off in the long run. The October Rat is also an emotional person who displays his or her feelings openly, and this openness may upset other people at times. This person also shows their concern for others and likes to be of service to other people. In many cases, this augments their business acumen and abilities. The October Rat loves good food and many born in this sign find employment in the food industry. They are also creative and many have an artistic flair.

The October Rat man is sexually active and has difficulty controlling his physical desires, leading to a life of pleasure seeking and promiscuity which can effect their standing in the community. An October Rat female may marry well and be devoted to her husband's needs, but she can become disenchanted and dissatisfied with her husband, and when this happens, she does not remain faithful. It is important for both men and women born in the sign of the October Rat to choose their mates carefully and to know the person well before marrying. The October Rat man finds love and compatibility with a woman born in the sign of April Monkey or May Dog. The October Rat female may find happiness with a man born in the sign of the August Cock or any month in the year of the Ox. Both men and woman born in the sign of the October Rat should avoid relationships with persons born in the year of the Horse or the year of the Rat. The October Rat is sensitive to others and makes every attempt not to hurt their mates, but they must guard against being tempted to stray.

November

Those people born in the month of November during the year of the Rat are independent and strong willed. This person can be decisive, opinionated, and somewhat unwilling to change his or her opinion once they have arrived at a decision. They have leadership qualities and work diligently to complete plans. This is not a person who seeks the advice of others, but moves forward in their own set direction. In many instances, this person is a loner who doesn't feel the need to socialize with others. This is an intelligent, self-disciplined person who prefers to make it through life on the merits of his or her own efforts. They may also be moody, temperamental, or preoccupied with their own thoughts to some extent. Many November Rat persons are possess some form of artistic talent which some decide to develop and others ignore choosing instead a different avenue of pursuit. There is some indication that the November Rat does well early in life then experiences loss in middle age only to regain their position again as they mature. It, of course, benefits this person to learn to listen to the opinions of others more and to be more tolerant of other people. Often, however, their energetic drive and inner strength impresses and earns them the respect of others who learn from them. This is an astute observer who often recognizes opportunities and makes the best of them. The November Rat's belief in their own ideas quite often sways the opinions of other people. This person likes to be presentable and is well dressed and neat in appearance even though they have the tendency to carry a few extra pounds.

The November Rat female does well in a relationship with a man born in the sign of December Tiger, November Tiger, or October Dragon. The November Rat man is compatible with a woman born in the sign of July Rabbit, June Rabbit, or any month of the year of the Monkey. Both men and women born in the month of November in the year of the Rat should seek to avoid serious relationships with other persons born in the year of the Rat or with those born in the year of the Horse. The ideas of the November Rat benefit them in life.

December

Those persons born in the month of December during the year of the Rat are intelligent, creative thinkers who use their abundant energies to put their ideas into actions. These people work hard and strive to attain their aspirations. strong-willed people, they are undeterred by obstacles or difficulties. This is an efficient manager at work who concentrates on productivity and appears to possess a tireless capacity for concentration. The December Rat can be relentless at pursuing his or her goals in life. But at times, this person can be overly concerned with details and the more insignificant aspects of life. They strive to exude an appearance of self-control but can also be temperamental on occasion. Those born in the year of the Rat are known for being extravagant, and the December Rat is no exception. They spend money freely on their own needs but are rarely the ones to loan money to others. However, in a romantic relationship, they can lavish attention and gifts on the other person. The December Rat can also be impulsive and make rash decisions or change his or her mind quickly. Inquisitive, curious, and sociable, this person also loves to collect tidbits of information about other people and are not above gossiping. They are open and direct about their sexual relationships which they talk about freely.

The December Rat man may discuss his conquests or even his sexual relationship with his wife. The female December Rat is also uninhibited about her sexual relationships and will talk about them openly. The female December Rat does best to guard against being overly extravagant and to learn to handle finances well.

The December Rat female does best in a relationship with a man born in the sign of the September Ox, the October Ox, or the October or November Dog. They may also have a successful relationship with a man born in the year of the Dragon. The December Rat man is well suited for a relationship with a woman born in the year of the Monkey, or in the sign of the October Ox, the September Rabbit, or the December Dragon. Both men and women born in the sign of the December Rat should avoid relationships with a person born in the year of the Horse or the year of the Boar.

Date With Destiny

OX

Share With Me Your Fantasy

THE OX

The Ox is gentle and patient with a quiet, somewhat reserved nature and a tendency to be precise and methodical. The Ox possesses a great deal of determination as well, and applies his or her efforts to accomplishing what they set out to do. They seldom meddle in the affairs of others, preferring to tend to their own affairs and working through any difficulties with a firm persistence. While well liked by others, the Ox can be a self-reliant loner who neither borrows nor depends upon the advice or assistance of others. This person is not extravagant and neither is the Ox comfortable with much debt or using credit unnecessarily. They don't always express their thoughts or ideas and may miss out on some opportunities. However, they are capable of inspiring confidence in others even though they can appear a bit eccentric themselves at times. This sign is considered neither lucky nor unlucky, but the Ox manages his or her resources well and generally finds financial security which may increase in later life. The Ox is a hard working and dependable person, many of whom find that they work better with a partner. Their business and finances improve with effort rather than through investments or speculation. This person has a tendency to move slowly but effectively, and their plans are usually well founded because of their powerful organizational ability. Their patience and confidence in themselves lead them to think ahead carefully and clearly. The Ox is a good manager and when in a position of leadership, this person becomes more talkative. But this is also a person who may resent the authority of others to some degree. The Ox takes care and always appears to be in control of his or her own direction in life. They rarely show-off, and many Ox gain the respect of others through their sincere efforts and concern.

The Ox can be so diligent, in fact, that there are times when they are stubborn, refusing to bend in any way or to accept the opinions of others. In this way, they can be intolerant of others and at times prejudiced, preferring to follow through in their way whatever they have started. They simple believe wholeheartedly in their own methods and ways of doing things. They can become nervous and develop anxieties about difficulties. And underneath a rather calm exterior is often found a hot-tempered person capable of becoming furious to the point of being violent or even dangerous to themselves or others. When this temper explodes, no one else can deter it. It is best for the Ox to be left alone to regain control of their emotions. It is advisable for the Ox to live and work in a quiet environment away from a lot of confusion. And it is preferable for this person to avoid other people who provoke him or her. The anxieties and temper of the Ox can even bring on illnesses and poor health.

The Ox is most generally conservative and conventional with a tendency to distrust change and to be hesitant to accept new trends or fashions. Many an Ox is more than a little introverted and makes every effort to avoid stressful situations. However, they form strong friendships and quite often it will be another person in a good position who helps the Ox the most in establishing a successful business built upon the hard work, diligent efforts, and careful planning of the Ox. This person may prefer not to travel much and this is true in business as well, and, actually, it is best for the Ox to take precautions and extra care when traveling. The Ox is not a person who aims for material possessions, per se, but there are strong indications that many of their hopes, dreams, and plans will be fulfilled. This person should always handle all legal problems has proficiently and swiftly as possible and should avoid any long drawn-out law suits. The Ox is benefited by most always finding any lost or misplaced items. Gains are also realized through real estate and other property.

Many an Ox must strive to overcome the hardships of a disadvantaged youth, but they are extremely well loved and cared for by their parents. This can make them feel secure throughout life, but it can also bring on jealousy tensions, or stressful situations among brothers and sisters who have a different nature from the Ox. Those persons born in the year of the Ox may not place a high priority on romantic relationships, but their gentle, caring nature makes them attractive to the opposite sex anyway. As in other aspects of their lives, the Ox can be cautious and especially careful in choosing a marriage partner, and many will wait until later in life to finally marry. The ox can be an adoring spouse who cares deeply for his or her partner. And they love and care for their children just as deeply. To understand the Ox well, the other person needs to pay attention to them because they want most to be listened to and appreciated. Unfortunately, in many cases, because of their quiet natures, friends and family members do not always listen well to what the Ox has to say. This can lead them to be somewhat dogmatic and stern with family members. The female Ox has a tendency to devote herself almost exclusively to her family, caring for them and making every effort to take care of their needs. She is a natural at managing and organizing a comfortable home for her family. Because of their concern for others, both men and women born in the year of the Ox can become overly concerned and anxious about the injustices and suffering in society. The Ox rarely tolerates unfaithfulness with their spouses, even though they sometimes place the blame on themselves for such infidelities. The Ox is tender, affectionate, warm, and good hearted, but rarely overly romantic.

Those persons born in the year of the Ox find their most ideal marriage partner is a person born in the year of the Cock because they are harmonious and develop a loving and lasting relationship. The Rat is also a sincere and loving marriage partner for the Ox, and together they are strong, fortunate, and truthful with each other. A person born in the year of the Monkey is sincere as well as kind and truthful, and this person would add a liveliness and light-hearted aspect to the relationship that would insure that both partners remained faithful and devoted to each other. Both men and women born in the year of the Ox should avoid relationships with those born in the year of the Ram, the Horse, the Dragon or the Dog. A relationship with a person born in these signs would be inharmonious and cause tensions and stress, resulting in poor health, discord, gloom, and numerous other problems.

The Ox finds the most relaxation with projects and hobbies that require working with the hands. They also enjoy music, quiet settings, and pleasant get-togethers with close friends. You choose your words carefully, and speak when you have something to say that is worthwhile. You find that you are successful in most of your endeavors.

JANUARY

A person born in the month of January during the year of the Ox is on the cusp of the previous sign, the Rat, and may possess some of the characteristics of that sign as well. The January Ox is a patient person with extremely strong powers of concentration who will not stop work on a task, no matter how difficult, time consuming, or repetitive, until it is completed. This same obstinacy, or stubbornness in some, may also make them appear to others as unapproachable. But, whatever the case may be, once this person has set a goal, he or she continues through with it until it is accomplished. The January Ox is creative and imaginative and many are graced with artistic ability, some even to the point of being considered a genius. Fortune also smiles on the January Ox granting those born in this sign the benefits of good luck. This is a self-reliant and independent person who has little difficulty relocating to a new place when it is necessary. They are adaptable and fit in easily in new situations and new people. They may have close, personal friends, but the January Ox is quite comfortable on their own. In fact, this person doesn't mind being alone. To other people, the January Ox may appear to not need other people, and, in many cases, this holds true with their marriage partners as well. The January Ox is strongly individualistic with particular likes and dislikes which can make them difficult to live with. This person doesn't always make the best marriage partner unless they find a very understanding and compatible mate.

The January Ox female is charming with a special allure that attracts the attention of the opposite sex. She is well suited to a relationship with a man born in the sign of the September Cock or the October Monkey. The January Ox man possesses a refined and sensual sexuality and uses subtle mind control techniques to control their relationship. This man finds lasting happiness with a woman born in the sign of the June Rat or the January Cock. Both men and women born in the sign of the January Ox should take precautions to avoid relationships with persons born in the year of the Sheep or the year of the Horse. Your independent nature and good luck bring you benefits.

February

The February Ox person is hard working, steadfast, and optimistic. They are sensitive to the feelings of others and this combined with their cheerfulness and sensual nature makes them especially appealing and attractive to members of the opposite sex. Added to that is their ability to forgive and forget easily which adds to their tendency to get along well with others. This is quite often an accomplished person whose slow but steady work habits and methods lead them to success on the job. And their generally optimistic nature helps them to brush off insignificant problems that worry or upset other people. But the diligent February Ox does have the tendency to believe he or she is right, and their outspoken and frank nature, however well meaning, can at times offend other less secure people. Those born in the sign of the February Ox are also imaginative people who can become preoccupied with their daydreams and personal thoughts. They love to read, appreciate good literature, and may also develop an appreciation for art and music. There is also a tendency among those born to this sign to collect books, music, art, or memorabilia. The February Ox is tolerant and accepting of other people and rarely judge others by their outward appearance. They are so accepting, in fact, that they possess the tendency to be somewhat naive and may find themselves taken advantage of by less well meaning people. In fact, even though this is generally considered a fortunate sign, the February Ox must guard against being cheated. Their amiable and socializing nature leads them to enjoy sports, games, cards, and gambling at which they are neither lucky nor unlucky.

The February Ox female finds that she finds love and happiness in a relationship with a man born in the sign of the August Tiger, the October Rat, or the June Monkey. The February Ox man does well in a relationship with a woman born in the sign of the September Sheep, October Rat, or May Cock. Both men and women born in the sign of the February Ox should avoid relationships with persons born in the year of the Ram, the Horse, or the Dragon. The February Ox forms lasting relationships, but it is advisable to hold off on marriage until they know the other person well.

March

Those persons born in the month of March during the year of the Ox are strong-willed, independent people who much prefer to think for themselves. They are patient and persevering, but they dislike others interfering with their plans or offering then advice on how to do things. In many instances, this is a talented, physically strong person who relies on personal ability to solve problems and to accomplish goals. They are congenial, active people with busy social lives, but it is often difficult for other people to feel close to the March Ox because this person has difficulty accepting criticism, suggestions, or directions in any form. Often these people prefer to either work by themselves or to be in charge of the situation. In fact, this person can be so over sensitive to the good intentions of friends and family that they perceive any form of advice as criticism. They can be so defensive that they will rarely admit to making a mistake no matter how insignificant the error may be. The March Ox is also hesitant to allow other people to get to know them well. They guard against this by keeping others at a distance and by keeping their most personal thoughts and feelings to themselves. At the same time, this person rarely gives up or admits defeat, but rather insists on completing what they started no matter what problems occur. The March Ox often faces numerous changes and problems in early life and succeeds only after much effort. It is important for them to find a direction in life or they can drift aimlessly. This person makes up his or her own mind and determines for themselves a code of ethics. For the most part, the March Ox is honest, honorable, and rarely takes advantage of or cheats others.

The March Ox female does best in a relationship with a man born in the sign of September Dragon, November Dragon, or in some cases the September Snake. The March Ox man does well in relationships with a woman born in the sign of the June Cock, the September Rat, or the December Monkey. Both men and women born in the sign of the March Ox should avoid relationships with those people born in the year of the Sheep or the year of the Dog. To avoid unnecessary mistakes, the March Ox must learn to listen to others.

APRIL

Those persons born in the month of April during the year of the Ox are intelligent and strong-willed. They are capable of applying a great deal of patience and thought to a situation and can even be obstinate and stubborn when need be. They are thorough and methodical with strong powers of concentration which allows them to complete their tasks no matter how difficult. They are not easily distracted once they begin something. The April Ox also has a tendency to desire fame, prestige, or positions of authority. They like to feel powerful and in charge of the situation. When they apply their strong powers of concentration and their will powers to gaining authority or fame, they most generally succeed. This same obstinacy and determination can at times lead them to miss out on other opportunities that come their way because they fail to recognize such chances until it is too late. However, the April Ox is not inclined to waste his or her time or efforts needlessly and if they are not convinced an activity is worthy of their attention, they do not apply themselves to it. In other words, this is not a frivolous person who wastes his or her time by being lazy or unmotivated. The April Ox acquires a degree of self-discipline which when combined with their natural patience can produce effective results. The April Ox can go so far as to be meticulous and overly concerned with details.

Those persons born in the month of April during the year of the Ox most prefer a relationship with a person who shares their values and characteristics. They seek another like themselves who they can admire and who is companionable. The April Ox female does well in a relationship with a man born in the year of the Ox or the sign of the April Rat. The April Ox man is well suited for a relationship with a woman born in the sign of the April Ox, the April Rabbit, or the year of the Rat. Both men and women born in the sign of the April Ox should avoid relationships with a person born in the year of the Horse or the Ram. April Ox females are sensible, well organized, and make good managers. Both men and women of the April Ox sign like a comfortable, well run home and a daily routine which allows them adequate time for their other interests.

MAY

A person born in the month of May during the year of the Ox is cheerful, optimistic, kind-hearted, and generally well liked. Most generally, they are also fortunate or lucky in some way, and many are born into wealthy or prominent families. This same luck can lead them to fail to have the experience necessary to face difficulties and obstacles which can come into their lives as adults. Perhaps for this reason, the May Ox has problems adapting to different situations, changes, or new people. While the May Ox may be financially successful in life, it is best if they do not test their luck by changing their careers or jobs because they are prone to misread the situation and to lose out in the long run. Their optimism can also lead them to be extravagant in their spending habits. Their appearance is important to them, and they like to be well dressed and fashionable and to live in stylish, comfortable homes. This congenial and outgoing person loves to entertain and to show off their possessions and homes. The May Ox can also be stubborn and willful and hates to lose in a competition or in a disagreement. This person can be a sore loser, but they rarely hold a grudge for long. The May Ox must learn to be more moderate in his or her lifestyle and to guard against over consumption of food and drink which can lead to health problems.

The May Ox can be ostentatious and superficial and has a tendency to be attracted to another person based on looks alone. Despite this tendency, there is a strong indication that the May Ox is happy in his or her relationships with others. The May Ox man especially admires physical beauty in a woman. He is best suited for a relationship with a woman born in the sign of the July Tiger, the August Rat, or the September Monkey. The May Ox female greatly prefers a man who is intelligent and well meaning. She does best in a relationship with a man born in the sign of the August Cock, the September Rat, or the September of August Rat. Both men and women born in the sign of the May Ox should avoid relationships with a person born in the year of the Sheep or the year of the Horse. You do well to seek skills and to develop attributes which enhance your lifestyle.

JUNE

Born in the month of June during the year of the Ox, this person is changeable and often impulsive. Their dual nature leads them to be introverted one moment and sociable and extroverted the next. They like change, meeting new people, and being exposed to different situations. They like variety and are involved in any number of activities and interests. They may have different groups of friends that they turn to as their moods change. They may establish a quiet, peaceful home where they can pursue their interests or studies, but they also seek large group activities where they can submerge themselves in the crowd. Many June Ox are unafraid of taking a risk and they often seek adventure or daring escapades as compared to their normally quieter lives. They develop particular likes and dislikes which they may change on an impulse and without explanation to others. This is not a person who does well in a dull routine which offers little opportunity for change or stimulation. The duality of their natures may lead them to change careers in mid-life or to seek a change of lifestyle or location during this period of their life. At times it seems as if the June ox has difficulty deciding which direction in life to take or what actually interests them enough for them to pursue one course diligently. The Ox, however, is known for being obstinate and headstrong, and the June Ox possesses these same tendencies even though they can be unsure of themselves. This person prefers to make his or her own decisions and doesn't really appreciate the advice of others, no matter how well meaning.

In a romantic relationship, the June ox can be either passionate or distant, depending upon their mood, and there are times when they are either late for dates or forget them altogether. This can make it difficult for them to get along with other people who don't really understand them. The June Ox female gets along best with a man born in the sign of the December Cock. The June ox man does best with a relationship with a woman born in the sign of the June Monkey or the May Cock. Both men and women born in the sign of the June Cock should avoid relationships with a person born in the year of the Dragon or the Ram.

JULY

Persons born in the month of July during the year of the Ox are intelligent and mild-tempered. They possess gentle, calm and caring natures. But like those born in the other months of the year of the Ox, the July Ox is stubborn and obstinate. This person is strong-willed with a firm belief in himself or herself and a strongly held belief that they are right. They resent others attempting to tell them what to do or how to do something. This can lead July Ox persons to believe too strongly in themselves and to over estimate their own abilities. One of their biggest challenges in life is to learn to listen to others and to take advice when necessary. The July Ox can also be devious in his or her attempts to get their own way. In many instances, the July Ox expends or wastes his or her energies in their efforts to be in charge. They also waste energy and get side-tracked in their efforts by their schemes and plots rather than working together with others to succeed at their endeavors. This, of course, can lead them to fail more often than they succeed. The July Ox can also be jealous of others and resent those who have the characteristics and stability which they lack. Often, it is not until after mid-life, that the July Ox person recognizes his or her own talents and abilities and then chooses to develop those skills. While this person is often aware of the social graces of other people, the July Ox often lacks these same attributes and fails to develop them, preferring to do things their own way no matter if others find their habits offensive. The July Ox must guard against developing nervous disorders due to worry and anxiety.

The July Ox female finds happiness and a lasting relationship with a man born in the sign of November or December Cock. The July Ox man does well in a relationship with a woman born in the sign of the June Monkey or the July Rat. Both men and women born in the sign of the July Ox should avoid relationships with persons born in the year of the Horse, the Dog, and the Dragon. Your inner strength and innate abilities lead you to develop your potentials and to seek worthwhile endeavors. Your relationships deepen as you mature and accept yourself and others.

August

Those persons born in the month of August during the year of the Ox are outgoing and sociable. Their extroverted nature makes them well liked by other people and especially attractive to the opposite sex. These people like to be the center of attention and often draw attention to themselves by their mere presence. They are open and outspoken and seldom keep their thoughts to themselves. They are honest and frank in their opinions which other people respect. The August Ox loves to entertain and to be included in social gatherings. And they are excellent conversationalists. In fact, if they have a fault, it is that they talk too much and every so often are guilty of putting their foot in their mouth. They can be demonstrative and entertaining, but often this person is not easy to get close to as they keep their most personal feelings to themselves. Their outgoing nature augments not only their busy social lives but their business endeavors as well. The August Ox is seldom at a loss for companionship and has many friends. In romantic relationships, the other person may become jealous of this person's friendliness and his or her ease at attracting the attentions of others. The August Ox is also sexually assertive and uninhibited about approaching members of the opposite sex. Their sexual nature, in fact, can also interfere with their family life or with serious romantic relationships. There is a tendency for the August Ox to become infatuated with a person and to idolize and place that person on a pedestal, only to become disappointed after they get to know the individual better. This person has a tendency to do better in a relationship when introductions are made through a friend or family member.

The August Ox female is well suited for a relationship with a man born in the sign of the May Snake, the April Rat, or the December Cock. The August Ox man finds lasting happiness with a woman born in the sign of the April or May Cock, or the October Monkey. Both men and women born in the sign of the August Ox should avoid relationships with a person born in the year of the Dog or the year of the Sheep. Your relationships improve when you get to know the other person well before making a commitment.

September

Those persons born in the month of September during the year of the ox are humorous, well spoken, and knowledgeable about any number of subjects. These are perceptive and intuitive people who read others well and use this ability in their interpersonal relationships. They are also honest, sincere and well meaning which gains them the respect of their friends, business associates, and family members. This is a diplomatic, congenial and well meaning person whose worst fault is probably having a soft heart and being overly generous with others on occasion which can cause occasional problems. The September Ox most generally succeeds both in business and in his or her social life. But in many instances this person must learn to be more decisive and assertive in their daily affairs and to guard against allowing others to take advantage of their better nature. This is a person who appears to know instinctively how to enjoy life to the fullest and they love to share this natural trait with others. Added to their other beneficial traits is the ability to handle their finances and business affairs well. They are good at managing and take the time to organize and run their lives accordingly. The September Ox is approachable and will help a friend or take the time to offer advice when asked.

In love, the September Ox is compassionate, gentle and caring. This is a person with good intentions who enjoys making the other person happy. Quite often their relationships are long lasting, pleasant and fortunate. The September Ox female is a helpful, attentive person who most generally finds happiness in a relationship with a man born in the sign of the December Rat or the January Cock. The September Ox man does well in a relationship with a woman born in any sign during the year of the Monkey, the sign of the February Rabbit, or the January Snake. Both men and women born in the month of September during the year of the Ox should avoid relationships with persons born in the year of the Dog.

Your kind-hearted ways and good intentions produce effective results, and this is most especially true in your love relationships.

OCTOBER

Those persons born in the month of October during the year of the ox are open minded, honest, and not prone to being distracted by unimportant issues. This is a confident and self-reliant person who sets high standards and high expectations not only for themselves but for others as well. They strive to achieve and because they believe in their own abilities they possess high aspirations and dreams which they often attain. But they are reliable enough to recognize they own limitations and most generally set goals which they are comfortable that they can succeed at. The obstinacy and endurance of the ox serves the October Ox well, and this person perseveres against any and all obstacles, overcoming problems and difficulties in life through effort, ability, and diligence. They are so determined and independent that they may choose to leave home at an early age to start work or to simply go out on their own to establish themselves. And often this very tendency adds to their ability to become successful in life. The October Ox person is self-involved and being independent can become an all-consuming effort. These are not people to follow the lead of others or to be overly impressed with style of fashion or what others are doing. This and their intent natures may hamper their relationships with others, and it can be an ongoing process for this person to improve their interpersonal skills. Be that what if may, the October Ox can be congenial and sincerely enjoys the companionship of others. This person, however, most enjoys other people who share his or her values, interests or opinions. They are also dependable and reliable and rarely make a promise they cannot keep.

When it comes to love, the October ox can be impulsive, and if they marry young, the marriage may not last. The second marriage of the October Ox is usually more successful and lasting. The October Ox female should seek a relationship with a man born in the sign of the April Cock or the May Monkey. The October Ox man is well suited for a relationship with a woman born in the year of the Monkey. The October Ox should avoid relationships with those persons born in the year of the Dragon.

November

Those persons born in the month of November during the year of the Ox are sensible, calm, and strong-willed individuals whose intelligence adds to their obstinacy. They are well-organized leaders who are naturally adept at managing their personal and business affairs. They appear distinguished and reserved and use their planning abilities to achieve their objectives. Generally speaking, they have a tendency not to listen to the opinions of others and at times can be selfish and self-centered. In some cases, they are also narrow minded, thinking their own opinions count more than what others may think. At the same time, the November Ox can develop an empathy for those persons less well off or in need of assistance. They are always ready for a challenge which allows them to put their organizational and leadership abilities into practice. Because the November Ox shuns the advice of others, they must guard against making errors in judgment or oversights which can result in costly mistakes or produce problems that then have to be dealt with. This person may not deal well with mistakes and errors which can cause them to lose self-respect. With many a November Ox person this fear of making an error in judgment makes them overly cautious particularly in relationships. And they may spend more time thinking about and considering an involvement than actually enjoying a relationship. For the most part, however, the November Ox makes a good friend and when they strive to become more congenial and sociable, they do well in interpersonal relationships. This person enjoys sports or physical activity.

The November Ox female is well suited for a relationship with a man born in the sign of the January Cock or the December Rat. The November Ox man does well in a relationship with a woman born in the sign of the January Sheep, January Monkey or the September Cock. Both men and women born in the sigh of the November Ox should avoid relationships with those persons born in the year of the Dog. With the right person, the November Ox is caring, understanding, and compassionate. They have the capacity of making others feel wanted and understood.

December

Persons born in the month of December during the year of the Ox use the full force of their obstinate natures in order to strive to establish a good reputation and to be respected. They set high standards for themselves and others and place honor and respect as a top priority. In fact, being well known and respected may be more important to them than even financial success or a prestigious position. At the same time, this is a competitive person who dislikes making a mistake or losing at anything. They make a diligent effort to succeed at their endeavors and to overcome any obstacles placed in their way. This person abhors idle gossip and rumors especially when it is aimed at them. They can become impatient and irritable when things aren't running smoothly or going their way. Taking all of that into consideration, the December Ox is a congenial person who is well liked and appreciated by friends and family alike. They enjoy an active social life and like to be included in social occasions. The December Ox is witty and their sometimes off beat sense of humor helps them to fit in to different situations, -making friends easily. But they remain sensible people who do not like for things to get out of hand or too racy. They also make every effort to put their talents to good use. The December Ox succeeds by making definite goals and then pursuing them gradually, one step at a time. They make excellent business people and often successfully establish their own businesses. They also do well in the entertainment field or in sports or the area of recreation.

The December Ox does best in relationships with others who share similar interests, values, and character. The December Ox female finds love and happiness in a relationship with a man born in the sign of the August or September Rat. The December Ox man is well suited for a relationship with a woman born in the sign of the June Cock or the July Rat. Both men and women born in the sign of the December Ox should avoid relationships with those persons born in the year of the Sheep or the Ram. You find success and happiness in love and romance when you take the time to get to know the other person well.

Date With Destiny

TIGER

Share With Me Your Fantasy

THE TIGER

The Tiger is an unpredictable mixture of intense caution and undisciplined assertiveness. You can be charming, imaginative and optimistic, but you are also highly likely to not follow directions and to openly disagree with your superiors. There are those who say you can be as cold-blooded as though carved from wood, and that you are capable of becoming either a great leader or a resistance revolutionary. You can be rebellious and most unconventional in your thinking and lifestyle. Your suspicious nature can lead you to be self-centered, selfish, and critical of others. You distrust your first impressions of people and can be slow to ever trust someone else. You rarely seek or take advice from anyone, and very few people ever confront you. In this manner, you often fail to listen to others, but you expect other people to listen to your opinions. It matters not to you if you fail because you will try again until you succeed or win at your endeavors. In this way, you refuse to abandon an idea or a project once you have started, and you can be quite obstinate when set on a particular course of action. You are hard working and determined and derive a certain amount of satisfaction from completing whatever you start. At the same time, you rarely waste your time arguing over what you consider insignificant or of little importance to your life. You are known for being strongly independent and courageous, and some perceive this as stubbornness. But it is wise for you to be cautious with friends and acquaintances because if you trust too much you can lose your position or lose financially.

You are fortunate in that the year of the Tiger is considered the sign of good luck. And with your adventurous streak, courage, and daring risk taking, you may need all the luck you can get. You are drawn to excitement and stimulating situations, dangerous activities, and the unexpected. You improvise well in such situations although your daring nature may lead you into life-threatening situations. You are admired by others for your courage and your leadership abilities as well as your strong will power which you use to succeed. But while others may admire you, there are those who are put off by your obstinate nature.

With all this courage, you can at times also lack self-confidence or self-esteem, and you are particularly offended and upset if you ever "lose face" in a situation. You remain an enigmatic character in that you can be humorous, generous, and magnetic, or then turn around and be hotheaded, rash, restless, and even reckless.

Your luck holds with you in other areas of your life although you prosper primarily because of your great determination to do so. Regardless of your profession or career choice, you strive to attain a leadership position and often accomplish this goal because of your assertive nature and because you are generally well organized. You would be wise to select a business partner who was born in the year of the Boar because they are honest and trustworthy. While you may be a risk-taker, you may want to invest wisely and avoid speculative ventures as there is indication for financial losses. It is also advisable to avoid court cases and to make every attempt to settle any law suits out of court. On the other hand, your luck brings you ease in acquiring property and possessions although you have a tendency not to hold on to possessions. Money and material possessions matter less to you than the challenge involved in any particular endeavor. Quite often, a person born in the year of the Tiger is helped in one way or another by parents, family, or an influential person. And you maintain a good relationship with family members especially your brothers or sisters who admire you even though you like to be the center of attention. You benefit from good health as well even though inclement weather can bring on minor illnesses. While you may at different times in your life be accident prone, travel is safe for you, but you should be cautious while driving. You can be honorable and sincere, and your curiosity and intellect leads you to seek knowledge which makes you a well informed person. But your strong, egotistical nature can also make you volatile and demanding and at times difficult to live with. Whatever your educational or financial background, you are astute enough to take advantage of any and every opportunity that comes your way, and this trait adds to your accomplishments and successes in life. Your independent and restless nature also leads you to be always on the move, looking for the next opportunity or new situation to become involved with. These opportunities also bring you luck when you are young and first starting your career or business activities.

You handle your love life in much the same way as the rest of your life. You are emotional, but you are restless and become bored easily which makes you always on the prowl for a new relationship or encounter. In this way, you may take your relationships as seriously as your partners, preferring not to become tied down. This can make you hasty and careless with the feelings of others. You are also somewhat impulsive and must guard against marrying at an early age or in haste. In this way, your love life can be unpredictable even to yourself. You can be undemonstrative in your affairs and are not the cuddly type. You have a tendency to want to dominate your partner and this is true in sex as well. You have a need to prove yourself sexually and like to be in control of the situation. In relationships, you tend to be argumentative and excitable. And you unconventional attitudes carry over into your relationships. With all that, once you find the right partner, you have a tendency to settle down and to be faithful and loyal. It is said that the Tiger is most compatible and finds a happy relationship with a person born in the year of Horse because this is also a strong sign, but the Horse is willing to be controlled by someone else. You would find a partner born in the year of the Dog to be protective and loyal, and you would be honest with each other. The Dragon is energetic and brings wisdom to a relationship with a Tiger. The Tiger would find it difficult to understand the Monkey who is cunning and may cheat the Tiger. The Tiger is too hasty for the Ox who prefers to take things more gradual, and the two of you would either learn to accept each other's faults or would eventually destroy each other. The Cat may set off your temper just for the fun of instigating a disagreement and then laugh at your response. The Tiger would find the Snake a very unforgiving partner and someone to avoid. You are not a good match for the Ram who is more conservative than the Tiger. While you may be reserved in showing your love, with the right mate, you can be most charming, compassionate, and intensely passionate. You take your responsibilities seriously and make a good spouse and parent to your children.

JANUARY

Those persons born in the month of January during the year of the Tiger are straightforward and direct as well as being honest in their dealings and relationships with others. They prefer that everything run as smoothly as possible and therefore are not overly assertive, and they may quite often appear calm and even reserved in many situations. At the same time, because of their honesty, they have a tendency to see things in black and white and can be found to be inflexible or a bit intolerant of the opinions and suggestions of others. They aren't the best communicators and once they have stated their opinions, they assume others agree or understand their point of view. This can, of course, result in disagreements. But their reserved manner combined with their direct statements or questions can give them an air of authority. This being the first month of the sign, the January Tiger can be eager to rush head first into life full of hope and anticipation of come what may. They are challenged to learn patience and to proceed at their endeavors gradually, one step at a time. They do best when they take their time and are not overly hasty in pursuing their goals. The January Tiger realizes the most benefits when he or she makes every effort to save their resources when they are young. You do well when working for large businesses or corporations where you can achieve your career potential, but expect steady promotions rather than a quick rise to the top. This is a person with an active social life who is energetic, likes a variety of activities, and loves the out of doors.

In romantic relationships, the January Tiger may have difficulty expressing his or her emotions. This leads others to misunderstand you, and you may find that your relationships don't develop as you had hoped. It might be advisable for the January Tiger to look for a partner whose personality traits balance out their own. A January Tiger female does well with a September Horse or an October Dragon. A January Tiger man is well suited for an August Dog or a March Horse. Both men and women born in the sign of the January Tiger should avoid relationships with a person born in the year of the Ram.

February

Those persons born in the month of February during the year of the Tiger remain calm and methodical but once they begin work on a task they move quickly and efficiently until completion. In other words, they prefer not to rush into things, but once they start, they finish their work. They may at times, in fact, take on more work or schedule more activities than they can complete on time. This tendency often leads them to be late for appointments or to miss them altogether. But the February Tiger doesn't become overly concerned when he fails to show up or has to break a promise to someone even though this can often make them miss out on opportunities. This is a person who learns best through experience rather than any kind of formal instruction, but this is also a person who easily acquires the skills for a job and who is adept at many things rather than specializing in any particular vocation. They work steadily on their goals, gradually accomplishing what they set out to do. The February Tiger is sociable and generally well liked and popular with friends and family, often enjoying a variety of social activities and a wide circle of acquaintances.

The February Tiger male is most likely very masculine. This man has a tendency to fantasize about women. The February Tiger female is neat, well organized, a good manager, and is capable of working just as hard as the male of this sign. She approaches her romantic relationships seriously and commits herself to the love of her life. The February Tiger may well marry their childhood sweetheart because they prefer a partner who they know well and who understands them well. This person much prefers a stable home life, even though there are times when they experience periods of anxiety with their families due to finances or illnesses. These difficulties can be dealt with.

A February Tiger female is well suited for a relationship with an October Horse, June Horse, or March Dog. The February Tiger man enjoys a relationship with a woman born in the sign of July Dragon or September Dog. Both men and women born in the sign of the February Tiger should avoid relationships with persons born in the year of the Monkey.

March

Those persons born in the month of March in the year of the Tiger are said to be extremely sensitive people who react quickly to their emotions, new situations, or beauty in any form. They may express their emotions through art, writing, music, or acting. This person is easily overwhelmed by sudden changes, surprises, or unexpected events. They are also sympathetic people who empathize with the needs of others and who can be generous and supportive. They develop interests in charities or community services where they can help other people who are less fortunate then themselves. They are well liked by other people because of their sensitive and caring nature. These are kind, friendly people, but they can also be somewhat reserved and protective, even a little secretive, about their personal lives until they know a person well. This is a person who makes every attempt to work wisely and diligently and who manages his or her finances well.

The March Tiger chooses friends and acquaintances who share common interests, values, and likes and dislikes. They also often prefer that a person is a friend first before furthering a relationship into a romance. There is a strong tendency for the March Tiger to develop a romantic relationship with a person who possesses a similar nature. And the problems in romance for this person is often associated with their sensitive and emotional nature which may lead them to have difficulties dealing with harsh realities. And while the March Tiger may marry someone with a similar nature, they quite often are attracted to people from a foreign country or different ethnic background. The March Tiger female finds love and compatibility with a man born in the sign of February Dragon, June Dragon, or January Dog. The March Tiger male is well suited for a woman born in the sign of the May Dog, December Tiger, or July Horse. Both men and women born in the sign of the March Tiger should avoid relationships with persons born in the year of the Snake or the year of the Ram. When you find the right partner, you enjoy a pleasant home and a busy social life. You are adept at social graces and many consider you refined and appreciative of beauty and good taste.

April

Those persons born in the month of April in the year of the Tiger are open-minded, inquisitive, alert, intelligent, and hard working. They have a tendency to be liberal in their thinking and tolerant of others. But they are independent thinkers as well and aspire to a leadership position. They also have a cheerful nature which makes them well liked by others. The April Tiger is considered the luckiest of the animal signs in the Chinese Zodiac. This luck often brings them success in their aspirations, for some before their thirtieth birthday, and in some cases this good fortune is in the form of fame and prosperity. Whatever difficulties come their way, the April Tiger appears to sort them out and to move on. This is an optimistic person who is comfortable and works well with others, but who likes to be in charge of the situation. They are happiest when in control of their own business. Work that requires routine or detail-oriented work may not be the most suitable type of occupation for the April Tiger. These people are good at making plans or coming up with new ideas, and they like to think and plan ahead. They also like to find solutions to problems and to explore and investigate new methods.

While they have a tendency to separate their business life from their personal life, in their preference for thinking ahead, the April Tiger can be somewhat cautious in romance and is prone to setting high standards for their marriage partner. And whether working or playing, this person loves good food and drink and enjoys sampling the cuisine at a new establishment, dining at a favorite restaurant, or planning a meal at home. The April Tiger female finds lasting happiness with a man born in the sign of the May Dog or December Cock. The April Tiger man finds love and happiness with a woman born in the year of the Dragon. Both men and women born in the month of April in the year of the Tiger should avoid relationships with a person born in year of the Ox, the Rat, and the Monkey. The April Tiger is generally lucky in love and marriage finding lasting love and companionship with the right person. Any difficulties are handled with ease.

MAY

A person born in the month of May during the year of the Tiger is kind, humorous, and clever. These are generally physically attractive people who are cheerful and congenial and well liked by their friends and acquaintances. A May Tiger is the perfect host or hostess and makes a lively, entertaining companion as well. This person is considerate of the feelings of others and doesn't like to impose on other people, and can even be shy about his or her acts of kindness to others. Neither does this person like to accept help or advice from other people, and in this way, the May Tiger can become quite obstinate and stubborn. But this is also a very determined person who sets his or her mind to something and then does just that. If the May Tiger suffers from faults, it is most likely pride and over confidence in their own opinions and decisions which can lead to mistakes. The biggest challenge in the life of the May Tiger is to learn to listen to others and to accept worthwhile advice when necessary. But they learn best from their own mistakes in this regard. This is a busy, energetic person who has trouble keeping to a set schedule and who often misses appointments. Also, this is not a person to take advantage of because the May Tiger can become angry and vengeful. This person becomes successful by learning patience and using their intelligence and acuity to increase their income. This personable person does well in the entertainment field.

With their charm, good looks, and kind-hearted ways, the May Tiger has no trouble attracting members of the opposite sex. These people can move with the grace, elegance, and regal bearing of the tiger representing this sign. The May Tiger female is well suited for a relationship with a man born in the sign of the August Dragon or the May Dog. The May Tiger male does well in a relationship with a woman born in the sign of the August Horse or the March Dog. Both men and women born in this sign should avoid relationships with people born in the year of the ox or the year of the Snake. When you allow others to get to know you and your good qualities, they trust and respect you.

JUNE

The person born in the month of June during the year of the Tiger is wonderfully congenial and honest in their dealings with others. They can be mild-mannered and even tempered and may even appear elusively shy. These people possess a special kind of luck which allows them to find success without working overly hard at doing so. There is some indication that the June Tiger is often born of a wealthy family and never experiences financial problems as long as they spend their money wisely, make sound investments, and limit expenses. There are exceptions to every rule, of course, and the June Tiger from unfortunate circumstances, no matter how successful they become, finds it difficult to emotionally overcome their earlier hardships. Whichever the case may be, this is a person capable of expressing themselves well and often creatively. They enjoy an active social life and a variety of activities and interests. The June Tiger is well liked and accepted by other people who respond to this person's honest and friendly nature. If they have a fault, it is that they can be forgetful and at times preoccupied with their thoughts which makes them appear flighty and somewhat fickle.

The June Tiger female is attractive and especially charming and alluring with lovely voices. Men are attracted to this demure and seemingly shy female only to discover later in the relationship that she can be assertive and determined. She finds the most happiness in a relationship with a man born in the sign of the July Horse or the February Horse because they are strong and loyal. A relationship with a man born in the sign of the June Ox is also a good match if they share common interests. The June Tiger male is attracted to sensual women and is usually very generous in relationships. He is most suited for a lasting relationship with a woman born in the year of the Rat or in the sign of the July Ox. Also, those people born in the year of the Horse match the honesty of the Tiger. Both men and women born in the sign of the June Tiger should avoid a relationship with persons born in the year of the Boar of the year of the Ram. Your luck in relationships leads you to choose carefully.

JULY

Those born in the month of July during the year of the Tiger are extremely energetic and physically active. They can be seemingly tireless. The July Tiger is methodical and systematic as well as well organized in their planning and thinking. This is also an understanding and empathetic person who is concerned about the feelings and welfare of others and who makes every attempt not to offend other people. They are for the most part optimistic, congenial, and cheerful, but are also hard working and serious about their work and their personal lives. They apply themselves to the task of acquiring the necessary skills for their career field. And they succeed in life on the basis of their skills and efforts and most generally do quite well financially, socially, and in their personal relationships. This is a person who knows how to pace their efforts in order to achieve their accomplishments and will use their strength, power, and abilities to persevere against any difficulties or obstacles. They are also adaptable to new situations and can change their mood or approach to best accomplish their goals. While the diligent July Tiger is most fortuitous, there is an indication that they need to be cautious after the age of forty and to guard against work related mistakes make by an employee or by a subordinate which can result in a loss. The July Tiger much prefers a well managed and comfortable home for themselves and family. They like to entertain and to be socially active and involved in community affairs or other activities depending upon their interests.

In love, the July Tiger has a tendency to select a romantic partner who is different in nature to themselves which often creates a pleasant balance of interests and ideas. The July Tiger female does well in a romantic relationship with a man born in the sign of the May Ox, the June Ox, or the June Dog. The July Tiger man is well suited for a relationship with a woman born in the sign of the May Ox, the March Tiger, or the December Tiger. Both men and women of this sign should avoid relationships with those people born in the year of the Snake or the Monkey. You succeed in love with choose your mate carefully.

August

Those persons born in the month of August during the year of the Tiger are friendly, congenial, and can be gregarious, sociable, and fun loving. But they are also strong willed and extremely independent people who do not like to be controlled or dominated by others. There is said to be a duality in the nature of the August Tiger. However outgoing and fun loving they appear to be, this is serious-minded, intent person whose strong independent nature can lead them to be somewhat selfish and self-centered at times. They like to have their own way. There is an indication for prosperity as long as the August Tiger handles his or her finances and investments wisely. There is also some indication for good fortune and luck which is dependent upon the August Tiger learning to be generous and to treat other people well. This person is intuitive and possesses some degree of precognition and psychic ability, but the August Tiger rarely relies on this ability alone. They prefer to combine this perceptiveness with their cognitive abilities and to think through their decisions carefully. But in their thinking, the August Tiger can be dogmatic and opinionated with a tendency to be critical of others. This person should guard against being jealous and talking in a derogatory manner about others.

The August Tiger may marry more than once, falling out of love, only to find happiness in a second marriage. The August Tiger females have a tendency to find lasting happiness with a man who is from a different part of the country or with someone with a background which is different from their own. They do well with men born in the sign of the December Dog or the February Horse. The August Tiger man likes to be in charge when in a relationship, making precise plans and decisions. This man is well suited for a relationship with a woman born in the sign of the March Ox, the February Ox, or the April Boar. Both men and women born in the month of August during the year of the Tiger should avoid a relationship with a person born in the year of the Snake, the Monkey, or another person born in the year of the Tiger. The August Tiger is a physically attractive person who prospers through their ingenuity and perceptiveness.

September

Persons born in the month of September during the year of the Tiger are energetic, diligent workers who are generous and giving of their time and resources. This person can be unconventional with a progressive and innovative outlook on life. They can also be impulsive and quick to make a decision or to take action, and in this way can be daring and unafraid of taking a risk. At the same time, they are serious, quiet, and productive workers who don't like to interfere with others. This person generally finds success in his or her career. The September Tiger is caring, kind, and sociable, but can be impatient and standoffish with conservative thinkers. They possess an keen appreciation for art, music, or beauty in any form. They also usually have a distinctive style of dress or fashion which they prefer, and they remain well-dressed and fashionable throughout their lives. This person must make every effort to take care of their health and to develop healthy habits and lifestyles. A good diet and an exercise program aids the September Tiger in preventing illnesses.

The September Tiger looks for a marriage partner who shares similar values and who is attractive. The September Tiger man has a strong preference for a voluptuous woman, but in marriage he generally chooses someone who is not only attractive, but confident, caring and attentive. The September Tiger man finds lasting happiness with a women born in the sign of the September Boar, the April Dog, or the December Ox. The September Tiger woman is generally more moderate than her male counterpart. She is a caring and considerate person who has a tendency to prefer a marriage partner who takes charge of the relationship, but who is also loyal, supportive, and equally attentive. She is a stylish and attractive woman who finds love and lasting happiness with a man born in the sign of the April Rat, the February Tiger, or the October Dragon. Both men and women born in the month of September during the year of the Tiger should avoid relationships with those people born in the year of the Ram, the Monkey, and the Snake. The September Tiger is most happy in a comfortable and well maintained home that is furnished tastefully with beautiful objects.

OCTOBER

The persons born in the month of October in the year of the Tiger find luck and good fortune in spite of themselves. You are strong willed and independent and often take impulsive actions without considering the consequences. This can cause the October Tiger difficulties, but they generally overcome any obstacles. You may feel that others take life much too seriously. The October Tiger can be vulnerable and easily taken advantage of or cheated, but they have a tendency to take effective action when this happens. This is a person who recognizes and takes advantage of opportunities and finds that their luck holds and that, although they may need to work hard, they succeed in life. They have an expansive nature and with their good fortune they find that their business grows profusely with little effort. Any loss, in fact, is usually due to over expansion and to the mistakes of others. Good opportunities benefit this person throughout life. The October Tiger is well liked by others and has a tendency to achieve social success and may become famous either socially or in their field of endeavor. Your optimism, cheerfulness, and attractiveness gains you the attention and admiration of other people.

The October Tiger is fortunate in love and marriage as well, finding the right partner and enjoying a happy relationship. The independent October Tiger female prefers a well run home and a partner who enjoys helping around the house and sharing household duties. She is especially fond of a man who likes to cook. She is loyal and faithful to her partner. The October Tiger female is best suited for a relationship with a man born in the sign of the August Horse, the December Horse, or the January Dog. The October Tiger man prefers an attractive woman who is a good hostess and who is socially adept. He also much prefers a well run and comfortable home. The October Tiger man finds love and happiness with a woman born in the sign of the June Ox, the November Boar, or the March Horse. Both men and women born in the sign of the October Tiger should avoid relationships with a person born in the year of the Ram and the year of the Snake. Your lucky star benefits you in love.

November

Those persons born in the month of November during the year of the Tiger are natural managers whatever their job or career position. These are energetic, active people who love the out of doors and many are best suited for jobs in which they aren't confined inside. The November Tiger is often in a big hurry to start things and then in a hurry to finish whatever it is he or she is doing. In this way, they can be extremely impatient which can in itself cause them problems and difficulties. The November Tiger is an attractive person who may have a rather quiet home life, but who generally has a tendency to have an extremely active and changing social life. They may move easily from one social circle to another always seeking some new experience or activity to keep their curious minds and precocious attitudes engaged. This is not a person who likes to waste his or her time, energies, or efforts. Even in relationships, if one doesn't work out, they are ready to move on to another.

The November Tiger may feel restrained and limited by one relationship and many times in marriage they tend to feel lonely and confined. This can lead to affairs outside the marriage if they are unable to come to terms with their partner in the marriage arrangement. There is also a tendency for the November Tiger to marry for financial security or for social prestige, especially the longer they wait to marry. The November Tiger female may find herself so preoccupied with her career that she doesn't marry, and is drawn to affairs with married men who are either older and well situated or who can offer her what she is looking for. The November Tiger female much prefers a strong, in charge type man. She may find the most suitable marriage partner with a man born in the sign of the July Tiger or the January Dragon. The November Tiger man is well suited for a relationship with a woman born in the sign of the April Sheep, the October Rat, or the December Boar. The November Tiger also does well with a person born in the year of the Horse. Both men and women born in the sign of the November Tiger do best when they avoid relationships with people born in the year of the Snake or the year of the Monkey.

December

Those persons born in the month of December during the year of the Tiger are calm and intelligent and very sure of themselves. They are especially self-reliant and confident in their own abilities to achieve whatever goals they set for themselves. These people seem to know where they are going and what they want to accomplish. They set high goals for themselves professionally and strive to acquire a comfortable home, possessions, and social standing in the community. This person seems to be unstoppable in getting what they want in life. They are hard working and willing to make any necessary sacrifice in order to achieve success. In their undaunted approach to attaining their objectives, they can offend others and at times will find that powerful people oppose them. This person strives, however, to be diplomatic and to use his or her negotiating abilities to handle any problems which occur without losing face. The December Tiger is not only highly motivated but capable of being successful in business. Often, they find that someone helps them in their endeavors. This is a person who rarely speculates or takes a risk, but seems to know intuitively the right track to follow. The December Tiger's biggest fault, though, is that this person has a tendency not to take money seriously. They love to spend and can be generous and giving with friends and family and especially in relationships. These are friendly and sociable people who like to make others feel good and can lavish praise and compliments on other people in their efforts. They possess an innate trait of making other people feel good about themselves which helps them in their business and social life. They are attentive and also like to give of their time and energy.

The December Tiger female finds love and happiness in a relationship with a man born in the sign of the January Rabbit, a May Sheep, or a December Dragon. The December Tiger man does well in a relationship with a woman born in the sign of the April Rat or the January Sheep. Both men and women born in the month of December in the year of the Tiger should avoid relationships with persons born in the year of the Ram, the Snake, or the year of the Cock.

Date With Destiny

Rabbit

Share With Me Your Fantasy

THE RABBIT

The Rabbit possesses a winsome character, that is, they are generally pleasing and engaging often because of a childlike charm and cheerful, innocent nature. They are outwardly attentive and hospitable, taking care to create an inviting atmosphere where others feel comfortable and relaxed. Although they are most likely conservative with a strong preference for the traditional, they are also adaptable and diplomatic, preferring to settle disagreements peacefully by avoiding arguments and quarrels. The Rabbit has a tendency toward a tranquil life, but their generous, accommodating, and sympathetic nature leads them to be always ready to help others. Their youthful energy gives them a lithe and gentle appearance which augments their ability to turn enemies into friends. And while they crave friendship and to be surrounded by people, they often have few close friends. Beneath their friendly, outgoing exterior is a person who is detached and analytical, always able to size up a situation or other people. In fact, they possess an almost clinical detachment regarding issues that do not effect them directly. In this way, they can be not only intuitive but circumspect with a serious, intellectual perception of situations. They can be meticulous and driven in their ambitions and work, and at times they like to display their knowledge on any number of subjects. But this may be more to impress others than it is an actual interest in the subject matter. Preferring to live and let live and to enjoy life in a peaceful, tranquil manner, the Rabbit may take little interest in becoming involved in social issues or politics, wanting to leave the everyday problems and difficulties behind them as they seek to find harmony. That is not to say that their moods are not effected not only by the world at large but by their environment and immediate surroundings. And the Rabbit will go out of his or her way to create a happy situation and circumstance where they and/or their friends and family can enjoy life. But, at the same time, their energy level leads them to be quick to react and driven to achieve what they want. In this way, no matter how patient they may be, the Rabbit is always on the move to accomplish what they set out to do. And they must make every effort to plan their futures carefully rather than rushing to get things done. This is an easy skill for the Rabbit to learn, fortunately, for they like to give careful thought to whatever it is they do.

Good fortune smiles on the Rabbit, however, and there is every indication that those born in the year of the Rabbit have a settled, peaceful life devoid of major problems, obstacles, and problems. Any financial problems are minimal as the Rabbit is well suited to any number of promising career choices. Their worst problem in determining a career choice is examining the job and determining whether it is what they want to be doing with their lives. Whatever career choice they make, they are well accepted and well thought of by their associates and colleagues, often becoming prominent in their fields and prosperous. They are most cunning in business and make the best negotiators. In fact, the Rabbit is capable of acquiring financial security with ease, and the chances are good that they will acquire property and assets just as easily. The Rabbit remains just as lucky in speculation and investment and should only be hesitant about high-risk investments which could result in losses. The wisest investments for the Rabbit are large corporations with promising dividends. The luck of the Rabbit holds in lawsuits as well, and if the individual is right, he or she will always win. Lost articles will seem to reappear almost as if by magic for the Rabbit, and they are able to maintain contact with friends and family, rarely losing touch with those close to them.

The Rabbit is considered an elegant individual with a suave and debonair appearance and a sophisticated outlook on life. They have a natural graciousness that endears them to others. Throughout their lives, they maintain a good and understanding relationship with their parents and siblings who are always supportive. They are honest and thorough and a sympathetic listener who enjoys other people. But, of course, there are any number of instances where the Rabbit can be overly hesitant and faint-hearted with a slight tendency to be superficial and perhaps too sentimental. Having a conservative streak, the Rabbit develops particular likes and dislikes, and however friendly and amiable they can be, they like things done their way--especially in their home environments. At times moody, the Rabbit can be unpredictable and at other times touchy and easily offended. At their worst, they are snobbish with a distinct, subjective dislike for what they perceive as intolerance in others. There are those Rabbit individuals who are strongly egotistical and others who are extravagant, hedonistic pleasure-seekers.

It is the strength and energy that most impresses other people though. And when it comes to romance, the Rabbit proves to be a person capable of the most sincere tenderness and gentleness. They are kind-hearted, generous, and loving with a tendency to want to make the other person happy and one of the ways they do this is by attempting to fulfill the other person's wishes. When the Rabbit finds the right partner, he or she falls deeply in love. And in love and marriage, the luck of the Rabbit holds foretelling a successful and happy relationship. The marriage partner should be a person who understands that the Rabbit-born individual has a tendency to keep their deepest feelings and emotions tucked away and well hidden. For the best relationship, the Rabbit should seek a partner born in the year of the Ram, the Dog, or the Boar. The Ram is most empathetic and understanding of not only the nature of the Rabbit but of their talents as well while the Dog and the .Boar are willing to help the Rabbit build a secure financial future. The Rabbit can also find love and happiness with another person born in the year of the Rabbit or with a person born in the year of the Dragon. Those born in the year of the Rabbit will find it more difficult to find the. harmony and peacefulness they seek when in a relationship with a person born in the year of the Rat, the Cock, or the Ox. A relationship between a Rabbit and a

Rat, Cock, or Ox produces idleness and problems and turns the Rabbit's good luck into bad luck. Again, it is most important for the Rabbit to take his or her time to find the right partner who returns their love and who is understanding. It is important they share the same goals and values in life. When the Rabbit takes the time to find the right person, he or she is promised a long and happy relationship in which the luck of the Rabbit holds.

The Rabbit is a talented and quietly ambitious individual who is somewhat detached but who remains well meaning and well liked throughout life.

January

Those persons born in the month of January in the year of the Rabbit are attentive and hospitable with an openhearted, sympathetic nature that is sincere and reflects their positive, optimistic outlook on life. The January Rabbit possesses a knowledgeable and quite often elegant manner which may in part be due to being well educated or self educated in matters of interest to them. Their pleasant and considerate ways may, in some cases, be an outward cover for a person who is quite dependent on others such as family or friends. But they are totally sincere in their concern and efforts to assist others whenever and however they can. At the same time, they may resent others taking advantage of this giving nature, and if they perceive that a friend or associate is doing just that, they will distance themselves from that person. For the most part, their attitudes win them numerous and long lasting friendships because they are admired and appreciated by many people with whom they come in contact. There is indication, also, that their natures lead them to be successful in their business endeavors although they may never achieve wealth or fame. In fact, the January Rabbit may care less about prestige or being the center of attention, preferring to remain in the background with a live and let live attitude. The January Rabbit is often drawn into the food industry, the hotel industry, entertainment, or other occupations which provide services to others. They work cautiously and quietly and seldom with an effort to take their time and do things right rather than to rush their business activities. Friends and family are often very supportive of the January Rabbit. Whatever the size of their income, they save wisely and always appear to have money on hand.

The January Rabbit female is well suited for a relationship with a man born in the year of the Dog or the year of the Dragon, especially the March Dog and the April Dragon. The January Rabbit male does well in a relationship with a woman born in the sign of the August Ram, the September Dog, or the November Rabbit. Both men and women born in the sign of the January Rabbit should avoid relationships with those born in the year of the Rat.

February

Those persons born in the month of February during the year of the Rabbit are delightfully pleasant company who make an effort to put others at ease. That being the case, they can be extremely cautious until they make up their minds and then they are direct and straightforward, exhibiting an ease at expressing their thoughts both eloquently and fluently. The February Rabbit possesses a flair for fashion and style and prefers to be well dressed and presentable at all times. At their best, they are elegant and sophisticated, but at their worst, they are superficial and overly sentimental and emotional. There are also times when their caution leads them to be overly hesitant and faint-hearted about matters which they may find displeasing. And they can become most displeased when faced with problems, mistakes, or misfortune (none of which they handle well). Their hesitancy in some matters may even lead them to miss out on opportunities in their efforts to avoid problem situations. Be that what it may, the February Rabbit are forgiving and most generally caring individuals who show a tendency to be sincerely good listeners with an ability to comfort others. This ability makes for many friends and others who admire them. The February Rabbit makes a distinct effort to separate career from family and home life, preferring to establish a home setting that is pleasant, relaxing, and hospitable. They most enjoy spending time with others who share their appreciation for music or art or other interests. There are times when the February Rabbit can be a bit moody and during these times, they are easily offended and can display an abrupt temper.

The February Rabbit female needs to feel appreciated and admired. She finds love and happiness in a relationship with a man born in the sign of the September Rabbit, the June Boar, or the July Dragon. The February Rabbit man is best suited for a relationship with a woman born in the sign of the October Dragon, the May Dog, or the October Ram. Both men and women born in the month of February during the year of the Rabbit should avoid relationships with those born in the year of the Cock or the Ox. The February Rabbit is serious and responsible toward family and spouse.

March

Those persons born in the month of March during the year of the Rabbit are enthusiastic, energetic, and kind hearted. They enjoy active social lives and are always on the go from one activity to another. They are usually too busy with their own active lives to bother with interfering with the lives of others. They are friendly, however, and this amiable disposition leads them to be able to make friends and acquaintances easily. They interact well with others, but may become bored with one group making them always looking for a new activity or a new group or new people with which to socialize. This tendency to become easily bored may stem from their impatience with others who do not agree with them. The March Rabbit is prone to form strong opinions and to be somewhat set in his or her ways. They also have a tendency to concentrate on their work or on achieving their goals in life and this too can lead them to be impatient with others. They can be cunning in business, but in their social lives they can also be unpredictable

45

and changeable. All the same, luck and good fortune seem to follow the March Rabbit, and they rarely loose possessions or friends. The March Rabbit loves to travel, meet new people, and see new places. They must however be cautious of accidents while traveling. Because of their need for new experiences, the March Rabbit may also move through a number of romantic relationships before finding the right person with whom to establish a lasting partnership. They like to pampered and catered to, but they may do best in a relationship with a strong person who can provide guidance and stability to their lives.

The March Rabbit female may find it difficult to make up her mind when it comes to marriage, and it may just be that a strong-willed man will hold her attention. She is best suited for a relationship with a man born in the sign of the February Boar or the February Ram. The March Rabbit man is well suited for a relationship with a woman born in the sign of the September Dog or the October Dragon. Both men and women born in the month of March during the year of the Rabbit should avoid relationships with persons born in the year of the Rat or the Ox.

APRIL

Those persons born in the month of April during the year of the Rabbit are proud, ambitious individuals with logical if somewhat idealistic thoughts. Their lofty perfectionist tendencies can lead them to be a bit snobbish and critical of others at times. But they are just as self-critical and will worry themselves to distraction over whether a task has been completed satisfactorily. Their perfectionism can also lead them to be indecisive and to have a great deal of difficulty deciding how to proceed. In addition, the April Rabbit can be more than a little selfish and there is indication that this tendency can result in unnecessary financial difficulties. Their self-centered and overly self-critical nature makes them prone to being loners who need time to themselves to fret, worry, and think through their decisions. Accepting others and learning to be more realistic as well as setting realistic expectations for themselves and others becomes a real challenge for the April Rabbit. These proud individuals, however, are excellent in careers which require precision and detailed work. They work best in free-lance situations or positions that allow for them to work by themselves as much as possible. The April Rabbit gets along best with others who share their values and who set high standards. They make few but close friends with whom they share common interests. And however subjective and moody they can be, once they accept another and establish a close relationship, they are faithful and supportive. Often their friends share their dry, witty sense of humor as well.

The April Rabbit female proves her efficiency in the manner in which she manages her career and her home. She is well suited for a relationship with a man born in the sign of the December Dog or the May Boar. The April Rabbit man finds happiness in a relationship with a woman born in the year of the Rabbit or the Ram. Both men and women born in the month of April during the year of the Rabbit should avoid relationships with persons born in the year of the Ox or the year of the Rat. April Rabbit individuals are quite discerning regarding their relationships, but once in love prove to be loyal spouses.

MAY

Those persons born in the month of May during the year of the Rabbit are courageous, energetic, and positive thinking. They are often wise beyond their years, and it may well appear that they were born with more than their share of lucky stars. The May Rabbit possess a strong humanistic tendency and will work diligently in their efforts to help others. Often this tendency leads them in to career fields such as counseling, social work, or teaching. But they are multi-skilled individuals who are often well suited for any number of professions. In fact, they may change careers with ease until they find the position which they feels best suits them. And their tendency to treat others with respect and in a manner in which they themselves want to be treated lead them to be well liked and accepted by many. In many instances, this makes adept at being managers or leaders. Their versatility combines with their energetic drive and physically healthy disposition to make them always ready for a new challenge or endeavor. The May Rabbit are best known for being generous and caring individuals who are never at a loss for friendship or companionship. At the same time, they maintain a strong sense of morality and values which earns them a great degree of respect from their peers. Their worst trait may be a tendency to make hasty or rash decisions and there is indication that this trait can result in difficulties, financial losses, and mistakes in their personal and business lives. They do best to choose stable friends and spouses who can offer the necessary advice and guidance to off set this unpredictability.

The May Rabbit female is a caring and loving individual. She is most happy in a relationship with another caring person, and she is most successful in doing this in a relationship with a man born in the sign of the September Dragon, the October Rabbit, or the May Ram. The May Rabbit man is well suited for a relationship with a woman born in the sign of the October Dragon or the year of the Dog. Both men and women born in the month of May during the year of the Rabbit should avoid relationships with persons born in the year of the Rat or the Cock.

JUNE

Those persons born in the month of June during the year of the Rabbit are warm-hearted individuals who are trustworthy, kind, and well liked by others. They are extremely adaptable to different groups of people and different situations. There is every indication for the June Rabbit to enjoy a peaceful and contented life because they are indeed lucky individuals. Because they get along so well with friends and family, they face few difficulties in their personal lives, and this holds true in their careers and in regards to their financial well being as well. But there are those June Rabbit individuals who are just a bit timid and indecisive which leads them to be forever seeking the advice of someone else, even regarding insignificant matters. And when the June Rabbit becomes overly critical with themselves and others which leads to them being discontented--and it is this discontentment which quite often turns their good fortune to misfortune and difficulties. Other June Rabbits take their good luck for granted rather than learning to work hard for what they want, and when this happens, they are unprepared for any difficulties which may arise. The truly lucky June Rabbit is more than satisfied with a peaceful life and finds contentment with his or her lifestyle, friends and family. This is the person who has learned the lesson of patience and diligent efforts. To be truly successful, the June Rabbit must also learn to focus on a goal, to determine the steps necessary for attaining that goal, and then to gradually reach their success step by step.

The June Rabbit female truly enjoys her good luck, though there are times when she doesn't understand why others have such difficulties in life. When this happens, she can be critical and gossipy. She is best suited for a relationship with a man born in the year of the Rabbit or the year of the Dragon. The June Rabbit man finds lasting happiness with a woman born in the sign of the September Ram, the June Dog, or the May Boar, Both men and women born in the sign of the June Rabbit should avoid relationships with persons born in the year of the Cock or the Ox. With the right partner, the June Rabbit proves a wonderful mate.

July

Those persons born in the month of July during the year of the Rabbit are quick thinking and honest individuals who rarely worry or fret or insignificant matters. In fact, they may be too busy to pay much attention to irrelevant matters. Rather, they are intent on pursuing their careers or their many diverse interests. They set high standards for themselves, but they are most just in their dealings with other people. They are easily offended by those people who take advantage of others or who seek to hurt others in any way. These are kind and well liked individuals who take pride in keeping their promises and in helping others when they can. The worst trait of the July Rabbit may be a tendency to be a bit timid about taking a chance, and this can lead them to miss out on opportunities. There is an indication that the July Rabbit is lucky in regards to inheritance of property or assets. In their careers, they do best when they specialize in a given field and avoid the temptation to change careers or positions. But it is difficult to give the July Rabbit advice, for they have a tendency to make up their own minds and then to stick to what they feel is best for them. And this often works out for the best, because the July Rabbit uses a logical approach to problem solving, collecting details and facts and using deductive reasoning in problem solving. The July Rabbit is adaptable and popular and this holds true in romance as well. They generally have numerous friends and date any number of people before deciding to settle down and establish a home.

The July Rabbit female is genuinely attractive and kind-hearted. She finds true love in a relationship with a man born in the sign of the May Ram, the April Dog, or the March Boar. The July Rabbit man is best suited for a relationship with a woman born in the sign of the September Boar, the November Dragon, or the May Rabbit. Both men and women born in the month of July during the year of the Rabbit should avoid relationships with persons born in the year of the Rat or the year of the ox. The July Rabbit does best in a relationship with a person who shares their outlook on life.

August

Those persons born in the month of August during the year of the Rabbit are attentive and hospitable individuals who are kind-hearted, friendly, energetic, and quick in thinking. Unlike many born in the year of the Rabbit, the August Rabbit are not in the least timid. Rather, they are strong willed and in some cases even obstinate. These individuals work hard and play hard, and whatever their endeavors, they apply themselves willingly and diligently. They are most agreeable to be around and are well liked by friends and associates, but they have a tendency to want to lead and to have their own way. There are those August Rabbit individuals who are more than happy to accept the leadership of another as long as that person, in their opinion, is making the right decisions. This particular type of obstinacy may lead the August Rabbit to face difficulties in establishing their careers when they are young, but when they persevere, they achieve success but rarely any great wealth. It is their strong wills and strong hearts that prevent the August Rabbit from abandoning hope, and with maturity, they find that their efforts produce good results. And it is with maturity that the August Rabbit becomes more contented, successful, and willing to accept the advice of others. The seemingly inexhaustible energy of the August Rabbit appears to serve them well throughout life. Whatever their situations, they prefer elegant and tasteful surroundings, preferable in an urban setting which offers plenty of social activities. These are not the sit-at-home type, and whether at work or socializing, they are busily engaged.

The August Rabbit female are actively in any number of social activities, and with her pleasing personality, is often the life of the party. She is best suited for a relationship with a man born in the sign of the July Dragon, the May Ram, or the June Rabbit. The August Rabbit man finds love and lasting happiness in a relationship with a woman born in the year of the Rabbit or the year of the Dragon. Both men and women born in the sign of the August Rabbit should avoid relationships with those born in the year of the Cock or the Rat.

September

Those persons born in the month of September during the year of the Rabbit are the type of people who are comfortable whether alone or surrounded by a group of people. The September Rabbit can be unpredictable in that they need time to themselves to think through their thoughts, but they can turn around and be a whirlwind of activity, mingling and socializing or working in a frenzy to complete a project. They may be capable and sensible with a tendency to take life too seriously, or at other times be rash and impulsive, doing whatever they feel like with little regard for the outcome. The worst trait of the September Rabbit, though, is a perfectionist streak which leads them to be always attempting to set and maintain a high standard for themselves and others. At their worst, they are overly critical of themselves and anyone else who doesn't meet these superficial standards. This leads them to be easily offended at which times they become upset not only with their friends or family but with themselves as well. Others appear to accept these personal traits of the September Rabbit probably because they are so impressed with their talents and sharp intellect. With maturity, the September Rabbit does, quite often, learn to relax more and to accept those facets of life which can't be changed. When they learn to accept other people as they are, the September Rabbit become more content and happier. With this acceptance, the September Rabbit also learns not to fret and worry quite so much about the future, but to concentrate on their present endeavors. That many of these individuals are talented in art or career fields which require detail work is to their benefit.

The September Rabbit female does well with an understanding partner. She is well suited for a relationship with a man born in the year of the Boar or the year of the Dog. The September Rabbit man is best suited for a relationship with a woman born in the sign of the December Dragon, the October Boar, or the January Rabbit. Both men and women born in the sign of the September Rabbit should avoid relationships with those persons born in the year of the ox or the Rat.

OCTOBER

Those persons born in the month of October during the year of the Rabbit reserved, dignified, and quietly intelligent individuals. They are quite often conservative people who adhere to a traditional lifestyle. They are also sentimental with a tendency to be nostalgic and to collect items which reflect their interest in history or their family heritage. They much prefer peaceful, refined and elegant surroundings which is exhibited in their preference for antiques or quality furnishing which reflect their good taste and high standards. The October Rabbit can also be a moody person who withdraws into a dream-like state to fantasize or to ponder memorable occasions. Their sentimentality also leads them to be true romantics who take their relationships seriously. They have a tendency to place their romantic partner on a pedestal, perceiving that person as the vision of their dreams.' When reality sets in and they have to accept the shortcomings of the other person, the October Rabbit can become disenchanted and discontented. At their worst, they will attempt to change the other person into their picture of perfection. But they are easily offended when others criticize them. They also easily become jealous of the success of others and are not above criticizing or gossiping. But these are determined individuals who rarely stop short of accomplishing their personal and career goals. And there is every indication that good fortune smiles on the October Rabbit, helping to make them successful. Their talents may lead them to become well known in the field of art, or they may pursue a career in research, teaching, or writing.

When the October Rabbit female falls in love, she becomes devoted to her partner and only her jealousy or possessiveness hampers her relationship. She finds true love in a relationship with a man born in the year of the Rabbit or the year of the Ram. The October Rabbit man is happiest in a relationship with a woman born in the sign of the April Dragon, the May Dog, or the June Boar. Both men and women born in the month of October during the year of the Rabbit should avoid relationships with those persons born in the year of the Cock or the year of the Rat.

November

Those persons born in the month of November during the year of the Rabbit are well spoken if somewhat quietly conservative with an intelligent, honest nature. The November Rabbit individuals are good listeners and sound judges of character. They are alert and energetic with an active social life and numerous good friends. They are respectful, trustworthy, and more than a bit modest with a tendency to want to be of service to others. Many times, others will come to depend on the November Rabbit and even take advantage of their generous natures. By the same token, they strive not to depend on others and even have difficulty seeking advice or assistance when needed. They also become impatient when others don't share their point of view or adhere to their advice. This produces some friction in relationships, but others continue to seek out the November Rabbit for companionship. There is indication that the November Rabbit must face the challenge of being thankful and contented with their lives or their good luck may diminish resulting in unnecessary difficulties. While they are not necessarily loners, the November Rabbit much prefers to develop interests that offer them time to themselves such as individual versus group sports. And their inherent independence makes them well suited to careers that allow them to work by themselves rather than in situations requiring team effort. Their worst trait is a slight tendency to be unpredictable and to change their minds on a whim or impulse. Because of this, their investments should be based on long term savings rather than short term or risky investments. And they must learn to gauge their spending carefully.

The November Rabbit female finds lasting love and happiness in a relationship with a man born in the sign of the May Dragon, the April Dog, or the June Boar. The November Rabbit man is best suited for a lasting relationship with a woman born in the sign of the August Dog, the September Ram, or the October Rabbit. Both men and women born in the month of November during the year of the Rabbit should avoid relationships with those persons born in the year of the Rat.

December

Those persons born in the month of December during the year of the Rabbit exhibit the graceful manners of the suave and debonair with an elegance that proceeds them. They are honest and straightforward, but this can easily turn to a brusque and abrupt manner in their dealings with others. At their worst, they are superficial and snobbish with little concern for the feelings of those close to them. At the best, however, they are kind, idealistic, and caring. But they most generally appear impersonal and non-demonstrative around friends and associates. It is only in the company of those closest to them that the December Rabbit will display their truer emotions to the extent of expressing their feelings and concerns. These are hard working and thorough individuals who set their mind to a specific task and complete it before beginning another. When they learn to be sensitive to the feelings of others, the December Rabbit may become less vain and ambitious and driven to succeed at any cost. They have more than a slight tendency to place the blame for any problems, difficulties, or personal failings on someone else, finding it extremely difficult to ever admit to a wrongdoing or fault. And despite the efforts of the December Rabbit to be independent and not to accept the advice of others, they find it extremely difficult to ever be by themselves--almost as if they fear to be alone. They are that type of comic who relies on poking fun at others, and they relish getting attention by criticizing someone else--but they abhor being the brunt of a joke and cannot tolerate personal criticism. This can make them difficult to live with except for the most accepting of people.

The December Rabbit female is attractive and feminine, but she must learn to temper her sharp tongue. She does well in a relationship with a man born in the sign of the May Ram, April Rabbit, June Boar, or December Dragon. The December Rabbit man is well suited for a relationship with a woman born in the sign of the April Dog, the March Ram, or the July Dragon. Both men and women born in the month of December during the year of the Rabbit should avoid relationships with those born in the year of the Rat, the Cock, or the Ox.

Date With Destiny

Share With Me Your Fantasy

THE DRAGON

The Dragon possesses a strong, at times almost arrogant, character, and the good luck bestowed by the stars. Success in life is one of the main attributes of the Dragon whether he or she strives for monetary gain, fame or fortune or to be of service and to help others. The strong will power of the self-confident Dragon faces any challenge and difficulty without trepidation. The Dragon can be a most obstinate person who follows their own lead and has little time for listening to the advice or opinions of others. And the Dragon has a short, fiery temper that quickly blazes out of control. But those born in the year of the Dragon are also known for their big hearts and generous nature. They get along well with others because they are accepting and agreeable, often keeping their opinions to themselves, and they have a sensible nature which often leads others to seek them out for advice. The Dragon thrives on the admiration they receive from others and can be demanding as well as ambitious and competitive. While they are comfortable around other people, the Dragon develops close relationships slowly and then becomes more talkative and animated.

The nature of the Dragon is highly unpredictable, excitable and impatient. Their friends and associates may have little idea what to expect from the Dragon next. Their ideas may lead them to any number of changes and from one job to the next. They are quiet, serious and dignified people with a charismatic appeal which attracts the attentions of others. This charisma combined with an honest, proud nature and a natural good fortune, leads the Dragon to be put his talents to good use and to succeed and be popular in almost any endeavor. They are talented and capable people, but they also require attention and admiration. They can be somewhat of the exhibitionist in this way, and when they become discontented when in a routine setting, they simple move on. The Dragon appears to be healthy, powerful and energetic with an active social life and a good humor that results in an abundance of friends and personal relationships. The Dragon takes his or her responsibilities seriously and is trusted and respected by others.

The Dragon can be an authoritative person who is decisive and confident in the decision making process and whose actions are strong and determined. others may think the Dragon overly obstinate and selfish in always wanting their own way, but this only covers a person with a strong moral code and a determination to do what is right. They are straightforward people who rarely turn to devious ways to benefit themselves or to get what they want. They have a curious nature which may lead them in any number of directions, but they do most everything they undertake in a thorough and complete fashion. They are somewhat distrustful and suspicious of the intentions of other people and this makes them wary of including others in the gains of their efforts. In other words, the Dragon, who is very proud, may feel that what ever results are gained were won on the merits of his or her own efforts--not because of any help or assistance from others with whom they work. And they want all the praise and credit for a job well done. The sheer vanity of the Dragon makes it difficult for them to work well or to follow orders from others. They prefer to be in the leadership position and to have others following their orders. Be that what it may, other people respect the Dragon and are accepting of this behavior, admiring their leadership abilities. This makes the Dragon successful in their careers and popular socially. But then too, the Dragon can be brave and courageous, and always blessed with luck, and they are often quite charitable and well meaning. And it is their sheer arrogance and pride that leads them to feel that they are better than others.

There is every indication that the good luck of the Dragon results in a comfortable and long life, and the Dragon lacks for anything. But they do have to take into consideration their excitable natures, and when choosing a business partner, it is best for the Dragon to look for a person who is calm and steady. A good partner can also help the Dragon to invest wisely and to control spending in order to realize a profit in business. The Dragon works best with another person who understands and appreciates their nature and who can calm them when they become too excitable. And their excitable, obstinate nature may lead the Dragon to have differences of opinions with their parents, but the chances are they get along well enough with brothers or sisters who may share their viewpoints and understand their thinking. The Dragon doesn't rely on family support, but works hard and acquires property and possessions on their own. But there is some indication that they may benefit at sometime in their life from an inherited asset. The Dragon overcomes his or her impatience by working diligently to succeed. This may require time, but it does produce results. Good fortune smiles on the Dragon when he or she is traveling, and they often make friends or meet new acquaintances through their trips. The energetic Dragon works hard and plays hard and must learn to relax more to prevent become overly tired or ill. The Dragon is lucky in lawsuits, preventing court cases, but must guard against pushing his or her luck too far. And luck allows the Dragon to find any lost items or friends.

In romance, like other areas of life, the Dragon is lucky and enjoys numerous romantic encounters throughout life. The Dragon can be the love them and leave them type, however, and is seldom as affected by a love affair as his or her partner. And members of the opposite sex fall easily in love with the Dragon. Because they are socially active and popular, they have little difficulty becoming involved in romantic affairs. But it requires a special kind of person to hold their attention for long. Remember, the Dragon needs most to be admired, and they require a great deal of attention. When they find the right partner, they enjoy a happy and long-lasting marriage. Certain partnerships produce more luck, success, honor, and monetary good fortune for the Dragon than others. For that reason, the Dragon does best in a relationship with a person born in the year of the Monkey or the Rat. The Dragon also does well in a relationship with another Dragon or a person born in the year of the Snake as the humor of the Snake can subdue the pride and arrogance of the Dragon. It is necessary that the Dragon avoid a relationship with a person born in the year of the Ox, the Ram, the Dog, or the Boar as these people may resent the Dragon's need for praise and affirmation and this can result in problems, separation and sorrow for both individuals.

The Dragon can be fascinating and full of vitality and energy with an enthusiasm for life and an impetuous nature that results in a person who is always on the go.

January

Those persons born during the month of January during the year of the Dragon are talented people who strive to be honest and hard working. They are studious and observant people who consider their decisions and actions carefully. They are more than willing to do what is necessary to succeed, even if it means taking on two tasks or jobs at the same time. They are stable people, but quite aggressive in getting what they want. They may be quiet and reserved, but they speak their minds openly without any reservation. They can be sociable and popular, but there is a tendency for them not to form many close, personal relationships. The good luck of the Dragon year persons stays with them throughout life and augments not only their business lives but their personal lives as well. But their successes are attributed to their ability to work hard and to continue diligently in their efforts, slowly but progressively succeeding at their endeavors. They accomplish slow, steady progress in all that they undertake. Their very obstinacy can make them difficult to deal with in business. Their determination and resolution, however, generally pays off in the long run, and the January Dragon finds success in life.

The January Dragon female is very feminine and enjoys her choice of suitors. They are rational and objective and while they may argue with their partners, they soon forgive and forget. They are exceptional wives who love to cook, sew, or decorate, and they enjoy making a comfortable home. The January Dragon female is well suited for a relationship with a man born in the sign of the November Monkey or the October Rat. They also do well with another Dragon year person. The January Dragon man can be an astute and driven businessman, but at home, he can be quite retiring and even lazy with a tendency to turn the -management of the home over to his wife. He does best in a relationship with a woman born in the sign of the September Monkey or the September Dragon. Both men and women born in the month of January during the year of the Dragon should avoid a relationship with a person born in the year of the Dog or the Ox. The January Dragon enjoys steady, continuous luck in all that they do and this holds true in their family life as well.

FEBRUARY

Those persons born in the month of February during the year of the Dragon make good leaders because they are analytical, logical thinkers with a sensible, good judgment which allows them to make decisions. They have a pronounced ability to concentrate, and they follow through on their ideas and decisions. Socially, they are accommodating people who are warm-hearted and congenial, but they can appear somewhat quiet and reserved. This apparent shyness, however, masks a confident, courageous spirit in a person who is more than willing to accept a challenge. The February Dragon is dedicated to whatever it is they are doing, but they don't function at their best in routine situations. They must feel challenged, and if the situation becomes unstimulating, they can become dissatisfied and discontent, For this reason, they may have a tendency to move on to a new idea or a new situation until they find a position that continues to interest them. one should never assume that they know the mind of the February Dragon as they can be quite unpredictable. While they can appear intelligent and dignified, the February Dragon is also gentle, kind, and most pleasant.

The February Dragon is the most romantic of those born in the year of the Dragon. The February Dragon female may find that men may hesitate to approach her because of her reserved and intellectual appearance for fear that she is aloof and snobbish. She is actually a most warm and accepting person who is a romantic at heart. Like the February Dragon man, she is attracted to persons with a strong character who are passionate, physical lovers. She does well in a relationship with a man born in the sign of the December Rat or the January Snake. The February Dragon man is also a romantic who is best suited for a relationship with a woman born in the sign of the December Monkey, the January Rat, or in the year of the Dragon. Both men and women born in the month of February during the year of the Dragon should avoid a relationship with a person born in the year of the ox or the Boar. The luck of the Dragon holds true for the February Dragon, and this persons enjoys life with the right partner.

MARCH

Those persons born in the month of March during the year of the Dragon are determined and stern, but they appear most kind and gentle. They are very observant and sensitive and responsive to their environments. They are easily excitable and respond to people, situations, and sensations. Their self-confidence, strong egos, and charismatic nature attracts the attention and admiration of others. Because of their congenial manners and kindness, they make friends easily. The March Dragon makes a good leader because they are capable of inspiring others to follow their direction. This is augmented by their reliable and responsible natures and their ability to make good decisions. The March Dragon find that their prestige and influence continue to increase as they go through life. They seldom resort to devious methods or deceit to get what they want, preferring to be honest, direct and outspoken. This gains them the respect and admiration of others. They move with ease from one social circle to another and possess the ability to influence society with their ideas. Socially, they achieve distinction and prestige and can become well known. The March Dragon can also become famous in their chosen career and many do especially well in acting, the arts, or sports.

The March Dragon female can be assertive and physically active with a tendency to dress for success. Beneath this ambitious demeanor, she is quite feminine and maternal and becomes committed to her partner in a relationship. In a relationship, she much prefers strong, physically fit and energetic men. She finds happiness and harmony in a relationship with a man born in the sign of the July Snake or the July Dragon. In romantic relationships, the March Dragon man is responsive and attentive. He does well in a relationship with a woman born in the sign of the June Rat or the April Monkey. Both men and women born in the month of March during the year of the Dragon should avoid a relationship with a man born in the year of the Dog. Because of their sensitive, alert natures and strong egos, it is important for the March Dragon to take the time to get to know the other person well and to be sure that their partner understands and admires them.

APRIL

Those persons born in the month of April during the year of the Dragon are sensible, kind, and understanding. They are perceptive, intelligent people with the ability to concentrate on important matters while ignoring much of the trivial problems that vex other people. They also have the ability to focus their impassioned beliefs and abundant energy into their strong determination to accomplish their goals. And they remain undaunted when they don't immediately get what they want, preferring to wait patiently for the next opportunity. They express themselves openly and often wear their emotions on their sleeves, laughing with joy when they are happy or crying when they are sad. While they are usually careful and always self-confident, their generosity and their tendency to be protective of others sometimes leads them to be naive about the intentions of other people. But the open friendliness of the April Dragon makes them well liked and accepted by others. There are indications that the April Dragon will encounter difficulties in life and that they must rely on their patience and fortitude to withstand such problems and to gradually overcome them. This ability will earn them the respect and admiration of their friends and associates. The April Dragon is a proud person who has a high opinion of himself or herself.

The April Dragon female is energetic and outgoing with a tendency to be assertive. This is especially true when she meets a man in whom she is interested, but she can be at times overly assertive which may cause her to lose out. The April Dragon man is openly friendly with everyone he meets, but he has a tendency to be overly generous with one and all. The April Dragon female is well suited for a relationship with a man born in the sign of the January Snake or the November Rat. The April Dragon man does best in a relationship with a woman born in the sign of the April Monkey or the July Rat. Both men and women born in the month of April during the year of the Dragon should avoid relationships with those born in the year of the ox or the year of the Dog. The April Dragon is lucky in love when they wait for the right person.

MAY

Those persons born in the month of May during the year of the Dragon are optimistic, cheerful and kind. They are outgoing and sociable and love to entertain others. They make the perfect host or hostess and others admire them for their personable manner. But they retain an inherent shyness which limits their close friendships to a few people. The May Dragon also possesses a great deal of determination and is always ready to stand up for a just cause. Their optimism fills them with hope, and there is every indication that opportunities will continue to come their way throughout life. But the May Dragon must guard against being overly confident and selfish--their greed can work against them. They love the finer things in life and are drawn to beautiful clothing, jewelry, and luxury items with a preference for a comfortable lifestyle and the best of everything. They can also be somewhat changeable with a tendency to vacillate between one decision or another. And their impatience leads them to have extremely short tempers. But for the most part they are pleasant, likable people who are considerate of others. The May Dragon is also brave and courageous with a willingness to face dangerous situations that others would choose to avoid. The May Dragon must also learn to relax as their tendency to work hard and to play hard can drain their energies.

The May Dragon female loves to lavish her money on herself, buying clothes, jewelry and make-up, and often spending freely to make herself over. often she gains the attentions of men who don't realize she has created her beauty herself and is often not satisfied with the attentions of one man. However, when she meets the right person, she falls in love deeply and becomes devoted to him. She finds the most fulfillment in a relationship with a man born in the year of the Snake or the Dragon. The May Dragon man is drawn to women who are lively, energetic, sociable and fashionably attractive. He does well in a relationship with a woman born in the sign of the April Monkey or the February Rat. Both men and women born in the month of May during the year of the Dragon should avoid relationship with those born in the year of the Ox.

JUNE

Those persons born in the month of June during the year of the Dragon are fiercely competitive and determined to attain their goals. They will spend freely to gain social status and prestigious positions, often stepping over others in their drive to succeed. They can be quite fascinating people who are full of vitality and energy and who easily impress others, but underneath this surface show, the April Dragon can be driven by selfishness and greed. At their worst, they are hot-blooded and hot-tempered with a tendency to be inconsiderate of others and self-centered. It does little good for others to criticize the pride and determination of the April Dragon because they often fail to heed the advice of others. They are most obstinate and willful.

They are skillful conversationalist who are adept at manipulating others in order to carry out their well laid plans. They also dress for success and are always presentable. Their fierce pride in themselves and their arrogant nature may stem not only from their inherent talents but also from their undaunted bravery. They know they are willing to attempt what others would not have the nerve to do. This blatant bravado often wins them the respect and admiration of others who are willing to follow in their lead. But the April Dragon can be most intolerant and impatient with those who oppose his or her plans. There are indications that good luck does not favor the April Dragon, and that they must overcome numerous difficulties and problematic situations in order to succeed.

The April Dragon female can be precocious and willful, but she is a romantic at heart who seeks out her true love and becomes devoted to him. She finds love and happiness in a relationship with a man born in the sign of the August Dragon or the September Monkey. The April Dragon man finds lasting happiness in a relationship with a woman born in the year of the Snake because the Snake can balance out his temperament with wisdom and humor. Both men and women born in the sign of the April Dragon should avoid relationships with those born in the year of Dog because the Dog can be too pessimistic and cynical. The April Dragon does best in a relationship based on understanding.

JULY

Those born in the month of July during the year of the Dragon possess an air of elegance and dignity. They are clever conversationalist and usually quite talented in some manner. They also have a tendency to be caring and kind and can be most gentle. They are not above caring for those in need and may do well in professions where they can be of service to others. The July Dragon man is most compassionate and considerate when in the companionship of women, but he can hold his own when with men becoming more determined and aggressive when necessary. Others admire and respect the July Dragon for their adaptability, intelligence, talent, and their ability to work hard. They easily inspire others to follow their lead. The energy level of the April Dragon propels them into constant activity, and it is difficult for them to take time off from their busy schedules for relaxation. Even at home, the July Dragon has a list of things that need to be done, and one can always find them busily engaged in one project or another. The July Dragon man expends his energies on numerous activities and at home will pitch in to help with household chores or projects to improve the home. There is indication that the July Dragon encounters numerous difficulties and financial problems in early life, but in later years becomes financially secure. A very full and long life is predicted for the July Dragon. These are personable and honest people whose worst fault may be a tendency to be overly curious about the affairs of other people which can cause them to become involved in problems other than their own.

The July Dragon female is very feminine and attractive. She does well in a relationship with a man born in the sign of the April Rat or the October Dragon. The July Dragon man makes an excellent husband, but to be happy, he must find a partner who appreciates him and who admires his talents. He is well suited for a relationship with a woman born in the sign of the February Rat or the March Dragon. Both men and women born in the sign of the July Dragon should avoid relationships with those born in the year of the Ram. The July Dragon needs to be recognized for their good qualities.

AUGUST

Those persons born in the month of August during the year of the Dragon continuously impress other people with their creativity. They are also gifted scholars with a clear intellect and ability to make good decisions. They appear to have unlimited amounts of energy combined with a strong self-confidence which makes them much admired by others. The August Dragon relies on their judgment, perceptions, and intuition to make decisions, and rarely bothers to ask for the assistance or guidance of others. Moving up the career ladder comes easily for the August Dragon as they are often offered promotions for their ability to work hard and to concentrate on specific tasks and objectives. There are times when the August Dragon can be too ready to accept new innovations or new ideas even though they have not been thoroughly researched. In this way, they can be influenced by their own enthusiasm and to some degree by their self-centered need to achieve success which can lead them to move too quickly. They must guard against allowing greed to influence their decisions. While the August Dragon is destined to meet with both success and failure, once they learn to balance out their natures, they are often highly respected for their ability to solve problems and to ferret out answers to difficult situations. There is every indication that this balance results in a secure future and that the August Dragon is both content and successful in their endeavors. Their adaptability makes them well suited for any number of professions. In their social lives, they are always available for their friends and will strive to help friends however they can.

The August Dragon loves family life and children, and, if at all possible, will have a large family. The August Dragon female finds love and lasting happiness with a man born in the sign of the February Rat or the October Snake. The August Dragon man is well suited for a relationship with a woman born in the sign of the March Monkey or the April Dragon. Both men and women born in the month of August during the year of the Dragon should avoid relationships with persons born in the year of the Ox or the year of the Boar. The August Dragon finds balance and harmony in love.

SEPTEMBER

Those persons born in the month of September during the year of the Dragon are natural born scholars who develop into talented intellectuals. They also have a practical mind and make sound decisions and judgments. They are adept at financial planning and comprehend easily the precepts of economics and financial planning. They often make wise investments and are skilled at handling their money. Their self-confidence is often well founded which earns them the respect of associates and colleagues. Money seems to flow their way, but to become truly well off, the September Dragon must curb an urge to be extravagant. And when the September Dragon becomes too greedy they are tempted to gamble or to make risky investments in hopes of realizing a quick return. Once they learn to control these urges, the September Dragon is capable of becoming extremely well balanced which leads them to control their thoughts and actions. The September Dragon also has a tendency to be pragmatic and to over rationalize emotions. They think too much to ever give of themselves freely in romance. If a person is looking for financial security rather than love and romance, the September Dragon is a good choice. Even the September Dragon female plans her relationships carefully rather than allowing herself to be moved by pure emotions. The September Dragon can be a bit demanding, intolerant and intimidating because they rarely consider the feelings of other more emotional people. In fact, they can be impatient with others who allow their emotions to cause problems for them. And like all Dragon-year born persons, the September Dragon has a stubborn streak and a quick temper when they become impatient. If the courageous September Dragon fears anything, it is the failure to be financially secure.

The September Dragon female is well suited for a relationship with a man born in the sign of the January Dragon or the December Snake. The September Dragon man is well suited for a relationship with a woman born in the sign of the May Dragon or the May Snake. Both men and women born in the sign of the September Dragon should avoid relationships with those born in the year of the Ram.

OCTOBER

Those persons born in the month of October during the year of the Dragon are cheerful, talented and honest with a determination and energy that allows them to achieve their goals. They are also unpredictable with a tendency to act before they think. This combined with an innocent naiveté leads them to make mistakes. Their changeable nature also leads them to be either too proud and obstinate or too emotional and intent. But the October Dragon is very active and enjoys a life filled with endless projects and activities at which they are generally successful. These are congenial, personable people who enjoy a type of instant popularity with who ever they meet. But the October Dragon can become so sold on themselves and so self-confident that they fail to listen to the advice and opinions of others. They can also be impatient with those who don't readily agree with their opinions. In some cases, the October Dragon can even develop an overpowering personality and becomes determined to have their way. They can be secretive and manipulative but appear persuasive and simply impetuous. Their tempers are not to be tampered with. The biggest challenge in life for the October Dragon is to learn to control his or her temper and prideful ways and to learn to listen to the advice of others. The October Dragon has an active social life and is always on the go from one activity to another. Anyone who seeks to have a relationship with an October Dragon, must be able to adapt to this busy lifestyle. The October Dragon prefers to be surrounded by people and doesn't like to be alone. The October Dragon man can be a philanderer who takes romance casually.

The October Dragon female may find that she does best in a relationship with an older man. She is well suited for a relationship with a man born in the year of the Dragon or the Snake. The October Dragon man does best with a woman who accepts him the way he is. He finds the most harmony in a relationship with a woman born in the sign of the November Rat or the December Monkey. Both men and women born in the month of October during the year of the Dragon should avoid relationships with those born in the year of the Boar or the Ox. The October Dragon becomes more settled with maturity.

November

Those persons born in the month of November during the year of the Dragon are the quiet loners of the Dragon year born. They are reserved and dignified with a sensitive and conservative nature. They shy away from boisterous, noisy people and from large group activities, much preferring the companionship of a few close friends. They have refined taste and a highly developed appreciation for art and literature. They function best in a career position which allows them to work by themselves. In close relationships, they can become more assertive and determined and will often get their own way in matters. The biggest challenge for the November Dragon is to learn to express their feelings. Because of this inability, they are often misunderstood, especially in romantic relationships. The November Dragon, however, can be his or her own worst critic and often will underestimate their own capabilities. Once they recognize their talents and learn to use them effectively, they can become quite successful. Their tendency to accept their fate in life or to be content with their position may prevent the November Dragon from acquiring wealth or fame. But quite often, they become financially secure and enjoy their own success in their own way. And when they overcome their own shyness, their success knows no limits. Those who fail to do this can lead lonely lives. They too often fail to realize their own lack of effort hurts the feelings of other people who may care about them. Be that what it may, the November Dragon remains a sophisticated person with a strong, inherent courage that sees them through difficult period times. But they must be careful of developing nervous disorders which lead to poor health.

The November Dragon female finds love and happiness in a relationship with a man born in the sign of the December Monkey or the October Snake. The November Dragon man is well suited for a relationship with a woman born in the sign of the March Rat or the March Dragon. Both men and women born in the month of November during the year of the Dragon should avoid relationships with those born in the year of the Dog or the Ram. The November Dragon finds that true love allows them to express their emotions fully.

December

Those persons born in the month of December during the year of the Dragon appear accommodating and most hospitable, but this outward appearance can mask a very determined and spirited inner nature. The December Dragon is strongly independent and capable of quick actions and decisions. Their ambitious determination also makes them capable of whatever action is necessary to achieve their goals. In this way, they can be fierce competitors. They can be tenacious in their efforts and will resort to overpowering or out maneuvering an opponent. When necessary, they can be quite stubborn and proud. There are also times when they exhibit an irritable nature and short temper. They most want to be in control of the situation and to have their own way in most everything. This obstinacy and selfishness can at times wreck havoc with opportunities that would otherwise result in good luck and good fortune for the December Dragon. They are challenged to be more considerate and compassionate toward others. When they accomplish this, there is every indication for success and good fortune in their lives. But they must take care in weighing the benefits of opportunities. When they learn to accept advice from others, the December Dragon is often aided by an influential person. Generally, the December Dragon makes profitable investments and knows how to use profits to generate additional income. There is every indication that the December Dragon is capable of achieving social prestige and financial success.

The December Dragon female possesses an independent nature and may have to learn to pay more attention to the man in her life. She does best in a relationship with a man who understands her nature such as a man born in the year of the Snake. The December Dragon man is well suited for a relationship with a woman born in the year of the Monkey or the year of the Dragon. Both men and women born in the month of December during the year of the Dragon should avoid relationships with those born in the year of the Ox or the year of the Ram. The December Dragon seeks admiration and affirmation in romantic relationships and is most happy with a person who greatly admires their abilities.

Date With Destiny

Snake

Share With Me Your Fantasy

THE SNAKE

To many, the snake is perceived as a bad symbol. But in Asia, the snake is considered very precious and represents kindness, intelligence and wisdom. Those persons born in the year of the Snake are appreciated not only for their intelligence but also for their artistic ability. This quiet, serious, and meditative person is often a profound thinker and philosopher who is physically attractive with his or her own distinct aura of glamour. The Snake is a friendly but sensitive person who can be quite decisive and original with dreams and ideas which are unique and different from others. They think out their actions carefully and rarely abandon a course of action or are deterred from their goals. In this manner, once the Snake has made up his or her mind, they will complete what they are doing no matter what difficulties may arise. Their sensible, self-confident and self-reliant character combined with their intuitive nature leads them to be a good judge of situations and of other people. The Snake is a skillful debater and loves to ferret out the facts and to investigate matters. Conversely, there are other times when they remain quiet rather than enter into conversations or small talk with others. With their sensitive natures they can be easily offended or hurt by other people, but they strive to conceal this pain and to bare it stoically. This sensitivity leads them to be somewhat defensive and at times even paranoid. At the same time, the Snake can be sympathetic, compassionate, and patient with other people and will make an effort to be of assistance as long as it doesn't require too great a personal sacrifice. They have a preference for helping their friends by giving personal advice rather than financial assistance.

Then again, at other times, the Snake can be selfish as well as stingy, self-centered and vain. However, many a Snake are charismatic people who are elegant and cultured with a taste for the finer things in life. They appear well bred and distinguished with a helpful and charming manner. The Snake can be most definitely decisive and his or her own worst critic, but the Snake is also capable of being lazy and even dishonest on occasion. And facing any opposition, this person can become indifferent and hostile, and once they have been offended, they will seek revenge.

The Snake patiently acquires the skills they need to be successful in life. And while it may be difficult to accumulate funds in early life, the Snake is fortunate in that they possess good self-control and an understanding of what needs to be done. They spend a lot of time thinking and planning on how to improve their business or finances. In the extreme, this can make them greedy and intent on acquiring possessions and money. But most generally, the Snake impresses other people and they are quite often helped by others in influential positions. In business, the Snake may get off to a slow start, and but they stick to their ambitions and can become well off and well known by middle age. Others respect the snake and their abilities, all of which helps the Snake to continue to prosper throughout life. This person is drawn to occupations which require precise thinking skills, study, and decisiveness. And while the Snake is usually never short of funds or good luck, their selfishness can lead them to be moody and susceptible to vain thoughts and an insufferable pride. Often, they are unable to relax, even in their leisure time or when traveling, and will spend their time pondering their future and how to make more money. But their confidence and good judgment does lead them to make wise investments, and they are particularly successful with short term investments and speculation. They are somewhat driven to be successful and quite often this dream is realized. The Snake must be careful that their greed doesn't lead to law suits, but in this regard they often win out in court. Their strong memories and retentive powers help them to locate lost items. In most cases, the Snake is an intent person with a clear vision of what they want, and they strive to acquire just that. They easily acquire possessions, property and financial success when that is their goal. The Snake rarely has financial problems because they work hard to overcome any difficulties.

In family matters, the quiet and reserved nature of the Snake makes them hesitant to openly discuss their personal lives with their parents, and they may find it difficult to accept their advice or assistance. The Snake, howeverdeeply they care for their families, remains a little distant, finding it difficult to even discuss personal matters with their brothers or sisters. Perhaps because of this tendency, they do attempt to choose a marriage partner who will meet with the approval and fit in well with their families. The Snake prefers to be around people who are agreeable and manageable. They do like a big family and after marriage will concentrate their attention on their families. With their own children, they become responsible and devoted and willing to make sacrifices for the needs of the child.

The Snake is romantically charming and alluring and very attractive to the opposite sex. They often find themselves the center of attention and sought after by others. Romantic affairs, however, can become complicated because the Snake has a tendency to lie and cheat. They are irresistible drawn to extramarital affairs. But the snake is possessive, and in love will jealously protect and cling to the other person. Even if they are unfaithful, the Snake will continue to attempt to control their marriage partner. In extreme cases, the Snake becomes over controlling and suffocatingly possessive, allowing the other person little personal freedom.

Those born in the year of the Snake are well suited for a relationship with a person born in the year of the Ox, in either marriage or business, because the Ox is hard working and responsible and will be tolerant of the selfish nature of the Snake. The Snake can develop a loving and companionable relationship with the Cock, as well, as long as they are both willing to cooperate with each other. A relationship is also advisable with a person born in either the year of the Ram, the Horse, the Dragon, the Rabbit, or the Rat because these signs respect and admire the Snake. The Snake should avoid a relationship with the Boar because the Snake will slowly suffocate the other person and the relationship will become unrewarding for both people. Neither is a relationship with either a Monkey or a Tiger advisable because these people will not tolerate the controlling nature of the Snake, and the relationship will be deteriorate into a series of arguments and disagreements. The Snake does best with other people who appreciate their good qualities and admire their nature.

January

Those persons born in the month of January during the year of the Snake are the most modest and reserved of the Snake year persons and the most content. They prefer refined and tasteful fashions and find anything the least bit flashy distasteful. The January Snake is also active, friendly and sociable with a happy, easy going nature which makes them popular with their friends. They desire knowledge and like to be well informed and well read on a variety of subjects. They have an open mind and are tolerant and accepting of new ideas. Their friends and associates admire them because of this and seek them out for advice not only on personal matters but in regard to business decisions as well. The January Snake thinks a lot, but they can be so reserved that they are overly cautious and this can lead them to miss out on what would have otherwise been a good opportunity. They become more successful in both personal relationships and in their business affairs when they learn to relax and to be more comfortable with themselves. But the January Snake does make excellent decisions and their judgment is well respected and followed by others. The January Snake is proficient and makes an excellent teacher, writer, doctor, counselor, psychiatrists, or advisor. This person is often hard working and diligent and will find that opportunities for advancement come after the thirtieth birthday. This person likes to save money and learns to make good investments, but they must guard against being too anxious or greedy which can lead to risky investments.

The January Snake is cautious in choosing a marriage partner and waits patiently for love and the right person. The January Snake woman is well suited for a relationship with a man born in the sign of the January, July, or October ox. She may prefer to marry an older man. The January Snake man does well with a woman born in the sign of the December Ox or the September Cock. Both men and women born in the month of January during the year of the Cock should avoid a relationship with a person born in the year of the Boar. Your love for your partner continues to grow with time as does your good fortune.

February

Those persons born in the month of February during the year of the Snake possess a dual nature. They can be quiet and reserved and then turn around and be energetic, active, and congenial. Their reserve gives way to an excitability which affects their sensitive and emotional nature. But then their mood can swing back to a cool, calm demeanor just as rapidly. They can become very enthusiastic about a new idea or new friend or relationship, and then lose interest and become impatient when the situation no longer holds their attention. When the February Snake finds a career for which they are well suited and which makes them happy, they work extremely hard. Otherwise, they can become discontent and may move from one position to another seeking something which satisfies their interests. When the February Snake makes a diligent effort to control their dual nature, they become successful and this success is often realized around the age of thirty. Otherwise, their duality can make them nervous which can lead to stress, anxiety and a loss of energy. Once they learn to focus their energies more effectively, they are more successful in their endeavors. The February Snake is generally fortunate financially, but must strive to also be content. Their dissatisfactions can result in them never fully realizing their potentials. The February Snake is especially intuitive, sensitive, and perceptive with a good memory and strong intelligence. They develop their intellect and have numerous interests which keeps them busy and preoccupied. The February Snake is also somewhat stubborn and may not take the time to listen to the advice of others.

Once the February Snake marries, they like a comfortable home and prefer activities centered around the home and family, but they continue their outside interests. The February Snake female finds love and happiness with a man born in the sign of the January Ox or the year of the Dragon. The February Snake man is well suited for a relationship with a woman born in the sign of the March Rabbit or the June Rat. Both men and women born in the sign of the February Snake should avoid relationships with those persons born in the year of the Boar.

March

Those persons born in the month of March during the year of the Snake appear active, energetic and outwardly confident. But, they also possess an innate curiosity and inquisitiveness which leads them to be overly cautious and suspicious of others. In extreme cases, they can become hostile and angry and will readily disagree with others, often refusing to take advice and most definitely disliking being told what to do or how to do something. This tendency to be insecure and to lack self-confidence can lead to them being unstable and unable to perform well on the job or in their personal relationships. Once this person recognizes this, however, they may discover that they do best when working by themselves on a job which allows autonomy and self-direction. The March Snake is drawn to research, study, and endeavors which require them to use their intellects and intelligence. The March Snake also develops their talents and this may help them to become successful early on, but they rarely become wealthy. This person does strive to be honest and just in their relationships. They admire beauty and the better things in life and may be drawn to marry for prestige or a better position.

Relationships may remain difficult for the March Snake, however, because they remain suspicious and distrustful of the other person. In some cases, this trait, even after years of marriage, will destroy the relationship. For that reason, it is especially important that the March Snake get to know the other person well and to develop a relationship based on friendship and trust before marriage. The March Snake has a tendency to expect a lot from a relationship, both emotionally and materially. The March Snake female does best with a man born in the sign of the August or February Cock. She has expensive tastes and does best with someone who appreciates this and can afford them. She also prefers to continue working or to have outside interests which keep her mind busy. The March Snake man finds happiness with a woman born in the sign of the March Rabbit or July Rat. Both men and women born in the sign of the March Snake should avoid a relationship with persons born in the year of the Tiger.

APRIL

Those persons born in the month of April during the year of the Snake are serious-minded and responsible with a confident nature that impresses other people. This may make success easier for them as they seem to progress easily through school, work, and promotions. The April Snake works hard to be achieve a position of leadership and prefers to be the manager or the person in charge at work. They are just as successful socially as well, and move easily from one group to another in their climb up the social ladder. They do have a tendency to be jealous and resentful of the good fortune or success of others which can cause them problems at times. And their own self-confidence and belief in their own abilities can lead others to be jealous of them. Given a choice, the April Snake is most interested in succeeding at business, and personal relationships may take a back seat to this preference. And the April Snake may discover that they do best when they follow they own lead rather than the advice of others. They have a natural instinct for making the right decision or judgment call, often like an inner feeling that leads them in the right direction. This is an idealistic person who pursues a field which is best suited for his or her talents. They must, at the same time, learn to handle their finances well in order to be financially secure. Then they find that success comes even easier and more quickly.

The April Snake does best in a relationship with a person whose nature balances and supports theirs. In other words, their partner may need to handle their social affairs so that they can more fully pursue their careers and endeavors. The April Snake female finds love and happiness in a relationship with a man born in the sign of the March Cock or the February Ox. Or she may be most interested in material gain through marriage, looking for a partner who can support her well. The April Snake man is well suited for a relationship with a woman born in the sign of the September Ox or November Rabbit. Both men and women born in the month of April during the year of the Snake should avoid a relationship with a person born in the year of the Tiger or the Boar.

MAY

Those persons born in the month of May during the year of the Snake appear friendly, considerate, and well meaning, but their true intentions can be selfish and inconsiderate of others. The May Snake has a tendency to be nervously high strung, excitable and especially stubborn. They have an idealistic sense of justice and what is right and wrong and are particularly opinionated in this regard. This is an intense person with strong cognitive abilities who applies all of their efforts and concentration on whatever endeavor occupies their mind. This ability to focus on their goals helps the May Snake to achieve career advancement and success in life. However, their impatience with others and intolerance of the ideas of others can hamper them socially. With close friends, though, the May Snake is loyal. They can be critical of other people, but they are always ready to be of assistance to a friend. But, then again, they may draw the line at lending money or helping in financial matters. The May Snake can be hard working and make diligent efforts to succeed and generally this pays off later in life. They must guard against missing out on opportunities and the chance to better themselves by being overly stubborn and resentful of the advice of others. The May Snake must also guard against drinking too much as alcohol can develop into a difficult habit to control.

In romance, the May Snake is intensely passionate and emotional. Generally, they desire an emotional involvement as well as a sexual one and seek to find true love in their relationships. The May Snake may marry more than once. Often, they become disenchanted with the faults of their first marriage partner and only find happiness in the second, or in some cases, the third marriage. It may be advisable for this person not to marry too young and then only after they know the other person well. The May Snake female discovers love and happiness with a man born in the sign of the April Cock or the May Ram. The May Snake man is best suited for a relationship with a woman born in the sign of the May or April Dragon. Both men and women born in the sign of the May Snake should avoid a relationship with a person born in the year of the Monkey or the Boar.

JUNE

Those persons born in the month of June during the year of the Snake are warm-hearted, friendly and easy going. They have a tendency to be cheerful and full of fun with a clever and imaginative sense of humor. They love socializing, good food and drink, and entertaining companionship. They like to keep busy, and within their social circles, they are the ones who plan and initiate the activities and are always ready to do something. They are intelligent and capable of quick actions and fast thinking. But, they are also quick to judge others. Despite this, they are well liked and popular with their friends and associates. The June Snake may often use their intelligence and capabilities to achieve success in life, but they are prone to spending money on a lavish lifestyle and social occasions. They prefer to dress fashionably and to always be presentable. They are attractive, well groomed, and have a flair for making a fashion statement.

The June Snake can also be overly generous with friends and lovers. In their careers and business, the June Snake is generally successful and makes money, but they must guard against over spending in this area of their life as well. And the June Snake may discover that extramarital relationships can result in losses. Be that what it may, the June snake enjoys a comfortable and busy life. They also enjoy their private moments and may have a home or place where they can retreat in privacy.

In romance, the June Snake looks for physical attractiveness, passion and love. But, they may be susceptible to those who seek to take advantage of their generous and carefree ways. The June Snake female is well suited for a relationship with a man born in the sign of the October Rat, December Cock, or in the year of the Horse. The June Snake man finds happiness with a woman born in the sign of the October Rabbit or the year of the Rat. Both men and women born in the month of June in the year of the Snake should avoid a relationship with a person born in the year of the Tiger or the Boar. The June Snake is in love with love and many marry early in life while others may enjoy life instead, choosing to marry later in life.

July

Those persons born in the month of July during the year of the Snake get along well and work well with other people. They are adaptable to new situations and tolerant of the ideas and lifestyles of other people. They are also honest, hard working people who strive to be trustworthy and reliable. They do, however, have a tendency to be independent persons who form strong opinions. But their opinions are easily influenced by others and may change depending upon their mood or situation. The July Snake takes time to carefully plan their lives and careers, but they are tempted to change their jobs and this can hamper their achievements and over all performance. The July Snake is talented and does well in a career which requires creative ability. While the July Snake may not be overly gregarious, they do enjoy an active social life and this is especially true of the July Snake female. At work, they are often the persons performing the important behind the scenes work rather than the ones managing the project. To be successful, the July Snake faces the challenge of learning to complete one task before moving on to another. They must also learn to accept the advice of others.

The July Snake female may decide to hold off on marriage and a permanent commitment until she feels she is ready or until she is sure she has found the right person. The July Snake man views romance as a conquest and may move from one to another until he meets a suitable female for marriage. once in love, the July Snake can become jealous and possessive, and it is best to find a person who sees this as admiration and an expression of their mutual feelings for each other. Extramarital affairs may also interfere with their relationships. The July Snake female does well in a relationship with a man born in the sign of the October Rat or the February Cock.

The July Snake man may find that he is a good match for a woman born in the sign of the March Horse or the February Ox. Both men and women born in the month of July during the year of the Snake should avoid a relationship with a person born in the year of the Tiger or the Boar. The July Snake may find that their relationships improve as they mature.

August

Those persons born in the month of August during the year of the Snake are cheerful, optimistic and outgoing. They are energetic and enjoy an active social life, but they are also kind, friendly and honest which earns them the respect and acceptance of others. Their adaptability and quick wit also helps them to fit in with any social group. They possess leadership abilities and are always ready to take charge of a group. These are people who are always ready with a new idea, but they can be somewhat hesitant to take unnecessary risks. The August Snake is perceptive, however, and usually sees a situation correctly, noting any chance for making an error in judgment. But they can miss out on opportunities by changing their mind too soon or moving on to a new idea before completing the last one. They like to be well dressed and for their clothing to fit the occasion. They may also have a tendency to drink or eat too much because of their busy social lives. But this is a person who interacts with others so well that their social lives continue to improve throughout their lives. Besides being able to liven up any group, they have the ability to remain humble and accepting of other people.

There is a tendency for the August Snake to meet a lost love later in a life and for the love to rekindled, even if only in an affair. These are attractive and fun loving people who may hold a picture in their minds of the perfect innate for them. This tendency may lead them to encourage their mates to continue on self improvement after marriage. But they must guard against being too critical. The August Snake female is strong willed and independent which makes her well suited for a relationship with a man born in the year of the Rabbit or the year of the Cock. The August Snake man finds love and happiness with a woman born in the sign of the April Rat or the May Dragon. Both men and woman of this sign should avoid a relationship with a person born in the year of the Monkey or the Tiger. It is important for the August Snake to seek a relationship with a person who enjoys the same lifestyle, interests, and activities in order that friction and misunderstandings do not hamper the harmony of the relationship.

September

Those persons born in the month of September during the year of the Snake are determined, calm and reflective, but they do exhibit opposition behavior when challenged. Despite this tendency, they are sometimes indecisive, but, again, when other's attempt to offer advice, they become paranoid and defensive. All of this adds up to a person who can be a hard worker, but who may lack the ability to establish strong interpersonal relationships. The September Snake female may be especially independent, preferring to do things her way regardless of the opinions of others. The September Snake also becomes easily discontented with their circumstances and may seek change just for the sake of something different. In this way, others may see them as selfish and self-centered and only concerned with what is best for their own well being. Once the September Snake finds a situation, social circle, job, or activity which holds their interest, they may become more satisfied with their situation. They do best with others who are nonjudgmental and accepting of their eccentricities. And the September Snake may be his or her own worst critic, forever examining their own faults, conversations, and direction in life. They must strive to become more self-confident and to build upon their own self-esteem in order to establish good relationships. They can then develop their own natures and become more empathetic toward others and kinder with a gentler nature. The September Snake often makes good money, but is susceptible to losing it through questionable investments and speculation. Often, these are attractive and appealing people who do well in the fashion industry or in some field where they receive recognition.

The September Snake female does well with an understanding man such as one born in the year of the Ox or the Horse. The September Snake man is well suited for a relationship with a woman born in the sign of the October Rabbit, the April Ox, or the November Dragon. Both men and women born in the month of September during the year of the Snake should avoid a relationship with a person born in the year of the Tiger or the year of the Monkey. Arguments persists in relationships with Boar year persons.

OCTOBER

Those persons born in the month of October during the year of the Snake retain their youthful appearance throughout life and also remain active, outgoing, and accepting of others. They are honest, straightforward people with a tendency to speak their mind, but who can at times be too blunt. They are most generally accepted by others, however, because they are honest, trustable and reliable. These people fit in well with others and their adaptability make them well suited for numerous occupations. Often their success in life is encouraged by an older friend or associate which results in them attaining their goals and becoming secure by middle age. The October Snake is known for a streak of carelessness, and may complain needlessly about insignificant matters. Learning to think before speaking or acting is a challenge to the October Snake as well. Neither do they always take the time to listen attentively to others, and this may lead them to miss out on opportunities. When these difficulties are dealt with effectively, many an October Snake becomes successful in his or her chosen field. Some 'may choose careers in which their youthful appearances are a benefit to them, and others enter creative fields such as acting or writing. The October Snake places as much value on accomplishment and self development as on material possessions. Be that what it may, the October Snake generally acquires property or possessions, but must guard against losses through being overly generous with friends and family.

The October Snake can be a gentle and suave romantic and may find that he or she falls in love with love. Romance is an important part of life to these people. The October Snake female finds love and romance in a relationship with a man born in the sign of March Rat, the April Dragon, or the January Ram. The October Snake man may fall head over heals in love with a woman born in the sign of the April Rabbit, the May Dragon, or in the year of the Rat. Both men and women during the month of October during the year of the Snake should avoid relationships with those born in the year of the Boar of the Tiger. The October Snake strives to be a good marriage partner.

November

Those persons born in the month of November during the year of the Snake are the most extroverted and social. They seem to need to be surrounded by people. But they can also be sensitive and emotional with a tendency to want their own way a great deal of the time. Their apparent self-confidence can at times be a cover for a person who is indecisive, has poor judgment, and lacks strong will power. These difficulties can be overcome, but the November Snake must guard against becoming too pleasure seeking. They must especially guard against developing a drinking habit and eating too much rich foods which can lead them to become even more insecure. The November Snake does best around people who are supportive and understanding. The November Snake finds that their luck improves as they resist the temptation to indulge and when they develop their talents and skills. They must also learn to recognize opportunities and to take advantage of them which at times means making the right decision at the right time. At the same time, these are industrious people who apply themselves to becoming financially secure. This goal is often realized by the age of thirty. The November Snake is also a bit competitive and can be jealous and envious of the accomplishments of others. Then too, they can be impatient with others at time and critical of not only themselves but of the other people in their lives. At the same time, they often make mistakes in judgment and are too trusting of others which can result in stress and even material loss.

The November Snake does best in romance when they marry for love and then work with their partner to establish a relationship that is secure and peaceful. The November Snake female is well suited for a relationship with a man born in the sign of the September Ox or the October Cock. The November Snake man finds love and companionship with a woman born in the sign of the September Horse or the October Dragon. Both men and women born in the month of November during the year of the Snake should avoid a relationship with a person born in the year of the Monkey or the Tiger because they will resent their jealous nature. True love brings the happiness this person desires.

December

Those persons born in the month of December during the year of the Snake are energetic and active, but they are also content with their lives. They are well bred and distinguished with an elegant air of grace and an aura of being in the right place at the right time. This self composed and self confident person, on the other hand, can be outspoken at times and prefers honest, straightforward remarks that express their opinions. Confidences are not their best suit, as they have a tendency not to keep secrets well or to bother to conceal their feelings about much of anything. Be that what it may, their generosity and adaptability makes them well liked and accepted by other people. The December Snake strives to improve their talents and with their intense powers of concentration are often quite successful. They can be hard working and will make diligent efforts to complete their endeavors. Many a December Snake realizes professional or career success at an early age. In relationships, the December Snake can be especially jealous and possessive of their loved ones, and this can make their partners feel smothered. It is most important that they take their time to get to know the other person before marriage in order to determine their compatibility. They do best in relationships with a person who appreciates their talents and hard work and who does not resent their possessive nature. The December Snake makes up for these short comings by being most expressive and demonstrative when it comes to love and romance. And then again, they are best suited for a person who responds in the same way and who appreciates this aspect of their nature. The December Snake is known to be determined and to have great staying power in all of their endeavors.

The December Snake female likes to give and receive the attention of her partner. In love, she does best with a man born in the sign of the September Snake or the May Rabbit. The December Snake man finds love and happiness with a woman born in the sign of the September Rabbit or the October Dragon. Both men and women born in the sign of the December Snake should avoid a relationship with a person born in the year of the Monkey or the Boar.

Date With Destiny

Horse

Share With Me Your Fantasy

THE HORSE

Those persons born in the year of the horse are gregarious, cheerful, fun loving, curious and strongly independent. The Horse thrives on attention, praise and flattery, and loves to be the life of the party or any gathering of friends and family. You are an attractive person who likes to be well dressed in a somewhat flamboyant style, attention getting style. You are active, energetic and alert, and you like the theater, music, movies, concerts, sports, and social events. You know how to turn on the charm and to be pleasing to others. Your charisma makes you a popular person with other people, and your outgoing nature makes you well liked by friends and family alike. You often find that your social life is enhanced by influential people. This is a strong willed person with considerable talent who makes every effort to improve himself. The Horse often knows how to make good use of his or her talents, but quite often doesn't like to take advice from others. The Horse is also known for being at times selfish and even a cold blooded, bitter person who easily loses his or her temper. You are wise, but often you are impatient with other people. You are, however, a hard worker who knows how to handle money. Generally, the Horse finds success because of being able to work hard to achieve goals. This person can be clever and adept at public speaking and does well in fields that require that type of skill and talent. There are times when you need to guard against talking too much rather than listening to others. The Horse finds that he or she must work hard before accumulating possessions, but the same hard work does result in successful business accomplishments. This person may not take the time to invest wisely for the long term which is necessary to ever realize a profitable return. The Horse should avoid taking risks in investments. You may find success in a small family-run business, but the chances are you will never be wealthy. Also, it is possible learning a new skill is beneficial. The Horse is not a lucky person in law suits. And you must guard against losing personal items or losing touch with family members because often they are not found again. Generally, the Horse is a person who was closer to his or her mother than father and often there are arguments with brothers and sisters. This is also a person who likes to travel, but often this travel results in disappointments and a discontented attitude.

The Horse has a tendency to attract many friends, but there are times when you don't keep personal confidences well which results in others being offended. You also have a tendency to become too involved in the personal lives of your friends and then become impatient when they don't follow your advice. Others, however, come to depend upon your friendship. You care about your appearance and spend a great deal of time looking at yourself in the mirror. You make every effort to make the most of life and to enjoy every minute possible. You are also intuitive and can anticipate what others are going to say and the feelings and intentions of other people. Your verbal skills and outgoing, pleasing personality combined with your leadership abilities may make you well suited for politics, public speaking, or acting. Generally, the Horse is an active, energetic person who is self-confident and who trusts in their own judgment regarding decisions and directions. Part of this self-confidence is based on a good impression of themselves and the Horse firmly believes they will succeed in their efforts because of this. The weakest point of the Horse if his or her impatience with others and a hot temper which at times cannot be controlled. This person can also be ambitious to the point of stepping over others in a selfish manner in order to obtain what he or she wants. You quite often are a skilled business person all the same who generally gets what you want. While you are capable of working as a team player, you may prefer to be the leader of the group and you much prefer to be praised, respected, and admired for your efforts. The Horse many times does better when he or she learns to be more tolerant of others. Whatever the case may be, the Horse quite often possesses an adventurous spirit which makes this person the center of attention. The Horse enjoys a fortunate life, and while this person may never be rich they do receive much attention and praise from others.

There are times when you offend your family or loved ones because you have so little time for home, preferring to be caught up in a social whirlwind of activity. A man born in the year of the Horse may find it difficult to take romance seriously. To a woman, romance is a very important part of life. Although this is a sensual sign, many persons born in the year of the Horse are not especially sentimental. You may find that your attention is drawn more to the physical aspect of love and romance than to the emotional side of caring for another. You much prefer attractive, sexy, and aggressive people for a romantic relationship. This is a person who may cheat or play the field for a long time before settling down to one person. You can be very honest in romance, but you may also be hesitant to make a firm commitment. Because you are so busy with your active lifestyle, you may want to find a partner who is capable of managing your household for you. At the same time, the Horse is quite capable of self-sacrifice to the point of giving up everything for the sake of a loved one. In some instanced, this can lead to problems for a Horse person because he or she will neglect everything else in their lives in order to win the attention of the person they love. Happiness in love is often found with another person who is honest and willing to commit to the relationship and to establishing a home and family.

Persons born in the year of the Horse may find their most suitable -marriage partner with someone born in the year of the Tiger, the Ram or the Dog. The Tiger or Dog person has a tendency to remain preoccupied with their own problems and may overlook or not pay a lot of attention to your coming and going. The Ram may be your ideal marriage partner, however, because with this person you feel safe to make a commitment which results in a long, happy and fortunate arrangement. You may find that a Goat is an ideal partner because that person is tolerant and patient with a Horse and will overlook your moody behavior and even your occasional selfishness. Then too, persons born in the year of the Horse can expect to find a good measure of happiness with someone from the year of the Dragon, the Snake, the Monkey, or the Cock. These relationships are often based on a mutual respect and understanding which can lead to happiness and fulfillment. Expect quarrels and arguments about money in a marriage with a person born in the year of the Rat, the Ox, the Rabbit, or another Horse. Your cheerful disposition results in you never being lonely.

January

Those persons born in the month of January in the year of the Horse are quite often known for their cleverness, way with words, and their diligent efforts. If this person fails to accomplish any goal, he or she will make an extra effort and try again and again to succeed. Often, this type of energetic spirit leads to success. These people are persistent and enduring and will face up to any and all obstacles and problems in their attempt to overcome them. There are times when other people resent this persistence, but that doesn't deter the January born Horse either. This person may find that even the boss doesn't always care for this attitude, but at the same time the boss's boss like the Horse just fine. And this of course can often lead to promotions. Many a January Horse will, in fact, pass the boss by on his or her way up the career ladder. These people also possess a good sense of judgment and make sound decisions. often they are found in careers associated with law or legal institutions. This is a systematic, well organized person who likes to make careful plans and then follow through with them. This person can, in fact, be overly precise with a strong preference for neatness and a firm belief that everything has its place. Another trait which the January Horse is well known for is the ability to effortlessly make other people feel like they are obligated to him or her. This is a person who can be overbearing and obstinate at times. The January Horse is also a proud and often powerful person. It is the inherent intelligence, talent, and independent nature of this person which may well lead to success and good fortune. You are unafraid of tackling the most difficult obstacles or problems. And this is accomplished with a dignity and regal bearing that draws the attention of other people who may both respect and admire you and your actions. Up until the age of thirty, this person may earn money easily and spend it just as easily. This person may receive unexpected money. A man January Horse may find the most luck in romance with a person born in November Boar or September Ox. A woman January Horse may do best with a May Boar of an April Dragon. You find good fortune in love and romance.

February

The February Horse is a person who is dynamically active, physically energetic, generous, honest and honorable. The men born in this sign are especially masculine and powerful. You are an attractive person who is socially popular, but you may be naive with a certain type of innocence that leads you to fail to understand the harsher ways of the world. This is, however, considered to be one of the luckiest sign, and you enjoy your good luck to the fullest. You find that success comes easily to you. And in competitions of any type, you often out perform and out-distance others. This is quite often the sign of the person involved in sports and athletics or other areas of physical endeavors. But then again, with your fine sense of humor, you also make a good comedian. The female February Horse may be more conservative and conventional in style preferring lasting fashions which may be expensive to trends and fads. She likes compliments and appreciates a person who understands her preferences and style, often making friends with such a person. The February Horse can be less tolerant of persons born in a lower social economic background or from another culture and will attempt to better them by changing their standards to their own. Pride can also be a fault with this person as they often know they are attractive and like to bask in the flattery of others. But most generally, this is an open-minded person who is sincere and honest with others and in relationships. They are deeply offended by others who lie to them. The February Horse person enjoys the good life and good food and drink and this can lead to weight problems or health problems associated with high blood pressure. The February Horse should not take unnecessary risks in investments, but this person can expect to earn a good living and even to be prosperous.

The male February Horse may discover a long and lasting relationship with a female born in July Tiger or April Dog. The female Horse born in the month of February does best with a man who is born in February Dog, September Sheep, or in November Sheep. Both men and women of this sign should avoid relationships with those born in the year of the Ox.

March

The person born in the month of March in the year of the Horse is a kind, generous day dreamed. But however absentminded these people are, they are also demanding of themselves and of others. They think a lot and have countless ideas. But failures can be caused by expecting too much of others and by not focusing intently on any one idea or pursuit. This person may often feel discontent or disappointed because they aim higher than their abilities. When this happens, they enjoy the escape of their dream world and fantasies. on the other hand, when this person agrees to do a job or a task for someone else, they will finish it no matter the difficulty or circumstance. Because of this, many times they do not like others asking them to do things. The March Horse is known for being brave and generous with friends and family, but often difficulties and even disasters are experienced. This person may worry a lot which results in mental and physical exhaustion and over exertion in attempting to succeed. This person may be an extravagant individual who finds it very difficult to save for the future or to invest in savings. It is important for the March Horse to learn to handle his or her finances better and to invest wisely. You enjoy the companionship of others with a good sense of humor because you may be that person who simply doesn't remember the punch line or a good joke. It is important to establish a good relationship with in-laws, and this may be especially difficult for the female March born Horse. The March Horse may be prone to a bit of clutter in their living environment and this may be due again to their daydreaming. This person may change employment often up until about age forty.

The men born in the month of March in the year of the Horse are well suited for marriage partners who are born in July Monkey or December Dog. A woman March Horse may find happiness and companionship with a man who is a December Sheep, July Cock, or a man born in the year of the Tiger. Both men and women of the March Horse sign should avoid relationships with another Horse or with persons born in the year of the Rat, December Ox, or September Rabbit. Once you find your love, you become more content and happy.

APRIL

These April horses are extroverted, outgoing, and sociable with a high energy level that keeps them on the go and moving from one social activity to another. These are logical people, however, who most enjoy a lively discussion or interesting debate. This person can also be proud and obstinate and takes a great deal of pride in their grooming, style, and clothing as well. They are very careful people but vain without meaning to be. They can also insist on having their own way to the point of being overly pushy or assertive. In fact, they are not above taking advantage of another person in order to gain or to get what they want. At times, this person doesn't realize when he or she is asking too much of other people. This can upset family members as well as friends. While respectful, the April Horse also talks a lot and can be somewhat dogmatic and opinionated which can offend others as well. This person must make an effort to learn to more polite and considerate, to accept criticism from others, and to listen more to what others have to say and to what they want. This type of consideration earns real friends. You find that your life is extremely busy, but it can be difficult as well and often times you spend as much as you earn. one of your main interests is traveling and if you can you choose to travel extensively or at least as much as possible. You are a person who finds that he or she may have many responsibilities and duties at different times in your life. But you have difficulty accepting changes and find it even more difficult to get over upsets and unexpected changes and occurrences. You do very well otherwise and have a strong character until something goes wrong.

Men born in the month of April in the year of the Horse find that they are most suitable for a woman born in August Dog, November Sheep, or September Monkey. The female April Horse is most compatible with a June Snake, December Monkey, or April Dog. Both men and women may find that they are compatible with a person born in the year of the Dragon. And both men and women born in this sign should avoid relationships with any person born in the year of the Rat. You learn to live well with another person.

MAY

This person is more often than not the brains of the operation, coming up with one good idea after another, and following through to success. You are exceptional at planning and detailing work to be done, but you may find that it best if others carry out the actual work because you have a tendency to move on to your next endeavor. You may be talented but there are times when you are slow to take advantage of your own talent and don't react to opportunities. You are, however, intuitive to the motives and intentions of other people. In some circumstances this makes you a good writer, but you improve yourself by listening to others and by observing the mannerisms of other people. You may also do well as an executive, manager, employer or in politics or public relations. You possess good judgment and are seldom wrong but you find that you often must strive to balance your ideas with the final outcome. You have many hobbies and move from one new interest to another easily. You like to travel and to see new places and rather than return to vacation spot, you prefer to go somewhere different. You also most prefer to appear reserved, studious and refined, but often you are quite the opposite. The male May Horse may have a nervous disposition and be prone to worry. The May Horse, however, is most passionate and caring and can be quite dedicated to their partners, but often they find that these feelings lessen after marriage. The May Horse has a tendency to put most of their energies into work often leaving play until later in life when there is time to enjoy leisure activities.

The men born in this sign may find that the most compatible marriage partners are born in April Ram or June Cock. These men may find happiness with an older woman. A female May Horse finds happiness and satisfaction with a man born in March Snake, April Dragon, or a man born in the year of the Tiger or Dog. Persons born in this sign may find it best to avoid marriage with another person born in the same sign. You are the type of person who a partner may not come to know well until some time after marriage. choose a marriage partner carefully and knowing each other well.

JUNE

Those persons born in the month of June in the year of the Horse are straightforward, direct, brave and clever. Often these people are born to be in the spotlight and glow when the center of attention. They are dramatic and whatever their professions they stand out from the crowd as if on display. They are of regal bearing and seem to draw attention to themselves by simply entering a room. These are passionate and socially popular people who others admire. Needless to say, this is one of the best signs for a career in show business. Their very sophistication spells success and helps them along the road to achievement. Other June Horses may best put their talents to good advantage by entering the fields of art, politics, diplomacy, public relations, managers, advertising, or by becoming an entrepreneur. This person finds that whatever they do when they are young brings success and good fortune. They should be aware, however, that after middle age a good business partner may help them be successful which could result in fame and wealth. The June Horse has a tendency to spend a great deal of money on extravagances which may result in difficulties and problems with others. Even when short on funds, they find one of their greatest faults is impulse spending and spending on unnecessary purchases.

This is an attractive person with a physically appealing body and plenty of sex appeal as well. Often this sex appeal is of the self-made type rather than being natural. This is a person who is willing to improve themselves and to make themselves over if necessary in order to create the image they want and which will bring them the attention they desire.

Those born in the month of June in the year of the Horse may find the most suitable romantic partner is a person born in the year of the Ram, the Tiger, or the Dog. A male June Horse finds love with a January Sheep, a February Dog, or a November Ram. The female June Horse discovers the most compatibility with a March Monkey, July Sheep, or November Dog. Both men and women of this sign should avoid other people born in the year of the Horse as well as those born in the year of the ox.

JULY

A person born in the month of July in the year of the Horse generally has a kind, honorable, and gentle character. This person is careful and understanding and has a large and giving heart. You take both work and play seriously, but you may have difficulty making a decision without seeking the advice of someone else whether it be friend, family, or spouse. This inability to be decisive may lead you to miss out on good opportunities. Because of this, you may feel unlucky, but if you are patient, you will lead a peaceful and content life. The July Horse experiences situations which he or she may perceive as being hopeless, but this person must learn to set priorities and to take things one step at a time in order to succeed. The July Horse likes to gamble and is at times extravagant which reduces his or her savings abilities.

Your willingness to make sacrifices carries over to your work place where you are known for your generosity and ability to take that extra step or to make the extra effort when it is necessary. And you find that you are appreciated and at times even rewarded for all this extra effort. And although you may feel like you are working harder than others or putting in extra time at the office, whether working for someone else or yourself, your effort does pay off in the long run. The July Horse male is an especially hard working person who is more than willing to go the extra mile in order to accomplish his goals.

July Horse females are known for being especially well liked and appreciated by everyone they know. Their families, extended families, and even their in-laws love them for their cheerfulness and optimistic attitude as well as their efforts and generosity to help others. This woman may marry early in life, some while still in school, and there are those who visit or live in foreign countries with their husbands after marriage.

The July Horse woman is well suited for the June Tiger or October Cock. The July Horse man finds a good marriage partner with the April Ram or the November Monkey. The July Horse should avoid marriage with the Ox. An influential person offers assistance.

August

The person born in the month of August in the Year of the Horse is sensitive and at times impractical because of a tendency to be overly emotional. At the same time, this person is generally a good judge of character when dealing with other people. This can be because they are intuitive and will often react to their first impression of another person. However, the August Horse can appear be an exhibitionist in that they consider themselves first in all matters, always thinking themselves the best. With many an August Horse, this self-confidence leads them to success in their chosen fields at an early age. This person can be an astute business man or woman who out distances others in the same age group. This is not the person who takes risks, though, but will only attempt what he or she feels they can accomplish. The August Horse most enjoys traveling and will collect souvenir to show friends and family of the many places they have been. They also like music and dancing. A woman born in August of the year of the Horse finds true love with a December Dog male or a man born in the sign of October Sheep. A man born in the sign of the August Horse is well suited for a woman born in November Dog or September Dragon. Both men and women of the August Horse sign should avoid relationships with people born in the Year of the Rat or other people born in the Year of the Horse. You are most compatible with another person who shares your interests, values, and hobbies. And you are most successful in a relationship when you take the time to understand the other person's likes and dislikes.

There are times when you are impulsive and make your plans based on the whims of your emotions. It is important for the August Horse to learn to think of others more and to be more considerate rather than focusing so intently on his or her own plans, desires and wishes. You overcome your mood changes by learning to plan carefully and by focusing on one plan at a time.

Those persons born in the sign of the August Horse are graceful in movement and gracious in manner. You achieve success based on the merits of your own hard work and diligent efforts.

September

The September Horse is a kind and caring person who can also be most generally straightforward, direct and honest with others. A self-reliant person, the September Horse doesn't refuse to be of help to others but at the same time rarely asks for help or advice themselves. This person gets along well with friends and family and is considered responsible and trustable. When others experience problems, the August Horse is always ready to help even if it is inconvenient or causes problems for them. But there are times when others will, for their own benefit, take advantage of the September Horses helping nature, and this person must learn how to avoid this. For success, this person most needs to learn to take advantage of an opportunity when it presents itself, and to listen to and take the advice of older or more experienced people. The September Horse may experience difficulties after middle age and should be cautious with new endeavors or new plans during this time. You may find that you have difficulty saving money and that you most enjoy little spending sprees which can become an extravagance for you. You are a person who loves to collect and this can become a compulsion with you even to the amazement of your friends and family. You are such a trustable and reliable person, however, that you make a wonderful friend and are well-loved by your family.

Persons born in the month of September in the year of the Horse are the romantics of this sign. They fall in love easily being people who are simply in love with love. But again they must be cautious because their lovers may take advantage of their loving and giving natures. The men born in this sign are wonderfully well suited for a relationship with a woman born in December Cock or October Rat. The women born in the sign of the September Horse find the best relationships with a man born in January snake or one with the sign of the March Monkey. Both men and women who are a September Horse should avoid relationships with a person born in the year of the Ox, the Rat, or the Rabbit as well as others in the year of the Horse. Once you meet the right person, you develop a strong and lasting relationship based on friendship, romance, and mutual interests.

OCTOBER

Those persons born in the month of October during the year of the Horse are kind and most humane, but they are also brave and self-confident with an enduring patience and keen intelligence. You enjoy a good conversation no matter how mundane the subject matter. The October Horse doesn't do as well in their own business as they do when they have an employer to give directions. This is because this person is most adept at following through and completing what others have started or initiated. More than a few October Horse persons are somewhat lazy and can tend to idle away their time. But at the same time, this person, who makes such an excellent assistant, is self-reliant, a good student, and a talented researcher who may decide to become an inventor. This is the person who picks up an intense interest in a hobby or particular past time which preoccupies their mind and their time. The October Horse may also have particular likes and dislikes and many of these people are notable neat and clean even as children. The one weakness of the October Horse may be a tendency to brag or show off which is a trait that can cut them off from others as they mature. This is a person who doesn't necessarily seek the advice of others but who learns that there are times when such advice is helpful. The October Horse discovers that once the right position in life is found, be it employer or business partner, ho or she receives guidance or assistance from an influential person.

A woman born in the month of October during the year of the Horse finds love and romance with a man born either in June Ram, October Snake, November Snake, or March Snake. A man born in this sign is well suited for a romantic relationship with a woman born in the sign of March Sheep, December Sheep, other months of the Sheep, and June Dog. Both men and women who are October Horse should avoid relationships with persons who are born in the signs of the Horse, the Rat, or the Rabbit. You do best in a relationship when their is a shared level of intellectual development. In other words, if you are well educated then you find the most compatibility with another person with the same degree of education. When you like a plan or idea, you follow it through to completion.

November

A person born in the month of November in the year of the Horse is an independent person who is content with life. They are kind, sociable, and fun loving with a tendency to be satisfied with their situation. This is a confident and self-assured person who is rarely influenced by the opinions of others. The November Horse relies a great deal on his or her own ideas and imagination for direction and in decision making. This is the loner of the sign of the Horse and may be a person with few family members or someone who chose to move away from the family setting at an early age in order to establish an independent lifestyle amongst strangers. Because they are easily satisfied with life, the November Horse may not work extra hard at succeeding which may limit their business advancement or career promotions. In selecting a business partner, the November Horse may choose a person who is outgoing and energetic to offset this limitation. This person is curious by nature and a natural researcher or investigator who loves to solve a problem or uncover a mystery. At ease with making friends and socializing with others, the November Horse does well with other people but at times may need to be more giving and considerate as other people may be more dependent on their friendships and relationships. This person is known for a good sense of humor and gracious manner.

There is every indication that the November Horse forms a relationship or marries someone from another area of the country or a person from a foreign country. And after marriage, those born in this sign develop a dependency on their partner based on sharing experiences and a mutual companionship. An November Horse man does well with love with a partner born in February Tiger or a person born in the sign of the February Ram. The November Horse woman finds love with a partner born in the sign of September Cock, April Dog, or May Snake. Both men and women born in the month of November in the year of the Horse should avoid relationships with those persons born in the year of the Rat, the Sheep, the ox, or the Horse. Your self-contentment with life makes you an easy person to get along with and to live with. You enjoy a harmonious relationship.

December

Persons born in the month of December in the year of the Horse are kind, open minded and personable. This is a person who likes to work hard and to play hard. It is said that December Horse persons have two sides to their personality. They can be sensitive, concerned and caring, or, when angered, they can become malicious. Although hard working and sociable, their dual nature shows in a tendency to like time alone and other times when they become somewhat lazy or indifferent to their pursuits and interests. This dual nature can, of course, make the December Horse changeable and moody, and there are times when this person displays too much pride and a selfish, uncaring nature. This is a person who desires change in life whether it be in activities, location, place of employment, careers, or groups of friends. Despite this tendency, the December Horse is a well-balanced person, but definitely a person who may find it difficult to establish stability in his or her life. This person's mind seems to drift from the job at hand, and they can be dreaming of faraway places. This yearning to be somewhere else can even lead to job transfers. This person does well in an occupation that offers changes such as in sales or the travel industry. They also make good writers. Others stuck in a routine lifestyle may turn to escapism through alcohol or drugs. This is a casually dressed person who finds it easy to fit in to a variety of different situations always appearing at their best. The December Horse is generally a good judge of character, but they can be too accepting of other people and there are times when this leads them not to face up to the reality of a situation. The December Horse becomes lucky once they learn to be more pragmatic.

The December Horse woman finds love with an October Monkey, a September Tiger, or a person born in the sign of the Dragon. The December Horse man finds love and companionship with a woman born in the sign of October Dog, February Rat, or December Snake. Both men and women born in the sign of the December Horse should avoid relationships with people born in the year of the Rat, the Rabbit, or the Horse. You discover luck and love when in love.

Date With Destiny

GOAT

Share With Me Your Fantasy

THE GOAT

Those people born in the year of the Goat are known for their sincerity and sensibility. They are considerate and capable people with a love of nature and an artistic flair or highly religious individuals. The Goat can be charming, friendly and good-natured who appreciates the simple things in life. But with their refined taste, they most prefer a comfortable lifestyle and are most secure when they possess what they perceive as an adequate amount of material possessions. The Goat is not particularly adventurous nor overly ambitious and often views the outcome of their life as fate. And their is a tendency for the Goat to never be quite happy about that fate which can lead them to be pessimistic and indecisive. They have the ability to make others unhappy with their moodiness, impatience and complaining. While the Goat may become accustomed to any type of lifestyle, they love attention and being taken care of by someone else. They are emotional and sentimental, and while they gripe about what fate has in store for them, they do little to change their situation.

All the same, the Goat is the most peaceful and gentle of people. It is their hesitancy and nervousness that leads them to worry and complain so much about insignificant matters. And their indecisiveness that leads them to leave their decisions to fate. The Goat has a distinct dislike for rules of any kind and is especially poor about making rules for themselves or others. They may like to tell others what to do, but in reality, they aren't very efficient at this, because of their hesitancy, and do much better when others are in charge and telling them what to do. Then too, they can be fussy and lacking in self-confidence, and they are always running late, and then complaining about being late, which makes them poorly suited for leadership positions. And they abhor tedious, mundane jobs, becoming easily bored with routine and sedentary jobs. Their creative imagination and adaptability make them well suited for any number of careers, but the Goat has a distinct distaste for menial or unpleasant situations and jobs.

After all, the Goat is dignified with a strong preference for elegance, and these individuals are well known for their good taste in all matters. The fact that they can be obstinate and stubbornly concerned with minor matters effects their personal relationships. And this is not helped by their hesitancy and indecisiveness. But the annoyance of others is often forgotten when the Goat turns around and displays their sincere concern for the welfare and well being of others. The Goat can be caring and attentive when called upon to do so. So others have a tendency to tolerate their shy and often timid demeanor and to overlook their constant complaining and discontentment with their fate. Of course, the Goat can be content when comfortable or wealthy and well provided for. That is, until any problems or difficulties arise, at which point the Goat becomes dissatisfied, moody, and prone to expressing their dissatisfaction to anyone who will listen.

Perhaps, the Goat is best described as a dreamer who envisions better things. They can easily become preoccupied with their dreams and fantasies, and it can be quite a jolt to be suddenly reminded of reality and the necessities of dealing with immediate problems and duties. The Goat does best when he or she learns to follow the directions and advice of others. And often the luck of the Goat gradually improves and changes for the better when they are in a situation that provides them with efficient leadership. They also function much better when their surroundings are comfortable and conducive to allowing their creative abilities to flourish. This creative day-dreamer simply has a pronounced need for the guidance of a strong and protective leader who knows how to elicit the most productivity from the Goat. The Goat, however, is intelligent and easily acquires an education and the necessary training or skills to do most any task well.

The hesitancy of the Goat can lead them to miss out on opportunities. But when they find direction early in life, they don't suffer as much from uncertainties and the difficulties of having to make decisions. The luck of the Goat varies, but more likely than not they become successful and financially secure and usually possess enough of the necessities of life to free them from excessive worries or want. After middle age, the Goat generally makes a good income and owns property and possessions. The Goat is careful with investments and does best with long-term investments rather than gambling, speculation, or short-term investments. Besides creature comforts, the Goat also likes to travel, enjoying both short and long trips. The Goat isn't overly independent, but because they don't always get along well with their parents, they often choose to live by themselves as soon as they begin work or are otherwise financially able to do so. They don't depend on their brothers or sisters either, and it may be that these relationships are either strained or not overly close. Their unpredictable luck may prevent the Goat from obtaining all their hopes and desires, and they rarely receive any unexpected help or sums of money. The Goat should avoid all lawsuits as they are as likely to lose as not. They do, however, find themselves luckier in finding lost items or friends who they may have lost contact with. The Goat may know intuitively that they are not lucky, but this awareness seems to grant them the ability to turn misfortune to their advantage. But while the Goat uses their talents, skills, training or education to earn a good income, they often spend lavishly on a busy social life rather than investing or saving as much as they should. Socially, the Goat is well liked because of their sincerity and their caring nature.

The Goat can be quite passionate, emotional, and romantic, but their shyness and inability to express their feelings may lead them to miss out on many opportunities. Because they can be shy with the opposite sex, they are often introduced to their romantic partners through friends or associates. And because of their awareness for a preference for a comfortable lifestyle, the Goat rarely marries anyone who is not equally concerned with possessions and finances. The Goat may wait until they meet someone who is financially secure and able to help take care of them. But once the passions and love of the Goat are focused on another, they become devoted to making that person happy.

The Goat are best suited for a relationship with the Rabbit, the Horse or the Boar because persons born in this sign will understand the needs and dependent nature of the Goat. Those born in the year of the Goat should avoid marrying any of the other signs; especially those born in the year of the Goat, the Ox, the Dog, or the Rat which would result in difficulties and misunderstandings. When the Goat is unhappy in love, they continue looking for someone to fulfill their passions.

JANUARY

Those persons born in the month of January during the year of the Goat are cheerful, active, and most independent of the Goat born. The January Goat has a tendency to be conservative, conventional, and especially truthful. They are just, honest, and willing to stand up for what they believe in. They find their luck holds during their youth and that their good fortune and financial success continues to improve gradually throughout life. But these are not people who like to tempt fate or take an unnecessary risk. Their financial security is important to them, and they avoid any gamble or speculative venture. Neither are they overly adventurous in other areas of their lives. Their basic insecurities and need to be secure leads them to fear any mistake or other decision which might cause them to fail in life. The January Goat is thoughtful and prone to planning their lives in great detail in order to avoid any errors in judgment. They may resist even good opportunities because of their pronounced caution and unwillingness to take a chance. All the same, they are extremely active and this is true in their social lives as well. They are drawn to social occasions and love entertainment, spending freely on having a good time. This extravagance may be one of the main reasons why the January Goat are rarely wealthy. They are persuasive speakers with a fun sense of humor who like to make others feel at ease and comfortable. All the same, they don't always trust the intentions of others and must learn to listen to the advice or opinions of other people.

They are passionate and romantic, but the January Goat may be so cautious that they prefer to wait until marriage before having a serious or ongoing relationship. The January Goat female is well suited for a relationship with a man born in the sign of the March Rabbit or the April Horse. The January Goat man finds love and happiness with a woman born in the sign of the August Rabbit, the September Rabbit, or the February Boar. Both men and women born in the month of January during the year of the Goat should avoid relationships with those born in the year of the Rat or the Dog. Your romantic nature comes alive when you meet the love of your life.

February

T hose persons born in the month of February during the year of the Goat are kind, friendly, mild-mannered individuals who are well liked and accepted by their friends because of their honesty and sincerity. The February Goat is most drawn to a peaceful life and is not overly ambitious. They are talented, however, with a tendency to be a romantic dreamer who may pursue art, music or a career where their talents can be put to use. They do not become particularly upset with difficulties and accept any failures as simply a part of life. With the right management, they advance steadily in their careers, becoming productive workers who are respected and well liked by colleagues. Without this management, however, they drift along becoming lazy and unmotivated. While they prefer not to have to struggle to achieve their goals, there is every indication that the February Goat is fortunate and easily acquires a prominent social position. They are gentle, compassionate and caring people which others appreciate, but they can also be self-indulgent and extravagant. When the February Goat learns to handle his or her money well, they become more proficient at business matters. Otherwise, they can become withdrawn and pessimistic always wondering why fate has brought them to their present dilemma.

The February Goat female is most feminine and romantic and falls deeply in love, becoming loyal and faithful to her partner. She stands by her commitment to her husband even through difficult times. She finds love and happiness in a relationship with a man born in the sign of the November Rabbit or the December Tiger. The February Goat man is passionate and caring and likes to create a romantic setting. He is best suited for a relationship with a woman born in the year of the Tiger or the year of the Rabbit, especially the June Rabbit. The Rabbit most likes a comfortable home and security and is good at managing financial affairs and organizing social activities. The February Goat should avoid a relationship with those born in the year of the Dog. The February Goat makes a protective and loving parent who may spoil their children, but the time they spend with their families makes for happy times.

March

T hose persons born in the month of March during the year of the Goat are congenial but reserved individuals who are conscientious and diligent hard workers. They can be creative, but may lack the initiative and assertiveness to develop or capitalize on their talents. These are the people who are discovered by others who then act as a catalyst for putting their talents to use. When this happens, the March Goat may become successful at an early age. But their greatest disadvantage is that while they may like to follow the ideas of others, they can be quite obstinate when it comes to taking directions or following orders. They definitely don't like to be bossed around, and they may have some difficulty putting their trust in others. This lack of trust only makes it harder for others to place their confidence on the March Goat. But perhaps, they are distrustful because others have a tendency to take advantage of the well meaning and honest Goat.

Once they find another person with whom they work well, the March Goat seems to become successful with ease. It helps that they know how to associate with the right people who are in a position to offer them opportunities. Whatever the case may be, the March Goat somewhere along the way experiences his or her fair share of difficulties and problems. They are logical rather than emotional people and when these difficulties do occur, they take them in stride rather than becoming emotional or angry. The March Goat is not overly extravagant and prefers not to waste his or her resources on social activities or on pleasure seeking.

The March Goat female is content with a domestic situation and after marriage becomes devoted and dutiful. She does best with a strong male figure who takes on the leadership role; otherwise, she can become discontented and may find companionship outside the marriage. she finds happiness in a relationship with a man born in the sign of the October Horse or the December Horse. The March Goat man finds love and happiness in a relationship with a woman born in the sign of the November Tiger or the December Horse. Both men and women born in the sign of the March Goat should avoid relationships with those born in the year of the Rat.

APRIL

T hose persons born in the month of April during the year of the Goat are quiet, meditative individuals who like time to themselves to explore their thoughts or other interests. And they find interruptions bothersome to the solitude which they seek. They are intelligent and scholarly and do well in careers which require detail-type work such as research, publishing, or teaching. Their biggest fault is that they can be obstinately inflexible and once they establish a routine, schedule, or particular lifestyle, they do not like to change. This can be the type of shy person who embarrasses easily in public places and abhors any type of public display which would draw attention to them. The April Goat is warm-hearted and compassionate, but they have a great deal of difficulty expressing their feelings to those they care about. And they can be hesitant to be charitable or giving toward friends or those in need. Their hesitancy and caution also leads them to miss out on opportunities which may have otherwise proved fortunate or lucky. Their are indications that the April Goat earns a comfortable income, but they can be extravagant in their spending. They possess refined tastes and have a preference for the arts and literature as well as for a comfortable home and possessions which reflect their good taste. They make diligent employees or do well in their own business, but they are not well suited for positions which require them to give directions to others.

The April Goat can be accommodating before marriage, but after marriage they can become outspoken and bold about getting their own way. They do best in relationships with persons who understand and appreciate their natures. The April Goat female is well suited for a relationship with a man born in the year of the Rabbit. The April Goat man finds suitable companionship in a relationship with a woman born in the sign of the May Rabbit, June Horse, or in the year of the Boar. Both men and women born in the sign of the April Goat should avoid relationships with those born in the year of the Dog. with the right partner, the April Goat learns to be more tolerant and accepting of others which leads them to be more compassionate and caring.

MAY

Those persons born in the month of May during the year of the Goat are elegantly attractive, friendly, sensitive, and caring. Their love of nature draws them to appreciate the out of doors and activities in natural surroundings. They are well liked and suited to any number of careers, but the May Goat does best in a position that doesn't require decision making or delegating responsibilities to others. Quite often, they are creative and talented in some area of the arts whether it be music, dance, design, or art. The May Goat must strive to overcome an impatience which arises over difficult situations. In such instances, they are tempted to abandon their projects altogether. This impatience can lead them to miss out on opportunities, but the May Goat is lucky in that there are indications that in time of need, they receive unexpected assistance often in the form of money. Like other Goat year persons, the May Goat has a tendency not to save adequately but to spend rather extravagantly on their interests, hobbies, clothes or furnishings. They much prefer a comfortable home setting resplendent with articles of worth which reflect their desire for possessions and their refined taste. They can also be pleasure seeking and will spend freely on entertainment. The May Goat can also be somewhat secretive and no one, not even their spouses, ever knows all that goes on in their minds. No matter how sociable they may be, the May Goat like their privacy and time to themselves.

The May Goat female is considered attractive and desirable with an alluring mystique which is reflected in her elegance and noble demeanor. She is at her best when she feels appreciated and is best suited for a relationship with a man born in the year of the Rabbit or the Tiger. A May Goat man is handsome with a boyish nature that they know how to use to their advantage. He is best suited for a gentle and accommodating woman and does well in a relationship with a woman born in the year of the Boar or the Rabbit. Both men and women born in the month of May during the year of the Goat should avoid relationships with those born in the year of the Dog or the Goat. The May Goat appreciates the attentions of his or her partner.

JUNE

Those persons born in the month of June during the year of the Goat are benefit from the luck bestowed on them by this month. There are indications for the June Goat to enjoy a happy and peaceful life with few difficulties associated with their friends, family, or financial affairs. It seems they can do no wrong and even when at lotteries and raffles. They are warm-hearted and optimistic people who are kind, honest, and well liked by others. The June Goat may be well off through an inheritance or by marrying well. And it may seem that his or her greatest obstacle is learning how to handle money well rather than allowing it to become the main focus of their lives. There are those June Goats, however, who become complacent and discontented with their good fortune, and this pessimism and sulkiness can turn their good luck into misfortune. And even the luckiest of June Goat individuals must remember to work hard rather than becoming lazy and taking their good fortune for granted. The June Goat also has a tendency to draw attention to themselves by always running a little late for appointments, meetings, or social events. This can be an annoying habit for their friends, associates, and family. Other June Goats are best described as followers who are adept at finding someone else to keep them comfortable and in style. But for the most part, the June Goat are cheerful people with pleasant, youthful appearances who know how to make a good first impression. They are accommodating and like other people, and, in turn, other people like them. At their worst, they can be stubborn and impatient with others and must learn to listen to the advice and opinions of other people. They do well in careers that allow them to work with groups of people.

The June Goat female is a lovely individual and a gracious hostess and is at her best when entertaining others. She is well suited for a relationship with a man born in the year of the Boar or the Horse. The June Goat man does well in a relationship with a woman born in the year of the Rabbit. Both men and women born in the sign of the June Goat should avoid marriage to a person born in the year of the Rat or the Dog.

JULY

The people born in the month of July during the year of the Goat are well liked by others because of their kind, just, and honest nature. They keep their promises and fulfill their obligations and responsibilities to the best of their abilities. They are hard working and diligent and rarely distracted by insignificant issues or minor problems. They are perfectionists, however, and whatever it is they are working on, they are not satisfied until it meets perfection by their own standards. In this way, they can be obstinate and most stubborn and impatient with others who prompt then to rush through a job. The July Goat is often talented and creatively imaginative. But they can be inflexible in that they do not like change, and once they are settled in one location or a particular job, they prefer not to relocate or change jobs. They feel most secure when they have attained a position which affords them a comfortable lifestyle. They become attached to their home and possessions and protective of their things. They make loving and protective spouses and parents as well. They enjoy caring for others and in this way can be generous with their time and efforts. But they do not like being taken advantage of and will abandon a friendship if this happens. Once married, the July Goat focuses his or her attentions on their families and are most loyal, devoted, and faithful. They make excellent husbands and wives when they find the right partner who understands and appreciates their natures. They are open, accommodating, and sociable people who express their compassion and caring. At their worst, the July Goat can be overly self-critical, indecisive, and weak willed with a need for someone else to lead the way.

The July Goat female finds love and happiness in a relationship with a man born in the sign of the June Rabbit or the May Horse. The July Goat man is best suited for a relationship with a woman born in the year of the Horse or the Boar. Both men and women born in the month of July during the year of the Goat should avoid relationships with those individuals born in the year of the ox, the Goat, or the Dog. The July Goat gets along well with a partner who is supportive and caring.

August

Those persons born in the month of August during the year of the Goat are friendly, kind, and compassionate with a quick, alert mind and agile body. They possess a logical, conservative nature and a well developed appreciation for beauty. The August Goat are generally well dressed individuals who look becoming in whatever they choose to wear, be it everyday clothing or designer wear. And they are the least likely of the Goat born to be extravagant, preferring to save and to invest wisely. They are well liked and easy to get along with, but a certain stubborn streak shows itself whenever things aren't going to the advantage of the August Goat. Be that what it may, they remain diligent, hard workers who are thorough and competent in all of their endeavors. There are indications that their abilities see them through difficult periods when they are young and that they become more established as they mature. When the August Goat comes to terms with his or her obstinacy in some matters, they become more content with their lives and more successful. From time to time, their easy going natures give way to a moodiness, but this too is overcome with patience as the August Goat learns to accept life and to be happy and content. For many an August Goat, a comfortable but conservative living style in a city or suburban atmosphere suits them well. This is a person who may be well suited for a career in some area of the fashion or cosmetic industry.

The August Goat female knows to save for a rainy day and is not prone to wasting her resources. She finds love and happiness in a relationship with a man born in the year of the Rabbit or the year of the Horse, especially the January Horse. The August Goat man does best with a partner who shares in his appreciation of beauty and refinement. He does well in a relationship with a woman born in the sign of the October Rabbit, the September Horse, the November Boar, or during the year of the Tiger. Both men and women born in the month of August during the year of the Goat should avoid relationships with those born in the year of the Rat, the Dog, or another Goat born person. The August Goat strives for a peaceful relationship with his or her partner.

September

Those persons born in the month of September during the year of the Goat are capable individuals who are sensible and intensely sincere and dedicated. But in many cases, they can also be perfectionists who set high standards for themselves and for others. The September Goat has difficulty delegating responsibilities to others, and they can become frustrated in their efforts to do everything to perfection themselves rather than asking for assistance. If any one trait can lead the September Goat to experiencing failures, it is this one. And the difficulties resulting from such failures lead the September Goat to be easily upset, worrisome, and anxiety-ridden, causing minor problems to escalate into larger ones. Without meaning to, the September Goat can also be demanding of others, wanting them to meet their expectations of what they perceive as perfection.

Often, the September Goat becomes mentally and physically exhausted without knowing why. When these individuals allow this pattern to continue, it can result in nervous disorders and even effect their health. It is wise for the September Goat to learn relaxation techniques and to recognize that their obsessive behavior is causing many of their difficulties. The September Goat does excel at financial management, and there is an indication that those born in this sign benefit from an inheritance. Their astute, detail-oriented minds and artistic ability may make them well suited for careers in art or design. Others are proficient in businesses such as insurance, accounting, or research or in teaching, or lecturing.

The September Goat female is often demure with the ability to make men feel in charge and superior. She is well suited for a relationship with a man born in the sign of the January Rabbit or in the year of the Boar. The September Goat man does well in a relationship with a woman born in the sign of the March Horse, the April Rabbit, or in the year of the Boar. Both men and women born in the month of September during the year of the Goat should avoid relationships with those people born in the year of the ox, the Rat, or with others born in the year of the Goat. The September Goat needs a person who adds fun to life.

OCTOBER

Those persons born in the month of October during the year of the Goat are calm and reserved but with a personable, cool manner that makes them well liked and accepted. They are intelligent, quick learners who can appear concerned about others, but who will use any means necessary to achieve their goals or to obtain what they want. In this way, they can be manipulative and self-serving, helping others when it benefits them in some way. Their personal achievements and self-gratification may even cause sacrifices and hardships to their friends, family, and associates. The October Goat can also be jealous of others and envious of the possessions, achievements, or happiness of others. In some cases, this jealousy leads them to gossip or to talk about others in a derogatory way. At the same time, they don't like to be the brunt of jokes and are easily offended by any criticism aimed their way. The October Goat is adept at turning misfortune into good fortune and in this way is quite often perceived as lucky. They are can be adaptable to new situations and people with a natural diplomacy and grace that enhances their interpersonal and romantic relationships. If they worry needlessly, it is usually about psychosomatic related illnesses, and they have a tendency for the sniffles to turn into a full blown case of influenza or for minor aches and pains to become so serious that they take to their beds. The October Goat does well in any number of occupations including art, music, politics, law, journalism, publishing, or business.

The October Goat female can be creative and elegant, but she may become overly concerned and preoccupied with the health of her family and loved ones. She is devoted to her spouse and does well in a relationship with a man born in the sign of the September Tiger, October Tiger, November Tiger, or in the year of the Rabbit. The October Goat man does best in a relationship with a woman born in the year of the Horse, especially the August or September Horse. Both men and women born in the month of October during the year of the Goat should avoid relationships with those born in the year of the Rat or with others born in the year of the Goat. The October Goat finds love and happiness when appreciated.

NOVEMBER

Those persons born in the month of November during the year of the Goat are personable, outgoing, and sociable as well as intelligent, quick thinking, and more honest than not. Their grace and eloquence are exhibited in their appearance and in their speech, making them excellent public speakers. Their worst trait is their impatience with routine situations. And they can be unpredictably impulsive which leads them to change their minds easily or to act on the slightest of whims. other people may like them despite these traits because the antics of the November Goat can be especially entertaining providing few dull moments. But some will consider them undisciplined and irresponsible with a weak will and erratic, emotional nature. It is advisable for the November Goat to establish firm friendships and relationships and to learn to listen to the advice of others as they develop self-control. There is also an indication that the November Goat should avoid speculation and short-term investments concentrating on savings and long term investment progGoats. The November Goat may also have a tendency to travel or to relocate to another area of the country, completely changing their lifestyles and adapting to the new location. It is perhaps advisable for these individuals to wait until these tendencies have been fulfilled and satisfied before they settle down and marry. The November Goat is a romantic who gathers relationships like stories to be remembered. When they find the right partner, they become devoted to the other person, falling deeply in love and wanting to establish a lasting relationship. If they marry too young, however, it may be on impulse and this relationship may not last.

The November Goat female finds lasting happiness in a relationship with a man born in the sign of the April or May Horse, the June or July Boar, or in the year of the Tiger. The November Goat man finds lasting love in a relationship with a woman born in the sign of the April or May Tiger, or in the year of the Horse. Both men and women born in the month of November during the year of the Goat should avoid relationships with another Goat as well as with persons born in the year of the Rat or the Dog.

December

Those persons born in the month of December during the year of the Goat are hard working individuals with kind hearts and a gentle nature. They are sincere and peaceful with a just and idealistic outlook on life. At their worst, they can be blunt and outspoken with a tendency to speak their mind, thinking others will not be offended because they are usually so easy going and likable. They can be generous to a fault, attempting to help others or becoming involved in causes to which they devote themselves wholeheartedly. In some, however, this charitable nature becomes a way of satisfying their vanity in an effort to make themselves look better or more important to others. The December Goat can also be overly ambitious, attempting to better themselves again in their efforts to satisfy their vanity. Their greatest fear is to be alone, and they have a tendency to cling to a friend or lover. The December Goat can also be sensitive to criticism and will rarely admit to any fault, preferring to place the blame on someone else even for minor mistakes. When the December Goat fails to achieve any overly idealistic hope or desire, they can become pessimistic, withdrawn, and even depressed, blaming society, the world, or other people for what they perceiveas injustices. Their inability to separate their emotions from realities can produce difficulties and misfortune which they often see as bad luck. Their best hope for financial security is to save when they are young. They should avoid get-rich quick schemes or speculation as they are easily influenced and taken advantage of by less scrupulous people.

The December Goat female does best in a relationship with a man born in the sign of the October Horse, the November Boar, or the April Tiger. The December Goat man is well suited for a relationship with a woman born in the sign of the March Horse, the April Horse, the November Rabbit, or during the year of the Tiger. Both men and women born in the month of December during the year of the Goat should avoid relationships with those born in the year of the Ox, the Dog, or others born in the year of the Goat. The December Goat finds contentment and happiness in a lasting relationship based on friendship and love.

Date With Destiny

MONKEY

Share With Me Your Fantasy

THE MONKEY

The Monkey is a most likable unique individual who is intelligent and clever with a good sense of humor. They are energetic and sociable and have an active social life because they adapt to new situations and fit in easily with others. But this accommodating adaptability covers nicely a tendency to be cunning, wily, and at times unscrupulous and mischievous. They will use any means available to have their own way, even lying and manipulating the situation if necessary. At the same time, the Monkey is wise with a warm-hearted curiosity about everything and everybody, and they are always ready to help another person. They are happy people with a loving good will and the intellect necessary to learn new skills and to become well educated. The Monkey is a logical thinking person who has a good memory and a tendency to be inventive and who is original not only in their thinking but in the way they carry out their plans. They are capable of solving the most difficult problems all by themselves. And with their strong characters, they like to draw other people to them by attracting the attention of others. The Monkey will go to extremes to prove a point, and will fight for a just cause even to the point of personal sacrifice. They are popular with other people who are impressed not only with the Monkey's personable manner but also by their strength and courage. But the Monkey possesses a double-sided character. one side is friendly and kind while the other side is crafty, vain and arrogant. The Monkey most generally puts themselves first in whatever they are doing. This may cost them some friendships, but the Monkey flits easily from one group to another, rarely staying in one spot long enough for others to become overly offended. And the families of a Monkey person often care so deeply for the person that they accept their behavior. There are, of course, those persons born in the year of the Monkey who are honest and trustworthy, but one may find that even they can become mischievous and that they have a curious bend to their sense of fun. Simply put, the Monkey can be a most obstinate person who is impossible to control and who resists any attempts to be controlled by others. They love their personal freedom and their independence. Despite these tendencies, the Monkey gets along well with parents, brothers and sisters, and other family members who are most willing to be of help when necessary.

In most cases, those persons born in the year of the Monkey find that their creative abilities and talents combined with their intellectual capabilities aid them in achieving financial success. They perceive problems and difficulties as challenges and work hard to overcome such difficulties by coming up with a workable solution. They are adept at learning new skills easily regardless of their training or educational level. But at the same time, they should not attempt to force themselves to learn skills for which they are not well suited because in these instances they can become discouraged by any subsequent failures. In other situations, the Monkey has the ability to look at the bright side of any problems and to make even bad situations appear hopeful. To this end, they are more than willing to stretch their imaginations and to misrepresent the situation, especially if they perceive the possibility of personal gain from doing so. The Monkey is a good organizer and does well in business ventures and in careers in big business often achieving a leadership position or becoming the manager of their operation. Their versatility and quick thinking abilities make them well suited for any number of career fields. They are so versatile, in fact, that they may find that they actually earn more from their second or part-time jobs, particularly if this is in an area of self-employment. While they most always earn a good income, the Monkey has a tendency to spend money quickly. The Monkey benefits from good luck and may find that he or she receives an unexpected amount of money at some time in life. They are also fortunate in investments as long as they consider the risks involved, but they should avoid speculation and most especially gambling. The Monkey easily acquires land, property and possessions by mid-life, but should avoid selling or mortgaging assets to cover expenses or to finance pleasurable pursuits. By using his or her intelligence and talents, many of the hopes and desires of the Monkey are fulfilled. They are benefited by both good fortune, diligent efforts, and determination. But the Monkey should guard against pushing themselves too hard as this can produce physical ailments in a person who otherwise enjoys good health. They Monkey loves to travel and finds that he or she often benefits from short or long trips. Their good luck holds after middle age, and the Monkey finds life enjoyable and pleasurable during this period of life. Quite often the Monkey becomes famous and wealthy.

The Monkey so enjoys romantic relationships and their own freedom so much that they may put off marriage for as long as possible. They can be unpredictable, selfish, and even immature in this regard, as they simple prefer moving from one pleasurable romance to the next, often with little regard for the other person involved. That is not to say that they don't care about the other person, but, as in other aspects of life, the Monkey has a tendency to put their own desires and pleasures first. When the Monkey does decide to marry, however, they do a complete turn around and become devoted and faithful to their partners even when this fidelity isn't mutual. In romance, their double-sided nature shows itself again. They can be either passionately sexual or non-demonstrative, cool and dispassionate, depending upon their mood. The female Monkey is attractive and alluring and enters into her first romantic relationship at an early age. The male Monkey is cheerful, pleasant, and understanding and knows how to charm women. It is advisable for those born in the year of the Monkey not to impulsively marry at a young age as this marriage may not last and, if the Monkey were to wait, there will be numerous romances throughout life from which to choose the right mate. The Monkey rarely needs an introduction or help meeting members of the opposite sex because they are quite adept at meeting people on their own. The Monkey is best suited for a relationship with a person born in the year of the Dragon, the Rat, or another person born in the year of the Monkey. The Monkey may dominate the Dragon, but they make good friends. The Rat is understanding and accepting of the Monkey. When the Monkey is companionable and works at the commitment, a relationship with a person born in the year of the Ox, the Rabbit, the Snake, the Ram, the Cock, the Dog, or the Horse works well. The Monkey should avoid relationships with those persons born in the year of the Tiger or the Boar which can be disastrous because these people are too easily tricked and dominated and come to resent the Monkey.

JANUARY

Those persons born in the month of January during the year of the Monkey are energetic and active, but of the Monkey year persons, they are the most mild-mannered and the least likely to show off or to be overly critical of others. They can be kind-hearted, humble and accommodating and they rarely interfere in the lives of others, but at the same time they don't appreciate other people interfering in their lives. They are strong willed, independent people who form strong opinions and particular likes and dislikes. The January Monkey is a good organizer and planner who does best when faced with a challenge, but they often ignore the opinions of other people and stick to their own preferences. Along with this obstinacy is a tendency to be selfish and self-centered and to always put their own preferences first before those of others. In business, they always seek the means by which to better themselves and most likely will only offer ideas if they feel it will in some way benefit themselves. Despite these tendencies, they form firm friendships and business associations and do well in life. Quite often, the January Monkey establishes a lifestyle which requires him or her to function more completely during the evening or night and to sleep during the day. They can be quite ambitious and at the same time impatient with those who -misunderstand their intentions. While they can be original and inventive, they can also be jealous, sly, and manipulative with an attitude that makes them preoccupied with their own desires and wishes. But the January Monkey is also intelligent enough to realize that good human relationships and interpersonal skills are important to becoming successful in life.

The January Monkey female finds love and romance with a man born in the sign of the September Rat or the December Dragon both of whom admire and understand the traits of the Monkey. The January Monkey man is best suited for a relationship with a woman born in the sign of the October Monkey, the January Ox, or the November Snake. Both men and women born in the month of January during the year of the Monkey should avoid relationships with those born in the year of the Tiger or the year of the Boar.

February

Those persons born during the month of February during the year of the Monkey are intelligent, kind, and extremely cheerful. These are popular, well-liked people who easily attract the attention of the opposite sex. They are often the comedians of their social circles. The February Monkey while being socially active also loves to gossip and can at times be critical of others. At the same time, they can be rather sensitive about any criticism directed at them and may lose their temper if they hear of any gossip about themselves. The February Monkey while amusing and shrewd can be vain and superior and not totally opposed to being vengeful on occasion. For the most part though, they do not hold a grudge for long and are the forgive and forget type. The February Monkey may for-m strong opinions, but they are not always the best judge of character when it comes to making a decision about another person or a particular situation. The February Monkey may mean well, but they may not always keep their promises, especially if it is not in their best interest. But friendships and being surrounded by other people are important to the February Monkey, and they like to be included in any and all activities that are being planned. In fact, they strive to be the leader in most situations. They are also somewhat secretive and may keep many of their thoughts or some aspect of their personal life to themselves not even sharing this information with their families or closest friends. They also like time to themselves. This person makes every effort to achieve their goals in life. They are more than likely casual dressers who prefer careers which do not require business suits or a professional appearance.

The February Monkey female discovers love and happiness in a relationship with a man born in the sign of the January Snake or the year of the Rat. The February Monkey man is best suited for a relationship with a woman born in the sign of the May Snake, the December Rabbit, or the March Ram. Both men and women born in the month of February during the year of the Monkey should avoid a relationship with a person born in the year of the Boar. Romance for the February Monkey becomes more serious as they mature.

March

Persons born in the month of March during the year of the Monkey are honest and straightforward with a tendency to say what is on their minds. They are also magnanimous with a good will toward others. They are active and socially involved, and the March Monkey most generally is fortunate in all of their endeavors. The March Monkey is hard working and industrious and even if they don't seek promotions, they benefit fro-m them anyway because of these traits. But the March Monkey is most interested in their personal past times and activities and is most desirous of pleasurable and enjoyable good times. In fact, they must guard against allowing their pleasures to interfere with their business endeavors. There is also an indication that there are those March Monkeys whose talents and creativity lead them to success and fame. The March Monkey possesses a refined appreciation for beautiful objects and luxurious surroundings.

They are also aware of their personal appearance and take pride in being well dressed, fashionable, and presentable at all times. And they are accepted, well liked, and respected by other people who seek them out for their friendship and companionship. To some extent, this is because the March Monkey is adroit at transforming bad situations into more pleasant and joyful occasions. Then too, the March Monkey is talkative and entertaining with an unquenchable thirst for new information and for meeting new people. Their adaptability and versatility benefits them in many situations.

There are good indications that the March Monkey enjoys a happy and peaceful marriage, but their relationship stands a chance of being destroyed if they become involved in an extramarital affair with another married person. The March Monkey female finds good fortune and happiness in a relationship with a man born in the sign of the October or November Rabbit, or in any month during the year of the Rat. The March Monkey man is well suited for a relationship with a woman born in the sign of the December Dragon, the September Horse, or the June Rat. Both men and women born in the sign of the March Monkey should avoid a relationship with a person born in the year of the Boar.

April

Those persons born in the month of April during the year of the Monkey possess a creative imagination combined with an energetic drive that enables them to be successful and to achieve their accomplishments. They are original thinkers with innovative ideas. And they are adept in financial affairs, rarely running short on cash. There are indications for fame and wealth for the April born Monkey, but even when this occurs, they continue to work on self-improvement and personal development. They are outgoing and cheerful and seldom lack for friends and associates. However, while they are comfortable in group situations, they may not have close, personal friends to rely on, and they may at times feel lonely even when surrounded by people. on the other hand, through hard work and diligent efforts the April Monkey makes a successful small business owner or entrepreneur which wins them the praise and recognition of others. In some situations, they find that it is an influential person who helps them along the path of success when it is most needed. The April Monkey is most fortunate throughout life in that they are likely to benefit not only from good health but from good luck as well.

The April Monkey female is an excellent care giver and may be drawn to careers in which they can serve others such as teaching, nursing, or counseling. They are willing to make personal sacrifices for their families and are good help-mates for their husbands. They are well suited for love and marriage with a man born in the sign of the May Rabbit, the July Ox, the August Cock, or the November Rat. The April Monkey man is considerate, warm and affectionate, and in love becomes devoted to his wife and family. They are well suited for marriage with a woman born in the sign of the February Rabbit, the April Ox, or the November Horse. Both men and women born in the month of April during the year of the Monkey should avoid a relationship with a person born during the year of the Tiger. The April Monkey does best in a relationship with a person who shares their enthusiasm, goals, and values in life. And they most admire another person with a strong sense of morality and a good character.

May

Those persons born in the month of May during the year of the Monkey are serious and intelligent individuals with a quick wit and a tendency to make on the spot decisions judiciously and accurately. They are often elegant, stylish people who possess a warm heart and are prone to stand up for what they perceive as right, Even though they have a strong intellect and alert mind, they are also aware of their limitations and this may lead them not take a chance even when a good opportunity comes along. others may become preoccupied with their intellectual thoughts, and their idleness may lead them to not face up to responsibilities or the reality of any given situation. There is some indication that the May Monkey may be the type of person who is so smart that they choose not to develop their talents or to use their intelligence effectively in order to become successful in life. There are some May Monkeys who may think that if they are so intelligent, then why should they have to work hard. And this attitude can lead them to look for the easy way of doing things or to become involved in unscrupulous activities which promise a big return for little effort. The May Monkey is a gregarious extrovert who loves to entertain and to be the center of attention. But they must guard against offending other people in their efforts to be noticed. Those May Monkeys who choose to apply their talents wisely become successful in any number of careers such as writing, journalism, photography, layout, design, and publishing. The May Monkey does like to read and study and is generally well informed on a number of topics. They possess good memories and retain details well. They have a tendency to be nervous and must guard against the temptation to turn to drugs, alcohol, or sex for relaxation rather than physical activities and sports.

The May Monkey female finds love and companionship in a relationship with a man born in the sign of the September Snake, the October Dog, or any month during the year of the Dog. The May Monkey man is well suited for a relationship with a man born in the year of the Horse or the year of the Ox. The May Monkey should avoid relationships with those born in the year of the Boar.

June

Those persons born in the month of June during the year of the Monkey are known for their alluring charm and charismatic appeal. At the same time, they can be easily affected by their moodiness, rapidly changing from excitement and happiness to discontentment and sadness. They are sensitive, emotional and easily excitable only to turn around and be calm and pleasant a few minutes later. They express themselves well, but they can cry almost at will over the most unexpected things. The June Monkey has a great need for other people in their lives, and it is very difficult for them to get along well when they are not surrounded by people. They love to socialize and crave the attention and admiration of other people. Needless to say, many a June Monkey develops a frantic, hectic lifestyle in their desire to satisfy their needs. They can be quite adaptable and accommodating in their need to be accepted by others, and they will often agree for the sake of agreeing even when they inwardly disagree about a subject or situation. At other times, this person may withdraw from the crowd and seek solitude in order to be alone with their thoughts. The June Monkey experiences the full range of emotions from happiness to anger, from fulfillment to disappointment, from gladness to loneliness. The June Monkey can become successful, however, when they learn to use their talents and intelligence and by observing and thinking of others instead of themselves. They must also guard against turning small problems into larger ones, learning to disregard insignificant issues while concentrating on more important objectives. The June Monkey often find that they benefit from good luck and the assistance of an influential person. Their attractiveness and ability to turn on the charm and alluring personality serves them well too.

The June Monkey female finds love and companionship in a relationship with a man born in the sign of the October Ox, the March Rabbit, or in the year of the Rat. The June Monkey man is best suited for a relationship with a woman born in the sign of the May Cock or April or May Rabbit. The June Monkey should avoid a relationship with a Tiger.

JULY

Those persons born in the month of July during the year of the Monkey are intelligent, clear-sighted, and confident. They are also responsible, just, and straightforward with a decisive attitude. These are adaptable people whose skills and talents make them adept in any number of professions. They easily learn what is necessary in order to move from one task to another. The July Monkey is seldom done in by problems or adverse conditions. They remain in control of themselves and express themselves well in any given situation. They are also innovative thinkers who respond well to a challenge and who seek new ideas and methods. The worst trait of the July Monkey is that they can be critical and impatient with other people, and they do not always take into consideration the wants or needs of others. When the July Monkey works hard and makes a diligent effort to be successful, they often become well off and most fortunate by middle age, but this requires patience and perseverance. The July Monkey often has more than one career in their lifetime, and many are drawn to the food industry where they are particularly skillful and talented. They can be talkative and entertaining and are most generally well liked and popular with other people.

The July Monkey is so drawn to new people from different backgrounds that they often marry someone from a different country or cultural background. The July Monkey female is attracted to physically handsome and well built men regardless of their financial situation or social standing. she finds fulfillment in a relationship with a man born in the sign of the March Horse, the February Rabbit, or in the year of the Dragon. The July Monkey man finds love and happiness in a relationship with a woman born in the sign of the August Dragon, the September Rat, or in the year of the Monkey. Both men and women born in the month of July during the year of the Monkey should avoid relationships with a person born in the year of the Boar or the Tiger. The July Monkey does best in a relationship with a person who understands their needs and who appreciates and admires their talents and skills. In such a circumstance, they are responsible and warm-hearted partners.

AUGUST

Those persons born in the month of August during the year of the Monkey take life seriously and strive to be good citizens and responsible members of their community. This is a person with good common sense who is decisive, practical, and responsible. They are also good planners and organizers. They are intelligent with a good mind for facts and figures, and they are often creative and talented in art or detailed work. The August Monkey can be friendly and accommodating with other people, but they have a tendency to have a stubborn streak as well and prefer to have things their own way. In this manner, they can be quite stern and obstinate and may use their creativity to devise means to turn a situation to their best advantage. They rarely miss out on an opportunity which they feel will benefit them in some way. They are persuasive speakers who can be quite cunning and crafty in manipulating other people, and they have a tendency to use this ability to incite jealousies and disagreements among others. Be that what it may, there is every indication that the August Monkey does well in his or her career, however, they do have a tendency to change their jobs or positions frequently. Their interpersonal relationships become more stable when they learn to take into consideration the opinions and feelings of other people. Their strong sense of duty serves them well throughout life and they do benefit from this responsible attitude. They also benefit from being ambitious and unafraid of working hard to overcome any obstacles.

The August Monkey female has an air of innocence about her that men find appealing. She is most likely to find love and happiness in a relationship with a man born in the sign of the May Monkey, the August Cock, the September Snake, or any month during the year of the Rat. The August Monkey man does best in a relationship with a woman who is supportive of his ideas and who admires his talents. He discovers loving companionship with a woman born in the sign of the November Horse, the March Ox, or any month during the year of the Dog. Both men and women born in the month of August during the year of the Monkey should avoid relationships with those born in the year of the Boar.

SEPTEMBER

Those persons born in the month of September during the year of the Monkey are emotional and most sensitive. This tendency leads them to be nervous, tense, and even paranoid. They develop a defensive attitude and take offense easily, always believing that others are criticizing them or trying to harm them in some way. They are distrustful and suspicious of other people, and while they may socialize and appear to be accommodating, they develop few close friendships. However, when they do form a friendship, they can become devoted and dependent upon the other person, and this often gets them into difficulties as they are easily influenced and lead by this person. At the same time, this very nervousness and paranoid attitude drives them to excel and to make gains for themselves. And they are not above lying or using unscrupulous means to obtain what they want. In this way, they are often perceived as selfish and self-centered with a preoccupation of obtaining as much as they can for themselves. The September Monkey derives satisfaction and feels most secure when he or she has accumulated possessions and financial security. Their negative traits are somewhat offset by their creative talents, imagination, skillfulness, and by their inventive streak. When they are at ease with themselves, they also possess a unique sense of humor and are adept at mimicking or acting the comedian. But most often the attention of others can produce a nervous reaction which causes them to shut down emotionally.

The August Monkey female has a nervous preoccupation with her appearance which often attracts the attention of men. She is best suited for a relationship with a man who understands her and admires her creativity. She does well in a relationship with a man born in the sign of the May Rabbit, the June Rabbit, or the August Dog. The August Monkey man must feel that his mate is totally devoted to him and admires him. He does best in a relationship with a woman born in the sign of the December Horse, the November Dog, or the September Ox. Both men and women born in the month of August during the year of the Monkey should avoid relationships with those born in the year of the Tiger.

OCTOBER

Those persons born in the month of October during the year of the Monkey are energetic, quick acting, decisive, and clever. They are most generally honest and well meaning with a generous, warm-hearted attitude. They lead active social lives and prefer to appear elegant and well dressed. They are drawn to busy, bustling places and events, and often can't tolerate quiet, confining settings for any length of time. Neither do they adjust well to a regular routine or a set daily schedule. They like change and meeting new and interesting people. Their idea of relaxation is to go somewhere to do something. Often, the October Monkey does best when their job requires them to travel or when they own their own small business and can plan and organize their own schedule. They are most independent and love their personal freedom to do as they wish. They can become impatient when faced with limitations or restrictions. The October Monkey has an insatiable curiosity, and they study and read becoming knowledgeable about diverse subjects. In conversations, they have a tendency to better talkers then listeners and they may tend to dominate the conversation. Despite this, they are often well liked and admired because of their versatility and adaptability. Their worst trait is that they can be a bit vengeful when angered or opposed, and they seldom forget a wrong doing, preferring to wait patiently for the opportunity to get back at what they perceive as a slight.

The October Monkey is a romantic at heart who loves being charming and sexual and who moves easily from one affair to another. After marriage, though, they become devoted partners who admire and respect their spouses. The October Monkey female does best in a relationship with a man born in the sign of the August Ox, the September Ox, or any month during the year of the Rat. The October man is well suited for a relationship with a woman born in the sign of the March Dragon, the February Monkey, or any month in the year of the Rabbit. Both men and women born in the month of October during the year of the Monkey should avoid relationships with persons born in the year of the Tiger or the Boar. The October Monkey is lucky when in love.

NOVEMBER

Those persons born in the month of November during the year of the Monkey are friendly, warm-hearted, and kind, but they are also unconventional day-dreamers who much prefer to things their own way. They can be quite passionate and energetic in their pursuit of innovative ideas and methods, but they have a tendency to start on something new before finishing their last project. Even so, the November Monkey is perceptive and intuitive, and once set on a definite direction, often becomes successful and prosperous in their careers and lives. The November Monkey is more persuasive than argumentative and has a knack for getting along well with others which makes them well liked and accepted despite their often eccentric ideas and lifestyles. They can be clever, ingenious, versatile, amusing, cunning, or sly depending upon what the situation calls for. They are most independent and freedom loving, hating restrictions and limitations, and in some situations can even become revolutionaries. At the same time, they develop particular likes and dislikes and form strong opinions. Their worst trait, however, is a tendency to gossip and to be critical of others. They can also become impatient with others who have different opinions from their own or who adhere to conventional and traditional lifestyles and trends. Because of their unique characteristics, the November Monkey is often helped by an influential person who is impressed with their talents and innovative perspective. Their friends and families are also supportive and this benefits them as well.

The November Monkey female is well liked and respected, and she becomes a devoted wife and mother. She is well suited for a loving relationship with a man born in the sign of the March Rabbit, the October Monkey, or in the year of the Cock. Women are attracted to the November Monkey man and many seek him out for romance or friendship. He is well suited for a relationship with a woman born in the sign of the May Dragon, the June Dragon, or in the year of the Rat. Both men and women born in the month of November during the year of the Monkey should avoid relationships with persons born in the year of the Tiger. The November Monkey does best when in love.

DECEMBER

Those persons born in the month of December during the year of the Monkey are intelligent and industrious. They are respected and admired because even though they enjoy pleasurable and social events and activities, they are also efficient hard workers who rarely waste time or energy on nonproductive pursuits. They prefer to make the most of every situation and seek to find balance and harmony in their relationships with others. They can be most diplomatic, accommodating, adaptable or flexible, depending upon the situation. They are energetic and active and like not only to work hard but to play hard as well. They are adept at separating these two areas of their lives and of also separating their work from their family lives. They are cheerful with an optimistic outlook and can be generous and charitable with others. At the same time, they can be persuasive and often inspire others to follow their lead. There are times, however, when they worry too much about insignificant matters, and they are prone to being selfish and self-centered. When others oppose their viewpoints, they can become argumentative, harsh, and critically outspoken. They benefit most from developing their more positive character traits and overcoming their negative characteristics.

The December Monkey is in love with love and is often involved with one romantic affair or another. Others find them attractive with an alluring sexual appeal. After marriage, they center their lives on their partner, families and children becoming devoted parents and spouses. The December Monkey female does well in a relationship with a man born in the sign of the December Rabbit, the November Cock, or the year of the Dog. The December Monkey man is well suited for a relationship with a woman born in the sign of the April Horse, the September Dog, or in the year of the Snake. Both men and women born in the month of December during the year of the Monkey should avoid relationships with those born in the year of the Tiger or the Boar. The December Monkey does best in a relationship with a person who shares their interests and who is also energetic and outgoing. Love and marriage brings happiness.

Date With Destiny

ROOSTER

Share With Me Your Fantasy

THE ROOSTER

The Rooster is an enthusiastic individual with an inner strength that grants strong will power and the aggressiveness to overcome obstacles. The Rooster is also intuitive, sincere, honest, and intelligent. You are a straightforward person who most generally expresses yourself well and without reservation. This tendency to speak your mind may offend others at times, but you are well liked and respected for your opinions. You are a brave person who readily accepts challenges and who remains undaunted by any problems which may occur. There are times, in fact, when you set goals or accept challenges which are beyond your abilities or means to complete. You are a diligent, hard worker, but you must learn to recognize when you are attempting the impossible or that for which you are unqualified. The Rooster is drawn to large groups of people and is often more comfortable in a large group than in smaller, more intimate circles. In a large group, you like to be the center of attention, and you manage to do this well. Because you seek the admiration of others, you also like to be well dressed and presentable.

The steady character of the Rooster lends itself to an innovative attitude. But while you are adaptable to changing situations, you remain cautious towards others who are too bold or adventure seeking. The Rooster has a tendency to distrust other people, and this may be one of the reasons why you don't often accept even the friendliest of advice. But while you are well liked, many times other people don't fully understand your independent nature. You know your own mind and develop your own opinions, and this too leads you to rarely seek or accept the advice of others. The Rooster also possesses a selfish streak, and there are times when you can become impatient and even rude to those people who don't agree with your opinions or your way of doing things. Despite that inclination, you enjoy an active social life, and you participate in numerous activities and develop diverse interests. While you are not a person with a strong preference for spending quiet evenings at home, there are times when you prefer your privacy and time to yourself. At these times, you become quite meditative and are comfortable alone. The Rooster can most certainly be a courageous, brave person who makes wise decisions, but in many circumstances you have a tendency to talk rather than to follow through on your actions.

The Rooster is a person who is neither lucky nor unlucky, and in many instances their hopes and desires are only fulfilled when they pursue them with a great deal of perseverance and diligent efforts. There are times, however, when you are helped by either a family member or an associate. The Rooster is well suited for managing his or her own business, but will face numerous difficulties. The ability to think through problems clearly and to use logical and innovative solutions augments your business acumen. Also, the ability to make good decisions gains you the respect of others and may lead you to be successful in business. There are times, though, that no matter how carefully you develop your skills and efforts, you are not fully appreciated. You should attempt to settle any law suits out of court because you are not particularly lucky in court. Your sensible and logical decisions lead you to acquire property and possessions easily, but you also suffer losses. When you use your intelligence, you can invest wisely, but you are not fortunate in short-term investments. Guard against misplacing or losing personal items and losing touch with friends and relatives, as it is very difficult for you to find or locate them again. You possess a natural business sense and you negotiate well which results in you always making money. But the Rooster loves to spend money and has a tendency to spend it as fast as he or she makes it. You must learn to handle your finances better and guard against wasting your money carelessly or being too extravagant in your efforts to receive recognition and attention from others. In many cases, it is advisable for the Rooster to find a business partner who manages finances well in order to prevent unnecessary losses. The Rooster, who is most generally honest, may be tempted to cheat to make up for losses, or can become interested in questionable or illicit businesses which offer quick profits. In some cases, other people may even consider you somewhat eccentric and this is true in your business practices as well. You choose business partners who are generally not from your friends or social circle. You may find your best business partner with a person born in the year of the ox, the Dragon, or the Snake.

Your family is supportive of you, and you have loving parents. But often your family relationships are disrupted by arguments over minor issues. You also frequently find yourself having disagreements with your brothers or sisters, but you and they manage to stay on good terms throughout life. Many of your quarrels with your family result from you taking their remarks as personal criticism. All of this arguing in your early life may lead you not to be overly critical of your own children and you make an effort to allow them to make their own decisions and mistakes.

The Rooster attracts the attention of the opposite sex by being not only well dressed but also an intelligent and witty conversationalist who has a knack for creating an intimate romantic atmosphere. The Rooster is a caring, compassionate and kind person who has a good heart and is attentive to his loved ones. Because you are a sociable person, you have many opportunities to meet romantic partners. You are passionate and fantasize a lot about sex. You have a tendency to be tempted to cheat in relationships. However, when you fall in love and make a marriage commitment, you become responsible and devoted to your spouse and family. Because of your carefree spending habits, you may want to choose a marriage partner who handles finances well.

The Rooster is quite often well suited for a relationship with a serious, conservative person such as a person born in the year of the Ox. A marriage to a person born in the year of the Dragon brings good fortune to both the relationship and the business activities of the Rooster because the Dragon is strong. You may find a harmonious relationship with a person born in the year of the Snake as the Snake likes to share knowledge with the Rooster. You should avoid a relationship with another Rooster or with a person born in the year of the Rabbit, Rat, or Dog as this can result in bad luck, misfortune or accidents. The Rooster is comfortable with either the Tiger, Horse, Ram, Monkey, or Boar, but a relationship with a person born in any of these signs would not be fortunate or financially prosperous. There is every indication that the Rooster finds happiness and fulfillment in love and marriage.

JANUARY

Those persons born in the month of January during the year of the Rooster are courageous individuals who possess a great deal of inner resolve and the ability to go the limits in order to complete their endeavors. The January Rooster may be somewhat reserved but they are also quick thinking, perceptive and intelligent. You are, at the same time, amiable and congenial with a flair for being diplomatic and for handling situations effectively. You are at ease when dealing with other people, but you have a tendency to say what you think others want to hear. There are times when you say one thing and then do something else. You may be drawn to seek a public position or to enter politics or become a diplomat. If you can, you like to travel extensively, meet new people, and be exposed to new situations. You seek knowledge and experiences and are also drawn to new experiences and situations which stimulate your mind and imagination. And you like to describe and reflect on these experiences in your conversations. You are active, energetic, and inclined to socialize, moving from one group to another. You develop diverse interests and have any number of activities which attract your attention. There are other times when you prefer time to yourself and you take precautions to protect your privacy. You are generally well liked and respected and other people seek you out for companionship. You enjoy good health, but you like good food and drink and must guard against gaining weight.

In romantic relationships, the January Rooster can be rather non-demonstrative, and you may have difficulty in expressing your feelings. But you care deeply for your loved ones and are protective toward them. The January Rooster female does best in a relationship with a man born in the sign of the December Ox or June Dragon. The January Rooster man is well suited for a relationship with a woman born in the sign of the November Ox or August Dragon. Both men and women born in the sign of the January Rooster should avoid a relationship with a person born in the year of the Rabbit, the Rooster, or the Dog. When you take the time to develop a mutually understanding relationship with the right partner, you enjoy a happy and fulfilling marriage.

February

Those persons born in the month of February during the year of the Rooster are independent and freedom loving. You are drawn to innovative, new ideas and ways of doing things. You are an unconventional thinker who prefers to establish your own somewhat unique lifestyle and fashion. You are most-comfortable in casual clothing and tend to shy away from formal attire and occasions. In your opinions and likes and dislikes, you rarely compromise or accept advice from others. In this way, you can be quite obstinate, and this stubbornness can at times lead you to make mistakes or to stick to a course of action rather than to change your mind about your decisions. The February Rooster, however, most desires to feel in charge of his or her own life and is reluctant to have anyone else be in control in any manner. This inability to compromise can lead to problems in your career and in your relationships. You may find that you do best when you work for yourself in a small business. Despite your innovative thinking and your preference for new methods, you can also be overly critical of yourself and others. You are forever chastising yourself for not doing something in a more productive manner, or finding fault with yourself over minor errors in judgment. While you can be friendly and likable, there are some people who avoid your friendship because of your headstrong and critical attitude. You do make friends though, and it may be that it is an influential friend or associate who helps you in some way to become successful. Your desire to be independent and self-reliant may lead some people to think you are selfish and self-centered. But your need and desire for freedom remains strong and deterred.

The February Rooster is a loyal and devoted lover who finds it difficult to end a relationship. The February Rooster female finds companionship in a relationship with a man born in the sign of the June Dragon or September Snake. The February Rooster man is well suited for a relationship with a woman born in the sign of the August Ox or in the year of the Snake. Both men and women born in the month of February during the year of the Rooster should avoid a relationship with a person born in the year of the Dog or the Rat.

March

Persons born in the month of March during the year of the Rooster are confident and self-assured. You are decisive, know your own mind, and usually make good decisions. You can also be bold and outspoken with an idealistic nature which seeks justice in all situations. You can be entertaining and a stimulating conversationalist who likes to make others feel at ease and comfortable. You possess a proud nature and there are times when you are opinionated, suspicious of other people, and selfish. At the same time, you are devoted, caring and responsible toward your family and loved ones. And you can be compassionate and caring toward other people as well, but there is a tendency for other people to take advantage of this aspect of your nature. You derive a good deal of pleasure from being positive, helpful, and generous, but, because of your sensitive nature, you can be deeply hurt and offended by those persons who you attempt to help. This also leads you to have difficulties with employees. Be that what it may, you are an industrious, energetic, hard working individual who remains active throughout life. You are knowledgeable and a logical thinker, but you also have a tendency to daydream and you love to tell a tall tale or two. You also love to give advice, but you can be a bit of a perfectionist in this regard. You are overly impressed with titles and awards and often strive to be the center of attention and to gain prestige for yourself.

In romance, you are a loving and caring partner, but you like to be the person in charge. You also possess a highly jealous nature, and can easily lose your temper on occasions. A March Rooster female is well suited for a relationship with a man born in the sign of the August Dragon or the September Snake. A March Rooster man does well in a relationship with a woman born in the year of the Snake or the sign of the August Ox. Both men and women born in the month of March during the year of the Rooster should avoid relationships with persons born in the year of the Dog or the year of the Dog. Other Rooster year people may make good friends, but in a romantic relationship, these two people may often disagree.

April

Those persons born in the month of April during the year of the Rooster are personable, easy going, and kind individuals who possess a witty sense of humor and a fun loving nature. You are active and energetic as well as open-minded and tolerant of other people. You have a creative imagination and lively sense of humor. You are a generous, helpful, and undemanding person who is content and happy whatever your station in life. At the same time, you can be an efficient manager and good organizer who is hard working and makes every effort to do a good job. You are fortunate and prosper because of your strong intellect and curious nature. You also have the ability to be self-disciplined, but this is most difficult for you when it comes to your financial affairs because you have a tendency to be generous and even an extravagant spender. You make every attempt to invest wisely, but you are tempted to speculate on short term gains in which you are neither lucky nor unlucky. You must also guard against becoming preoccupied with the pleasures in life including food, drink, and sex.

The April Rooster is romantic and passionate, and once this person finds the right partner, he or she becomes devoted and loving. They develop a happy and lasting relationship and in most cases are loyal and faithful to their partner and responsible toward their families. The April Rooster female finds lasting love and companionship with a man born in the sign of the December Snake or any month during the year of the Dragon. The April Rooster man is well suited for a relationship with a woman born in the year of the Snake or the September or December Dragon. Both men and women born in the month of April during the year of the Rooster should avoid relationships with persons born in the year of the Rabbit or the year of the Rat. In a relationship with another person born in the year of the Rooster, you may discover that the two of you become too critical of each other or develop definite differences of opinions. You most prefer to have your own way, but your gentle and kind nature leads you to be well liked by both friends and family. Common interests produces lasting relationships.

MAY

Those persons born in the month of May during the year of the Rooster are bold and energetic, but they possess a kindhearted, understanding, and mild nature. You can be most straightforward and outspoken when expressing your opinion, but you are also conscious of the impression you are making on other people and you do not like to be offensive. In fact, you may be overly concerned and worry about what others may think of you. When this becomes a concern, you can withdraw and prefer to spend time to yourself. But with your creative imagination, you easily entertain yourself with your own daydreams. You are most generally well liked and admired, and with your strong management abilities you are often chosen to be the leader. But you remain a Rooster, and you can be at times a little too proud and obstinate, always believing your own ideas are the best. This can lead you to be self-centered and even selfish on occasion. While you are usually a good judge of character, you can also be cautious and distrustful of others and at times unforgiving and overly critical. You may prefer the companionship of one or two close friends or associates rather than mixing with different social groups. The April Rooster has a tendency to develop not only self-control but self-discipline and is often adept at money management and financial affairs. You possess diverse interests and an intellectual nature that may lead to an interest in literature or the arts. Because you can become preoccupied with your own interests and thoughts, you may not be especially attentive to your loved ones. Some may feel you are cool and indifferent, but when your thoughts turn to romance you can become passionate, caring and compassionate.

The April Rooster female finds lasting happiness in a relationship with a man born in the year of the Snake, especially the November Snake. The April Rooster man finds love and companionship with a woman born in the sign of the January Dog or the June Ox. Both men and women born in the sign of the April Rooster should avoid a relationship with a person born in the year of the Rabbit. Relationships with another Rooster can also be trying. You are best suited for a relationship with a person who shares your interests.

JUNE

Those persons born in the month of June during the year of the Rooster are perceptive but overly sensitive. It is easy for this person to become defensive and to have his or her feelings hurt by the remarks and actions of other people with whom they are dealing. Many times, you are incapable of admitting to even the most insignificant errors or mistakes. This preoccupation with your own feelings leads you to be moody, easily depressed, and even angry with other people. You are creative and many innovative ideas, but you can become impatient with others and may also become distracted. You do best when others are available to carry out your plans and ideas. And no matter how quiet or reserved you may be, you are also obstinate enough to usually see that in one way or another you get what you want. For such a sensitive person, you can also be overly critical of others and sharp and sarcastic in your speech. You also like to collect and repeat gossip which may cost you friends. You are most likely close to your family and your parents were over indulgent with you as a child. But you enjoy being spoiled and indulging yourself as well. You have a tendency to worry about the future and this leads you to be selfish, self-centered, and driven to having as much as you can get for yourself. In fact, your security appears to be based on how much and what you own. You are conservative and either superstitious or religious, but often self-righteous. But as far as you are concerned, you are never wrong. You strive to be accepted, however, and have the ability to fit in and to adapt to different people and situations. You like to be socially active and involved and you may attempt to do more than you can manage well. This too, can lead you to be disappointed with yourself, despondent and moody.

The June Rooster female does well in a relationship with a man with a strong nature such as a September Dragon or an April Ox. The June Rooster man does best with an understanding and devoted partner. He is well suited for a relationship with a woman born in the sign of the August or September Snake or Ox. Both men and women born in the sign of the June Rooster should avoid relationships with another Rooster.

JULY

Those persons born in the month of July during the year of the Rooster are well organized and methodical people who strive to be successful. You are an innovative thinker with the ability to make good plans and to carry them through to completion. You are a loyal and sincere friend and a good judge of character. You enjoy being popular and well liked by both friends and associates. You are open to new ideas, adaptable to new situations, and tolerant of other people and different lifestyles. The July Rooster is generally well loved and protected by his or her family but has problems and difficulties after leaving home. However, the luck of the July Rooster improves steadily and they are often prosperous and fortunate in business and personal affairs. And if you don't lose hope or your ability to be diligent and persevering, you will most always rebound from any difficulty. You discover that in life, when difficulties do arise, you are often assisted by an influential person, associate, or friend. You can become obstinate and stubborn and this is especially true when you are angry. It may take you days to recover from your angry moments and to forgive the offender. This can produce a strain and unnecessary stress on relationships. You are hard working, energetic, and active, but you have a tendency to become tired and at these times you require additional rest in order to recuperate and regain your stamina. The July Rooster must guard against being tempted by the ideas of persons who are less than scrupulous in their business practices. These types of speculative and/or questionable activities can result in financial loss and additional hardships.

In love, the July Rooster is sensitive, compassionate and caring with a deep and abiding affection for his or her partner. You have an appreciation for beauty but you seek someone who is a good person as well. The July Rooster female finds lasting love and happiness with a man born in the year of the Dragon or the sign of the March or December Ox. The July Rooster man is well suited for a relationship with a woman born in the year of the Snake. The July Rooster should avoid relationships with those born in the year of the Rabbit or the Rooster.

August

Those persons born in the month of August during the year of the Rooster are elegant and graceful with a natural flair for the dramatic. You have an appreciation for beauty, the arts, music, and literature. You are socially active, popular and well liked, but others may criticize you for being vain and conceited. You have a tendency to talk a good story, but in many cases, you don't fulfill your promises to others. Until you learn to correct these less than admirable traits, you suffer from the numerous difficulties which they produce. However, you possess a quick and alert mind, and quite often you manage to recover from difficulties with little effort. Your congenial and charming nature makes others willing to help you even though they are aware of your faults. You are also open-minded and accepting and tolerant of other people. Your liberal attitudes adds to your popularity. You desire most to enjoy life, and you seek pleasurable activities and past times. At times, your lack of self-discipline leads you to be ,easily tempted by pleasure and a good time. In some cases, the August Rooster is eccentric with unusual ideas and an unique lifestyle that adds to his or her mystic. But this same eccentricity can result in difficulties in relationships. You develop particular likes and dislikes and can be opinionated and stubborn in this regard. You have a tendency to feel that you are always right in your beliefs, decisions, or opinions. You prefer comfortable, relaxing and even luxurious surroundings, and you enjoy entertaining your friends and acquaintances.

The August Rooster is attracted to the opposite sex, and the August Rooster female knows how to be especially alluring and attractive. This woman does well in a relationship with a man born in the sign of the September, October, or June Dragon or Snake. The August Rooster man is well suited for a relationship with a woman born in the year of the Ox or the Snake. Both men and women born in the month of August during the year of the Rooster should avoid relationships with another Rooster or with a person born in the year of the Rat or the Rabbit. You do best in a relationship with a person who appreciates and admires your personable nature.

September

Persons born in the month of September during the year of the Rooster are strongly emotional and individualistic. They are independent thinkers who rarely follow others in any manner whether it be opinions, fashion, or lifestyles. They develop distinct preferences and personal tastes which suit. heir nature. The September Rooster is an expressive person who likes to be admired for his or her individualism and creative ideas, unique plans, or bold actions. Other people admire their inner strength, resolve, and charismatic personality. You are outgoing, personable and sociable, but you are also impatient with others who may disagree with you. And this impatience can grow beneath your calm demeanor until you are no longer capable of holding in your anger. You can then become abrupt and rude. You are also courageous and unafraid of taking a risk if you think it will produce the desired results you are seeking. You like to take advantage of opportunities and are willing to create your own opportunities if necessary. Your obstinacy And stubbornness is channeled into your energetic drive to complete your endeavors. Whether you succeed or fail, and you experience both success and failure in your endeavors, you are always ready to try again at something new. You may become prosperous and then suffer losses, but you work hard to regain your good fortune. You do well in a career which requires you to be involved with other people and which allows you to express your creativity. Other people are often willing to help you complete your projects, and you may realize profits due to the aid of a benefactor. You do best in a relationship with a person who understands your nature and appreciates your talents. The September Rooster female is best suited for a relationship with a man born in the year of the Ox because they can be supportive, compassionate and caring. The September Rooster man finds love and admiration in a relationship with a woman born in the sign of the November Snake, the June Dragon, or the May Ox. Both men and women born in the month of September in the year of the Rooster should avoid relationships with those born in the year of the Rabbit, the Monkey, or the Rooster. You do best when you can express yourself openly.

OCTOBER

Those persons born in the month of October during the year of the Rooster are friendly and kind individuals who take pride in their honesty. You are a strong, independent person who can be stubbornly determined. But others accept your obstinacy as part of your nature because you also possess a good heart and are gentle and understanding. Those people who are close to you, admire your strength and courage and you are well loved by friends and family who trust you to be responsible and devoted. You are capable of taking charge of a situation are often sought out to be the leader whether socially, at home, or in your career. others seek and respect your advice. You also have a tendency to be somewhat secretive about your personal affairs even with your close friends and family members. And you are drawn solving problems or getting to the facts in a mystery. You may find yourself well suited for detective work. Whatever career you decide upon, you apply yourself diligently and through your strong will power, gradually acquire promotions and success. While you are not one to depend upon luck, you come to realize that your own good fortune is steady rather than good or bad. You depend upon your strong will power even more than your physical abilities and there may be times when you have to endure minor illnesses and even injuries. But endure you do even when your energy is drained and you know you need to rest. When you do rest, your body has recuperative powers and you regain your energy and drive. You overcome any difficulties experienced in early life and experience success generally in middle age.

The opposite sex is attracted to your mysterious, alluring nature. In romance, however, you become devoted to your partner. And once you marry, you prefer for your family life to be stable and you strive to be respected. But there are times when you complain too much. The October Rooster female is best suited for a relationship with a man born in the year of the Dragon or the Snake. The October Rooster man does well in a relationship with a woman born in the sign of the July or August Dragon. Both men and women born in the sign of the October Rooster should avoid relationships with other Rooster year persons.

November

Those persons born in the month of November during the year of the Rooster are forever young at heart. They retain a childlike innocence and naiveté throughout life and are capable of enjoying life with the gaiety of a child. Children's literature and fantasy-type movies may appeal to this-person even into adulthood. They have a simplistic, uncomplicated outlook on life. They are also intelligent and have good memories which helps them to retain facts and makes them good at debating or negotiating. The November Rooster loves a good mystery and will investigate and ferret out facts. They will also work hard to accomplish whatever goals they set for themselves. The November Rooster can be obstinate as well and will become angry when others oppose his or her viewpoints. They may also respond in anger to what they perceive as injustices in any form. You must learn to control this anger so that it does not cause you misfortune. In fact, the November Rooster can usually turn .difficulties to his or her advantage as long as their anger doesn't prevent this. The November Rooster is lucky in so far as opportunities intermittently present themselves, but this person must learn to recognize and take advantage of those opportunities when they happen. And it is the very innate innocence of the November Rooster which may bring about these lucky opportunities, seeing them safely through life. There is indication that the November Rooster lives a contented and happy life and enjoys old age.

Your simplistic nature may lead you to enjoy romantic relationships as well and you may even be a bit promiscuous in this regard. When you decide to marry, however, you build a warm and loving relationship and are devoted to your partner. The November Rooster female finds happiness in a relationship with a man born in the sign of the July Ox or the December Snake. The November Rooster man is well suited for a relationship with a woman born in the sign of the April Dragon or the year of the Ox. Both men and women born in the month of November during the year of the Ox should avoid relationships with those born in the year of the Dog or others born in the year of the Rooster. You are well liked for your good-hearted ways.

December

Persons born in the month of December during the year of the Rooster are confident and persuasive. These are cheerful, optimistic people who can be generous and warm hearted with others. They are so well liked and received by other people that the December Rooster must guard against taking advantage of the good natures of others who are willing to do things for them. While the December Rooster is most generally an honest person who can appear calm and in control, they are also strongly emotional and hold firm beliefs in what is just and right. And in this regard they can be quite obstinate, stubbornly insisting that they are right. But the December Rooster is known for making good decisions. There are times when the December Rooster may feel like they have set impossible goals for themselves, but when they persevere, they generally achieve their goals. You are an intent person who may prefer to work alone rather than on group projects. And while you can be outgoing and sociable, .there are times when you prefer your privacy. You are an intelligent and knowledgeable person who desires to be well informed and who is open to new ideas and methods. You are neither lucky nor unlucky, but when opportunities do come your way, you are perceptive enough to recognize them. In many instances, in fact, you may bring about your own good fortune. But you must learn to handle your finances well so that you do not suffer losses. It is easy for you to get the best of a situation or of other people, but you must guard against using other people as this can cause losses in the long run.

You are open and receptive to meeting new people and this is true in romance as well. In romantic relationships.. you must learn to give as much pleasure as you receive. The December Rooster female finds love and happiness with a man born in the year of the ox, especially the July Ox. The December Rooster man is well suited for a relationship with a woman born in the year of the Dragon or the sign of the August Snake. Both men and women born in the month of December during the year of the Rooster should avoid a relationship with another Rooster or a person born in the year of the Rat or the Rabbit.

Date With Destiny

Share With Me Your Fantasy

THE DOG

Those persons born in the year of the Dog win instant popularity and friendship throughout life. And this popularity and attention is an aspect of life that the Dog actually thrives on and needs. These individuals are kind and trustworthy and make the best of companions. They are willing to sacrifice for their friends and family and remain loyal, supportive, and devoted throughout relationships. The Dog is constantly alert and aware of his or her surroundings, and is an astute observer of situations. This can make these individuals somewhat cautious. They may have a tendency to avoid large groups or crowded gatherings, preferring smaller gatherings or open spaces. And the Dog likes to roam a bit, traveling or sight seeing, and gathering information. That is to say, the Dog is a person who is content with life, but, at the same time, these individuals possess a natural curiosity that leads them into new situations or to meeting new people. And while they are the kindest of friends, their tendency to be observant results in them also being natural critics. At its worst, this leads them to be fault-finding and somewhat cynical, but generally in a friendly and well meaning manner. Untempered, this tendency can develop into a nagging pessimism and a stubborn introverted perspective. And those individuals born in the year of the Dog can be most stubborn when they set their minds on what they want. They have their own brand of defensive maneuvers designed to get their own way and are not above turning on the charm or manipulating another in order to get what they want. The Dog is not so much antisocial as he or she is suspicious or cautious of those who they may not know well. Once this initial parlay is broached, the Dog responds, when deemed appropriate, in a positive manner.

What do they want most? A contented, peaceful life in pleasant settings surrounded by friends and family with plenty of time to pursue their interests or endeavors. They cherish their families and dote on their children. They enjoy an active social life, developing strong friendships, and whenever in need, they can depend on their friends. They have strong bonds with their parents and siblings, but they may spend time away from both--either traveling or due to relocations--and generally speaking their siblings are not helpful to them in life. Problems may come and go, but these individuals gradually and persistently attain a secure financial position and a profitable professional or business situation. The rewards of their own hard work and personal merits benefit them, and the Dog learns to invest wisely and to save. There is indication that these individuals should begin their savings when they are young, gradually building their financial resources. The Dog may find it difficult to hold on to assets or possessions, and therefore should look to savings instead. The cautious nature of the Dog may lead these people to steer clear of speculative ventures, or it may be that they learn from experience that they have a tendency to lose money if they take a chance or risk too much. Their luck and good fortune is augmented by their trustworthy natures and dependable character. They have a knack for finding whatever it is that they may misplace, whether it be a lost item or a friend with whom they have been out of touch. And while their luck may hold with lawsuits, it is probably best to avoid them. These individuals rarely strive for fame or wealth, much preferring to establish and maintain a peaceful, settled environment.

These active and energetic individuals can be clever and amusing with a love for life and other people. They appear to instinctively know how to inspire confidence in others, and other people just seem to take to them. But they are also emotional with a tendency to become moody and restless. At times, they can appear distant or coolly impersonal, but, considering their emotional natures, the Dog also has quite a bit of difficulty expressing their feelings and deeper thoughts. These they hold to themselves, and it is a very understanding friend or mate who understands and respects this aspect of their nature. The Dog may appear indifferent in that they do not particularly want anything special from life and are not driven by money. This is a good hearted person with a good soul whose honesty speaks for itself. And these individuals are known for their dedication, perseverance, and hard work. They are considered by some the most sincere people in the Chinese Zodiac. While they can be impatient, bad tempered, and at times cantankerous, for the most part, they are good listeners who care about others. The Dog never betrays a friend, partner, or associate.

The caution of the Dog may actually be a reflection of their shyness, and this becomes most evident in romantic relationships where they may find it difficult to create the right mood. That they may also have problems displaying or expression their emotions and true feelings doesn't help them either. In love and romance, the Dog may experience some difficulties, but these are individuals who keep trying until they discover that special person with whom they can build a long and lasting relationship. In love, the Dog is as devoted, loyal and dedicated as anyone could ask.

Luckily for the Dog, they get along so well with so many people, their chances of developing a close friendship and lasting relationship are good. They are especially lucky when in a relationship with a person born in the year of the Tiger, the Rabbit, or the Horse, which can result in a prosperous and happy union. The Tiger is brave and romantic and especially impressed with and supportive of the good qualities of the Dog. It is said, in fact, that the Dog and the Tiger may be a case of love at first sight. The friendliness of the Rabbit also attracts the Dog, and the

Dog learns that the Rabbit is also loyal. Then too, the Rabbit may best understand the critical nature of the Dog. A lasting relationship can also be developed with the Horse, but the Dog must allow the Horse adequate freedom and independence. The Dog can also develop a long lasting, prosperous, and happy relationship with a person born in the year of the Snake, the Monkey, the Rat, or with another person born in the year of the Dog--as long as this relationship is based on a mutual friendship, respect, admiration, and love. There is every indication that those individuals born in the year of the Dog should avoid relationships with persons born in the year of the Dragon, the Cock, the Ram, or the Ox which could result in disasters, difficulties, adversities, financial loss, and general misfortune.

The Dog is a dedicated and loyal friend and partner or a formidable foe. It is, of course, best to stay on the good side of the Dog who can be an unselfish champion of the right cause. They possess a yearning for an uncomplicated life and a passion for fair play.

January

Those individuals born in the month of January during the year of the Dog are mild-mannered, honest, kind and caring, and quietly refined. They are most agreeable and friendly toward those with whom they come in contact, making friends and acquaintances easily. They work steadily and diligently, finishing their tasks on time and often to perfection. They are content with their lives, have a preference for a comfortable setting, a peaceful home life, and good food. The January Dog must watch a tendency to enjoy life too well as they gain weight easily and must plan into their schedules time for exercise and daily activity. These individuals possess an alert intellect which makes them good at ferreting out information, solving difficult problems, or settling disputes between co-workers. At their worst, their stubborn streak can develop into an obstinacy when it comes to their likes and dislikes or their particular ways about doing things. They much prefer an orderly routine and well organized atmosphere whether at home or at work. Theirs is the type of luck based primarily on the merits of their own efforts. There are indications that their luck can be hampered by their stubbornness to adhere to their own wishes. Friendly and good natured, the January Dog can also be observant, outspoken, and not above offering criticism and advice whether it is asked for or not. There are also those times when the January Dog becomes moody, irritable, and temperamental. They best express their feelings for others in actions rather than words, and their emotions can be guarded in a rather cool, impersonal manner.

The January Dog female is a pleasant, attractive and kind-hearted person who makes a sincere effort to please her partner. She finds lasting love and happiness in a relationship with a man born in the sign of the March Tiger, the May Horse, or with someone born in the year of the Rabbit. The January Dog man finds contentment and lasting love in a relationship with a woman born in the sign of the April Horse, the May Rat, or the September Tiger. Those persons born in the sign of the January Dog should avoid relationships with those born in the year of the Dragon.

February

Those persons born in the month of February during the year of the Dog are friendly and outgoing individuals with pleasant, cheerful natures. They sincerely wish to be of help to their friends and family, and this combined with their generosity wins them the good wishes of those they know. They are thorough, hard workers who set their minds to a task and face any difficulties or obstacles bravely. They even win over luck and good fortune, finding that more often than not, their efforts and good intentions pay off making them successful at their endeavors. At the same time, they are not conservative in the traditional sense but are open and liberal-minded thinkers who are receptive to new ideas. The February Dog pride themselves on being good judges of character, and once they make a friend, they are more than willing to make any necessary sacrifice in order to help that person. Their good intentions are repaid in kind by an influential person, and this help often comes when the February Dog most needs it. The February Dog may prefer a casual lifestyle as well as fashionable but comfortable clothing, but they also possess a knack for always appearing well groomed and presentable--no matter what they are wearing. The February Dog attends to family matters, and both men and women make an effort to establish a warm and secure home environment. They provide well for their children and are supportive in the efforts of their children. They can be, at times, stubborn and cynical, but this is often off set by their good intentions. Superficial people and relationships don't appeal to them as they seek that which is true and meaningful in their lives.

The February Dog female makes an extra effort to create a pleasant home environment for her spouse and children. She is well suited to a relationship with a man born in the sign of the June Tiger, the May Snake, the April Monkey, and the July Dog. The February Dog man finds lasting happiness in a relationship with a woman born in the sign of the March Horse, the October Rabbit, the July Rat, or the June Monkey. Both men and women born in the month of February during the year of the Dog should avoid a relationship with those persons born in the year of the ox or the Cock.

March

Those persons born in the month of March during the year of the Dog possess the energy and alertness to face any challenge. These are creatively talented individuals who are at times impulsive and rash and quick to act. They are bold and courageous, but while they have good judgment, their impulsiveness can make them unpredictable and even eccentric at times. They have a tendency to departmentalize their lives, separating their family and home lives from their professional or business lives. The March Dog is always ready to befriend another person., but their changeability and moodiness often leads others to question their true intentions. Their quiet dignity, however, combines with their talents and energies to produce a prodigious business mind and quite often a talented artist, writer, or designer. There is some indication that the March Dog may benefit from an inheritance, but these individuals must guard against any loss brought on by speculative or risky investments. Which ever the case may be, the March Dog who pursues and develops their talents find that with endurance, patience, and hard work, they attain success sometime after middle age. The worst trait of the March Dog is a slight tendency to be outspoken and to offer their opinion even when it is not sought. They can be critical to the point of fault finding, and if they are not careful this becomes a part of their habitual nature. The March Dog may change his or her style on a whim, suddenly appearing with a new hair style or changing their clothing style. They love to have make overs almost as if in an effort to recreate themselves in a new image.

The March Dog female is charming and attractive, but she is also sensitive and emotional. She does best in a relationship with a man born in the sign of the September or October Rabbit, the July Tiger, or the February Snake. The March Dog man is well suited for a relationship with a woman born in the sign of the December or January Horse, the October Tiger, or the February Rat. Both men and women born in the month of March during the year of the Dog should avoid relationships with those born in the year of the Dragon, the Ram, or the Cock.

APRIL

Those persons born in the month of April during the year of the Dog are sensible, logical individuals with a sincere determination to be honest and accommodating toward others. The April Dog is kind and caring, but they also possess a cautious if not suspicious nature, making them at times unsure of the intentions of others. This tendency can make them overly critical, and they will attempt to logically pinpoint and deduce the meanings of the words and actions of other people. At the same time, they can be manipulative themselves in an effort to get others to do things their way or to accept their point of view. The April Dog appears to know how to effortlessly win others over with compliments and flattery, and, if necessary, they may also stretch the truth in some instances. But this is always done within the boundaries of their determination to be as honest as possible and still get what they want. Included in their better points, are their ability to work hard and to overcome obstacles. They have an emotional quality to their natures which can drain their energy on occasion. But these are energetic individuals for the most part who continue on in their efforts until they complete the task at hand. The April Dog is popular and well accepted, enjoying an active social life, but they most desire a quiet and pleasant home atmosphere where they can relax and spend time with their families or friends. Much of the luck and good fortune of the April Dog is dependent upon their own actions and decisions and the life they choose to live. Their luck improves when they come to recognize the difference in their dreams and their own talents and abilities.

The April Dog female often seeks a partner who will be financially secure and who wants to provide well for a family. She is well suited for a relationship with a man born in the sign of the January Dog, the February Rabbit, or the May Horse. The April Dog man desires a partner who wants a peaceful home environment. He does well in a relationship with a woman born in the year of the Dog or the sign of the May Rabbit. Both men and women born in the sign of the April Dog should avoid relationships with a Cock.

MAY

Those persons born in the month of May during the year of the Dog are cheerful, easy going individuals with a tendency to be calm but personable. They have a clever sense of humor, are accommodating, and enjoy making others feel at ease. They are most generally kind, honest, and direct in their speech. The May Dog is well accepted and popular with friends, associates and family members. And they seem to bask in this popularity like a favorite son or daughter. They are often contented and peaceful individuals who are secure within themselves and who came from a family environment that enhanced that feeling of well being. Their sincerity and sympathetic natures allow them to listen attentively to others and to empathize with their needs. They possess a desire to help others no matter what the sacrifice to themselves, and this inability to ever say no may reduce the effectiveness of their management ability. So sensitive are they, that the plight of others and the problems of the world seem to wear down their better natures at times. And even difficulties in their own lives can lead them to become distressed. Their challenge is to learn to accept and face those problems in life that cannot be changed. Then too, at other times, the May Dog can show a stubborn streak and become moodily obstinate. Once they learn not to be critical of others, they become more contented with life in general. Their friends and family, however, are generally supportive and understanding of their moodiness, knowing that soon enough they will pop out of it and become their happy selves once again.

The May Dog female is an attentive partner, a creative homemaker, a loving wife and mother, and she may also pursue a career with all her energy. She is well suited for a relationship with a man born in the year of the Tiger, the sign of the October Monkey, the February Dog, or the July Snake. The May Dog man finds lasting happiness in a relationship with a woman born in the sign of the April Tiger, the June Rabbit, the November Rat, or in the year of the Horse. Both men and women born in the month of May during the year of the Dog should avoid relationships with those born in the year of the Ram.

JUNE

Those individuals born in the month of June during the year of the Dog are quick in actions and thought. They are intelligent, brave, and well spoken with a tendency to be sociable and outgoing. Their popularity benefits them as friends or associates are more than willing to help them when problems are confronted. The June Dog is fortunate in that luck benefits them as well, and their problems in life are generally minor and easily handled. Their sincerity and willingness to help others is only hindered by their seemingly inability to take advice from others. Then too, they find it difficult at times to work for someone else because they don't readily take directions from another person. once they learn to work well with others and in group situations, however, their careers flourish. The June Dog enjoys a comfortable lifestyle and the best of food and drink. They are active and athletic but must also guard against weight gain. While they most like the simple things in life and a peaceful life, they do like to dress well and to always be presentable.

The June Dog makes every effort to thoroughly learn a skill or to become well educated in order to pursue their careers. They are hard working and make diligent efforts to complete their tasks. There is every indication that the June Dog benefits from a long lasting and happy marriage with the right partner. They quite often realize most of their hopes and desires in life as long as they don't become overly greedy or ambitious. These are well liked people and their efforts at work often pay off in promotions. The June Dog also likes to travel and enjoys meeting new people and seeing different lifestyles and locations.

The June Dog female is a delightful companion who most desires to make her partner happy and content in their home setting. She is well suited for a relationship with a man born in the sign of the August Tiger, the October Dog, the July Horse, or the December Monkey. The June Dog man finds lasting happiness win a relationship with a woman born in the year of the Tiger or the year of the Rabbit. Both men and women born in the sign of the June Dog should avoid relationships with those born in the year of the Dragon.

JULY

Those persons born in the month of July during the year of the Dog are open minded and outgoing with an accommodating nature that adds to their popularity. They have a tendency to have untraditional ideas and liberal viewpoints. And they often have a unique lifestyle and casual manner of dress. But their pleasing natures and quick sense of humor make them well accepted by others. At work, they benefit from their serious efforts, and they are generally well liked by employers. Their intelligence is often enhanced by a creatively artistic flair which can be subtle or dramatic depending upon their mood. The July Dog does have an emotional, moody side as well, and this can be a distraction from their usual congeniality. often their pride and their beliefs in their own opinions lead them to take whatever steps necessary to obtain what they want. This determination and head strong nature serves them well in some instances, but in other cases they often miss out on overlooked opportunities because of this single-minded tendency. Which ever the case may be, the July Dog faces problems head on, rarely stopped or intimidated by obstacles. Their strong wills and prideful manner may be a by-product of a somewhat lucky youth, but the June Dog learns that luck can fade and obstacles in maturity must be endured. The July Dog must also save when they are young because although money may come easily at that time, they enjoy spending. The July Dog should guard against the temptation of risky investments as these can result in loss and usually do not pay off as expected. The July Dog is often a free and easy spirit who is not materialistic, but they must learn to save for their later years.

The July Dog female is cute, energetically perky, and a natural hostess. She makes a loving and devoted partner when in a relationship with a man born in the sign of the November Rabbit, the November or September Tiger, the May Horse, or in the year of the Rat. The July Dog man is well suited for a relationship with a woman born in the sign of the August Tiger, the year of the Rabbit, or the year of the Monkey. Both men and women born in the sign of the July Dog should avoid relationships with the Ox and the Cock.

August

Those persons born in the month of August during the year of the Dog are strong willed and resourceful. Their clever abilities, intelligence, and good memory serves them well in their endeavors. The August Dog is adaptable to various situations and people, but these individuals often choose to remain somewhat cool and aloof and withdrawn from the group. They are calmly analytical which in certain circumstances can become a critical, fault-finding perceptiveness. When they become moody, their tempers flair and their pessimism makes them even more introverted. That they are persevering and enduring individuals who make diligent efforts to complete their tasks and be reliable leads them to be successful at what they undertaken Difficulties appear to fall by the wayside for the August Dog who takes life in stride for the most part. The August Dog prefers to think things through before they act and to apply a logical if somewhat pedantic approach to all that they do. They are also somewhat defensive, preferring not to make mistakes, but when they do, they accept the blame and work to correct their mistakes. They can be a bit antisocial at times, wanting to do things their own way and becoming impatient with those people who don't agree with them. When the August Dog develops the ability to interrelate and communicate well with others, they do better. For all of that, the August Dog can be caring and kind, and they are generally well accepted and respected by their friends and family. There are indications that if the August Dog becomes too cantankerous or discontented with his or her life, it can result in disastrous situations.

The August Dog female likes to establish a well run home environment and a predictable daily routine. She is well suited to a relationship with a man born in the sign of the December Rat, the January Snake, or the year of the Horse. The August Dog man does well in a relationship with a woman born in the sign of the September Tiger, the August Horse, the December Dog, or the year of the Rat. Both men and women born in the month of August during the year of the Dog should avoid relationships with those born in the year of the Ram or the Dragon.

September

Those persons born in the month of September during the year of the Dog are dependable and reliable individuals, but they are also extremely cautious and contemplative. They give all of their decisions the same degree of intent concentration, choosing their direction carefully, and rarely taking a chance. Because of this calm deliberation, they rarely fail at what they are doing, but by the same token, they miss out on what would have otherwise been good opportunities. They appear to sense that their luck will hold if they maintain this cautious approach to life rather than becoming involved with others. This is true to some degree because there is indication that any questionable activities could diminish their luck resulting in losses. The September Dog are devoted friends and responsible family members, always wanting to do what is best in any given situation. But they adhere strongly to their thought processes and are extremely self-disciplined, rarely compromising when they have made up their minds. They are open to listening to others and to being accommodating if possible, but they do not change their own opinions or their direction. This is simply the approach to life that works best for them. There are those August Dog individuals who must guard against over estimating their abilities. The August Dog desires a peaceful lifestyle and a comfortable home setting as well as a productive career. They do best when they are content with this type of lifestyle and with their friends and family. They are detail-oriented and analytical which can lead them to be critical of others and themselves in their efforts to understand a situation or other people.

The August Dog female feels secure when surrounded by family in a comfortable environment. She does well in a relationship with a man born in the sign of the June Tiger, the July Horse, or the December Rat. The August Dog man is well suited to a relationship with a woman born in the sign of the August Monkey, the September Rabbit, or the July Monkey. Both men and women born in the month of August during the year of the Dog should avoid relationships with those born in the year of the Cock or the Ox.

OCTOBER

Those individuals born in the month of October during the year of the Dog are energetic and socially active and popular. They are honest and kind with a clever personality and sense of humor. They most like to be accommodating and diplomatic, but they can use this ability to maneuver others to get what they want. They have the ability to read other people and can sense the intentions and weaknesses of others. They are quick thinkers and can be impulsive at times which can lead to difficulties. But the October Dog appears to handle difficulties with their own style of aplomb, moving on to other situations while handling their problems as best as possible. There is a tendency for the October Dog to daydream and they can get lost in their imaginary world. When they are working, however, they apply themselves with diligent efforts expecting to be judged on the merits of their abilities and hard work. They are not particularly luck, but seem to be adept at making their own luck. And when in time of need, luck seems to come through for them at the last minute. For all their difficulties, the October Dog often becomes quite successful if not famous and wealthy by later life. The October Dog must learn to think decisions through and to make careful plans in order to avoid unnecessary losses or disasters. Their genuine kindness and sincere regards for others win them the gratitude and respect of their friends. The October Dog is a romantic at heart and desires to find the love of his or her life. Then they become devoted to that person, but they can also be possessive and jealous. Their amiable ways are at times offset by a quick temper, moodiness, and faultfinding. They like people, but can be distrustful and cautious of the intentions of strangers.

The October Dog female finds love and happiness in a relationship with a man born in the year of the Tiger, the sign of the September Rabbit, the May Snake, or the June Horse. The October Dog man is best suited for a relationship with a woman born in the sign of the September Rabbit, the August Tiger, or the February Monkey. Both men and women born in the sign of the October Dog should avoid relationships with those born in the year of the Dragon.

November

Those persons born in the month of November during the year of the Dog are talented and self-confident. Their reserved, dignified and noble mannerisms cover a kind heart and well meaning nature. They get along well with others and are often sought out to be leaders. Their impersonal nature can often lead them to keep their deeper feelings and thoughts to themselves. These are strong willed individuals who face obstacles and difficulties courageously. Their is a tendency for them to encounter people who challenge their authority by questioning their opinions and decisions which result in continuous problems that the November Dog must deal with. There is indication that these well meaning people must learn to be cautious of adversaries who want to compete with them. The November Dog, however, does well when he or she continues to be straightforward and honest in their dealings with other people. They enjoy an active social life even though they also desire a quiet time to themselves to pursue their own interests. There may be a tendency, in fact, for the November Dog to separate their careers and social lives from the home life. They most desire a peaceful and calm setting where they can relax with close friends and family members. They are responsible family members who want the best for their spouse and children. But they have a tendency to stray if they become discontented at home. And their usual good nature can turn moody and stubborn leading them be critical and demanding of others. When they find contentment, this moodiness is often dispelled leaving this dignified and discreet individual as attentive and generous as anyone would wish.

The November Dog female has a tendency to over analyze her romantic relationships although she has difficulty expressing her concerns. When she finds lasting love, however, she learns to relax. She is best suited for a relationship with a man born in the year of the Tiger or the Dog. The November Dog man is best suited for a relationship with a woman born in the year of the Rabbit or the Monkey. Both men and women born in the month of November during the year of the Dog should avoid relationships with those born in the year of the Ram or the ox.

DECEMBER

Those persons born in the month of December during the year of the Dog have a flair for being the center of attention. They are energetic, sociable and outgoing with an active lifestyle, but they can become easily bored when they are not receiving enough attention. They are honest and outspoken, but may maneuver the topic of conversation or the situation back to themselves. This flair for the dramatic is also noticeable in their romantic lives. The December Dog can not be alone and will seek romantic relationships, moving from one to another until they find the most perfect partner. They have a tendency to idealize their partners and may attempt to make them over into what they perceive as the perfect image. The December Dog is well groomed and fashionable with a taste for the latest in styles. They most prefer a comfortable setting that is conducive to entertaining. They strive to be entertaining and to insure that others are entertained and enjoying themselves. They are competitive both at work and in their social lives, wanting to succeed and excel at what they are doing. They are particularly concerned with their image and how they present themselves to others. When they do not feel they are receiving enough attention, they can become moody and withdrawn and particularly impatient and critical of others. They can be willful, discontent, and bad tempered at these times. Despite these traits, the December Dog is quite often successful because of their competitive nature. And once they learn to control their weaker points, they find that others trust them more and their relationships become more dependable.

The December Dog female is charmingly attractive and knows how to use her good looks to her advantage. She does best when appreciated and admired. She is best suited for a relationship with a man born in the sign of the April Monkey, the March Rabbit, or the April Tiger. The December Dog man desires an attractive and personable mate. He is well suited for a relationship with a woman born in the sign of the July Tiger, the August Monkey, or the September Rabbit. Both men and women born in the sign of the December Dog should avoid relationships with the Ram or the Dragon.

Date With Destiny

BOAR

Share With Me Your Fantasy

THE BOAR

The Boar, or the Pig as sometimes called, is honest, well-mannered, and virtuous. You are always ready to help another. You are a trustable person who never breaks a promise or deceives or betrays a friend. There are many times when you are too good for your own benefit. Because you trust others, you are at times gullible, naive, and susceptible of being cheated. But you seldom blame others for their actions and you rarely waste time feeling sorry or guilty about what has already happened. The Boar person is most generally brave, careful, affectionate, and cooperative with a special innocence and a willingness to be of service even if it results in you being dutiful but lonely. You are extremely open and straightforward in your relations with others. The Boar is most generally studious and loves to read and to be well informed, but you also like to socialize and you have a good sense of humor. There are times, though, when you prefer small groups to large groups of people where you can become withdrawn. A Boar is either well educated or has the appearance of being well educated, but either way a Boar can be a psuedo-intellectual or somewhat superficial. Your calm nature makes you a natural authoritarian. You must guard against a tendency to push your business too hard or to look for the easy or fast way of doing things. You succeed best when you take things one step at a time and follow conventional methods to success. Your intelligence and skills make you suitable for a number of career fields, but you may find that you do best in a quiet work environment and in a career such as writing, art, research, or business. You love to be well informed about a variety of topics and this keeps you busy and at times preoccupied. You have a tendency to complete your tasks and are good at many different things. You may find that you generally get what you want, but you must spend a great deal of time looking for what it is you do want or working toward it. There are times when you can change into a negative person when it seems to you that things aren't going your way. And when in this mood, as if "rolling in mud", nobody seems to be able to change your disposition. Most of the time, however, you are kind, generous, affectionate, and honest, but you find that others cheat on you and take advantage of your susceptible nature. And more aggressive people may not always fully appreciate your low-key and laid back mannerisms. However, gallant you prefer to be, you can also be impulsive at times and others learn it is best never to make you angry. You have a strong will and inner strength which can't be changed, broken, or overcome. With your tremendous staying power, you work at a task until it is completed. Neither lucky nor unlucky, you succeed on the merit of your own hard work. Gambling or speculation may not be profitable, but with patience you realize profits through long-term investments. Generally, you are healthy although you may suffer from minor illnesses. Your lofty ideals may lead you to be impatient with others and a bit selfish or overconfident in a leadership position. But you can also be reserved, hard-working and warm-hearted. And you make every attempt to be cooperative, brave, careful, and affectionate. Your frank and outspoken nature may offend some people, but you are also generous. You find yourself concentrating on your own personal endeavors and are not overly competitive. You can be self-protective, but you don't attempt to harm others. You at times find it difficult to trust others because you grow wary of being taken advantage of and used. Boar persons who become famous can be easily angered if someone threatens their fame. Although you are capable of self-sacrifice and you can suffer pain and difficulties, you do not like having your good reputation questioned or threatened. You strive to improve yourself and can become a perpetual student. You are also responsible in all your duties. You become quite skilled in your career or profession with a sensible approach to your organization abilities. Your parents are loving, kind, and morally supportive, but chances are they don't offer financial assistance. You enjoy a good relationship with your brothers and sisters, but they are not of help financially either and you do not expect them to be. You may sell your property at a loss. While you find that very few of your dreams are fulfilled, you remain steadfast in your efforts and eventually realize success. In law suits, you find it best to settle out of court.

You may find that you are somewhat uncommunicative with strangers, but with close friends you are sociable, warm and friendly. You most enjoy a rather quiet life, a few close friends, good food, pleasant surroundings, and financial security. And you manage your money well. Unfortunately, in love relationships you are also vulnerable and others take advantage of your naivete. You are an emotional person and when you become involved with another person, you don't take a sexual relationship lightly. Sexually, you derive as much satisfaction in giving pleasure as receiving it. While your lovers may cheat on you, they also love you as well. You are no more competitive in your love relationships than in other aspects of your life.

You may grow impatient because it seems difficult to find the right partner, but when you do, you fall deeply in love. You must guard against becoming discontent after marriage. A male Boar is attracted to women born in the year of the Rabbit or Ram. A female Boar may prefer not to readily show her affections until she understands the intentions of her partner. She too is attracted to those born in the year of the Rabbit or Ram. Those born in the year of the Snake, Monkey or Pig may not make the best partners for the Boar. With mutual respect and common interests you can develop a good relationship with someone born in the year of the Rat, Ox, Tiger, Dragon, Horse, Rooster, or Dog. You don't like to argue, and you may find that a person born in the year of the Cat shares this trait with you. Although there is every possibility that you have several unsuccessful relationships before you find the right partner, but when you do find that special someone, you are loyal, trusting, and faithful. Your partner will discover that you can be depended on, loving, caring, and kind. You make an excellent marriage partner. With your family and children, you are protective and supportive. A male Boar is a loving father who is extremely fond of his family and willing to work hard for them. A female Boar is just as protective of her family, and likes to care for them. She is a loving mother who manages her home well.

With many Boars, youth is either not advantageous or problematic, but later life is happy and peaceful. The Boar can be a clever person who uses his or her talents well and is not afraid of hard work in order to succeed. Kindness in life brings the Boar people their luck and good fortune. A Boar person often finds fame and success in life.

JANUARY

Those persons born in the month of January during the year of the Boar are steady, straightforward, and prefer to stay on any given path until the accomplish their goals. Because this month is a cusp sign, there are some traits of the Dog, the previous sign, found within the personal characteristics of the January Boar.

Financial security is usually not overly important to the January Boar person as they are good in money matters. You prefer to pursue one career field or to stay with a job as long as possible. Careers related to the field of law are good for you. Female Boars born in the month of January prefer to keep their jobs and to further their careers even after marriage. You may find that you feel more mature than other people your own age. You are a sincere person who is intelligent and determined to see each task through to completion. You are ambitious but you can become preoccupied which at times makes you slow to react and even careless. But you much prefer to approach every task with completion and success in mind. In fact, you have a great fortitude and use all your strength to accomplish your goals. You seldom stray from a task until it is completed, and you have a tendency to never give up.

Your marriage or love relationships do best with a partner who shares common interests and hobbies. You may meet this person through social activities. Because of your tendency to mature early, your first love affair may be in your adolescence. The female January Boar, especially, matures early, both physically and emotionally and is excellent at setting the mood for romance. Men born in January in the year of the Boar may find that they do well with a partner who is born in February Rat or April Dog year. A January female would do well with a person born in March Dog year. Also persons born in any month of the year of the Rabbit do well with the January Boar. Persons born in this month should more than likely avoid anyone born in the year of the Monkey.

The January Boar will find success in life but must guard against greed which will bring loss, failure or misfortune. Your fortitude sees you through life.

February

The person born in February of the year of the Boar is known for being a helper. They are often drawn into a career field where they can be of service to others in some way. A field such as religion, medicine, counseling, teaching, or social work is good for persons born in this month as are positions in management in areas that cater to other people. You are honest, determined, and financially savvy which allows your business endeavors to succeed. Your strong relationships with others also furthers your career. often you find that people in influential positions offer to help you in your endeavors. But while your plans may initially run smoothly, you may encounter numerous difficulties and obstacles. Follow your initiative and do not be tempted to abandon your projects before completion because through endurance and patience your projects reach their completion with success.

You can be vulnerable in that other people attempt to take advantage of your good nature. You must be especially careful in your choice of friends and associates. You must also guard against depending on others but rather learn to make your own decisions and to persevere and forge ahead on your own efforts. You are more successful when following your own ideas and intuitions. You must also guard against a tendency to be drawn to pleasures including sex and to overspend on material possessions such as clothing. You also have a tendency to grow tired of other people and are drawn to spend time by yourself. Although you dislike arguments and disagreements, you can be short-tempered at times. Because you usually hold your anger in, you discover that when another person pushes you too far you can become angry and even prone to violent reactions.

You are well meaning, but after marriage you are tempted to cheat. Male Boars born in this month are attracted to woman born in March Ram or to November Monkey females. Women of this sign may find that they do best with October Tiger or with November Rooster year men. Both men and women Boars born in this month should be careful in relationship with those born in year of the Snake.

Do not be overly tempted by fame or fortune.

March

The person born in March of the year of the Boar can be a perpetual student with a thirst for knowledge. You may know a little about a number of different subjects. You must watch that this tendency doesn't lead you to be preoccupied or to dream away your days rather than working toward your goals. Thinking about what you want to do is fine, but you must also do them. You must also guard against being impulsive and reckless, and you must take others into consideration when making decisions. You have a tendency to make your own decisions and to go your own way. Generally, you don't seek the advice of others but find that others ask for your advice. You have a trust in your inner feelings and often times are correct in your judgments. But that isn't to say that you can't make disastrous decisions on occasion which lead to problems. You may take over the family business or inherit from your family in which case you further your business endeavors successfully. Stay on good terms with relatives and influential people. You may feel that your luck is against you, but it is actually in your favor.

The male Boar born in March has a tendency to be passionate and caring. This can be a reserved man who prefers to watch and listen, but when he does talk he has plenty to say. When angered, this man can verbally defend his point of view and make a strong impression on others. He can be overconfident at times and should avoid quarrels if at all possible. But he is brave and straightforward in all of his relationships both personal and business. Women of this sign have a great sense of color and fashion and a modern taste all of which benefits them personally and at times in their career choices.

Men born in March of the year of the Boar may find a happy relationship with a November Rabbit or a February Tiger or a person born in any month of the year of the Rat. March Boar year women may find their most suitable romantic partner in the October Dragon or with a man born in the year of the Rabbit. Persons born in this month should perhaps avoid relationships with other Boars.

Pursuit of gradual success brings good fortune.

April

Persons born in April of the year of the Boar are most generally eloquent and the sophisticated. They possess exquisite good taste and can become connoisseurs. They also have a good character, are quick-witted and sociable and often quite popular with friends and family. A kind, accommodating and open person, the social life is active and fulfilling. This is a determined and ambitious person who seeks to carry out his or her goals, but who, at the same time, can be somewhat nervous and worried about personal endeavors. Although money is easily earned, it may be spent too readily on social events. Females born in this month are good conversationalists, attractive, and talented.

Some born in April will marry young while others marry later in life because of being hurt by an early love affair. But they always have numerous friends and acquaintances. There is a tendency to lose friends because the April Boar has trouble making decisions and can become irritable. Male Boars born in April are energetic and apply this to working hard. They have a tendency to be persevering and not to give up easily no matter how difficult the tasks becomes. These people rarely ask for help from others because they much prefer to do everything themselves and to prove their abilities. This person may even avoid doctor visits until it is absolutely necessary. The April Boar is a well-mannered, considerate person who is honest and well meaning. While they must control other traits, these tend to lead them to be successful.

April Boar year men do well with a May Ox woman or September Horse year females. A female April Boar finds romance most fortunate with a May Rabbit, October Ram or a March Tiger or with a man born in the year of the Dragon. Both men and women born in April of the Boar year should be most careful in relationships with those born in the year of the Monkey or the year of the Snake or with other Boars. You must be careful not to be too trusting or to allow others to take advantage of you in love because your good nature makes you vulnerable.

You are most generally well liked and respected because of your chivalrous nature and good heart.

May

Persons born in the month of May in the year of the Boar are kind, caring, affectionate and known for their honesty. But this person may also have a tendency to be stubborn even when following the lead of others. This makes them not open to compromise in any given situation. Although they are good listeners, the chances are they do not listen to the advice of others but prefer to make their own decisions. However, this person makes a much better person to carry out instructions rather than to be a leader. This person finds success on the merit of his or her own hard work and often success does not come easily and they may have to work harder than others. This person is trustworthy, straightforward, and honest, but at times has difficulty working closely with others. However, once a May Boar learns to be more patient with others, they appear to offer help and assistance in tasks and goals. The May Boar must also learn to use his or her talent well and to be more self-confident. You may find that you are not well suited for leadership positions or positions of power, but you find that you are capable of earning a good living. This person is also drawn to outdoor activities, sports and an active lifestyle. While this person may not benefit from inheritances, they do well in their own businesses. It is best not to speculate. Success and security are usually attained by middle age and life is happy and prosperous.

In love, the May Boar may find happiness with a person from a different social or cultural background. The men born in March of the year of the Boar are attracted to females who are January Horse or March Sheep. May Boar year females do well with August Ox or August Tiger men. Both men of women of this sign find problems with people born in the year of the Snake or the year of the Monkey. Those born in the year of the Dog or the Dragon are found to be compatible and well suited to the Boar. You are a kind and compassionate person who is also passionate when in love. You make a good mate who is caring and supportive of your spouse. Your honesty and trustworthiness earns you great respect from both friends and family. And your ability to persevere and to work hard brings you eventual success.

June

Persons born in June of the year of the Boar are the most sensitive of this sign, and they allow small things to worry and to upset them. This can lead to pessimism and making decisions become difficult. But once they make a decision, they are competent and hard working. This person may be talented but may lack the self-confidence to develop or pursue his or her talents. Once talents and capabilities are developed, the June Boar does better. You may prefer to be a slow starter until you find your direction and make up your mind about what it is you want to do. Then you stick to your plans with tenacity and apply diligent efforts to your tasks. You are responsible and have good intentions. You like to be knowledgeable and women of this sign may prefer to further their careers even after marriage. This person must guard against being overly emotional, abandoning projects and taking off impulsively in a new direction rather than using good judgment. This person learns to succeed at endeavors by recognizing his or her weak points and making every attempt to overcome them. This type of self-improvement leads to good results. There is a tendency to rely on good luck, but although you may feel lucky or even seem to be lucky, you must strive to not misplace your efforts or to waste your talents and energies. Seek gradual accomplishments rather than shooting for the moon and watch attempting to over expand your business. Also, this person should avoid unnecessary quarrels and arguments.

The June Boar may find that many personal relationships are filled with arguments or quarrels, but this also has another effect. It allows this person to blow off and release pent up steam and emotions. This is a forgive and forget person, and in no time the arguments are forgotten. A June Boar may have a tendency to have more than one marriage or serious relationship. The men are very masculine and aggressive in their relationships, but can be indecisive which often results in losing out.

Men and women born in June of the year of the Boar do well in relationships with a September Rabbit, September Rooster, or a person born in the year of the Tiger or the Ox. Success in life comes in middle age.

JULY

Persons born in the month of July during the year of the Boar for the most part live in harmony with the other people in their lives. They establish close ties with family members and friends, and may feel obligated or responsible toward them. However, the July Boar can also be argumentative at times which results in the loss of some friendships. This person can be restless and impulsive and often bores easily with routine jobs or tasks, becoming easily distracted. They are best suited for careers or jobs which excite them or otherwise attracts their attention because they otherwise become listless, discontent and exhibit a certain amount of laziness or a poor attitude toward work. This tendency leads them to not complete projects or to abandon tasks that they have started, failing to follow through on their efforts. To be happy, it is important for them to find a challenge in life. They do well in their own business or in the entertainment or art field. Generally, this is a person with good intentions who is well mannered and well thought of by others. They have little interest in gossiping or spreading tales about other people, and, in return, other people seldom talk harshly about them. Being well thought of may bring them favor with other people, but usually this person doesn't overly benefit from their relationships with others. It is necessary for the July Boar to be patient and to persevere, facing any obstacles or difficulties in order to achieve in life. And while great wealth is usually not indicated, this is a person who attains a good position. The July Boar must choose friends and associates carefully in order not to experience financial losses.

There is some indication for the July Boar to lose at love early in life only to be reconciled years later with this same person. The July Boar man does well in a relationship with a woman born in the sign of the November Ram or the April Rabbit. The July Boar female finds happiness in a relationship with a man born in the sign of the May Ox or the April Dragon. Both men and women born in the sign of the July Boar should avoid relationships with others born in the year of the Snake or the Boar.

August

Persons born in August of the year of the Boar are elegant, strong willed, and high spirited, but they are also charitable, hard working and find themselves well respected and popular with friends and family. They have good judgment and once they make a decision, they stick to it. They prefer to finish what they start. They are honest and straightforward and usually say what they mean. They can also be too blunt when pointing out the errors in others. However, this person makes a responsible, respected and energetic leader who makes good decisions and has good intentions. They make good employers and managers, and they do not misuse their authority or power. This person is generally financially secure and often may receive an unexpected amount of money some time in life. But, this person must also guard against becoming greedy which could result in problems and unnecessary arguments with others. You must also guard against over exertion no matter how physically strong you may be. This person has a tendency to take on extra responsibilities and duties and may find themselves doing too much. If necessary, they care for elderly relatives or other family members out of a feeling of duty, and they are not afraid of self-sacrifice in this regard. Even on the job, they don't hesitate to take on the work of others if they feel it is necessary. This person finds that he or she must work hard in order to overcome obstacles and problems, but generally they are capable of doing just that. Career success is equally challenging but this person finds benefits through hard work and diligent efforts on their part. The August Boar may discover that the necessary help comes from an influential person who is impressed with this person.

Women born in August of the year of the Boar do well with men who are a July Rooster or a January Tiger. August Boar year men find happiness in marriage with an April Rat, November Sheep, or December Ox. Both men and women born in this sign should avoid other persons born in the year of the Boar or those born in the year of the Snake.

This person finds happiness through developing his or her kindness, charitable attitude, and understanding.

September

Born in September of the year of the Boar, you are sensitive and kind with a refined taste in all things. You apply yourself diligently and completely to all of your endeavors. This person has a tendency to take a serious approach to life and important matters as well as to smaller matters and makes every attempt for things to succeed in an organized and well thought out manner. One problem, in fact, may be mental fatigue brought about by thinking seriously and considering all that must be accomplished and done in a day. It is easy for this person to expend all of their energies in their endeavors, and they may find it difficult to restore their energies when this happens. Luckily this person is stimulated by other people and is socially active as well. They excel is social groups as well as in business gatherings. They develop numerous hobbies and past times and find that they are socially appealing and popular with friends and family. This may be the person who takes up a hobby such as a craft in order to relax and unwind. This person possesses much self-respect and strong morals which combined with a resolute attitude enable him or her to work efficiently and diligently without complaints. You must, however, guard against over eating which easily results in weight gain and can effect your health. This person finds that luck has smiled on him and her and this is especially true in business. It appears as if good fortune has been bestowed on this person throughout life. This person is most generally financially secure throughout life and this is especially true after middle age. Your nature is gallant, true and morally pure. And you are a person respected for your honesty, trustworthiness, fortitude and good effort. others are fortunate to have you as a friend.

Women born in September of the year of the Boar may find that they do best with men born in the year of the Rat and with a July Rat. Men find the best success with a woman who is a June Sheep, February Rooster or January Rabbit. Both men and women born in this sign should avoid relationships with people born in September of the year of the Boar and those who are born in the year of the Snake or Monkey. With the right partner, you find success and much happiness.

OCTOBER

Persons born in October in the year of the Boar are honest, trustworthy, kind, and friendly. They possess an outstanding character and enjoy warm and long-lasting friendships. These people are the least introverted and reserved of the Boar sign and can be cheerful, sociable, and passionate in their affairs. They are straightforward and direct in their personal relationships with good intentions and uncomplicated motives. This person is generally moderate in his or her opinions while also being active and considerate of others which brings them much respect and admiration from friends and family. While strong willed and determined this person is generally not overly stubborn in his or her relationships. These people have a tendency not to like anything that is overdone of distasteful in any way. They prefer casual dress, and uncultured and simply decorated homes and lifestyles. You must develop the ability to not be stingy or absent-minded however. While this person is well-liked, there is a tendency to prefer a few good friends and compatible companionship. They like social functions and find gambling interesting and fun, but are not drawn to becoming overly involved in such pursuits. This person must make every effort to control spending which has a tendency to get out of hand. Many times this person will have a good income, but spending or debt deters from the enjoyment of it. You most prefer people who share your interests in life and you are capable of developing deep friendships and relationships which last throughout life. You must guard against being taken advantage of by fast talking friends or lovers who mean well but can't follow through on promises or whose intentions are not sincere. You are most attracted to home-loving types.

In love, the October born Boar female may find the best success with a man born in either November Rat, July Rabbit, or April Tiger. Men of this sign may find the most compatibility with a woman born in June Sheep or November Rabbit. Both men and women born in October of the year of the Boar should avoid those born in the year of the Monkey, Snake or other Boars. You make an excellent partner and a loving mate who is faithful and sincere in your efforts.

NOVEMBER

The person born in November of the year of the Boar is drawn to seeking justice and what is right more than others born in this sign. This person is sensitive but courageous, alert and self-confident. You can be outspoken, frank, and direct and this tendency may at times offend others. You are extremely interested in new ideas, innovative thoughts, trends and fashions, but you have a tendency to stick with your own ideas. In this manner, you usually are not one to compromise when you believe you are right or you have a strong preference. You are not overly sociable and on the job you have a tendency to ask questions and to seek explanations for what you don't understand and this can find you in disfavor with a boss or employer. With an employer looking for an inquisitive and seeking nature, you do quite well. You also do well when you have autonomy in the work situation and can have a say in your own area. You have a strong fondness for your friends and family, but you do not become overly active in social affairs or family gatherings. This person finds that he or she enjoys a certain amount of luck throughout life and this can produce a worry free disposition that is intent on other matters. This persons also enjoys a youthful appearance throughout life. You can be a very patient person, but you discover that your patience is often tested by other people or family members. Persons born in November of the year of the Boar find success in any number of careers often in fields such as teaching, research, science, writing or publishing.

You do best in relationships with people who are mentally and physically strong and emotionally secure. A man born in this sign may find that he is attracted to and does best with a woman born in November Rat. Women born in this sign may find the most happiness and compatibility with a January Horse or October Rabbit. Both men and women born in November of the year of the Boar should avoid relationships with those born in the year of the Snake, the Monkey or with other Boars. You must be on guard in your relationships or you will find that your luck disappears.

December

Persons born in December in the year of the Boar are self-confident, cheerful and optimistic with an outgoing nature and open character. This person finds that as long as he or she feels self-confident, they do well in their endeavors and efforts. This is an enjoyable and pleasing person to be around. You are a decisive person with good judgment who makes sound decisions and possess a tendency to act on your decisions. You must guard against hasty decisions, however, which can lead to problems or disagreements with friends, family or business associates. Also this person must guard against becoming over confident and letting his or guard down in relationships both personal and professional. This is not a person who avoids problems or obstacles, however, and when they encounter such difficulties they make every effort to be diligent and to work through any problems, facing them head on. They possess a great deal of fortitude and perseverance in this regard and are known for making diligent efforts to succeed which they do quite often. The December Boar may not be sensitive or perceptive to the thoughts or wishes of others and this is especially true in relationships. This is a trait which must be consciously developed in order to get along well with a mate or lover. Patience and understanding must be developed as well. November is not a good month for a Boar person to go into business on his or her own, but with a good partner this becomes more advantageous. This is a person who quite often finds that an outspoken approach must be softened as well as a tendency to exaggerate. An open mind, tolerance, and patience brings success with other people. While this person enjoys good luck, quite often this does not help to attain leadership positions. This person must also guard against trusting others too much which can result in unnecessary difficulties.

Women born of this sign do well in relationships with June Rabbit or November Tiger men. The December Boar man finds success in a relationship with a May Dog, September Rat, or August Sheep. Both men and women born in this sign should avoid relationships with other Boar persons or with those born in the sign of the Snake or Monkey.

The Influence of The Planets

The study of Astrology dates back to a time when the Earth was considered the center of the Universe, and it was believed that the celestial bodies orbited it. Thus it was described by Ptolemy, and it wasn't until 1543, that Nikolaus Copernicus initiated a new idea that had been floating around for centuries. That is, that the Sun is the center of the Universe and all the planets, including Earth, orbit it. Scholars and theologians alike opposed this new idea presented by Copernicus, and it was considered heresy. The fact of the matter is that Astrology along with astronomy and the other sciences has evolved over time to encompass new information. Today, it is accepted that new information will continue to influence our lives, and astrologers continue to study the planets and the influence of the celestial bodies on horoscopes. And by ascertaining the positions of the planets at the time of birth of an individual, a person's birth chart is formed.

The word planet in Greek means wanderer (stars), and it was these wandering stars that attracted the attention of astrologers from the earliest of times to the present. Is it that the planets possess inherent qualities that effect our lives, or is it the motion of the planets as they make their transit through the signs of the Zodiac which influences lives? It is accepted that the Moon has an effect on the Earth's tides as well as other aspects of nature. Do these other planets effect our daily lives as well? The study of these questions is the basis for the study of astrology.

In astrology, those planets orbiting the Sun inside the Earth's orbit are referred to as inferior. The inferior planets, therefore, are Mercury and Venus. The planets outside the Earth's orbit are the superior planets, Mars, Jupiter, Saturn, Uranus, Neptune, and Pluto. When compared to the Moon which takes 28 days to complete one cycle, Mercury requires 88 days to complete one orbit of the Sun, Venus requires 224.5, Mars takes 22 months, Jupiter spends 12 year in each orbit, Saturn's orbit is between 28 and 30 years, Uranus requires 84 years, Neptune takes 165 years, and Pluto's orbit is 248 years. Thus, the time each planet spends in each sign of the Zodiac varies.

Each planet rules one, sometimes two, signs of the Zodiac. When the planet is located in the sign it rules, it is said to at home, or in Domicile. When the planet is placed in the opposite sign to the one it rules, it is in Detriment. Each planet is also most harmonious or most powerful when it is located in another sign, referred to as its Exaltation. When located in the sign opposite where it is exalted, the planet is in its Fall. A planet has the most influence when located in its Domicile, and next its Exaltation. Its influence is weaker in its Detriment and the weakest in the location of its Fall.

Ancient astrologers classified the Sun, Mars, Jupiter and Saturn as masculine. The Moon and Venus were referred to as feminine, and Mercury was androgynous. Jupiter (the Greater Fortune) and Venus (the Lesser Fortune) were considered benefic while Saturn (the Greater

Infortune) and Mars (the Lesser Infortune) were malefic. The Sun, Moon, and Mercury, being neither benefic nor malefic, were referred to as common. Modern astrologers in many instances ignore many of these ancient classifications, preferring to think of the influence of the planets as being either positive or negative.

The ancients also assumed that these wandering stars moved in a fairly circular pattern, but it was later determined that planetary orbits could better be described as ellipses. Then too the retrograde movement of planets was observed, that is, what appears to be the backward movement of a planet in its orbit. Retrograde occurs when an inferior (faster) planet overtakes Earth, or when a superior (slower) planet is overtaken by the faster moving Earth.

Each planet is felt to influence a particular aspect of the lives of individuals, whether emotions, love, temperament, intellect, subconscious, or luck. Neptune and Pluto are also believed to have a generational affect because of the length of their orbits and their length of stay in each sign. By planning a birth chart based on the time and date of birth, the influences of the planets on an individual life can be determined.

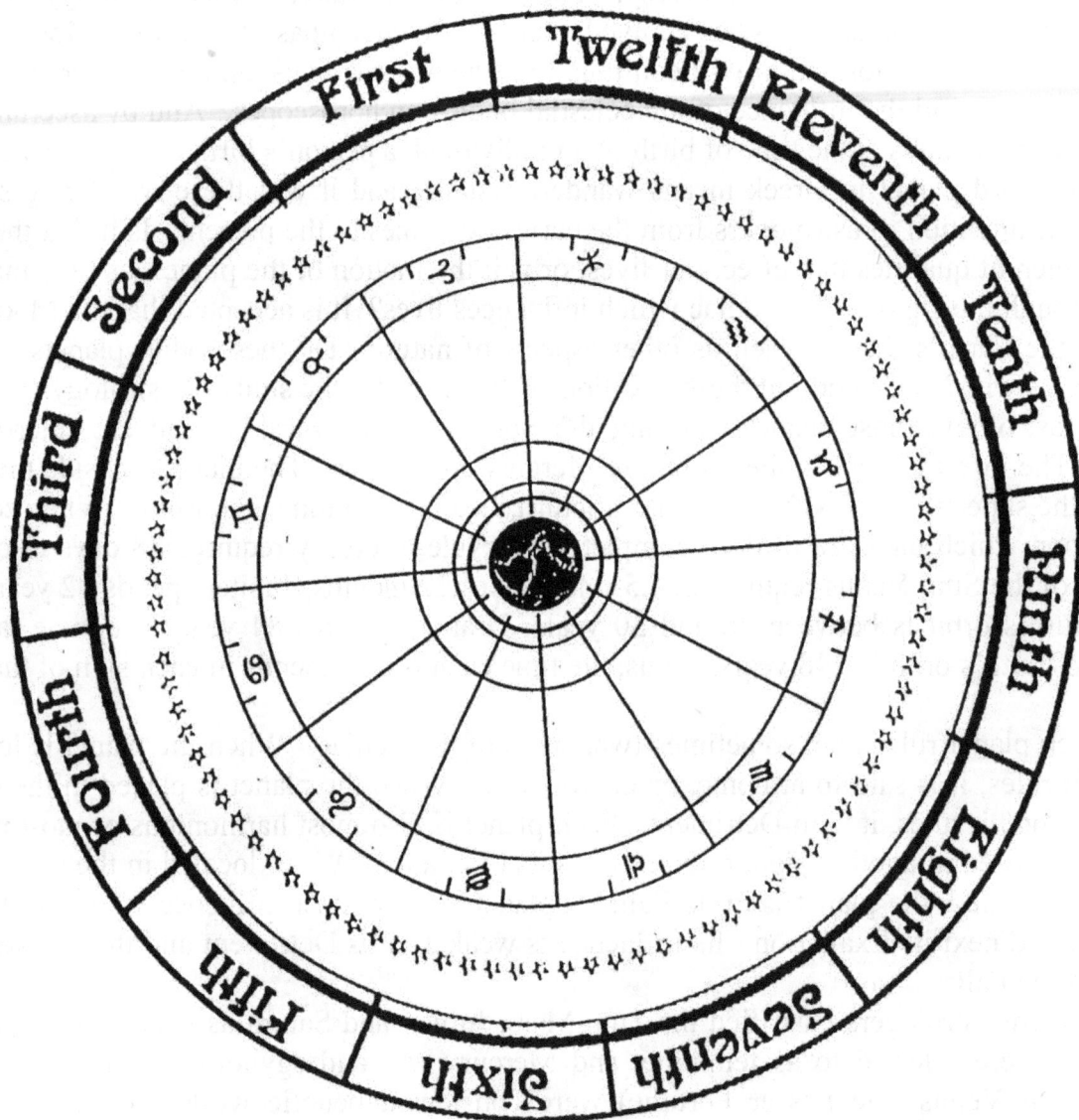

MERCURY

Rules: Gemini and Virgo

KEYWORDS: Ideas, the intellect, communication, conversations, mental activity, facts, statistics, sensory perception, mental processing, mental interpretation, logic, short trips, siblings, neighbors, colleagues, health, diet, medicine, employment, skills, transportation, surroundings, critical thinking, inquisitiveness, documents, writing, teaching, worry, anxiety, nervousness. Mercury rules Gemini and Virgo; is exalted in Virgo; is in detriment in Sagittarius; is in fall in Pisces.

The task of Mercury, the wing-footed messenger of the gods, is to serve as an emissary for the Sun. Mercury's orbit is closest to the Sun, and since there is always a twenty-eight degree distance between them, Mercury is either in your Sun sign or in the sign preceding or following your Sun sign. Orbiting back and forth with wave-like motion in its short orbit, it collects and weaves the sense impressions into practical and meaningful connections.

Mercury is variable, convertible, neutral and dualistic. It represents the practical mind with its intelligence enhanced by matter and is influenced by whatever the Moon is reflecting, thus it suggests the quality of changeability. Mercury collects the sensory perceptions and indicates a person's reactions to sights, sounds, odors, tastes and touch reflecting the person's response to the material or physical world. In other words, it is the messenger through which your physical body and your spirit reacts to the world. The benefit of Mercury is the ability to analyze, and it signifies not only how a person perceives the world but how the mind works. Therefore, it communicates and channels the will of the person allowing man to connect the perceptions with the intellect. In modern terms it is the central processor which collects and deals with the sensory input, channeling it into thoughts and actions. Mercury is the planet of mental activities, communication and intellectual energy which rules over intelligence, perception and reason, memory, speaking and writing. It has influence over the nervous system, bowels, hands, arms, shoulders, collar bone, and tongue as well as vision, perception, comprehension, interpretation and expression. Mercury enhances both the Sun's ability to act and the Moon's ability to react. Mercury also has some influence over daily travel, short trips, and it rules over transportation.

The rulership of Mercury influences natives to be clever and shrewd, quick in speech with an alert mind, fast to discern, eager for knowledge, likes to investigate, inquire, research, and explore. They are good speakers, singers and actors.

If afflicted, Mercury inclines its native to be overly nervous and worrying, hasty, irritable, forgetful, controversial, critical, sarcastic, argumentative, and to possess a tendency to be sly, to lie and commit fraud, forgery, or scams. It is associated with mischievous, childlike and amoral behavior.

Ancient scholars referred to Mercury as Thoth or Hermes, the messenger of the gods and was depicted as a youthful person flying with wings on his heels. Mercury represents duality, speed and wisdom. Mercury's nature is neutral and dualistic. Its colors are blue and gray and its metal is silver.

The influence of the planet Mercury is neutral, dualistic, cold, moist, sexless and convertible. Mercury is associated with the mind and the intellect. It expresses according to its position, aspect and location in the Zodiac. It can either be good when in favorable aspect to other planets or malefic when adversely aspected. When well-aspected Mercury influences persons to acquire knowledge and to communicate it to others. It bestows good reasoning and intellectual powers and the ability to be perceptive, clever, versatile, superior in debate or argument, and with good attention to detail. If poorly-aspected this planet produces a person who is inconsistent, critical, cynical, sarcastic, questioning, argumentative and lacking in purpose or direction. Nervous energy results in stress.

Mercury in Aries

Mercury in Aries produces a person who is aggressive and brilliant but also impulsive and fiery. You possess a quickness in thought and speech which produces a faculty for public speaking, debating, or argument, but at times you consider only your own opinion. If afflicted, there can also be a tendency to be contentious, antagonistic and disputing. You can be expressive, earnest, demonstrative, liberal, inventive, unique, interesting and clever, but with a tendency to exaggerate. And whether funny or sarcastic, you are clever, outspoken, enthusiastic and most often original and quick in thought. Changeability is seen in opinions and projects. There may be a restless nature that at times lacks stability of purpose or a sense of order. However, there is an inclination for your first decision to be the correct one. You possess sharp and alert reactions to sense impressions. Often, music harmonizes and balances your nature.

Mercury in Taurus

Mercury in Taurus makes for a happy and pleasant person who may prefer to ponder and think before making a decision. Then this person sticks obstinately to that decision. You have a practical, stable, and constructive mind which is determined. Good judgment and reasoning ability is seen and this person is persevering, discreet and diplomatic, however, with strong likes and dislikes. There is a strong desire to for financial success and material possessions, and a talent for handling money. Sociable and refined, there is seen a fondness for the opposite sex, pleasure, recreation, music, art and cultural amusements. This person is a strong visual learner who perceives the world best through the sense of sight. There will be a strong preference for harmonious and pleasant surroundings. Likewise, the sense of touch is well developed, and this person can see how a material will feel or feel how something will look. While taste and smell are adequate, hearing may not be as strong. This person can tune out sounds and may not correctly hear the sound of his or her voice, sounding at times a bit flat or unfeeling. Learns through practical experience.

Mercury in Gemini

Mercury in Gemini grants a quick and ingenious mind that is logical, clever, inventive, resourceful, and well informed. There is a preference for travel, change, novel experiences and speculation. This person longs to be free and while lucid and expressive in speech and thought with a rational, curious and seeking mind, at times there is seen no need to be overly practical. This person is sympathetic and generous but not given to sentimental and emotional inclinations. Being perceptive, observing, and shrewd, there is a tendency to categorize and organize ideas. The outlook is unbiased with a lack of prejudice. This is the phrase maker, whose quick wit, humor and charm is noted in his or her clever remarks. This person has strong senses and reacts to variations and shades of light, color, texture, and sound. Failing to tune out all of the sensory perceptions in the environment can leave a person at times tired, and there is a need to seek solitude in a relaxing and quiet setting.

Mercury in Cancer

An individual with Mercury in Cancer is diplomatic, tactful, and sociable with a strong desire to please others. Readily adaptable to their surroundings and other people, they can change their opinions easily. There is a tendency to be influenced and swayed and at times this person tries too hard to please others by saying what he or she thinks others want to hear. This person is sympathetic and tolerant of others and appreciates the kindness and admiration of others. There is seen a keen imagination, but at times this person is sensitive, melancholy or moody. The mind is at times restless and reflective, requiring knowledge or informative material. The intellect is clear, sensible and reasonable, and the mind is retentive with an interest in historical times or family histories. This person enjoys social occasions, pleasures, picnics and family reunions. There is a spiritual inclination for this person. Being sensitive, this person personalizes the sense impressions seeing and hearing either beauty or at other times noise and distractions. There is a found a fondness for music and a creatively artistic streak.

Mercury in Leo

With Mercury in Leo, a person is ambitious, confident, and articulate and possesses a warm and magnanimous personality that can sway the opinions of others. Persistent and determined, the mind is lofty if not fiery and quick tempered. Able to concentrate and to communicate ideas and concepts well, this assured speaker and thinker may become a bit conceited and dogmatic lacking an appreciation for anyone who doesn't readily agree or accept his or her ideas. Distinctive, aspiring and persuasive, this person seeks to be understood, admired and accepted. This person is kind and sympathetic with an intuitive intellect and noble ideals of honesty and justice. There is seen a positive mind which tends to govern, control and organize. This person is known for a strong will power, an expansive personality, and a progressive mind. Outgoing and fun-loving, this person shows a preference for children, animals, music, arts and culture, sports and pleasures, and may be prone to self-indulgence. This person is more attuned to his or her own image, voice or projected self-impression than to the sights and sounds of the environment. Having fixed opinions, this person most perceives those images which agree with these opinions producing a person who is unobservant. Because of a lack of interest, this person may be slow to react.

Mercury in Virgo

Mercury both rules and is exalted in Virgo. This is a pragmatic person with a mind that is comprehensive and practical. There is a tendency for this person to be cautious, prudent and discriminating while at the same time being versatile and inventive, but less innovative than adaptive. Studious with a good memory, this individual is perceptive as well as intuitive. This person is precise in observing, organizing and categorizing details and information. There may be a compulsion to be correct. This person can be persuasive in speech but may lack confidence, and being self-critical, this person may be unwilling to take a firm position. The nature is quiet and serious and can at times be critical and skeptical. This is not an easy person to convince without facts, as he or she will want to see, know and understand before making a decision. There is an inclination toward math or literature. This individual has a tendency to intellectual the senses preferring to name, describe and put in a slot every sight and sound thus diminishing the pleasure of simply feeling. With effort, learning to appreciate feelings can be developed.

Mercury in Libra

Mercury in Libra grants a flair for being expressive in both speech and thoughts. This person is refined and graceful of mind and spirit with a preference for being easy going. In an effort to balance, this person's mind is less concerned with details than it is with how the details relate to the overall pattern of the situation. The intellect and the emotions combine to assess the value of perceptions. This is likely to be a quiet, impartial, caring and judicial person who prefers to compare and contrast before arriving at a reasonable decision. Wanting to weigh and consider differing viewpoints, there may be a tendency to be indecisive and vacillating. This is a favorable location for intellectual pursuits, and the person may possess natural abilities for music, math, or inventions. This person prefers to apply discerning balance and judgment to all of the sensory perceptions preferably from the standard of art or of good taste. This person is sensitive to the environment preferring sights to be neat and attractive rather than offensively messy, odors to be appealing, and sounds to be pleasant rather than loud or offensive. This sensitivity may be appeased by the natural beauty of the wilderness setting, and offended by the man-made clutter.

Mercury in Scorpio

Mercury in Scorpio influences the individual to be bold but obstinate with an inclination toward research and solving mysteries. This person is ingenious and shrewd possessing the ability to size up people, situations and deceptions. This perceptive mind is curious and seeks knowledge, but is at the same time critical, suspicious and adept at discovering secrets and solving puzzles. This person can be the persistent investigator or may prefer to delve into the mysteries of life. This person can be sarcastic and reckless and if involved will be vindictive, holding a grudge for a life time. The nature is fond of the opposite sex, friends, and pleasures. There is indicated a mental resourcefulness, a practical ability, but problems with disappointments and troublesome friends and relatives. This person's sensory perceptions are filtered through a critical filament of discernment. The observations are astute, comparable to the eye of an eagle. This person is constantly aware of what is going on as if there were an eye in the back of the head. These critical observations, while not always kind, are more often than not accurate. This person may well develop an interest in the occult, mysticism, or other unsolved mysteries of mankind or the universe.

MERCURY IN SAGITTARIUS

The person who has Mercury in Sagittarius will find that the nature is purposeful and ambitious, the mind is quick and bright, and the spirit is expansive and loves freedoms of expressions and movement. This is a sincere and generous person who is independent and, if freedom is restricted, can become rebellious. Changeable, restless and impulsive, the progressive mind is constantly seeking stimulation, new information and knowledge. This person is both physically and mentally active and can become prophetic, wise and philosophical. At times, this person is outspoken and blunt. There is a fondness for nature, travel, home, family, sports and animals. In regards to sensory perceptions, this person has a tendency to scatter the attentions in many directions rather than paying close attention to any one stimulation. There may be a tendency to overlook things, to not hear all of a conversation, as if the person is thinking about something else.

MERCURY IN CAPRICORN

Mercury in Capricorn precludes patience, prudence and practicality focused through a sharply acute and penetrating mind. You are tactful but curious, diplomatic but critical. You are thoroughly careful, economical, and will systematically finish what is started. There is a sensible and dignified demeanor about you with an ability to speak clearly. There is a tendency to be rational, calculating and methodical in decision making. You have a dry humor which at times is pessimistic. If afflicted, this Capricorn native can become narrow-minded and rather stiff and opinionated. You appear to see and hear all with sense perceptions that cut to the basics presenting a realistic and accurate portrayal of your surroundings. Your mind is seldom in the clouds, and you don't linger on abstractions when you could be focusing on details and facts. You solve difficult problems with practical solutions. At times you can be disapproving of less serious-minded folks.

Mercury in Aquarius

Mercury in Aquarius produces a person who is refined and original, comprehensive and inventive, with a penetrating intuition. You possess a good reasoning ability and powers of concentrations. You are observant and perceptive of human nature, and being sociable and kind, you make friends easily. You are forward looking and futuristic in your approach to life. You most like to circulate freely, with no restrictions on your thinking and freedom, and you enjoy new and stimulating ideas and experiences. Innovative and progressive approaches appeal to you, but you may find that many of your ideas, especially about humanity, are too progressive to be accepted. Your senses are strong, alert and sensitive, and you may find beauty where others don't see anything. Your values lead you to possess a realistic approach to experience that results in you appraising all things based on their utilitarian or artistic merits.

Mercury in Pisces

Mercury is in detriment in Pisces, and this individual resists using a purely logical approach to life. After all, what fun are hard facts? You much prefer to be kind, charitable, flexible, imaginative and then again at times impressionable. Much of your knowledge is based on perception and intuition, and you appear to be able to memorize and absorb your environment. You can choose to be emotional and secretive or bubbly, fun-loving and talkative. You have a fondness for pleasure, travel, and socializing with friends. While not materialistic, you may love beautiful things, and you know how to save for your future wants. There are times when you choose to isolate yourself in order to retreat into your private world. Chances are you possess psychic tendencies. You can be analytical but diplomatic, cautious but versatile, ingenious and above all else you most probably possess numerous creative talents. You are selective and personal in your senses and prefer to pick and choose what you want to see, hear, taste, or smell, and perhaps you simply have a strong desire not to see the world as it is.

Venus

Rules: Taurus and Libra

KEYWORDS: Love, beauty, desire, attraction, partnerships, personal relationships, relationships with the opposite sex, romance, art, music, adornment, luxury items, valuables, power to attract, acquisitiveness, values, harmony, vanity, refinement, appreciation for beauty, indulgence, laziness, ostentatiousness, retention. Venus rules Taurus and Libra; is exalted in Pisces; is in detriment in Aries; is in fall in Virgo.

Venus, named for the goddess of love and beauty, governs emotional responses. In the influence of Venus, the Sun is found to energize the emotions and relationships. The principle of attraction is emphasized which brings feelings into an artistic pattern, and procreation is channeled into artistic creation. The Venus influence tends to bring balance or harmony to the basic sexual desires, helping to create cooperation and acceptable forms of behavior. The Venus influence is generous, kind, humorous and loving.

In mythology, the goddess Venus, known for her beauty, instructed mortals in the ways of love and seduction and possessed legendary skills in this regard. This alluring goddess who taught the pleasures of life was also known to be temperamental and sensitive, and at times vindictive.

The orbit of the planet Venus falls between the Earth and the Sun, and is never more than forty-eight degrees from the Sun. In your horoscope, Venus is either in your Sun sign or falls in one of the two signs before or after your sign.

Venus influences our emotional reactions to other people, ruling over our personal relationships and our love affairs as well as art, beauty, manners, decor, affections and friendships. Venus is a feminine planet which rules the sense of touch and influences our disposition. This influence inclines toward the higher attributes of the mind such as music, poetry, art, and literature. In its elevated aspect, it is the higher quality of love reflecting devotion, happiness, and grace. In the body, it rules circulation, nerves and impulses, the throat, kidneys, veins and ovaries.

Venus in relation to your birth sign indicates your attitude toward emotional experiences. If Venus is found in the same sign as your Sun sign, emotions are important to you and effects your actions, making you graceful and congenial. If Venus occurs in your Moon sign, again emotions are important and they influence your inner nature making you more self-confident but sensitive with a loving heart. With Venus in the same sign as Mercury, your mind and emotions work together so that you idealize the world through your senses and interpret emotionally what you perceive.

Venus in Aries

Venus in Aries influences this native to be ardent, affectionate and to attract. You are demonstrative and fall in love easily with a powerful erotic compulsion. More intense than sympathetic and tender, you are sensual, idealistic and imaginative. You are inclined toward love and admiration while being warm-hearted, passionate and attracted to the opposite sex. You are persuasive with a tendency to be controlling, but you are a natural at making friends. Often, this person will have an impulsive attitude toward money and will dream up money making schemes. You are known for being generous and charitable. Venus, a feminine sign, is in detriment in Aries which may give you a tendency to seek pleasures and gratification. Fiery, intense Aries becomes impulsive in emotions under this influence. If Venus is afflicted, there are arguments, restlessness and emotional upsets in relationships. Your sensory perceptions are personal and you see the world as how it relates to you and your pleasures. You are warm and sociable, but sensitive and you show more concern for those who admire you than you do for the world at large.

Venus in Taurus

Venus in Taurus produces an affinity between the planet of love and the sign of material possessions which more than likely will prove fortunate to affairs of the heart and to security. This person may be inclined to marry for money finding it easier to love in comfort. However, in your emotional relationships you are passionate and affectionate but also possessive. You desire to enjoy the best of life and all it has to offer. Both Venus and Taurus are closely associated with money and art, and while you work toward financial security, you also cultivate an appreciation for the arts. At the same time, you have a desire for pleasures and the best that money can buy. In love, you have deep, lasting feelings, are loving and faithful if not tenacious. Sociable, friendly, and generous, you possess fixed and stable feelings and opinions. Earthy and sensual, you believe in love, both physical and spiritual and look for both.

Venus in Gemini

Venus in Gemini may refine the intellect making it imaginative and poetic, but it leaves Gemini flirtatious, lively and lighthearted. Your emotions are elusive while you remain discerning of beauty and virtue. You are naturally friendly, intuitive, inventive, and original with a good sense of humor and an appreciation for light, airy and mental recreations. You may be rather cool and indifferent to the idea of marriage and obligations until you find that partner who shares your intellectual outlook. You most enjoy the fun and excitement of romance and a social life that is filled with other fun-loving companions. A thinker, you prefer ideas to being overly emotional. Your duality and amorous, easily aroused nature gives you an inclination toward more than one affair at a time. You are emotionally aware, but you are not particularly sensitive or touchy. In fact, you may be able to tell or write about emotions better than you actually feel them, and you seldom care to argue about them. This is a positive position for speculation in trade.

Venus in Cancer

Venus in Cancer grants an affectionate, charming and sympathetic nature, and feminine Venus bides well in Moon-ruled Cancer with both being concerned with reproduction and growth. You are sentimental, sympathetic and affectionate, cherishing those you love especially your family, and you may have a strong relationship with your mother. You are loyal and devoted, and strive to provide a good and secure home for your family and children. If well-aspected, Venus influences this person to be charitable and kind and to desire to be cherished and admired. Your emotions can run high and at times you are sensitive. If afflicted, you may be possessive, susceptible to flattery, shy, and excessively emotionally. Receptive, imaginative and responsive to emotions, you may indulge in several love affairs of a secretive nature and perhaps a relationship with an older person. While not selfish, your need for home and security gives you a deep sense of self-preservation and there is a tendency to maintain a protective shell. You may find that your are more emotional than physical in your perspective of life.

Venus in Leo

Venus in Leo leaves this powerful nature wanting to adore and exalt a lover while dominating at the same time. Venus grants Leo a graceful and artistic desire for self-expression. Leo, the sign of personal creativity, becomes endowed by Venus with an affinity for beauty, good taste, and fashion. You are sympathetic, charitable, kind, and often generous with a tendency to be sincere and ardent in love. Strongly attracted to the opposite sex, you can be demonstrative, pleasing, entertaining and amusing. You enjoy friends and pleasures. You lean toward the dramatic with a flair for excitement and romance, wanting life to be significant and meaningful. You love to be loved, appreciated and admired. You have a tendency to idealize love, and can become disappointed when reality sets in. Not spontaneous or hypocritical, your emotional responses can be honest but somewhat calculating. You are capable of turning your emotions on or off at will. You are not easily swayed, and in your affairs you have a tendency to listen to your own words and to do most of the talking.

Venus in Virgo

Venus in Virgo produces a discriminating and restrained influence that applies rational analysis to emotions and love. You have a quiet nature and seldom expressed deep sympathy. However, while you are sensitive, at times you can be overly critical of others in your life, avoiding involvement by finding faults and imperfections. Virgo works hard to produce tangible results, but may overlook sentiment and emotions. Venus and Virgo, both Earth signs, possess a love of nature which in this position is practical and realistic rather than awe-inspiring. You are generous, but hold your emotions in check perhaps with a concern for your reputation or standing in the community. You often shine in the business world, and may enjoy gains through business or investments, but emotions may be experienced through a well-organized code of conduct. Cautiously, you hold your emotions in check perhaps fearing someone will take advantage of them. You most desire someone who will appreciate you and consider you unique and deserving.

Venus in Libra

Venus rules Libra endowing this native with kindness and sympathy, affection and a love for nature. You are graceful with refined manners and taste and an alluring magnetism that appeals to others. Friendly and sociable, you are popular and well liked and know how to captivate an audience. You love music, poetry, the arts and cultural entertainment. You are a romantic, in love with love, but you may turn love into an ideal complete with rules of conduct, ethics and good manners. Being tolerant but versatile, you may breeze through a number of relationships before finding the love you desire. There is luck in finances and business for this position. Emotionally, you are direct, to the point, and crystal clear with a youthful but sometimes sensitive outlook on life. The struggles of life don't dampen your regard for beauty, emotions, and love, and you usually bounce back happy as ever. You prefer to think the best of others and attempt not to hold a grudge. And you seek emotional experiences of life with no limitations, clarifying your responses to suit your whims.

Venus in Scorpio

Venus in Scorpio is at its sexiest increasing the passions and emotions. You are ardent in love and demonstrative in affection, loving sensations, luxury and pleasure. Venus is in detriment in Scorpio, meaning that you who are already preoccupied with sex become even more seductive and sensual and your emotions are susceptible to temptations. You may have a sultry charm but find that you make every effort to channel this emotional energy into more productive pursuits. Even so, the emotionally indulged judgment of this person knows infatuations, love affairs, and endless flirtations. Relationships are highly emotional, and you are at times jealous and possessive in love. You may experience trouble and disappointments and numerous problems with your love life, even attacks on your reputation. You make every attempt to hold yourself to high standards in love, and your vivid emotional personality elevates love to a glorious plane. Your sincerity and willingness to sacrifice earn you respect and admiration.

Venus in Sagittarius

Venus in Sagittarius makes this person eager to experience emotions, but with a desire to remain free as long as possible. You idealize love, turning it into an experience of the mind, and you are unafraid of loving and being loved. You are impulsive but sincere and loyal in your affections. You possess a refined nature, loving beauty, but are fun-loving and impressionable while remaining intuitive and imaginative. Charitable and generous, you may enjoy travel, art, romance, sports and amusement all with the same degree of enthusiasm. You also equally pursue knowledge and intellectual and spiritual enlightenment. Sociable, you have many friends and acquaintances and enjoy a busy and active life. Loving to travel, you may find yourself developing attractions and affections for members of the opposite sex in far away countries. You possess a reverence for life, freedom, philanthropy and may express love as goodwill towards others. You may choose freedom, or a long leash, over long-term affairs or marriage. In marriage, your partner must understand your need to explore and experience life. You are sensitive, but no matter how many times your heart is hurt, you will continue to seek that new experience.

Venus in Capricorn

Venus in Capricorn influences this native to be wise, trusting, responsible, and seldom found in compromising situations. Your affections are serious but restrained. Emotionally, you are ambitious, diplomatic and honorable, and you may choose to marry for convenience or social position. If you have emotions, you hold them in check and refrain from displays preferring to use tact and instinctively saying or doing the right thing at the right time. When well-aspected, this person is loyal, trustworthy and dependable. Your first love may be your career. Venus turns your personal affections into emotional discipline, and for you true love may need time to grow and blossom. If afflicted, this is a person driven by physical pleasures with a self-serving ego. This can be a productive position for Venus granting a sense of order, discipline and proportion and success in endeavors. You seldom struggle between emotions, love, desire and ambition and duty. They all blend together quite naturally for you.

Venus in Aquarius

Venus remains unemotional in Aquarius finding an individual who is kind, charitable and an humanitarian. Your emotional responses have a tendency to be frank and open serving your aesthetic outlook on life. You have a love for ideals and may idealize love as well. You are intuitive, philosophical, and generous. You possess a flair for the exotic and may find yourself experimenting with love and emotions but settling in life. Your lover must be a friend as you hold friends and friendships in high regard. Your tendency is to love humanity as much as any individual and you may not devote your attentions exclusively to one person. You may be unconventional in love and attracted to stronger, more independent types who aren't as demanding on your time and affections. You have a strong fondness for pleasure and social life, easily make friends and acquaintances from all walks of life. While you are faithful as long as the affair lasts, you may find that you encounter sudden or unexpected experiences, secret affairs, and a love for a person either older or younger. You are sensuous but not necessarily sensual, and are perpetually pure of heart seeking freedom of expression and personal independence.

Venus in Pisces

Venus brings out the best in Pisces compelling a person to respond deeply to emotions, perhaps more so than you can express. Venus elevates and refines your sentiments bestowing you with a nature that is charitable, sympathetic, and inclined toward relieving suffering and pain in others. You are found to be compassionate and sensitive, emotional, psychic, idealistic and inspirational. You are ruled by the emotions, and in a relationship can be overly sentimental but genuine. You are willing to sacrifice for your love and for your family. You receive genuine joy from giving. You love all things that are beautiful including poetry, music and art. You are cheerful and congenial and prefer peaceful settings and comfort and luxury. If you aren't careful, you can become a bit lazy and indolent, and must strive to overcome this tendency in your nature. You are faithful and devoted with a tendency to idealize your partner, and you most desire to share the pleasure you find in life with another person.

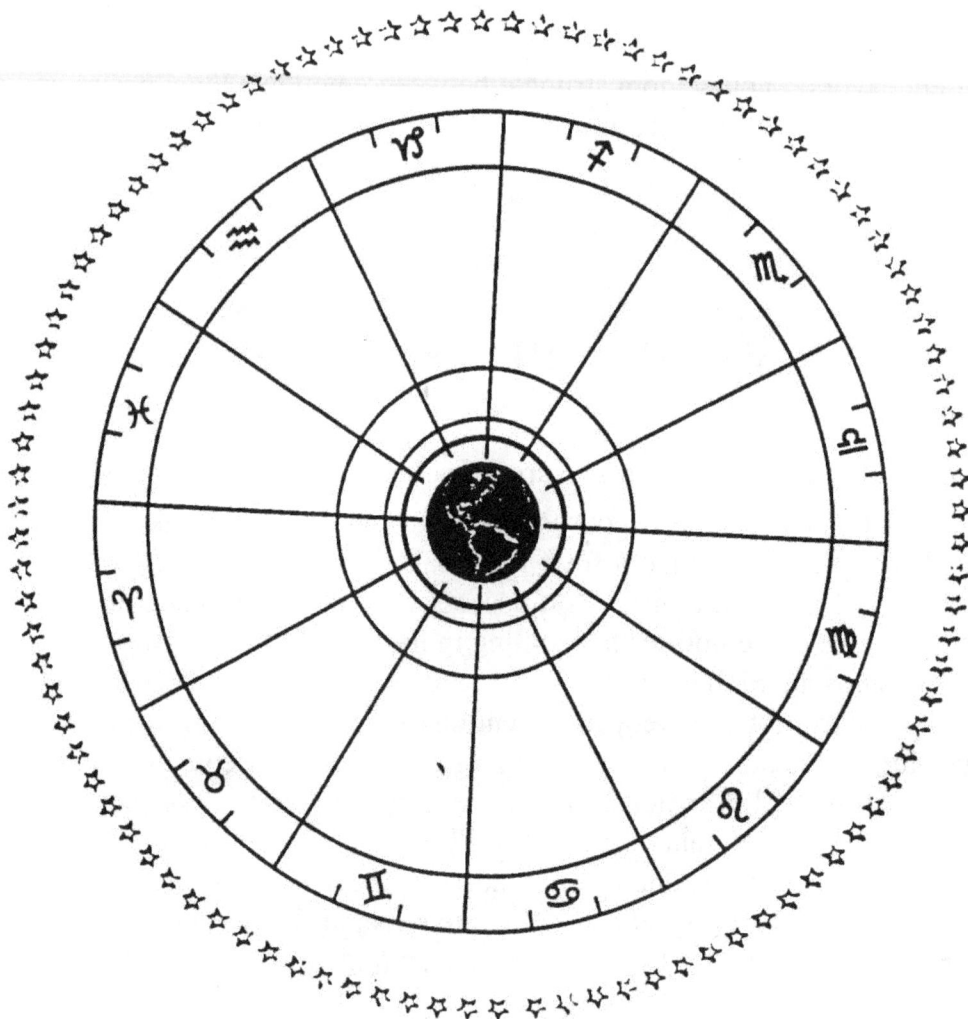

Mars

Rules: Aries

KEYWORDS: Energy, initiative, desires, new beginnings and ideas, motivation, physical exertion, forcefulness, ambitions, effort, competition, combativity, courage, bold, daring, impulsiveness, direct action, adventure, aggressive, fearless, sexual drive, temper, violence, destructiveness, passion, fires, quarrels, pain, war, weapons, accidents, cuts, bruises, fevers. Mars rules Aries; is exalted in Capricorn; is in detriment in Libra; and is in fall in Cancer.

Mars, named for the Roman god of war, is indicative of physical energy, the natural flow of energies into channels, and the most effective uses of energy. Mars manifests as hot, dry, masculine, and pertains to ambitions, initiatives, desires and the animal nature. The planet of Mars, because of its red color, is considered fiery.

Mars is referred to as a superior planet, orbiting outside of Earth from the Sun, and is not as closely associated with the Sun as the inferior planets, Mercury and Venus. Mars remains in each of the twelve Zodiac signs for a little over two months, taking about two and a half years to travel through all of the signs. It represents the energy or actions which express the reactions of the mind (Mercury) and the emotions (Venus).

If Mars is in your Sun sign it endows you with abundant energy which is at times directed into a fiery temper. In the same sign as your Moon sign, Mars grants the ability to use and realize your most inner nature and desires. In the same sign as Mercury, Mars sharpens your senses and makes you quick in speech, thought and actions. In the same sign as Venus, it makes your emotional responses quicker and causes you to act on them, making you ardent and passionate with an earthy awareness.

Mars rules over energy, ambition, actions, boldness, tenacity, and courage. Mars, the warrior of the Zodiac, was known in Greek mythology as Ares, the god of battle. This planet represents the union between matter and spirit, and it is the energetic influence of the planet Mars that sets out to conquer the limitations of the human experience, refusing to submit to the necessity for struggle and pain. Positive traits of Mars include ambition, strength, decisiveness, independent, direct and forceful, leadership, and the defender of others. It is strongly sexed, passionate, and forceful. The negative traits include aggressive, irritable, rash, impulsive, brutal, selfish, argumentative, rude and boisterous. Mars rules over fire, earthquakes, violence and war, and the negative influence can cause accidents, injuries or illnesses. Mars is associated with the muscular and sexual system, adrenal glands, red blood cells, the kidneys, bruises and burns, the male hormones in both sexes, heat, action, weapons and sharp tools.

Mars in Aries

Mars rules Aries and when it is positioned in this sign, it grants the native initiative, abundant energies, and powerful force. It influences you to be positive, self-assured and active but combative. You are originally enterprising and probably have mechanical ability as well. You can be fiery and electrically inspiring. You are frank and outspoken, know what you want, and love your independence and freedom. Your nature is at times impulsive and if you choose you can act rashly, but always on your own initiative. You possess a love of enjoyment, pleasure, sports, adventure and excitement. You are intellectual and love knowledge of all kinds, but at the same time you are sexually active. If Mars is afflicted, you have a fiery temper. Courageous, you are at your best when not limited or restricted by others. Your self-interest is strong, and you will push your ideas forward with or without encouragement from others. You are abrupt and aggressive in speech, and a formidable adversary reacting with strength and vigor to any challenges.

Mars in Taurus

Mars in Taurus can turn an otherwise stubborn person into an obstinate one who sees 'red' on occasion. But usually, your energy is more tempered and is used in practical and persevering ways. The impact of Mars compels you to force your way through obstacles and any resistance to attain your achievements. You have a quiet ambition, quick wit and plenty of plans for the future. You possess good executive ability and excel at organizing and directing. You work tenaciously and with determination to turn plans into practical applications. Mars in Taurus can alternate from being peaceful and aggressive, stoic or indulgent, generous or self-seeking. You are ardent and stable in love, and focus your energies on your security. You have a strong preference for doing things your way. You have the strength to achieve your basic objectives in life and love. Your passions are earthy, passionate, but possessive. You gain financially, materially and in pleasure, and you spend freely. If afflicted, there is a tendency toward loss, and the temper is hasty and at times violent. This position of Mars indicates opponents and difficulties, but you persevere.

Mars in Gemini

Mars in Gemini and you project your ideas forcefully with sharpened insight and perceptions. You are intellectually quick and energetic, but mentally combative and forceful. You are direct and plain spoken. You can be restless and changeable with a tendency to expend your energy in many pursuits. You are a nimble talker who prefers to change words into action as quickly as possible. Your sudden inspirations and impetuous ideas propel others to actions as

166

well. You have a way with words and can talk yourself out of most difficulties being convincing or at other times sarcastic and cutting. You love the adventures in life and have a need to circulate, your mind requiring action both mentally and physically. Rarely bored or boring, you can wear others out with your energetic outlook on life. If Mars is well-aspected you like information, travel, and science, and you are inventive, ingenious, and practical. If afflicted, you can be disagreeable and critical with a restless and indecisive mind. Your energy flows naturally to your speech, and in love you are a sensualist who likes the physical expressions of love.

Mars in Cancer

Mars in Cancer focuses energy on the emotions, and this person must constantly strive to rule the emotions rather than having the emotions rule. You want security, and you strive to acquire and to protect what you have. You may find that you hold your emotions in check and then periodically, you suddenly blow up. The well developed native will strive to direct this energy into productive causes and accomplishments. You are tenacious, ambitious and boldly industrious, but those temperamental outbursts cause you problems. Sensuous and fond of luxury, you are prone to nursing bad feelings for those who have offended you. You have a rebellious streak, but can be originally creative and strongly independent. If Mars is afflicted there is an indication of an early death of mother or separation from; trouble in home life and marriage; many worries and sorrows; changes in residence; trouble through inheritance of property; accidents to home through fire, storm, earthquake, theft; danger through water; stomach disorders and problems related to sight.

Mars in Leo

Mars in Leo grants enthusiasm and a strong sense of purpose. You are candid, fearlessly independent, free and enterprising and strive to be honest and conscientious. Your energy flows through positive, self-centered channels insuring that you always make a good impression. You are ambitious, active, trustworthy, industrious and capable of responsibilities and positions of authority. You are sociable, friendly, warm and kind, and ardent in love and pleasure. You have a tendency to be hasty and impulsive in love, emotions, and passions. There are times when you are aggressive and defiant, even forcefully so. You have a preference for projecting ideas in a dramatic manner, and will take an action regardless of consequences. There is a tendency for you to be impatient with less courageous types. A fiery planet in a Fire sign makes you doubly assertive. Mars makes this individual motivated enough to push forward purposefully with style and presence. Mars endows you with strength, courage and generosity. Your ego is in control of your energy. If Mars is afflicted, there are disappointments in love and fortune.

Mars in Virgo

Mars in Virgo makes you original, bold, hard working and enterprising. While you are usually successful, Mars places difficulties and struggles in your path. Your energies are directed into systematic and orderly ways in. Logical and precise, you may do well in scientific endeavors. Mars and Virgo are a vigorous combination, and you are shrewd and adroit with mental and physical finesse. Through specialization and skill, you rise in your profession, at times spending more time on your career than you do on your love life. But whether you are ardent in love or at work, you strive for perfection and results. The problem with this position of Mars is a tendency to focus on details rather than the bigger picture, losing your energy for more important things. If Mars is afflicted, there is a tendency to be irritable, rash, proud, stubborn, reserved, secretive, revengeful, and you suffer from loss of friends. Otherwise, Mars in Leo grants financial gain in later life.

Mars in Libra

Mars in Libra leads you to float between being languid and passionate. Either way, you are perceptive and observant with a clear vision and refined taste. You are intuitive and idealistic, but enterprising. You may have a love for science or fondness for law or business. In love you possess a strong attraction for the opposite sex, and can be ardent but rash, passionate but impulsive. You believe in moderation, however, and your energies are controlled by judgment. By nature peaceful, you are exposed to opposition and adversity which requires your strength, wit and skill to overcome. In spite of yourself, you become involved in arguments and can be belligerent. You are sociable and magnetic, but emotionally impressionable, and you strive against being overcome by stronger personalities. You are sensitive to love, and your energies flow effortlessly in that direction. If Mars is afflicted there are many separations and difficulties between. Your energies and your emotions are always working either with each other or not, and you express your energies both emotionally and physically.

Mars in Scorpio

Mars in Scorpio makes you deeply emotional. This position of Mars can either be the best or the worst location. If positive, these energies make for personal security both financially and in love. You are magnetic with a forceful ego which attracts others. You are passionate, determined, practical and relentlessly striving toward accomplishment. You possess executive skill and mechanical and inventive ability. You may appear cold to more emotional types, and can be rash and impulsive, selfish and even revengeful. But at the same time, you are diplomatic with a sharp mind and acute awareness that produces results. At its worst, Mars indicates a fear of not attaining security, producing a person who seeks escapes and can be timid or quarrelsome, self-indulgent and weak, dangerously cruel, jealous and revengeful. This position of Mars requires a great degree of strength of character and self-control. If afflicted, this native is unsociable and ungrateful, overbearing and selfish with little regard for others. Otherwise, this is a practical person with the ability to work hard and accomplish much.

Mars in Sagittarius

Mars in Sagittarius and you are youthfully boisterous and energetic, mentally and physically. The fiery planet in this Fire sign makes you open and free, generous and enthusiastic, ambitious but impulsive. With your love of freedom, you are uncontainable and balk at discipline forever seeking the future and what it will bring. Your energies flow into physical channels, and this vital position of Mars, grants you an athletic ability. You can also be studious and prefer generalizations to details. You are charmingly witty and humorous with an original and independent outlook on life. You are brave, daring and love a good adventure, full of excitement. Then too, you can be inspirational and prophetic with strong intuitive abilities which may develop into metaphysical interests. You are the seeker of quests, or the questioner of fate. Your ideas and antics can be outrageous, regardless of the opinions of others. If Mars is afflicted, it signals danger through travel, risk, or miscalculations. You are the dreamer of dreams forever peering into the clouds.

Mars in Capricorn

Mars in Capricorn begets a person who puts raw energy to work with sustained efficiency. Mars is in exaltation in Saturn-ruled Capricorn, joining the drive of Mars with the organization of Saturn. Your energy is controlled through containment and compression making you ambitious, enterprising, industrious and intent on acquisition. You direct your energies toward your objectives in life. You are courageous, brave, bold, self-reliant, self-assured and are not opposed to adventure or excitement. Danger is an afterthought for you, which at times means unforeseen accidents. However, your practical self-control shows in your speech and actions. You are as comfortable with details as generalizations, and you exhibit executive and leadership abilities. You are proud but tactful, subtle but intuitive, and have a love for duty and responsibilities. If Mars is afflicted, it indicates conflicts with authority figures and friends, a quick temper and irritability. This is a favorable position for Mars for financial success.

MARS IN AQUARIUS

Mars in Aquarius and you spend your energy freely, independently and inventively. You are the reasoner looking for a unique and original point of view. You are impulsive but determined, whether idealistically or in personal pursuits. You are forceful and convincing in speech, arriving at conclusions and summations quickly. You have a strong intellect, a quirky sense of humor, and a scientific bend to your mind. You can be impulsive, rash, head strong and abrupt, but you are an humanitarian at heart. You make friends easily, are faithful in love, and most capable of quick decisions in emergencies. There is a tendency for you to become set in your ways, and you must guard against becoming introverted, preferring your own thoughts to the ways of the world. Then too, you can just as easily become the humanitarian leader of social revolution and change. If a writer, this produces an interest in expressing socially significant messages. If Mars is afflicted, this native is too independent, rash, abrupt, and outspoken; indicates separation from friends. You are ambitious, enterprising, and fixed in opinion but capable of abrupt changes. Gain is seen through your career.

MARS IN PISCES

Mars in Pisces and your energy is diffused in numerous directions. This can leave you feeling either tired or emotionally drained. Your emotional desire is so strong, and you are so generous, you are capable of great sacrifices for others. At the same time, cautious in financial affairs, you strive to accumulate money. You can be timid or bold, cautious or outspoken, but you are easily influenced by others. There is a tendency for you to deflect the vital forces inward, to worry yourself into becoming indolent and moody. This location of Mars leads one to be either sensual, temperamental, and pleasure-seeking, or, when self-control is exhibited, passionately focused on a purpose. If Mars is afflicted, the native sees misfortunes and difficulties, suffering from theft, scandal, slander, and false accusation. Your nature is sympathetic, affectionate, and sensitive. At times, you must slow your pace so as not to lose your energy.

Planets are the Daughters of the Sun by Science, Pain, Love, or by Death

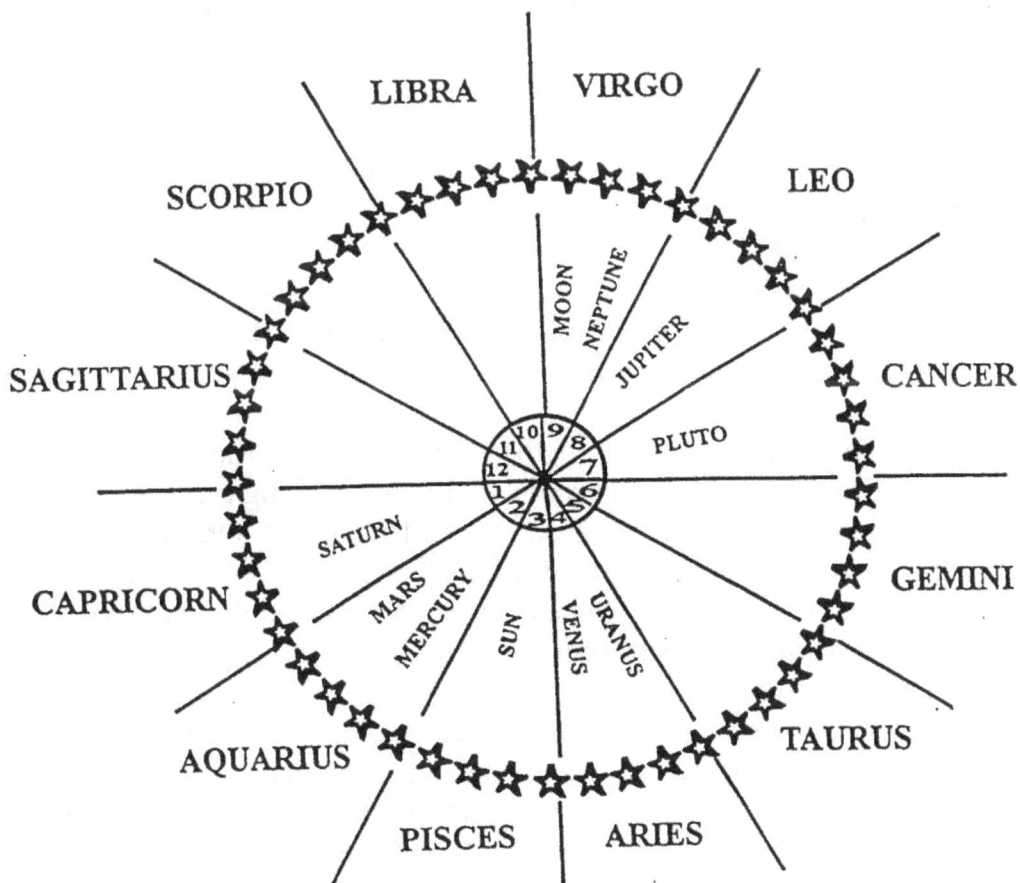

LIBRA VIRGO

SCORPIO LEO

MOON NEPTUNE JUPITER

SAGITTARIUS PLUTO CANCER

11 10 9 8
12 7
1 6
2 3 4 5

SATURN

CAPRICORN MARS MERCURY SUN VENUS URANUS GEMINI

AQUARIUS TAURUS

PISCES ARIES

JUPITER

RULES: SAGITTARIUS

KEYWORDS: Good luck, knowledge, philosophy, faith, religion, justice, law, wealth, opportunity, expansion, speculation, understanding, higher education and universities, compassion, congenial, growth, optimism, jovial, enthusiasm, new experiences, wisdom, legal affairs, long distance travel, foreigners and foreign countries, cultural institutions, abstract mental processes, ethics, social order, overconfidence, conceit, indulgence, freedom-loving, resentful of restrictions. Jupiter rules Sagittarius; is exalted in Cancer; is in detriment in Gemini; is in fall in Capricorn.

Jupiter, named for the Greek god Zues or the Roman god Jove, is the planet that represents opportunity and your responses and reactions to it. Jupiter is associated with the characteristics of joy, friendliness, hope, benevolence, compassion, justice, honesty, spirituality, and calculations. Jupiter is referred to as the greater fortune, and it is often associated with expansion, growth, abundance, wisdom, luck, optimism, success and generosity.

In mythology, Jupiter or Zeus, the ruler of the gods, was as a child nurtured on the milk of a goat whose horns overflowed with food and drink. The planet of Jupiter pours outs it gifts through the opportunity of experience. Jupiter can be thought of as coordinating the personal experiences of the Moon and Venus.

If Jupiter is in the same sign as your Sun or your Moon sign, it grants a direct and effective response to opportunity and is at its most fortunate or luckiest. If Jupiter is in the same sign with Mercury, you perceive opportunity optimistically, but with some degree of stubbornness. If Jupiter is in the same sign with Venus, you perceive emotions in a way that allows you to turn them to your advantage, harmoniously. In the same sign with Mars, your energy, drive and courage are directed toward opportunity and you aren't above taking a chance in life.

Jupiter indicates whether you are sociable and friendly or not, how you will respond to opportunities in life, and what career choices are best for you. Jupiter represents that aspect of life in which you are the luckiest, or most fortunate, or where you find the most opportunity, and it is that stroke of luck that arrives when you most need it. Jupiter is warm, moist, sanguine, temperate, social, expansive, masculine and moderate.

If well-aspected, Jupiter grants a logical mind, confidence, and a self-assured determination. If Jupiter is afflicted, it can indicate restlessness, uncertainty, misjudgment, unfortunate investments and speculations, and a tendency toward extravagance, overspending for luxury items, and an overly optimistic outlook. In regards to health, Jupiter rules the thighs, blood, liver, veins, the pituitary gland and arteries.

Jupiter in Aries

Jupiter in Aries makes you progressive, ambitious, and aspiring with an urge to expand that inspires direct and forceful action. You possess grand ideas and the enthusiasm to carry them out. Opportunity becomes personal and ego driven. Jupiter adds fire to a self-assured person or self-assurance to one who is not as outgoing. You have the ability to be responsibility and to handle situations of authority well. You prefer to be in a leadership position or working for yourself. Either way, you like to be the person making the plans. You may find yourself changing your career, following different objectives, or having two occupations in life. You have a strong intellect and are attracted to science, literature, philosophy, travel, sports, and nature. Success may be realized through personal merits, domestic relationship, social standing or influential friends. You find yourself naturally attracted to work that is satisfying to you, but you must watch against being rash, impulsive and over-optimistic. You possess an ability to exert yourself and the courage of your convictions.

Jupiter in Taurus

Jupiter in Taurus grants a love of justice, and to you, opportunity and security are perceived as one in the same thing. The chances are, you hold back from taking a chance, preferring to protect your security. Your tendencies toward expansionism is related to utilitarian endeavors in that you understand the value of money. At the same time, you are affectionate, generous, peaceful, reserved and firm in your opinions. You have a strong love for your home, and not being restless, you have little interest in change. With Taurus being the sign of material possession and Jupiter the sign of abundance, you may do well in careers related to finance and investments, minerals and oil, ranching, banking, construction, and merchandising. You find that you benefit through the opposite sex, society, the church, philosophy, your intellectual endeavors, gifts, investments, speculation, and inheritance. If afflicted, you suffer through the same aspects. You are most likely a person who believes that positive thinking brings profitable results. Not a gambler, you take life seriously and consider opportunities cautiously.

Jupiter in Gemini

Jupiter in Gemini and you are a person who is versatile and open minded with an adventurous attitude toward opportunity. Sympathetic and benevolent with humanitarian tendencies, you are friendly and well mannered with a love for novel situations, traveling and intellectual pursuits. You may find the most opportunity in exciting adventures in far away places, and others may think you impractical at times. Although clever, your interests may be scattered. You are more than willing to try anything once, and opportunities seem to come to you through social contacts and friends. Sometimes restless and changeable, you want to see and experience life, to travel, and to meet other people. You may find opportunities in the fields of math, communications, literature, education, law, diplomatic positions, consulting, inventions, and the travel industry. If afflicted, there are problems in marriage, some difficulties with adversaries, an unprofitable profession, separation from relatives, and problems associated with writings. You may find your fortune success by accident while you're not even looking for it.

Jupiter in Cancer

Jupiter is exalted in Cancer making you a kind, generous person filled with good humor. You are ambitious, enterprising, popular and sociable with a friendly disposition. You tend to see opportunity along lines that promote security and prosperity. You aren't grasping, but you have a tendency to hold on to your money and to strive to satisfy your wants. You have a close association with your mother, but you yearn to travel and to improve you insight and knowledge of cultures and places. You are intellectual, intuitive, and imaginative with an appreciation for the arts. You are patriotic and possess an interest in public welfare and an insatiable curiosity about psychic matters. You do well in investments, real estate, inheritances, public work, or gains through marriage. The chances are you will travel far from where you began. To you, home, family, spouse, children and friends are important, and you strive to maintain your security in life. You follow traditions and find comfort in attending religious services. Your knowledge of life develops out of a sympathetic response to feelings and emotions.

Jupiter in Leo

Jupiter in Leo grants you public appeal. You are good-natured, magnanimous, loyal, courteous, generous and compassionate. More than likely you see opportunity as a chance to show your virtues and abilities and to expand results, preferably on a large scale. Intelligent and ambitious, you have a love of display with a flair for the dramatic. The expansion of Jupiter working through this sign of self-expression enlarges the ego, making you dream of conquering the heights of success. Your emotions are deep, sincere and honest, and you are endowed with your own special wisdom, judgment, and strong will power. Capable, honest and trustable, you do well in positions of responsibility, and you enjoy honors and recognition. Intuitive, at times inspiring and diplomatic, you do well in acting, politics, advertising, public relations, government, and positions of prominence. You benefit through investment, speculations, travel, sports, education or diplomatic positions. Guard against easy success making you boastful or power hungry. You are most satisfied when others are aware of your success.

Jupiter in Virgo

Jupiter in Virgo makes you kind and conscientious but also skeptical and with a factual, sometimes critical, outlook. Jupiter is in detriment in Virgo making you intellectual but overly concerned with details. You can systematically accumulate knowledge and perceive progress in material aspects rather than as an expansion of consciousness. Chances are, you are cautious, prudent, discreet, and discriminating with a tendency to put all of your efforts into one endeavor, and find success by building on that endeavor, however humble your beginning. Practical minded, you are also analytical, persevering, and endowed with common sense. You probably prefer ethics to theology. You may find success in literature, investment, commercial and speculative dealings. You have the ability to live by your wits, but others trust your judgment and will probably back your business interests. If afflicted, there is seen loss of benefits, illness, and a tendency not to apply yourself through lack of concentration. Not rash or impulsive, you gain through your own endeavors.

Jupiter in Libra

Jupiter in Libra makes you kind, charming, magnetic, sympathetic, and hospitable with an artistic flair. Well mannered and sincere, you enjoy being conscientious, compassionate, imaginative, and perceptive with a judicial outlook of acceptance and harmony. You gain much wisdom through an understanding of relationships, but at times you are easily influenced. And while you may be talented in a number of areas, you may find yourself pursuing a course that isn't what you're best suited for, or which was suggested by someone else. You love to entertain and to attend to the needs of others, and being diplomatic, you like to inspire and encourage others. You do well in art, music, business, commerce, law, science, medicine, or public institutions. Jupiter influences your ability to plan functions and grants you social consciousness with a capacity to interact with others. This is also a good location for marriage. If afflicted, there are legal problems, problems through friends, laziness, self-indulgence and conceit. You must become firm to find success in any one area and worry less about the opinions of others.

Jupiter in Scorpio

Jupiter in Scorpio effects personal growth and transformation, and grants you a strong and powerful will power, deep emotions, enthusiasm and perseverance. You are ambitious, generous, proud, aggressive, and self-confident without being ego driven. Opportunity to you is associated with deep personal needs, and you follow instinctively that which satisfies these needs. You have an active and analytical mind and an intense desire to live life fully. You have a flair for business, finance, and positions requiring locating information such as investigative work or research positions. Opportunities may come from associations with the opposite sex, and gains from litigation and investments. You travel, experience strange adventures, and meet questionable people. If afflicted, you have jealous friends, loss through speculation, danger through travel and social or political associations, and if Mars overrules Jupiter's restrictions, you seek sensations through pleasures and experiences. You can be forceful, farseeing, and driven to complete your tasks, possessing a sublime intuition for finding the right information and making the right choice. You are ardent in life.

JUPITER IN SAGITTARIUS

Jupiter in Sagittarius brings a wide range of interests and generally good fortune in financial matters. You are attracted to opportunities related to ideals, knowledge, philosophy, and adventure. You are courteous, friendly, tolerant and accepting of others, kind, and generous. Your mind is liberal, progressive, and compassionate. You are the sincere humanitarian wanting what is best for all. You may well find that you are successful in your endeavors, receive honors, and become a leader in your field. You prefer to live well and to spend freely, turning your ideas into profit. You find success in sports, literature, law, government, politics, scientific, or spiritual matters. You may gain through speculation, marriage, inheritance and travel. If afflicted, you experience problems with social affairs, sports or loss by speculation. You can be prophetic and inspirational and often impress others. Not particularly materialistic, you find success when it relates to an ideal such as duty or security. You prefer to please the crowd, and must guard against this desire for popularity.

JUPITER IN CAPRICORN

Jupiter in Capricorn and you are prudent and conscientious with a deliberate, thoughtful, ingenious, serious and well-organized mind. You seek opportunity for positions of authority and dignity. You are the perfect mixture of daring and caution but with your strong will power and ambition you are unafraid of hard work. You can be either economical or frugal, and this doesn't slow you down in the least. You gain success in positions of authority, exhibiting natural leadership abilities. You may have an interest in philosophy, politics, science, or theology, and do well in government service, business, finance, trading, mining, construction, real estate and development, or the petroleum industry. You gain by commercial or foreign affairs, and an inheritance. There is indication for travel, a marriage affected by family, and a concern for others. If afflicted, there is an unorthodox attitude, career difficulties, and public discredit. You must account for all that you do. But for you opportunity is seen practically, with no inclination toward speculation, and you abide by traditions and well-tried practices.

Jupiter in Aquarius

Jupiter in Aquarius grants you the ability to win friends and to realize opportunities through social, artistic, or political matters. However, you are as content to be a drifter as a success story along as you have your principles and other people to keep you company. Naturally cheerful and good humored, you are compassionate and sympathetic with a giving and congenial manner. You strongly dislike discord or disharmony in your affairs. You possess a love for the ideal and may idealize love in an impersonal manner. You envision expansive dreams that may be realized or may just be fanciful thinking. You prefer careers having a higher goal than simply making money, and you seem to bring luck to others. You are intuitive about human nature and prefer to consider new concepts of social, spiritual or humanitarian causes. Careers that attract you include the media, computers, electronics, aviation and the space industry, and you have an interest in science, literature, music and the arts. Gain is seen through the career and perhaps pertaining to the government, and expect travel and unusual experiences. Nonmaterialistic, you view money as a means of doing what you want in life.

Jupiter in Pisces

Jupiter in Pisces grants compassion, benevolence, and good humor with strong ideals and an interest in charitable organizations. You prefer to expand consciousness by broadening relationships, seeing opportunity in how it relates to your private dream of self-justification. Chances are, you will pass up financial gain if it is associated with unworthy causes, having a personal code by which you judge all things. Your are attracted to unusual interests including the mystical. Original, independent and progressive in your thinking, you prefer professions in which you deal personally with others such as social work, counseling, teaching, politics or government services. You also like nature and do well in careers dealing with animals and wildlife. If afflicted there is seen loss through speculation and deception, and an attitude that is unambitious and pleasure seeking. You may find success in your secret ambition and will always find that you attract popularity through your personality.

Saturn

Rules: Capricorn

KEYWORDS: Discipline, restrictions, responsibil-ities, limitations, patience, long term relationships, authority, leadership, career plans, older people, groups, influential relationships, public reputation, social standing, political outlook, ambitions, serious social affliations, government, legalities, math, science, time, separations, difficulties, obstacles, constriction, sorrow, delay, self-discipline to overcome problems, wisdom through effort and experience, crystallization, caution, perseverance, endurance, stability, reserved, dignity, contraction, organization, seriousness, duty, obligations, prudent, economical, suppression, suspicion, pessimism, selfishness, indifference. Saturn rules Capricorn; is exalted in Libra; is in detriment in Cancer; is in fall in Aries.

Saturn represents limitations, self-preservation, hard work, and the burden of responsibility and duty. It is basically a protective influence providing growth and survival through restriction. Saturn provides us with the lessons in life, the trials and tribulations, and the training needed to become self-disciplined. Saturn is exacting, but once a person has endured the hardships, it can be expected that Saturn will then grant acceptance, justice and achievement through personal ambition and aspiration. Mars is referred to as the lesser malefic, and Saturn as the greater malefic with its influence granting patience and a mature stability as seen through realism. Saturn rules the end of life.

Saturn in mythology was Cronus, the father of Jupiter, Neptune, and Pluto; Cronus was dethroned by Zeus. Saturn's symbol is the sickle that Father Time uses to harvest the fruits of a man's life.

Those with a positive Saturn influence are serious, practical, and can be profound thinkers with an economical, prudent, and conservative outlook. The further Saturn is located from your Sun, Moon, or Ascendant, the more objective and extroverted you are. If Saturn is in your Sun or Moon sign, you are defensive and at times prefer to seek isolation. If Saturn is located in the same sign as Mercury, you are a profound thinker with a preference for serious reactions to sensory impressions. With Saturn in the same sign as Venus, you may be cool in emotions and have problems with interpersonal relationships, being luckier in finances than with love. Saturn combined with your Mars sign makes you either defensive, indecisive and aggressive, or if the Sun and Moon are stronger, well-balanced, calm and moderate. With Saturn in the same sign as Jupiter, you take opportunities seriously.

Saturn rules our destiny and fate, deciding the price we pay for what we have received in life. It rules the skin, teeth, bones, gall bladder, spleen, some nerves and joints, and hearing, and is associated with maturity, perseverance, tenacity, inhibitions, and intolerance. Poorly aspected, it produces adversities, delays, disappointments and sorrows. Saturn's nature is cold, dry, melancholy, earthy and masculine. It teaches us not only to respect life, but that through

179

introspection and reviewing our lives, we can see how the lessons learned in life benefit our choices in the future.

SATURN IN ARIES

Saturn in Aries and you are resolute, determined and ambitious, but this planet spells delays and obstacles in your path that can cause you some confusion. You realize success through hard work and perseverance. You find yourself contemplating your plans with sound thinking, but you are easily angered when opposed and you forcefully assert your control over your situation and over others. Ruled by Mars, you are aggressive, but Saturn provides a restraint that leaves you a disciplinarian of military precision, and this combination of forcefulness and organization dictates limits and restrictions. You can be reserved but are striving to acquire success or possessions. At your best, you are self-reliant, dutiful, and persistent with the strength of character to face hardships and obstacles. In its worst, this location of Saturn influences a person to be narrow-minded, self-centered, stubborn, defensive, and feeling that the world is against him or her. You may experience jealousies in life from a partner or friends, difficulties in marriage, and numerous obstacles and setbacks in your early life. Life becomes more settled as you mature.

SATURN IN TAURUS

Saturn in Taurus and you are a kind, thoughtful person who is cautious and prudent, but at times easy to anger, stubborn, resentful and unforgiving. You take responsibilities and duties seriously, working tenaciously with capable and determined effort. Your self-preservation instinct is satisfied by security and material possessions, but you aren't necessarily selfish or materialistic as much as striving for a sense of well being and security. Once you have acquired enough in life to make you feel secure, you become generous with others. You can be either purposeful or obstinate, but you have a strength of character and morals that see you through most situations. Economical and thrifty, you are not one to speculate on the future, but choose to follow a conservative path, working hard and gaining through your own merits. You may experience loss and sorrows associated with relatives, but gain through savings and investments. With a strong love of nature, you prefer to abide by nature's laws.

Saturn in Gemini

Saturn functions in Gemini to concentrate the mind on facts, organizing your energies constructively. This is considered one of the best locations for Saturn, granting you an ingenious and observant nature with a mastery of mind and perhaps of life. You have intellectual abilities and an interest in scientific subjects, research, education, literature, or math. You possess a shrewd business mind, and justify your actions forcefully and articulately. You can be versatile, adaptable, and resourceful with a steady, impartial mind, cautious reasoning, and profound insights. Unfortunately, Saturn and Gemini can combine to produce a person who is controlling, cynical, stubborn and pessimistic who doesn't see through obstacles to the future. A realist, this person becomes overly concerned with the present and is at times gloomy and depressed. You prefer travel associated with your career rather than for pleasure alone, may experience problems legal affairs, accusations, and problems which develop into restrictions and limitations with relatives and friends. You exhibit an interest in innovative thought and intellectual pursuits.

Saturn in Cancer

Saturn in Cancer finds a person whose practical concerns for security restrict emotional responses. You are shrewd, ambitious and tenacious, make a reliable partner, but may be emotionally dependent on others. Saturn is in detriment in Cancer which can make you somewhat changeable, fretful, discontented or jealous. You may worry yourself into being introverted, or on the other hand use your insecurities to force yourself into the world. If afflicted, this native becomes defensive, builds protective emotional walls, and turns irritable and self-absorbed. The challenge here is to learn to use the lessons learned from Saturn's discipline and Cancer's naturally protective instinct to overcome any obstacles. And your obstacles may be many including domestic problems, sorrows with parents and friends, and worries with home and children. You develop an interest in psychic affairs, prefer hobbies and subjects related to the home, and are very concerned with social issues. Take the intuitive to overcome obstacles in preparation for maturity.

SATURN IN LEO

Saturn in Leo grants you powerful characteristics. You are determined, ambitious, strong-willed and bold, but quick tempered. You are self-assured, well organized, and authoritative. Dramatic, you excel when in the public eye, finding justification but perhaps not much fun. This love for attention predominates your life, whether successful or not. Saturn's restrictions and Leo's expansive nature aren't necessarily in harmony. You may find problems expressing your emotions and troubles in relationships of love. And your authoritative nature may make you appear cool, haughty and controlling. Your strength is solving problems, organizing, and establishing order, and you may do well in government, politics, or leadership positions. Rigid and conservative, your sense of personal responsibility oversees your life, and at times you prefer this to more pleasurable past times or relationships. Your challenge is to learn to be more accepting of others and less suspicious of the good will of others. You experience sorrows and obstacles, and may attain success in your career, but must watch for reversals.

SATURN IN VIRGO

Saturn in Virgo is advantageous and compatible. The earthy, practical nature of Virgo does well under the limitations placed by Saturn. You are reserved, discreet, and cautious with a preference for sound methods, rules and regulations, precision, and good judgment. You are found to be most conscientious and responsible, finding expression and a sense of worth in work and duty. You are most happy when things are going well in your career, but you must guard against stress and depression associated with over working. You realize gains through investments, but encounter obstacles in marriage and partnerships particularly in the early adult years. If afflicted, this person is stubborn, intolerant, and too detail-oriented. You possess the ability to work well by yourself, and no doubt will find success in your career. However, in order to find harmony in your personal life, you must develop an understanding and acceptance of other people whose methods and opinions differ from yours. Real vision in life is found when one pulls all the details in life together to find a larger picture.

Saturn in Libra

Saturn is favorably placed in Libra making you kind, pleasant, lovable, and well-adjusted. You are patient with a spiritual nature and a reasonable perspective on life whose balanced judgment inspires others to trust you. You possess refined tastes, an interest in scientific subjects, and a good intellect with a tendency to play the advocate, loving a good discussion on controversial topics. Your nature inclines you to be faithful in love and marriage, honest in your transactions, and reliable in your friendships. You are responsible, flexible, diplomatic and know when to be assertive, and you naturally make a good impression on others with little effort. Saturn is exalted in Libra, allowing you to seek harmony and balance in life and relationships. If afflicted, this person becomes intolerant, insincere and impractical, becoming lonely and depressed. Obstacles in life may include sorrow and separations, delays in marriage, domestic problems, and reversals in the career which are unfortunate. With your grace, tact and charm you develop an overpowering interest in the world at large and other people.

Saturn in Scorpio

Saturn in Scorpio and you are a passionate and complex person, subtle, forceful, independent, willful, inquisitive, but somewhat reserved and at times secretive, jealous, and resourcefully cautious. Your shrewd mind combined with the seriousness of Saturn results in executive ability and business sense. You are capable of making sudden resolutions, having deep feelings, being masterful and difficult to fathom. It is almost like, at times, your primal instincts and urges are driving you, rather than conscious thought. You have a great sense of purpose and strong reserves. At times you are inflexible and have a tendency to be moody or to brood over circumstances. Your rather dry humor can turn everyday situations into insightful ones. Excellent detectives or secret agents, persons with Saturn in Scorpio prefer to work secretively, even deviously, rather than taking a straightforward path. Cautious, this person doesn't avert danger. You may experience sorrow through love, secret alliances or intrigues, or difficulties in marriage. Persistent effort against obstacles produces success.

Saturn in Sagittarius

Sagittarius warms the austerity of Saturn, and Saturn influences this native to be serious about goals and objectives, bringing into reality abstract ideas and plans. You are candid, show little fear, and are kind, obliging and somewhat philosophical. You are self-reliant and confident in your abilities, but your aim in life is to master it not merely to acquire position, success, or material possessions. You are dignified and can be grave, even at times moralizing, with an intuitive perception and insight into social welfare and scientific innovations. Strongly motivated, you excel at directing others and do well in administrative positions. You can be thrown off course by a sensitivity to criticism or opposition, but otherwise you gain through your own merit and hard work. If afflicted, this person can be ostentatious, cynical, harsh, and insincere. Problems are indicated from public, political or government affairs which may harm your reputation. You may find that you have more than one career, and you realize gains from investments. Guard against not watching your health, worrying too much, and nervousness. You develop intellectually and spiritually finding success in leadership positions.

Saturn in Capricorn

Saturn in Capricorn influences this person to be self-disciplined, practical and persevering. You are serious and at times become melancholy, apprehensive or suspicious. You are ambitious, and your persistent diplomacy and tact allows you to succeed in your career. You are more than willing to start at the bottom if it offers an opportunity for advancement. You possess the ability to turn abstract concepts into concrete terms. This position of Saturn allows a person to attain achievement or defeat, depending on personal actions in that he or she follows a path to its conclusion. You most desire success either financially, materially or through fame. If afflicted, this person is pessimistic, miserly, selfish, and arrogant. There is a tendency for unreliable friendships, difficulties in marriage, illnesses associated with stress and nerves. Your ambitions are important to you, and you are willing to sacrifice for them, and you prove you are capable of responsibilities. Success is based on your own merit and ability to face obstacles.

Saturn in Aquarius

Saturn in Aquarius and this independent, original and inventive person, once a decision is made, sticks to plans and objectives. You are courteous and friendly but somewhat reserved and thoughtful with a humane and serious nature. You have a tendency to ponder and think deeply, and with your penetrating intellect and reasoning ability, your thoughts are insightful. You are an impressive speaker when you care to be, and then too you are deliberate in your actions. When the intellect is developed and refined through study, observations, and experience, this person's inclinations toward humanitarian and scientific subjects produce profound achievements. Success is indicated in a career choice utilizing practical application of knowledge, and you are capable of leadership positions. Guard against your tendency for inhibitions, frustrations and isolation. If afflicted, this person is obstinate, indifferent, cunning, and sly. This person has numerous friends and acquaintances and a lasting romantic relationship. You may find that success is attained during maturity after encountering obstacles in early adulthood and learning to deal with them.

Saturn in Pisces

Saturn in Pisces and you are imaginative and intuitive, ingenious and aspiring, but sensitive to the disharmony in the world. Saturn's influence forces you to come to terms with your own tendency to brood and give up and to withdraw to safe territory. Only through self-discipline do you gain a spiritual ability and the perseverance to withstand the many obstacles Saturn places in your path. Called to sacrifice for others and to watch less deserving and talented achieve when you don't, it is left up to you to find the courage to utilize your creative abilities and strive for the success which Saturn eventually endows, that is if you don't give up by accepting existing conditions. You are sympathetic, flexible, creative and many times blessed with talent. If afflicted, this individual is moody and hypersensitive, lacks courage, worries excessively, and is sloppy and untidy. Misfortunes and sorrows are experienced, at times caused by the native's own actions, and success is realized only with the development of hope, resolve, and effort.

Uranus

Rules: Aquarius

KEYWORDS: Insight, originality, illumination, invention, genius, friends, groups, organizations, sudden events, changes, surprises, sudden destruction, the unexpected, unusual occurrences, revolution, disruptive events, advanced technology, electronics, electricity, independence, personal freedoms and liberties, ideals, humanitarian efforts, astrology, occult, corporate finance, intuition, reforms, ingenuity, eccentricities, rebelliousness, unconventional, new rules, futuristic visions, revelations, aeronautics, aerospace. Uranus rules Aquarius; is exalted in Scorpio; is in detriment in Leo; is in fall in Taurus.

Uranus, the seventh planet from the Sun, is referred to as one of the three modern planets. It was discovered in 1781 by William Herschel, British court astronomer. Uranus rules over sudden or disruptive changes and unpremeditated actions, signifying dramatic flashes of insight and of overcoming barriers. Then too, it influences liberty, equality, curiosity, inventions and investigations, innovations, unexplained occurrences, originality and modern art. Uranus is associated with advanced science, electricity, social and humanitarian changes and movements, and revolutions. This planet signifies the unexpected, upheavals, and the unorthodox: change, new or different, and original. Those people influenced by Uranus are often reformers, pioneers, inventors, geniuses, or are creative and intuitive.

Uranus was named for the Roman god of the sky, Ouranos, the personification of the heavens and ruler of the world. Ouranos was the father of the Titans and of Cronus who dethroned him.

Because of its distance from the Sun, Uranus remains in each Zodiac sign for seven years, taking eighty four years to pass through all twelve signs. The nature of Uranus is considered cold, dry, airy, positive, magnetic, unusual, occult and malefic. It rules the ankles and the intuition, and relates to your inner will and inherent abilities and powers. It is that energy which lends itself to will power and purpose.

In the same sign with Mercury or the Moon, Uranus grants acute awareness, quick reactions to perceptions and experience, and a quick mind. If in the same sign as the Sun, there is found nervous activity, creativity, originality and at times an eccentric nature. In the same sign as Mars, it produces fast actions and fearlessness. In the same sign as Venus, there are found unusual reactions to emotions, and sensual, idealistic, and original perceptions of love and human relations. With Saturn in the same sign, the person must balance both creativity and practical tendencies. In the same sign as Jupiter expect opportunity, inventive, daring and executive skills.

If afflicted, Uranus inclines one to be too forceful, abrupt, erratic, willful, obstinate, sarcastic and overly sensitive. These natives must strive to improve by combining effort, aspiration, and self-control, and by avoiding any risks associated with those things ruled by Uranus.

URANUS IN ARIES

Uranus in Aries and you love independence and freedom and are positive, forceful and at times impulsive with an original, imaginative and inventive mind. You are mentally and physically energetic and resourceful. You prefer to produce your own plans, be in charge of them and take charge of other people involved in your plans, and to receive recognition of your successes. You can be impatient with the ideas of others, preferring to develop your own, and fearless of following through with them. You like innovative new devices either mechanical, electrical or electronic. Travel inspires you and you change locations when your curiosity gets the best of you. If afflicted, this person can be impulsive, tactless, sharp-spoken and blunt with radical ideas and a lack of self-control. Disputes and disagreements are frequent, but your strong will power, self-reliance, and belief in yourself continues to compel you. Forceful and often inspiring, you convince others of the worthiness of your plans.

URANUS IN TAURUS

Uranus in Taurus influences a person to be determined and headstrong with fixed and immovable opinions. You are resourceful and ingenious with a tendency to plow through any obstacles in your way. You strive patiently and with effort to consolidate your work using will power and determination to make plans for future endeavors. You are good humored and friendly and enjoy the companionship of others. You possess a tendency to have an active mind that focuses on wanting newer and different material possessions, and you may come across a particular find that gains in value. You possess an inner drive to be constructive and to achieve through your efforts. You may be lucky in finances and prosperous in marriage. Uranus, however, brings gains and reversals in life, including sudden losses and a domestic difficulty caused by jealousy. You gain through partnerships, marriage, friends and associates, inventions and creative endeavors. You possess a natural intuitive ability that brings insights and inspiration into your pursuits. Tenacity, hard work and the results of your own efforts see you through life.

URANUS IN GEMINI

Uranus in Gemini and you are versatile and inventive with a flair for writing, speaking, invention or scientific skills. You are an original and imaginative thinker whose ingenuity impresses others. You feel drawn to innovative ideas, revolutionary concepts, and unusual subjects. You love ideas and your active mind draws the attention of other people, and you possess the ability to influence the thoughts, opinions and ideas of others. You are humorous and possess an amusing and quick way with words often turning dull phrases and subjects into a light-hearted discussion. You love to create a new and impressive outlook, an original slant, or an insightful perception that you can express in writing or speech. If afflicted, problems with nervousness, estrangement from family, undue criticism, problems with education, tests, letters, and with trips. Your enjoyment of travel and meeting new people brings you exposure to new people and ideas.

Uranus in Cancer

Uranus in Cancer produces a person who possesses a love for home and family. You are highly sensitive to other people and may at times choose to isolate yourself finding comfort in your personal creativity, imagination, art and self-made fantasies. You possess a quick sense of humor often turning ordinary incidences into funny, quirky happenings. You can be unpredictable or at times uncertain, but you are also intuitive and insightful having sudden flashes of inspiring thought which at times borders on mysticism. A bit restless, you love to travel, to be exposed to new places, scenes, and tastes. You have a love for food which at times you find difficult to control. Because you are so sensitive, the opinions and remarks of others offend you, making you sometimes nervous, eccentric, cranky, and impatient with friends and family. Changes occur in the domestic situation, and loss and difficulty may be experienced in regards to possessions and property. Guard against health problems associated with nerves and stomach. Your creative and original nature finds expression through seeking security and comfort in pleasant home situations.

Uranus in Leo

Uranus in Leo grants this native an industrious and aspiring nature with a forceful personality and strong leadership skills. You like being in charge and take to this position naturally because you dislike taking orders from others. Expect this person to be either independent, arrogant, headstrong, fiery, eccentric or defiant. You can be unconventional and even rebellious and worry little about the opinions of others. You have an interest in electricity, machines, and inventions. When your creative flair exerts itself, you are original in thought and like to see your ideas carried out, with you receiving the deserved recognition. Leo, as usual, stands out in a crowd. At the same time, being daring, you love a good adventure, excitement, and unusual distractions. You may encounter sudden chances through romantic relationships which bring either opportunity, odd experiences, danger or sorrow. If Uranus is well aspected, you achieve a distinctive position in a professional or public career.

Uranus in Virgo

Uranus in Virgo and you find that your originality most asserts itself in the areas of methods, organization, routine which allows you to analyze and use facts and figures. Your mind is subtle, independent and original while your nature is somewhat quiet and reserved but at the same time stubborn. You have a general fondness for science and unsolved phenomenon. You can be eccentric in your curiosities and determination. You are concerned with the social welfare of society, but sometimes become bogged down with the details which doesn't allow you to draw a general conclusion. If you are rebellious, it is when your plans, ideas and independence are restricted by others. If afflicted, there are disappointed ambitions, problems with either employment or employees, and a restrictive career. Generally, however, Uranus brings this native gains through opportunities and changes in the career at which time this person must set aside his or her usual caution and accept the change for the better. With your tendency toward weight gain, it is helpful if you become knowledgeable about health and nutrition.

Uranus in Libra

Uranus in Libra inclines this person to be a good reasoner who loves harmonious situations. You find yourself at times restless, and you enjoy the opportunity to travel and meet new people. Being open and tolerant, you are considered somewhat eccentric in your choice of friends and associates. Difficulties can make you quick-tempered, but generally you are ambitious with a creative imagination and insightful intuition. You possess a refined taste and love aesthetically pleasing surroundings and situations. You find opportunities through your associates and friends, and possess a fondness for artistic, literary or judicial careers. Your strong magnetic personality and inventive flair draws the attentions of others. Guard against a hasty marriage which brings separation or divorce. If well aspected, Uranus in this location favors partnerships for Libra natives. If afflicted, there is an indication for an unsympathetic or unaffectionate nature, oppositions, rivalries and criticisms resulting in broken friendships and problems with plans. You are capable of always making a good first impression.

Uranus in Scorpio

Uranus in Scorpio makes you an angel or a little devil with the innate ability, on some deep, subconscious level, to use your genius and power for either good or evil. Uranus is exalted in Scorpio granting you a powerful personal magnetism that compels others to your control. You possess a strength of mind, determination, and persistence that drives you through your daily life. You have a strong power of concentration and a will power that is un deterred by opposition. Resistance may leave you unable to express your emotions because your emotions, working under a force of their own, focus on gathering strength and empowering your will power and determination. You can be boldly stubborn and at other times quite reserved, forceful and if necessary sharp-spoken and direct. You can also be shrewd and secretive, but aggressive to the point of being rebellious. If afflicted, there is danger of accidents. You have a mesmerizing quality and a strong magnetic appeal especially to the opposite sex.

Uranus in Sagittarius

Uranus in Sagittarius and expect the scope of your originality to be drawn from pure vision. Uranus increases the already natural imaginative, inventive and creative nature of Sagittarius inclining you toward dreams, insights, intuition, premonitions and travel. The planetary influence of Uranus also increases any rebellious nature leading to recklessness. But generally speaking it drives your freedom-loving nature to pioneer in new endeavors and to risk adventurous schemes. You love all areas of science but have a curious interest in innovations in the travel industry. You are proud and courageous with a seeking, progressive mind. You can be daring and bold, loving excitement, and you are just as enthusiastic in your beliefs and opinions. You are known for being generous, congenial, and well liked among peers and friends. If afflicted, strange occurrences related to adventures, partners, scientific endeavors, foreign affairs or travel. If well aspected, these same circumstances may bring you opportunities and gains. You often know intuitively that moment when something is about to occur, and your prophetic visions serve you well.

Uranus in Capricorn

Uranus in Capricorn grants you leadership abilities and excellent organizational skills, sound reasoning faculties and a penetrating mind with perceptive and profound insights. You are thoughtful with a serious nature that is marked by self-discipline. Your strong will power compels you to positions of authority, and you strive to succeed in all endeavors. Uranus intensifies your already ambitious and persevering nature making you even more independent with a will to be in charge and not under the dictates of others. Uranus influences you toward bold enterprises, radical and innovative changes in methods and procedures, and to take strong initiatives in your efforts. You display foresight, intuition and a propensity for correct hunches and guessing trends. If afflicted, there is an indication for family problems, oppositions from authority figures, reversals in career, and public criticism. Guard against periodic restlessness and feelings of uneasiness.

Uranus in Aquarius

Uranus rules the sign of Aquarius influencing these individuals to be original, ingenious, inventive and independent. You are resourceful as well as intuitive with comprehensive mental abilities and a creative imagination. You are pleasant, friendly and sociable with humanitarian tendencies and a desire for freedom in thought and actions. Your love of people and ability to make friends plus that special stroke of creative genius allows you to sway others. Not tied to present day thinking, your unique outlook jumps light years ahead giving you what others consider an eccentric or different set of beliefs. You are drawn to science, novel subjects, and unusual endeavors not to mention other people with unconventional interests and tastes. You hold firmly to your personal opinions sometimes not listening well to others. You interests may lead you to a successful career in science, mechanical endeavors, with partners, in public life, in transportation fields, or in radio, television, or the computer industry. If afflicted, this person has difficulty through friends, partners and travel. If well aspected, friends and partners bring you luck.

Uranus in Pisces

Uranus in Pisces can grant an individual an unique idealism with pure visionary insight. You are emotional, sensitive, and subtly insightful with an interest in philosophy, theology, astrology or psychic subjects. At the same time, you are logical and take a realistic viewpoint, and are able to research and investigate the subject matter that interests you. Your creative flair expresses itself in sensual expressions of humanity and the human condition. Being sensitive and perhaps psychic yourself, you may find that you experience strange occurrences or have remarkable dreams and visions. Your intuition leads you to act or begin an endeavor at the most advantageous time, granting you a special luck in opportunities if you listen to your own counsel and follow it. Your visions can bring you insights into the future, but this ability may also tends to dampen your spirits at times. You discover friends who are also gifted with unusual abilities and others who are public figures. If inclined toward a career in science, you may discover an affinity for developing new techniques, treatments or medicines for man's welfare.

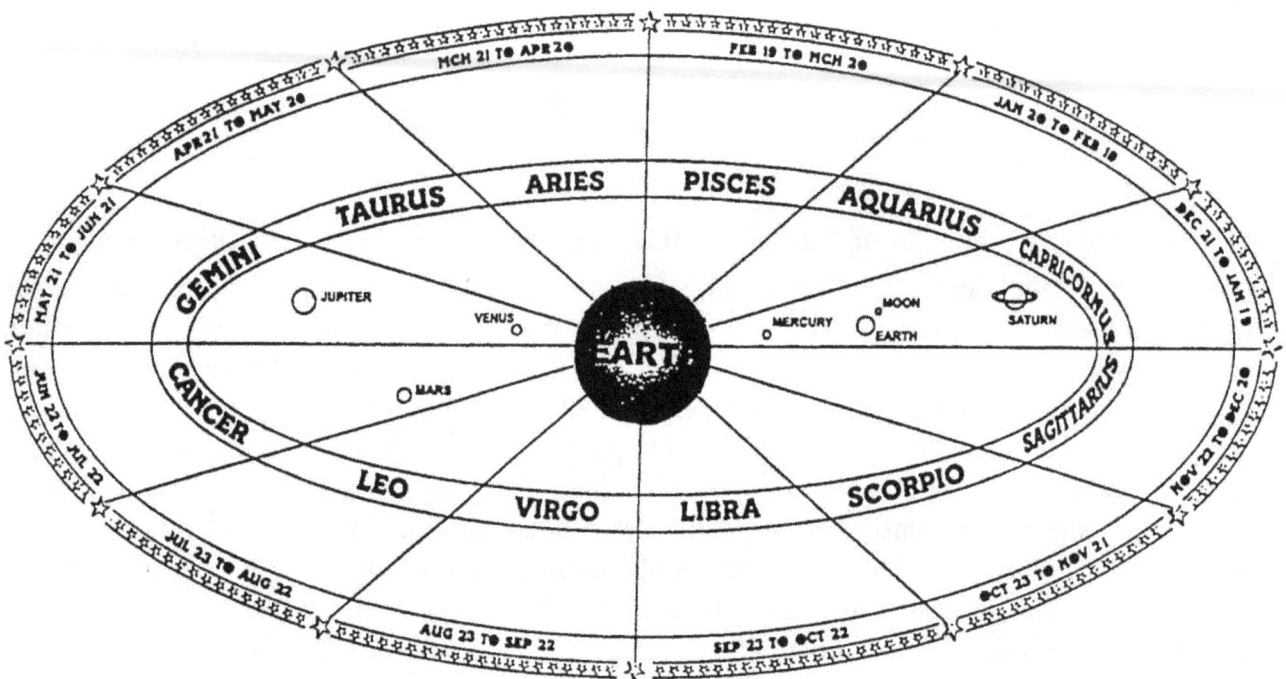

Neptune

Rules: Pisces

Keywords: Intuition, imagination, sensitivities, psychic abilities, dreams, unreality, shadows, oceans and seas, cosmic consciousness, artistic creativity, sentiments, journeys by water, idealism, past memories, self-sacrifice, sympathy, confusion, illusion, magic, universal love, charity, romance, alcohol, drugs, indulgence, perfumes, oil, airplanes, movies, television, chain stores, business mergers, subliminal thoughts, secrets, hidden thoughts, the subconscious mind, mysticism, the astral plane, confinement, prisons, hospitals, institutions, hallucination, hypnotism, neurosis, psychosis, chemicals, poisons, deceit, self-deceiving, impressionable. Uranus rules Pisces; is exalted in Leo; is in detriment in Virgo; is in fall in Aquarius.

Neptune is indicative of love, beauty and joy found on the subliminal plane of higher consciousness. It is that mysterious and elusive desire to experience what is beyond the confines of man, material possessions and earthly desires. Because most people prefer the distractions of everyday life, few individuals respond to this influence of Neptune, causing it to be diffused, like light passing through the depths of the ocean. Neptune's influence can either be indicative of human compassion or personal confusion, of a spiritual tendency or devilment, insightfulness or a susceptibility to indulgences in alcohol, drugs or sex, and to an optimistic visionary outlook on life or to delusions in life, dreams and visions. Neptune relates to the deepest part of our subconscious, allowing human kind to see pass the everyday delusions in order to contemplate the essence of being.

Neptune, Roman god of the sea and identified with the Greek god Poseidon, ruled over the watery depths of oceans, seas, rivers, streams and fountains. The planet was discovered in 1846 and spends approximately fourteen years in each sign, resulting in it producing strong generational effects. Under the auspices of Neptune, the planet of idealism and spirituality, man is entranced by the mysteries of phenomenon. It rules over the subconscious awareness of memory, intuition and visions, and represents inspiration, spirituality, inner feelings and artistic imagination. It rules over drama, theater, dance, poetry, and religious inspiration; hospitals and anesthetics; maritime endeavors, institutions, prisons, gases, poison, and drugs; and the nervous system, mental processes, and the thalamus. Its negative influence leads to pessimism, apathy, carelessness, and to impractical worries and weaknesses that can develop into tendencies to lie, steal or to commit fraud or scams.

Neptune in the same sign with the Sun or Moon, relates to intuition and insights, or delusions, and to a need to stay grounded in reality. In the same sign as Mercury, there is creativity with a sensitive mind which responds with intensity to sensory impressions. Located with Venus, this idealistic person is romantic and sentimental or overly fond of exotic pleasures. In the same sign as Mars, your energy and intuition work harmoniously. With Jupiter, Neptune influences intuitive, practical responses to opportunity which produces security. Located with Saturn, either your intuitive self-defenses protect you or you suffer from delusions and unhappiness.

Neptune in Aries

Neptune was last in Aries from 1861 to 1874/05 and will reenter this sign in 2014 to 2028. Neptune influences the fiery energy of Aries with leadership qualities and an introspective intellect. Neptune intensifies the emotions and sensitivities, elevating the nature of this person to be inclined toward sympathy and benevolence with a spiritual perception and attuned inner understanding. Aries innovates his or her personal ideas and plans and this will be noted in the areas of science, medicine, social institutions, international affairs, and political traditions. You have a strong imagination, creativity and self-awareness combined with a desire to travel and see the world. Your impulses lead you to improve and reform with opportunity seen in a public career in business, the media, politics, institutions, science or medicine. If afflicted, this individual distrusts his or her own abilities, seeks self-gratification and pleasures, promotes personal ideas and plans out of egotism, and experiences unusual occurences and notoriety. Your strong will power and inexhaustible energies will see you through the completion of your plans.

Neptune in Taurus

Neptune was last in Taurus from 1874 to 1887/89 and will reenter this sign in 2028. Taurus feels most secure with the acquisition of material possessions, and Neptune influences those born during this period to reevaluate the value of life. You are inclined toward artistic, musical and spiritual endeavors with a progressive outlook regarding the natural sciences and the arts. This is a good position for Neptune for financial profits, perhaps through speculation, and in business endeavors. You have an appreciation for history, antiques, and the legends and traditions of the past. You possess an aesthetic appreciation for beauty and nature, and enjoy travel and sight seeing. You are enthsiastic, patient, compatible, soft-hearted, and fun loving with a good sense of humor. This location of Neptune favors love and marriage. If afflicted, the individual over indulges in fun, pleasure and materialism. You are professional in business and social in your private life, preferring to simplify methods and procedures

Neptune in Gemini
(1887/89--1901/02)

Neptune in Gemini and you are inclined toward innovations and novel subjects. You are intuitive, sensitive, imaginative and creative with a taste for music, drama, or poetry and an inclination for prophetic dreams. You have a personal magnetism that is expressed in your speech and communications. And you have an aptitude for science, math, inventions or mechanical skills. You are sympathetic and congenial with quick perceptions. You are resourceful, flexible, and versatile with a tolerant attitude toward others. If afflicted, the individual can be somewhat flighty, irresponsible, overly friendly and interested in the opposite sex, preoccupied with worries, restless and changeable; experiences difficulties with relatives, friends, associates, and with deceptions, and broken promises. Under affliction this native can become narrow minded, gossipy, and lack vision. This was a period of time of invention, development and change requiring individuals to adjust to new lifestyles and changing trends.

Neptune in Cancer
(1901/02--1914/16)

Neptune in Cancer and you discover that your formative years at home deeply influenced your perceptions, morals, ethics, and sense of responsibility. You Possess an inner strength and spiritualism that is refined and idealistic. You have a strong imagination, powerful impressions, and intense emotions with a love for outdoors and the grandeur of nature. Sensitive, compassionate, and congenial, you are fond of sports, social activities, and the arts. You possess the ability to escape the everyday world with the imaginative world you create through dance, music, theater, and art. Your are fond of travel, science, and new and innovative ideas, gadgets, inventions, and other items that add to the home life. You my feel that you have experienced unusual or strange occurrences in your life as well as psychic phenomenon. If afflicted, the individual is overly sensitive and impressionable, discontent, worried, and restless with a desire for change. You love harmonious settings and feel most secure when settled in your home.

Neptune in Leo
(1914/15--1928/29)

Neptune, exalted in Leo, and you are ambitious but reserved and dignified, benevolent and warm-hearted with courage and leadership abilities. You possess intuitive insight and are sympathetic, charitable and generous. There is an idealistic and creative artistic ability and high aspirations for bringing dreams and innovations into reality. You are sensitive to human emotions which is expressed in your spirituality, making you conscientious, dutiful, and responsible. You possess a love for sports, the companionship of others, social gatherings, literature, music and art. You are receptive and responsive to your inner feelings and perceptions and openly tolerant of change and unconventional ideals. If afflicted, you are drawn to pleasure and your emotions are easily swayed, you are impulsive and restlessly seek change, you can overly generous and spend too freely on luxury items. You have a love for glamour, innovative schemes, and self-expression of freedom and ideals.

Neptune in Virgo
(1928/29--1942/43)

Neptune in Virgo and you may find that you have a tendency to be discerning, questioning conventional and accepted thoughts, principles, and institutions. Practical and capable, you have a concern for the welfare of humanity. Your nature is gentle and patient, and you are reserved with an inclination toward science, math, and a curiosity about psychic phenomenon. You have a tendency to develop your own thoughts and ideals about lifestyles, change, and innovations. Your analytical nature compels you to critically observe and this may deter you from your powerfully inspirational and idealistic tendencies. You possess a drive to serve, and being creative, this is expressed either in art, music, or public endeavors. In some aspects, these individuals appeared to have been born with a special wisdom, taken for granted, which granted a direct and warm-hearted nature and concern for others. If afflicted, the person is selfish, deceitful, confused, and prone to addictions. Your intuitive insights are more powerful than you accept.

Neptune in Libra
(1942/43--1955/57)

Neptune in Libra grants a compassionate and caring nature while intensifying the imagination in literature, music and art. You are easy-going, preferring to be pleasant, congenial and to relate well to others. You are strongly and prophetically idealistic particularly about issues related to social welfare, humanity, war and peace. Neptune in the sign of love, beauty and balance of Libra predisposes a strenuous striving for establishing harmonious institutions and personal situations. You find yourself offended by injustices and inequities against people, humanitarian efforts, nature, the environment and social institutions. Along with your idealism, you possess strong powers of influence and persuasion, and your patience allows you to wait for the results. If afflicted, this person is overly sympathetic and emotional with a strong attraction to the opposite sex. You experience the mystery of meeting new, different and eccentric individuals who expose you to unique thoughts, perceptions and occurrences.

Neptune in Scorpio
(1955/57--1970/01)

Neptune in Scorpio intensifies the emotions inclining an individual to be either subtle or sensational. You find yourself fascinated by innovation, change, advanced technology and science, and are particularly interested in unlocking the mysteries of the universe. You can be persistent in your efforts, possessing a secretive nature that can be reserved but that at times flairs in anger. You have a love for the unknown whether it be unsolved phenomenon or delving into the secret lives and unsavory deals of public officials and celebrities. You are somewhat skeptical of accepted practices and policies, questioning the soundness of decisions being made by politicians and business interests. You have a strong attraction toward sexual relationships and this is reflected in your tastes in movies and entertainment. If afflicted, the individual is drawn to sensational activities, luxury, pleasures, drugs, alcohol and is given to lies, business deceptions, and an unscrupulous nature. You have a strong interest in what is unanswered.

Neptune in Sagittarius
(1970/01--1984/05)

Sagittarius in Neptune influences natives who are open-minded, direct and idealistic to seek higher spiritual and philosophical values. Freedom loving Sagittarius, you possess an inherent reverence, sound reasoning ability, determination and ambition. Left to your own desires, you would prefer to travel seeking knowledge and insights. You are farsighted and inspired by your dreams and visions. You prefer to explore the powers of the intellect, revising laws and mores, and searching for a truer and more meaningful outlook on life. You are curious and insightful regarding business affairs, art, science, theology, international affairs, literature, politics, and education. If afflicted, this individual is vague, indefinite, distrustful, overly sensitive, and restless seeking aimless change and travel; experiences unusual dreams, problems in travels, discord through religion, politics, or following the wrong cause and ideas. Your strong ideals, high purpose, and humane nature leads you to possess an utopian outlook on life.

Neptune in Capricorn
(1984/05--1998)

Neptune in Capricorn makes for a well-disciplined, practical outlook on life with an inclination for turning ideas into real achievements. You possess a strong faith in your sound reasoning ability and can be cautious and prudent but decisive. You are meticulously thorough in your effort to carry through in your endeavors, exhibiting strengths of courage and fortitude and the ability to see your plans through to completion. You find yourself questioning outdated practices, methods and accepted social institutions. You are insightful in professional endeavors and discover gain through the areas of art, music, large businesses, science, chemistry, medicine, and institutions both public and private. If afflicted, the individual is indefinite, secretive, and scheming, and experiences family difficulties, and numerous discords and complications in business affairs. You turn ideas and inspiration into standards and practices.

Neptune in Aquarius
(1998--2012)

Neptune in Aquarius enhances the intuition and innate perceptual abilities. Humanitarian Aquarius under the influence of Neptune is idealistic and desires social fairness. You possess an appreciation for nature, wildlife, and the environment which you would like to see preserved and protected for future generations. You are friendly and sociable while accepting others and being tolerant of different lifestyles and circumstances. You would like to see society protect and benefit the less fortunate. You have an independent outlook on life, not accepting conventional ideas and practices. You are creative and would prefer to build a better tomorrow than to destroy what exists today. If afflicted, the individual is overly independent to the point of being eccentric, suffers from problems in love and marriage, and is involved in endless problems with friends. Your ideas are progressively humanitarian and sympathetic with an interest in social, philosophical and political policies.

Neptune in Pisces
(2012--2026)

Neptune rules Pisces and, if well aspected, it lends dignity to these inspirational, spiritual, sensitive and even prophetic individuals. You are creative but with a serious, contemplative outlook on life that lends itself to profound thought and perceptions. You are accepting, sympathetic, and charitable, benefitting through giving to and receiving from others. You would love to see the world, and you are inclined to travel, seeking exposure to different cultures and practices. And through travel you seek to experience and to gain insight from the various and different auras found in natural settings throughout the world. You are creatively talented in music, art, drama, science and medicine and insightful in business endeavors. If afflicted, this individual suffers from personal psychic experences, and endures loss and misfortune in plans and endeavors. You are an understanding person who is not necessarily materialistic, and you may well lead the way in new cultural concepts.

Pluto

Rules: Scorpio

Keywords: Change, mutation, transformation, regeneration, reproduction, death, catastrophes, dictators, death, endings, resourcefulness, creativity and acceptance of change, enforced changes; a curiosity regarding the unknown; the unconscious/subconscious, abnormalities, unnatural events, natural disasters, the underworld, Hades, the Mafia, clairvoyancy, advanced science and technology, atomic energy, corporate enterprises, insurance, taxes, alimony, funerals, secrecy, mysteries, espionage, cruelty, sadism. It rules Scorpio; is in detriment in Taurus.

Pluto, the farthest planet from the Sun, was discovered in 1930 by Clyde W. Tombaugh at the Lowell Observatory, and was named by a young British girl whose letter was the first to arrive at the Observatory. Unlike the other planets, Pluto's orbit tilts, bringing it at times closer to the Earth than Neptune, and its complete orbit is larger and therefore slower than the other planets. It requires approximately two hundred and forty-eight years for Pluto to travel through the Zodiac, staying in each sign for somewhere between twelve to thirty-two years. Pluto has a generational influence as well as an influence on personal horoscopes.

Pluto was the Roman god that ruled the underworld and Hades, overseeing the spirits of the dead. The planet Pluto rules the subconscious workings of the body, change and transformation and regeneration. It represents both the reforming and the destructive forces on Earth. It is associated with the creative and regenerative influences, the reproductive system of the body, volcanoes, earthquakes, big business, mass media and communication, world finances, and the beginning phase of life as well as the last phase of life. The positive aspects of Pluto include the ability to make a new beginning under unfavorable circumstances and an understanding and ability in big business. It grants an analytical intellect and financial security. It influences the discover of mistakes and detection of wrong doings. Pluto instills new life into ideas and ideologies. If afflicted, the negative traits include a troubled subconscious, low morality, indecency, secretiveness, criminal tendencies, treacherousness and cruelty. Pluto, it is felt, represents the highest and lowest of which man is capable.

Pluto influences the dramatic changes that takes place in the lives of people, either as a group influence or as experienced by individuals. Sudden changes or shifts in location of groups of people, for example, or the sudden relocation of an individual. Upheavals, revolutions, and revelations are influenced by this sign as are the personal plans of an individual that get off to a quick start or are suddenly disrupted and brought to an end.

Pluto may well be that influence that bridges the material and spiritual world, but it will require more time for astrologers to fully understand the influence of this planet.

Pluto in Aries

Pluto in Aries indicates exploration, reformation and a drive for power or, if adversely affected, revenge. These individuals possess much daring, energy, and drive. Changes in political, social, scientific and economic situations are indicated with revolutionary and innovative ideas marking this period of endeavor. The individual possesses courage, will power, a strong self-reliance, individuality, and a powerful belief in one's ideas and abilities. Energetic and driven to succeed, this individual is imaginative and resourceful. Pioneering new thoughts, innovations and directions will be undertaken with foresight and an energetic drive to succeed. This person will find it necessary to strive to develop a self-discipline in order not to diffuse the natural energies and strengths. The individual is expansive, progressive and broad-minded, striving for a leadership position in all endeavors.

Pluto in Taurus

Pluto was last in Taurus from 1851 to 1883. Pluto in Taurus grants endurance, perseverance, sensuality, and an obstinate stubbornness. This individual sticks to his or her goals and objectives seeking completion of the tasks at hand. There is an over powering need for stability, permanence and security as realized through material gain and possessions. In historical times, it was noted that the influence of Pluto lead the wealthy to exploit and take advantage of the working poor. Pluto is in detriment in Taurus, and while Pluto in Aries brought pioneers and adventurers, Pluto in Taurus promoted industrialists, investors and financiers who built complicated and complex social and political institutions. Pluto located in Taurus grants determination, drive, and the endurance to succeed at all endeavors, but it influences extremes in ideas and emotions, promoting either wealth or poverty, success or failure. Taurus is steady and tenacious, but Pluto indicates changes and abrupt starts and endings to plans, endeavors, relationships, and encounters. Taurus brings a desire for permanency and lasting values and relationships, but these are gained only through hard work and great effort.

Pluto in Gemini
(1882/84--1912/14)

Pluto was last in Gemini from 1882/84 to 1912/13 seeing a need for change, and indeed change was noted in many aspects of life including communication, transportation, and technologies. The Air sign of Gemini saw the realization of mass communication through newspapers, radio, telegraph and the telephone. New ideas and ideologies were promoted, and old ideas, customs and habits were put aside. Pluto grants Gemini natives a depth of character but a restless, seeking and curious nature which wants to experience and to explore. And Gemini individuals influenced by Pluto want the freedom to express their ideas and thoughts. This person loves family and friends, but can be changeable and restless, seeking new experiences and forms of expressions. There is a desire to expand the intellect and to feed the mind. This is a sensual person who appreciates the beauty found in nature, and is forever pulled to see and meet new people, and to experience new places and different settings.

Pluto in Cancer
(1912/14--1937/39)

Pluto was last in Cancer from approximately 1912 to 1939. Cancer possesses a love for home and family, and this period saw a drastic rise in the world population as well as the drastic changes brought about by a sudden relocation of people from country to cities. This individual possess a need for security through home, family and through personal and supportive friends and neighbors. There is indicated a strong social awareness, an appreciation for wildlife and nature, and a fondness for artistic pursuits and creative past times. Cancer is instinctive, intuitive, and highly creative and imaginative. This native is also intensely emotionally and craves the attention and recognition of friends and family. If afflicted, this person tends to isolation, pride, selfishness, overly sensitive, and is jealous of the success, possessions or attentions of others. Persons born during Pluto in Cancer are bound to effect traditions, society, and social institutions preferring conservative politics that favor the home and family. Warm-hearted with a quick sense of humor, this native creates a home atmosphere that pleases.

Pluto in Leo
(1937/39--1956/58)

Pluto in the forceful sign of Leo saw WWII and its after effects. There was an unprecedented number of rise and falls of governments throughout the world. The formation of the United Nations brought about an institution that allowed a forum for the discussion of ideas, social order and the development of third world countries. This native is self-confident, authoritative and possesses a good business ability, but loves power, prestige and leadership positions. The native has a strong sense of personal pride as well as a pride in his or her country and beliefs. The domineering ego of Leo brings austerity to the drive for success and completion of endeavors. If afflicted, the individual can be selfish, arrogant, prideful, ego-driven and pleasure seeking with an obstinate and over bearing nature. Persons born during this period have experienced many changes, successes, and reversals in their lives both personally, socially and economically. Pleasures, enjoying life, and making heroes out of personalities marks the influence of persons born during this period. It can also be said that this generation produced any number of important leaders who influenced the world.

Pluto in Virgo
(1956/58--1971/72)

Pluto in Virgo influences the individual to be analytical, inventive, technical, and detail-oriented. You are practical in the sense that you acquire all the details necessary to complete your projects. You possess an interest in matters related to science, medicine, mental and physical health, diet, and the upkeep of the body. You posses the ability to bring order to a situation. There is a tendency to over analyze and to be critical and discerning. Afflictions bring dependency on alcohol and illicit drugs, a self centered outlook on life, concern for self coming first above all others, a suspicious and selfishness nature, fault-finding and critical, and pleasure seeking to a destructive degree. You most seek perfection in all that you do and prefer that sensible ideas also be reflected in society, institutions and in government and politics. Your obstacle overcoming an overly cautious, detail-oriented nature that may prevent you seeing the bigger picture and from sharing in life with others.

Pluto in Libra
(1971/72--1983/84)

Libra seeks balance and harmony, and Pluto in this sign sees persons who are strongly aware of injustices and inequalities. These individuals prefers that people are treated as equals, including themselves. You want laws, rules and regulations to be fair, and not misused, and you find that many laws actually limit the freedoms of individuals. You exhibit empathy and concern for other people and for groups of people and abhor the misuse of power and laws for personal gain and to restrain others. You strive for new approaches to relationships, justice, prison reforms, politics, and especially to international relationships. You are indeed adaptable and flexible with an appreciation for beauty and harmony that transcends into art and music. You prefer to be direct and honest in relationships and expect the same from others. Becoming upset when you discover dishonesty, you are capable of leaving and seeking harmony elsewhere. You are also capable of ignoring the harsher, more distasteful aspects of life, preferring to shut them out of your life and to avoid stressful and painful situations. You discover that the greatest challenge in your life is resolving conflicts in your personal relationships.

Pluto in Scorpio
(1983/84--1995)

Pluto rules Scorpio and moves rather rapidly through this sign in only eleven to twelve years as compared to up to thirty years in other signs. In fact, for a period of time in 1989, Pluto, in its tilted orbit, came closer to the Earth than Neptune. The Pluto/Scorpio influence grants an individual strong will power and a curious, penetrating intellect. Your relationships are intense, sensual, and passionate, and you are not sexually inhibited. The conflict in your life centers around jealousies, and rivalries, and you have a tendency to be private even to the point of being secretive. Your curious, fact finding nature leads you to want to uncover the truths in situations, politics, and institutions. You are imaginative and insightful and may discover that you have some psychic abilities. You possess a strong interest in the environment and appreciate the beauty of nature. You can be aggressive when necessary, and will push forward your ideas and objectives forcefully even ruthlessly at times. Life to you is a mystery, and you feel that mystery must be examined and understood.

Pluto in Sagittarius
(1995--2008)

Pluto in Sagittarius finds a person who loves his or her personal freedom and wants to preserve it. It is expected in this period, that individuals will strive for new ideals, new philosophies, and new ways of dealing with laws, rules and regulations that would enhance the freedom and independence to think and to move about freely. You are humane, expansive, friendly and open to new ideas and innovations. There is found a spiritual inclination to your nature, and you seek a greater awareness of being. You are energetic, enthusiastic, flexible, versatile, and tolerant of different ideas and lifestyles. The negative aspect is the tendency to become eccentric in ideas and lifestyle, to follow an idea or thought to extremes, and to be indecisive in judgments. You may possess revolutionary, new ideas, but must guard that your wanderlust doesn't deter from the implementation of your ideas.

Pluto in Capricorn
(2008--2024)

Pluto in Capricorn finds a person who is persevering, ambitious, efficient, and who seeks to establish order by organizing and managing well. Capricorn finds security through establishing order, by attempting to preserve systems that work well, and by innovating new systems to replace outdated ones. Capricorn seeks to protect and even to insulate or isolate in self-defense. You possess a conservative tendency and indeed would like to conserve energies, preferring to utilize what is necessary to produce the best end results. You are materialistic in the sense that you want a secure home, financial security, and possessions to enhance your personal situation. You also possess a strong spiritual nature, preferring traditional and sensible religions and doctrines.

Pluto in Aquarius
(2023--2044)

Pluto was last in Aquarius from 1778 to 1798, and this period saw the American Revolution, the French Revolution, and the revolution of new ideas which produced the Constitution and the many French writings which called for instilling freedoms into everyday lives, a casting off of the old ways and an inaugurating of new ways. You are the intellectual who seeks to understand and to instill new humanitarian means and methods into our daily lives which promote freedom, understanding and acceptance. You are unconventional, ingenious, and you love your personal freedom. You are capable of inspiring others to great thoughts, and when called upon, you become a great leader yourself.

PLUTO IN PISCES
(2043--2068)

Pluto was last in Pisces from 1798 to 1823, which was the Romantic period in art and literature. This is a compassionate and caring person who is willing to sacrifice for the good of others. You are intuitive, creative and imaginative, and if left to your own tendencies will produce even more innovative changes in art, music, literature and the philosophies of this period of time. You are adaptable, flexible and tolerant of the ideas, thoughts and lifestyles of others. You adapt easily to new situations and new people, and have a strong desire to meet new people and to make friends. You must guard against be easily impressed, influenced and led by others. Intuitive and introspective, you must also guard against procrastination and living much of your life in the fantasy world of your mind. You are a spiritual, insightful person who may discover great gifts of psychic and prophetic abilities.

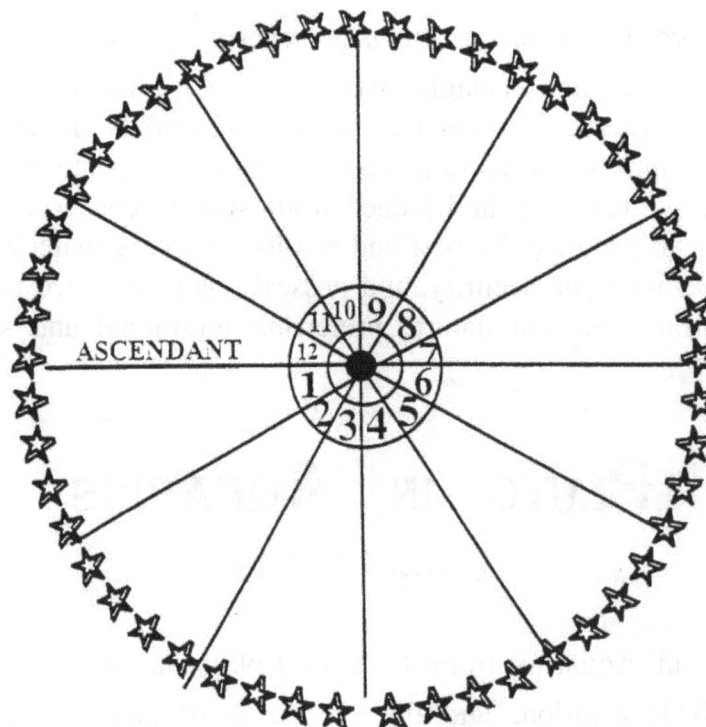

THE MOON SIGNS

The Moon reflects the light of the Sun, and in your horoscope it reflects your inner feelings and urges. The Moon sign is considered the second most important aspect of the horoscope next to the Sun sign. The Moon influences the individual's habits, reactions, emotions, moods, instincts, and the subconscious. Your hunches and intuitions come from the influence of the Moon. The waxing and waning Moon represents duality and change as in the ebb and flow of the waters of the Earth which produces two tides a day in the oceans. It exerts a strong influence over the daily and ordinary affairs of life as well as the cyclical nature of life.

In ancient times, the Moon was called Isis, Esses, Luna, Eleusis and the Virgin Nere. While the Sun remains unchanging, the Moon is constantly changing but is predictable in its orbit. The Moon completes its transition through the Zodiac faster than any of the other Celestial bodies, making a complete cycle in twenty-seven days, seven hours and forty-three minutes, staying in each sign for two to three days. As it orbits the Earth, the Moon's appearance changes, and in astrology the New Moon is in conjunction with the Sun, the First Quarter is ninety degrees of the Sun, the Half Moon is illuminated, the Full Moon is in opposition to the Sun, and the Third Quarter is ninety degrees west of the Sun.

The Moon is receptive, sensitive, passive, and feminine, representing women, mother, and child birth. It is nocturnal, cold, moist, phlegmatic and fruitful and rules the digestive system, stomach, sympathetic nervous system, breasts, and the left eye of a male and the right eye of a female. The Moon rules the public, commodities, liquids, nutrition, the home, family, ancestors, responses, and the memory.

Your Moon sign indicates where the Moon falls on the date of your birth, and the Moon sign indicates your instinctive behavior. The positive Moon traits include being passive, patient, tenacious, imaginative, sensitive, and protectively maternal, while being sympathetic and receptive but shrewd in business and possessing a good memory. The negative Moon traits shows an inclination to be moody, overly sensitive, stand-offish, changeable, exhibiting a poor reasoning ability, unreliable, gullible, narrow minded, and unforgiving.

Your inner self, your dreams, fantasies and private feelings are all indicated by your Moon sign.

Moon in Aries

The Moon in Aries influences an individual to have an instinctive and noticeable "me first" tendency. The quick temper can be erratic and fluctuating with an irritable and irascible nature. While Aries is fiery and energetically active, the Moon is passive and cool, resulting in nervous tendencies. There is found, however, a strong imagination and a nature which is positive, masterful, independent, and self-reliant. You can be courageous, practical, and sincerely enthusiastic in your endeavors, but there is a tendency to be restless and changeable with an unwillingness to take direction from others.

This individual possesses a strong inclination to be aggressively persistent, but at the same time can be impulsive, occasionally making rash decisions. Your impatience marks many of your decisions, and rather than being tied down with considering details, you will follow your own ideas and impulses. Your energy and independent nature leads you to attempt at all cost to achieve through action. You are known for being enterprising, original and inventive. Your plans are no one else's but your own, and you pursue them with all the energy and drive you possess. At the same time you are implementing your own ideas, you strive to be in charge, and will relentlessly take charge of whatever you become involved in. You can be head strong in this manner, and you have a tendency not to listen to the advice of others.

You are vigorous and direct in your methods, protecting what is yours, especially your emotions. You love travel, meeting new people and becoming involved in new plans, many of which are of your own design. You move through people collecting friendships and relationships as are necessary to augment your plans, your ego, and your feelings. In a relationship, the other person must reflect what you most admire and desire, and that is making a lasting impression on others. It would appear that you use other people in your attempt to complete your plans, but it may well be that you simply see your plans as all important to the situation at hand.

If well-aspected, the individual is energetic, enterprising, optimistic, open to new ideas, and possesses strong values and ideals; will attempt to succeed at his or her own ideas and will attain a leadership position. If afflicted, this individual is rash, opinionated, domineering, impatient and egotistical; encounters danger through travel and water, problems with the opposite sex, and numerous changes in his or her career. The nervous disposition can bring on health problems related to headaches, indigestion, and blood pressure. Physically active, the native can suffer from accidents.

You are socially uninhibited, and possess a strong preference for being in the middle of a social scene, being noticed and admired. Outgoing, decisive and self-confident, you can delight and inspire others. Socially, you are congenial, generous, witty and humorous, well-informed, entertaining and more than likely you stand out in a crowd. You have a great love of adventure and for personal freedom, and you desire new beginnings, new experiences, and new endeavors.

Moon in Taurus

The Moon in Taurus indicates a person who is sensual but possessive, loyal and determined, and obstinate with a strong need for security. Your emotional security depends upon possessing and protecting that which you value whether it be material possessions, home, or ideals. You are courteous and sociable with a strong preference for friends and family, love and marriage. You have a strong determination and prefer not to be undermined in your efforts to achieve your goals. You are probably conservative, and if not you still hold strongly to your own ideas and concepts and strongly resist any obstacles or forced attempts to change your mind. Ambitious and determined, you aspire most to excel at your endeavors. This is a stable position for the Moon, and the ever changing Moon influence becomes more stable and reliant, lead the native to be persistence rather than impulsive. You outlook on life is positive, energetic and upbeat. And the Moon in this sign influences the native to be financially successful.

You possess a strong self-image and project it to others. This native can be either enterprising or lazy, and either way, the person thinks well of himself or herself. Friends and companions are easily made because the native is as accepting of others as he or she is of self. You are considered sympathetic and good hearted with a sincere concern for others. You possess a strong appreciation for pleasure, luxury, music, art, and enjoying life.

Not always a self-starter, once this native begins on a path, the natural tenacity of Taurus takes over and this individual becomes determined to follow through to the end of the endeavor. You are intuitive and generally speaking, you exhibit sound judgment making reasonable and sensible decisions. However, not being easily influenced, you are known for your stubbornness, and you face down any opposition fearlessly.

The Moon in Taurus influences the native to be concerned with family and home, and there is often found a strong interest in relatives, ancestors and the family history. You also possess a strong business sense, and you make, save and invest money purposefully and sensibly. You appear to possess a rather special intuition in business being able to judge and to perceive coming trends and fashions.

If well aspect, this individual is caring and compassionate and will succeed in endeavors connected to nature and related products, chemistry, business, water-related industry, speaking or singing careers, real estate, and the food industry, and there is some indication for gains through a relationship with the opposite sex. If afflicted, the native is overly possessive, stingy, selfish, lazy and concerned with self-indulgence and earthly pleasures. There is some indication for health problems related to the throat.

The Moon sign of Taurus indicates a person who is trustworthy, devoted and tenaciously determined. You are known for being warm and affectionate with a strong appreciation for nature and the beauty found in art and music.

Moon in Gemini

The Moon in Gemini signifies a strong urge for this native to strive to communicate effectively. You are energetic with a creative and imaginative intellect, and a tendency to be restless and to seek change. You are best known for being agreeable, sociable, sympathetic and warm-hearted with a caring and humane nature. You can be reserved at times, but you are a progressive, ingenious and innovative thinker who has a well-developed intellect and who is well informed. You love information, literature and science and seek out new information from books, the Internet, or other sources. The duality of your nature is expressed in your changing interests and preferences.

Your great need for variety and your curiosity leads you on any number of adventures, some daring and some simply complying with your natural love of other people, places and lifestyles. Your quick responses to ideas adds to the variety in your life, making you flexible, ever changing, and fluctuating in your ideas and activities. You are seemingly most capable of reflecting any number of ideas with the same enthusiasm and intensity. You respond intimately to the moment at hand, but you possess a strong desire for the novel and new situation, losing interest in situations requiring any depth of emotion, long-sustained feelings or undivided loyalties. While you exude a fantastically warm personality, you can appear at times insensitive or uncaring.

At the same time, you are perceptive, and you know from experience that your perceptions and intuitions are often correct. You innately respond to situations and to people impersonally through your intellect rather than through your emotions. You accurately and rapidly size up situations, people and ideas, and with a critical and analytical perception, you determine for yourself which direction or decision is best suited for you. Your ability to analyze and your curiosity leads you to self-examination, and you make every attempt to understand your actions and your thoughts. The Moon in Gemini produces excellent writers, psychologists, artists, or other professionals and tends toward the fields of communication.

You are a free spirit in your relationships as well as in your intellectual pursuits. You have a strong desire to maintain your freedom of movement and thought. And your outgoing and personable energies propel you into ever new and changing situations. You generally prefer discussions to quarrels and upsetting situations.

If well aspected, this native is an amusingly witty, charming, lively and entertaining conversationalist with a curious and strong intellect and the ability to succeed in his or her career. If afflicted the native is overly indecisive, changeable and restless, disorganized, inconsistent, superficial, cunning, manipulative, not cautious, and has a tendency to be drawn into questionable situations.

The Moon in Gemini inclines you to numerous changes and frequent travel, and your dual nature leads you to be involved in more than one endeavor at a time. This strong preference for being involved in numerous activities should be accepted as it is a part of your nature. You are a caring and warm individual who responds well to your immediate family and you possess a love and affinity for nature and for children. You appreciate others who are intellectually challenging.

Moon in Cancer

The Moon rules Cancer granting the individual strong maternal feelings and a sympathetic nature. Passive, affectionate, gentle, and peaceful, you are a romantic at heart, preferring good feelings to the harsh realities of life. You are conscientious, humane and emotionally sensitive. You can be at times too passive, preferring to follow the path of least resistance, and there are times when you naively trust your own feelings and judgments rather than following the advice of others. You have a strong need to cherish and protect what is yours whether it be family, home or possession. You have a strongly developed domestic urge and prefer the stability and security of the home. At the same time you love to travel, to visit and see new places and peoples.

You possess an innately strong desire for supportive friendships and relationships, and become devoted to friends, associates and to family members. Being sensitive, you are often defensive and have little appreciation for criticism no matter how constructive. To you, being admired is to be loved and appreciated, and that is what you seek in life. While you are strongly devoted and supportive, you can become dependent on the other people in your life--needing others in order to feel emotionally secure and whole. You can be somewhat territorial, building walls of defenses and protecting your personal space from outside invasion. You appear to need your private space which you withdraw to for your own personal time.

Your mind is meditative, and you feel that need for your own place to think. Also, you are creatively imaginative and your personal place is where you prefer to dream away your own time in the fantasy world of your own creation--safe from the intrusions of the harsh and demanding world. Intuitive and sensitive, you pick up negative feelings as if from the air, and can become moody. Along with this strong intuitive ability, you also possess psychic abilities which you may or may not choose to develop.

The Moon in Cancer favors occupations that deal with the public, science fields, the food industry, shipping, antiques, art, music and acting. If well aspected, the native is creatively imaginative, sympathetic, protective, tenacious, loyal and devoted. If afflicted, the native is overly sensitive, emotionally unstable, critical, selfish, possessive, moody, and overly self-pitying with a tendency to nag others.

With the Moon in Cancer, you perceive your surroundings and other people through your emotions rather than your intellect. Romantic, intuitive Cancer is in harmony with the sensuous and receptive qualities of the Moon, making you a gentle, devoted and loving individual.

Moon in Leo

The Moon in Leo and you are ambitious, self-confident, and with your naturally good disposition, you reflect with radiance the attentions of others. You are exuberant, energetic, and self-reliant with a strong tendency to be loyal, trustworthy and honorable. You call attention to yourself through your personal creativity and your magnanimous personality. You are open-minded and warm-hearted with a generous and caring nature, but you have a tendency to be most concerned with those matters which directly apply to your life and well being.

You have a dramatic flair and stand out in a crowd, loving to attract attention and to be the center of attention. You excel in the adoration of others. You have a love of pleasurable situations but also an inherent need to control your personal situation. You are persevering with a lively and penetrating intellect. You possess strong leadership abilities and the capacity to organizing and to apply sensible methods to your endeavors. You make quick, accurate and intuitive judgments regarding other people and their motives.

Adventuresome and daring, you have a natural appeal with the opposite sex and enjoy much attention, praise and adoration from others. In love, you are sincere, loyal and devoted, generous and affectionate, but there is a tendency to hold your emotions in check and you may find it difficult to express your emotions well. And when you feel unappreciated by your loved one, you will stray. You are known for being particular in dress and style, preferring a tasteful and fashionable appearance. You have a fun-loving nature with a preference for sports, social gatherings, pleasure, music, and art.

If well aspected, the influence of the Moon grants you trust, respect and a responsible nature that does well in leadership positions. If afflicted, the native is ostentatious, domineering, self-indulgent, self-centered, and conceited. Generally speaking, the Moon in Leo brings positions of authority and responsibility which you handle seriously.

Moon in Virgo

The Moon in Virgo enhances the native's intellectual abilities, and the steadiness and practicality of Virgo stabilizes the Moon's changeable nature. The Moon influences Virgo to be more flexible, and these natives use their natural resources to produce better and more efficient results. You are receptive of new information, careful of details, and possess a good memory. You are analytical and discriminating with a strong preference for arranging your life the way you want it, generally neat and orderly. You prefer to put to good use the knowledge you acquire through your intellectual, critical, and analytical process, but you may have a tendency to worry needlessly.

You are not the ostentatious center of attention, but rather the reserved, unpretentious individual who strives ambitiously to achieve on your own merits and talents through hard work. You do respond to appreciation and praise for your responsible and sensible work methods. Your executive abilities make you a natural manager of your career, business or home life. You are steady, reliable and ingenious with an innate interest in diverse aspects of life such as science, nature, politics and world affairs.

Although reserved, you like other people and are generous with your time and attention, wanting to help others as best you can. You love to talk and to share ideas with others, gather information, and listen to other viewpoints during lively discussions. But you stick steadfastly to your opinions, and at the same time, you apply your detail-oriented outlook to the lives of others and can't quite understand why people don't apply common sense to their lives. You may try to enlighten others to the best way, and can sometimes be insistent that you know what's best. That aside, you collect friends easily, especially of the opposite sex, because other people know you are dependable and can be relied on.

If well aspected, this native is ambitious, intellectual, imaginative, meticulous, responsible, and likes changes, travel, investigating and collecting information. If afflicted, this native becomes obsessed with details, not being able to arrive at a sensible conclusion, is overly critical and fault-finding, argumentative, snobbish and hypochondriac tendencies. There is indicated numerous short trips and changes in the life of the Moon-Virgo native, and proper diet becomes very important to the health.

MOON IN LIBRA

The Moon in Libra favorably influences partnerships and popularity. You possess a charming personality, and you are naturally courteous, congenial and diplomatic with a sincere desire to please and to be accepted by others. The Moon's romantic inclination is enhanced granting you an appreciation for beauty, nature, art, poetry, and literature. You delight and enjoy social gatherings, pleasurable activities, and the companionship of others. You have a tendency to put your best foot forward and win appreciation through your considerate, warm-hearted and affectionate nature. You prefer to be agreeable, easy-going, friendly and kind with a joyful and fun-loving outlook on life.

The most important experiences in life for you are those that are shared with others. The balance and harmony of Libra leads you to be open minded. You attempt to rationalize both sides of a conflict or even a discussion, remaining neutral regarding various opinions, outlooks, or even lifestyles. As far as you are concerned, other people must decide what to do with their lives. You are more than capable of gracefully using tact and diplomacy to maintain a peaceful situation. Accepting and tolerant of others, other people naturally gravitate to you for you are an easily likable person. Your greatest challenge is not to be easy influenced by others or to be susceptible to flattery and praise. You must learn to follow your own direction, and on occasion to say no to others. Your most important challenge is to develop self-reliance and independence. If afflicted, this individual is whimsical, changeable, erratic, flighty, dependent, self-indulgent, indecisive or lethargic and overly critical.

You have a strong inclination toward love and marriage, an appreciation for pleasant surroundings, elegance and refinement, and you like nice things whether it be fashionable clothing or stylish furnishings. You also prefer that a certain amount of convention and formalities are observed in everyday life. You would, in fact, prefer if all the world was a pleasant and congenial place for everyone. That not being the case, you prefer to dwell on the more pleasant aspects of life yourself. You may find that you have a tendency to live for the moment not giving much attention to long range plans and goals. This works for a time, but then you must settle and make decisions. Intellectual, you experience beauty and sensations through the mind, and are capable of evaluating, criticizing, analyzing, or appreciating what you sense.

Moon in Scorpio

The Moon is in its fall in Scorpio, and this individual is intensely emotional, imaginative and forceful, but sensuality is sublimated. A reserved and somewhat reticent individual, you have a tendency to hide your deep emotions and feelings, preferring to hold them in check and to control them. While others discuss and display their emotions freely, yours are personal and you keep your innermost thoughts to yourself, even your dreams and fantasies. Then too, you are enterprising, determined, practical and possess a strong will power. Your forcefulness, magnetism and self-confidence allows you to succeed in your endeavors.

You are observant with a keen intellect and perceptive mind which allows you to size-up other people and situations accurately and shrewdly, and to use that information to manipulate or control the situation. Your determination and stubborn persistence makes you a reliable and hard worker who is willing to take on difficult tasks and follow them to completion. You may prefer your private space in which to work, away from the comings and goings of other people. And generally, you aren't easily influenced or swayed by the opinions of others.

You are a highly sexual person with a healthy libido, and this aspect of your life is very important to you. You also appreciate pleasures, fun, comfort, and the companionship of others, and possess the capacity for pure enjoyment of life which only others of a like mind realize. For the most part a positive individual, you don't appreciate opposition from others, and with concise and cutting words can put a contender in his or her place. You have a tendency to be unforgiving and even revengeful, and these become traits which you strive to control throughout life. Energetic, masterful, and aggressive, you are also known for your courage and daring willingness to face adversity. If afflicted, this individual is moody, impatient, jealous, revengeful, secretive, intolerant, obstinate, domineering, self-indulgent, with a weakness for drugs and alcohol; problems are indicated with the opposite sex and in marriage.

You have a strong interest in the mysterious and are adept at locating information. You don't appreciate being inconvenienced, but at the same time you are known for being kind, generous and willing to make sacrifices.

MOON IN SAGITTARIUS

The Moon in Sagittarius and there appears to be no boundaries to the heights or realms of your intellectual pursuits. Forever a philosopher, you seek ever more information and will either travel frequently seeking new adventures or experiences, or if tied to one place, your mind will travel great lengths. You are emotionally idealistic and are forever exploring and expanding your wealth of knowledge, experiences or ideas. Accepting and tolerant, you are open and friendly with others, and can merge gregariously within the group, accepting possibilities and knowing few limitations, especially in thought. You are freedom loving, searching, and restless. You are optimistic, cheerful and an inspiring and well informed speaker.

Sociable and kind-hearted, you have a generous nature and a willingness to help a friend in need, no matter the inconvenience to yourself. You are the benevolent humanitarian whose charitable instincts know no limits. Optimistic with a good sense of humor, you can be jovial and exuberant, but are also frank and outspoken. When others attempt to limit your freedom, you become easily angered, but you are quick to forgive. You are an energetic person, either physically or mentally, and this is apparent in your restless mannerisms. You prefer to have some form of physical outlet for your energies and usually have a favorite sport or activity.

You have a love of nature, beauty and harmony, and may find that you prefer the wide open spaces that allow your mind and soul to soar. Your perceptions are clear and concise making you insightful, intuitive, and at times prophetic. You find your strength in your independence, and you want no limitations placed on your freedoms. If afflicted, the individual is careless, reckless, overly restless and changeable, too optimistic and extravagant, irresponsible, and indecisive.

You can be enlightening, informative, persuasive and inspiring, and may be called upon for public leadership or speaking. There is an indication for involvement in spiritual, educational, or political reform, and an inclination to receive benefits from the opposite sex. Your dramatic flair can make you as excellent sales person, speaker or leader. More serious and sensitive than your outgoing nature indicates, you may well possess psychic abilities and premonitions.

Moon in Capricorn

The Moon in Capricorn is in its detriment resulting in the emotions being well disciplined and restrained. You are a reserved, cautious and prudent person who prefers to conduct himself or herself in a dignified manner. You display common sense and practical abilities which are exhibited in your dutiful manner. In fact, your responsiveness is displayed through your responsible nature, and you express your thoughts and ideas clearly and rationally. Poised, charming, and ambitious with strong administrative abilities, you are drawn to positions of leadership and authority and do well in public positions.

Your reserved nature may be a reflection of a certain shyness or reticent nature on your part, but you react quickly to your perceptions and intuitions. Your strength lies in your inherent desire to obtain success through power and leadership positions. Your drive and intent can appear austere, cold and calculating to others, but it is simply a reflection of your inherent nature and abilities. And it may be that your fear of failure drives you to success. At the same time, your ambitious nature is focused through caution and prudence with the security of the home remaining important to you. You marry judiciously preferring a spouse who is ambitious and socially aware, economical, responsible, and who likes a well-run home and family routine. You are not adverse to marrying in order to improve your social status or to benefit your business or career. At times self-doubting, you desire the appreciation and admiration of your family, friends and associates.

If not well-developed, the native's natural ambition is overridden with fears and anxieties which lead to indecisiveness. If afflicted, this person is overly sensitive, antagonistic, fanatical, obsessive, morbid, brooding and melancholy, insecure, cold, austere and unsympathetic; self-indulgence and problems with the opposite sex are noted. Enemies cause problems whether deserved or not. If well aspected, leadership and administrative ability is augmented with the native receiving awards, honors, and positions of prominence. This native achieves either public popularity or public notoriety.

You accept responsibility, apply yourself to your goals, and may earn success through hard work, commitment and persistence. You find that you desire practical knowledge that you can put to use effectively.

Moon in Aquarius

The Moon in Aquarius finds a person who can be unconventional and who values personal independence and freedoms. Creative and intuitive, you have unusual, original, and progressive ideas. Sensitive and perceptive, you are a humanitarian with an open and tolerant acceptance of others. You are rational, logical and altruistic in you outlook on life. Now, add to all that your visions, ingenuity, imagination and inventive abilities. Insightful people accept you in spite of your outspoken frankness and any misdeeds or mistakes you've made along the way.

Friendly, sociable, and kind, you can be a witty, charming, and expressive companion who sincerely and genuinely likes other people. Gregarious, your friendships are important to you, but at the same time your sense of security is based on your independence and you can appear emotionally detached. Energetic and active, you divide your time between socializing and spending time by yourself either in contemplation or pursuing one of your many interests. Male or female, a friend is a friend until that person attempts to place limitations or restrictions on your independence, freedom or privacy. Chances are you'll back away from that person for awhile, but will accept him or her back in your life after time has erased the threat of restrictions.

You react almost simultaneously with your intellect and emotions to perceptions and sensory input taking into consideration the human element in a situation. Visionary and future thinking, you gather information to you whether it be scientific, artistic, musical, mathematical, or literary. Your interests appear unlimited. You are also interested in the mysteries of life and more importantly the mystery that encompasses the future. Most of all, you would like to be a beneficial influence on humanitarian or social causes. You are drawn to teaching, social work, counseling, science, inventions, politics, communications, the arts, astronomy and astrology. If afflicted the native can be overly independent leading to loneliness, aloof, indifferent, suffering from nervous tension and depression, opinionated, erratic, unpredictable and tactlessly harsh and outspoken; indicates problems with friends, the opposite sex, and needless wandering.

You are the dreamer of dreams, capable of producing ideas and philosophies, and your success is discovered through your magnetic charm and insight.

Moon in Pisces

The Moon in Pisces makes you a highly responsive, compassionate and considerate person who responds to the needs of others. Sensitive and intuitive, you respond emotionally to perceptions and may be easily hurt or offended by indifference or insults. A loving person who is loyal to others, you are easily disillusioned when you discover the faults and failings of others. You can be benevolent and quiet or charmingly outgoing and gregarious with a strong appreciation for beauty, comforts, luxury, harmony and change. Your nature is somewhat restless in that you are seeking, but you are not always sure what it is you seek. You appreciate new sights, sounds, experiences, and meeting other people.

You are naturally talented, imaginative and creative with an insightful and intuitive intellect. You enjoy time to yourself to lapse into your creative, fantasy mode, but you are drawn to other people. You are so intuitive and perceptive that you can feel the emotions of others which has a tendency to drain you emotionally. The sometimes cold, objectivity of the human condition also confuses you because you prefer harmony and an easy-going life for one and all rather than difficulties and strife. You are at heart an optimist and romantic.

You can be hard working and industrious or easy going to the point of being lazy and restless. You have an appreciation for beauty, nature, art and music, and your works of self-expressions are prolific. You are not afraid of working diligently for a worthwhile cause or goal, but the outcome or goal must be attainable and in clear sight. You must guard against being easily influenced by other people into changing your direction or leading an unproductive life. If afflicted, this native is indecisive, moody, discontent, easily discouraged, overly sensitive, gullible, pleasure-seeking, self-indulgent, vague, secretive, and easily lead and confused. There is a tendency for sorrow, obstacles and self-undoing brought on by the actions of the native or by indecisiveness or self-doubts. You must guard against allowing the criticism of others to undermine your creative efforts and abilities.

You are inspirational and receptive with a tendency to be gifted with psychic or mediumistic abilities. Utilizing your many talents and creativity is your greatest aspiration.

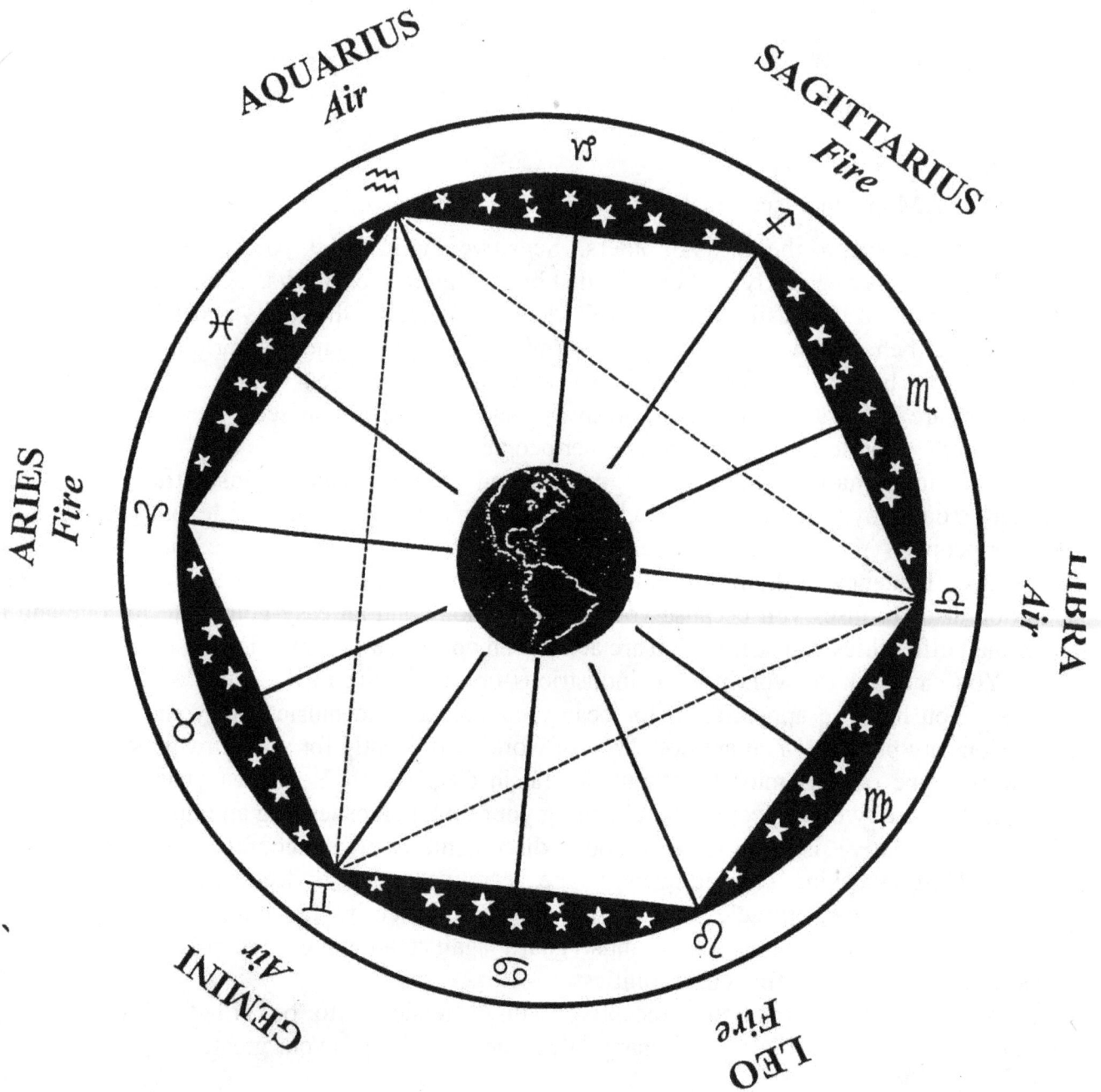

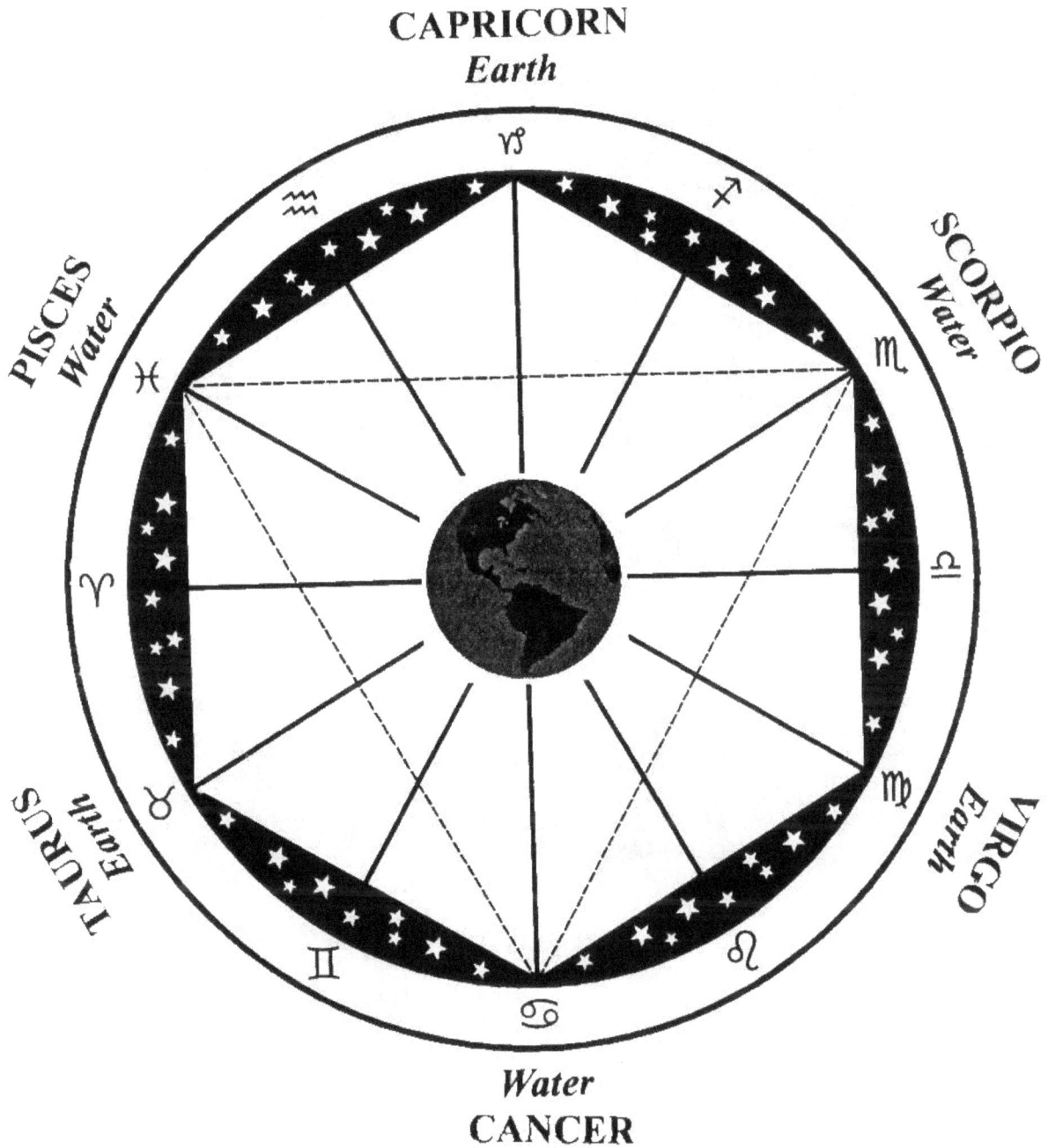

CAPRICORN
Earth

SCORPIO
Water

PISCES
Water

VIRGO
Earth

TAURUS
Earth

Water
CANCER

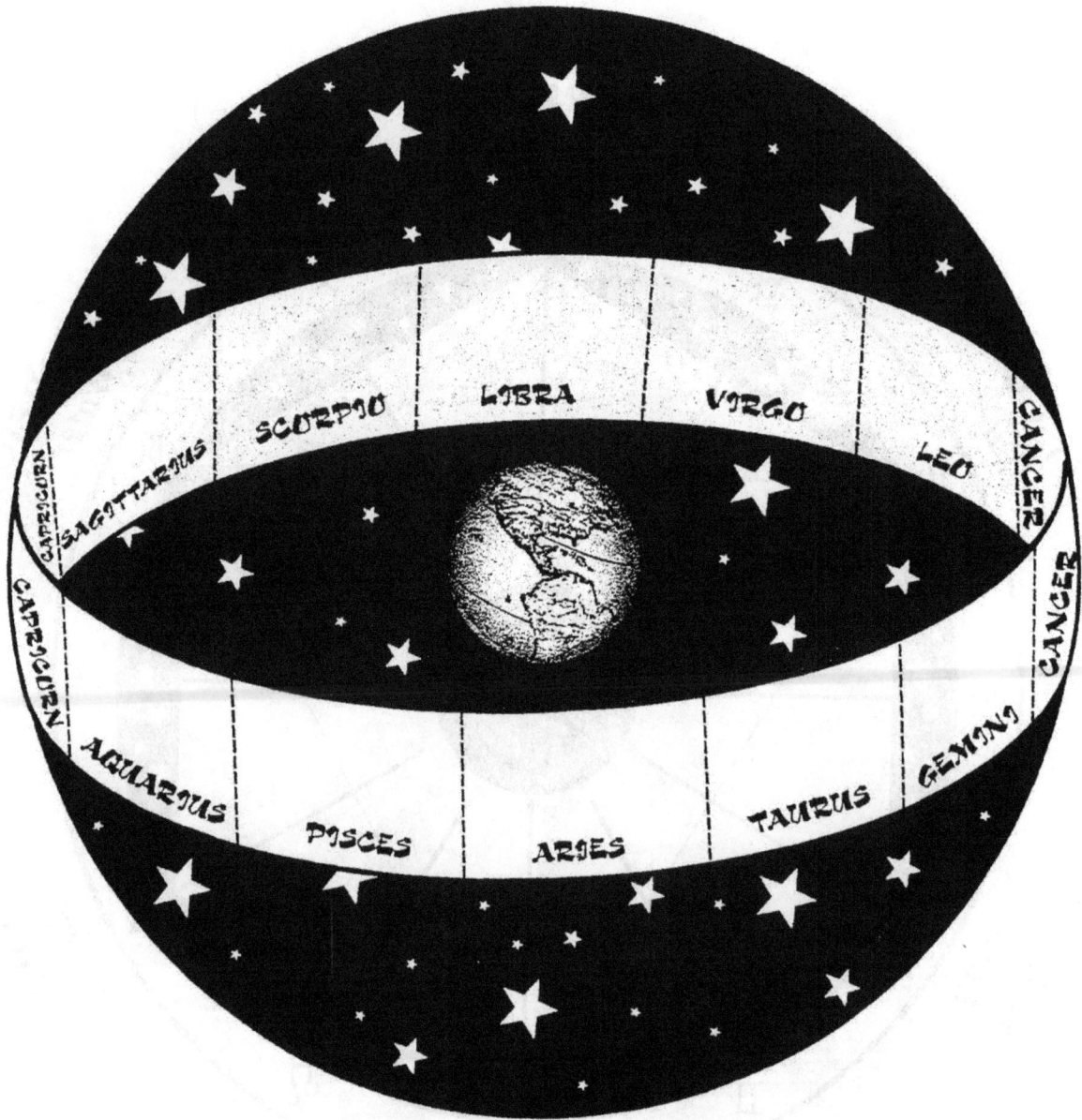

1900 MOON SIGNS

JAN	FEB	MAR	APR	MAY	JUN
3 AQU	1 PIS	2 ARI	1 TAU	2 CAN	1 LEO
5 PIS	3 ARI	4 TAU	3 GEM	5 LEO	3 VIR
7 ARI	5 TAU	6 GEM	5 CAN	7 VIR	6 LIB
9 TAU	7 GEM	9 CAN	7 LEO	10 LEO	8 SCO
11 GEM	9 CAN	11 LEO	10 VIR	12 SCO	11 SAG
13 CAN	12 LEO	14 VIR	12 LIB	14 SAG	13 CAP
15 LEO	14 VIR	16 LIB	15 SCO	17 CAP	15 AQU
18 VIR	17 LIB	19 SCO	17 SAG	19 AQU	17 PIS
21 SCO	19 SCO	21 SAG	20 CAP	21 PIS	19 ARI
23 SCO	22 SIGT	23 CAP	22 AQU	23 ARI	22 TAU
25 SAG	24 CAP	25 AQU	24 PIS	25 TAU	24 GEM
28 CAP	26 AQU	28 PIS	26 ARI	27 GEM	26 CAN
30 AQU	28 PIS	30 ARI	28 TAU	30 CAN	28 LEO
			30 GEM		

JUL	AUG	SEP	OCT	NOV	DEC
1 VIR	2 SCO	1 SAG	1 CAP	1 SIP	1 ARI
3 LIB	5 SAG	3 CAP	3 AQU	3 ARI	3 TAU
6 SCO	7 CAP	5 AQU	6 PIS	5 TAU	5 GEM
8 SAG	9 AQU	7 PIS	7 ARI	7 GEM	7 CAN
10 CAP	11 PIS	9 ARI	9 TAU	9 CAN	9 LEO
13 AQU	13 ARI	11 TAU	11 GEM	12 LEO	11 VIR
15 PIS	15 TAU	13 GEM	13 CAN	14 VIR	14 LIB`
17 ARI	17 GEM	16 CAN	15 LEO	17 LIB	16 SCO
19 TAU	19 CAN	18 LEO	18 VIR	19 CSO	19 SAG
21 GEM	22 LOE	21 VIR	20 LIB	22 SAG	21 CAP
	24 VIR	23 LIB	23 SCO	24 CAP	24 AQU
26 LEO	27 LIB	26 CSO	25 SIG	26AQU	26 PIS
28 VIR	29 SCO	2SIG	28 CAP	29 PIS	28 ARI
31 LIB			31 AQU		30 TAU

1902 MOON SIGNS

JAN	FEB	MAR	APR	MAY	JUN
3 SCO	1 SIG	1 SIG	2 AQU	2 PIS	3 TAU
5 SIG	4 CAP	3 CAP	4 PIS	4 ARI	5 GEM
8 CAP	6 AQU	6 AQU	7 ARI	6 TAU	7 CAN
10 AQU	9 PIS	8 PIS	9 TAU	8 GEM	9 LEO
12 PIS	11 ARI	10 ARI	11 GEM	10 CAN	11 VIR
15 ARI	13 TAU	12 TAU	13 CAN	12 LEO	13 LIB
17 TAU	15 GEM	14 GEM	15 LEO	14 VIR	15 SCO
19 GEM	17 CAN	17 CAN	17VIR	17LIB	18 SIG
21 CAN	19 LEO	19 LEO	20 LIB	19 SCO	20 CAP
23 LEO	22 VIR	21 VIR	22 SCO	22 SAG	23 AQU
25 VIR	24 LIB	23 LIB	24 SIG	24 CAP	25 PIS
27 LIB	26 SCO	26 SCO	27 CAP	27 AQU	28 ARI
30SCO		28 SAG	29AQU	29 PIS	30 TAU
		31 CAP		31 ARI	

JUL	AUG	SEP	OCT	NOV	DEC
2 GEM	2 LEO	1 VIR	3 SCO	2 SAG	1 CAP
4 CAN	4 VIR	3 LIB	5 SAG	4 CAP	4 AQU
6 LEO	7 LIB	5 SCO	8 CAP	7 AQU	6 PIS
8 VIR	9 SCO	8 SAG	10 AQU	9 PIS	9 ARI
10 LIB	12 SAG	10 CAP	13 PIS	11 ARI	11 TAU
13 SCO	14 CAP	13 AQU	15 ARI	13 TAU	13 GEM
15 SIG	17 AQU	15 PIS	17 TAU	16 GEM	15 CAN
18 CAP	19 PIS	18 ARI	19 GEM	18 CAN	17 LEO
20 AQU	21 ARI	20 TAU	21 CAN	20 LEO	19 VIR
23 PIS	23 TAU	22 GEM	23 LEO	23 VIR	21 LIB
25ARI	26 GEM	24 CAN	25 VIR	24 LIB	24 SCO
27 TAU	28 CAN	26 LEO	28 LIB	26 SCO	26 SIG
29 GEM	30 LEO	28 VIR	30 SCO	29 SAG	29 CAP
31 CAN		30 LIB			31 AQU

1901 MOON SIGNS

JAN	FEB	MAR	APR	MAY	JUN
1 GEM	2 LEO	1 LEO	2 LIB	2 SCO	1 SIG
3 CAN	4 VIR	3 VIR	5 SCO	5 SIG	3 CAP
5 LEO	7 LIB	6 LIB	7 SIG	7 CAP	6 AQU
8 VIR	9 SCO	8 SCO	10 CAP	9 AQU	8 PIS
10 LIB	12 SIG	11 SIG	12 AQU	12 PIS	10 ARI
13 SCO	12 CAP	13 CAP	14 PIS	14 ARI	12 TAU
15 SIG	16 AQU	16 AQU	16 ARI	16 TAU	14 GEM
18 CAP	18 PIS	18 PIS	18 TAU	18 GEM	16 CAN
20 AQU	20 ARI	20 ARI	20 GEM	20 CAN	18 LEO
22 PIS	22 TAU	22 TAU	22 CAN	22 LEO	21 VIR
24 ARI	25 GEM	24 GEM	25 LEO	24 VIR	23 LIB
26 TAU	27 CAN	26 CAN	27 VIR	27 LIB	26 SCO
28 GEM		28 LEO	29 LIB	29 CSO	28 SIG
31 CAN		31 VIR			

JUL	AUG	SEP	OCT	NOV	DEC
1 CAP	1 PIS	2 TAU	1 GEM	2 LEO	1 VIR
3 AQU	4 ATI	4 GEM	3 CAN	4 VIR	4 LIB
5 PIS	6 TAU	6 CAN	6 LEO	7 LIB	6 SCO
7 ARI	8 GEM	8 LEO	8 VIR	9 SCO	9 SIG
9 TAU	10 CAN	11 VIR	10 LIB	11 SIG	11 CAP
12 GEM	12 LEO	13 LIB	13 CSO	14 CAP	14 AQU
14 CAN	14 VIR	16 SCO	16 SIG	17 AQU	16 PIS
16 LEO	17 LIB	18 SIG	18 CAP	19 PIS	18 ARI
18 VIR	19 CSO	21 CAP	20 AQU	21 ARI	21 TAU
20 LIB	22 SIG	23 AQU	23 PIS	23 TAU	23 GEM
23 SCO	24 CAP	25 PIS	25 ARI	25 GEM	25 CAN
25 SIG	27AQU	27ARI	27 TAU	27 CAN	27 LEO
28 CAP	29 PIS	29 TAU	29 GEM	29 LEO	29 VIR
30 AQU	31 ARI		31 CAN		31 LIB
31 LIB			31 AQU		30 TAU

1903 MOON SIGNS

JAN	FEB	MAR	APR	MAY	JUN
3 PIS	1 ARI	3 TAU	1 GEM	1 CAN	1 VIR
5 ARI	4 TAU	5 GEM	3 CAN	3 LEO	3 LIB
7 TAU	6 GEM	7 CAN	5 LEO	5 VIR	6 SCO
9 GEM	8 CAN	9 LEO	8 VIR	7 LIB	8 SAG
11 CAN	10 LEO	11 VIR	10 LIB	9 SCO	10 CAP
13 LEO	12 VIR	13 LIB	12 SCO	12 SIG	13 AQU
15 VIR	14 LIB	16 SCO	14 SAG	14 CAP	15 PIS
18 LIB	16 SCO	18 SAG	17 CAP	17 AQU	18 ARI
20 SCO	19 SAG	20 CAP	19 AQU	19 PIS	20 TAU
22 SAG	21 CAP	23 AQU	22 PIS	22ARI	22 GEM
25 CAP	24 AQU	25 PIS	24 ARI	24 TAU	24 CAN
27 AQU	26 PIS	28 ARI	26 TAU	26 GEM	26 LEO
30 PIS	28 ARI	30 TAU	29 GEM	28 CAN	28 VIR
				30 LEO	

JUL	AUG	SEP	OCT	NOV	DEC
1 LIB	1 SAG	3 AQU	3 PIS	1 ARI	1 TAU
3 SCO	4 CAP	5 PIS	5 ARI	4 TAU	3 GEM
5 SAG	6 AQU	8 ARI	7 TAU	6 GEM	5 CAN
8 CAP	9 PIS	10 TAU	10 GEM	8 CAN	7 LEO
10 AQU	11 ARI	12 GEM	12 CAN	10 LEO	9 VIR
13 PIS	14 TAU	15 CAN	14 LEO	12 VIR	12 LIB
15 ARI	16 GEM	17 LEO	16 VIR	14 LIB	14 CSO
18 TAU	18 CAN	19 VIR	18 LIB	17 SCO	16 SAG
20 GEM	20 LEO	21 LIB	20 SCO	19 SAG	19 CAP
22 CAN	22 VIR	23 SCO	22 SAG	21 CAP	21 AQU
24 LEO	24 LIB	25 SAG	25 CAP	24 AQU	24 PIS
26 VIR	26 SCO	27 CAP	27 AQU	26 PIS	26 ARI
28 LIB	29 SAG	30 AQU	30 PIS	29 ARI	28 TAU
30 SCO	31 CAP				31 GEM

1904 MOON SIGNS

JAN	FEB	MAR	APR	MAY	JUN
2 CAN	2 VIR	1 VIR	1 SCO	1 SAG	2 AQU
4 LEO	4 LIB	3 LIB	3 SAG	3 CAP	4 PIS
6 VIR	6 SCO	4 SCO	6 CAP	5 AQU	7 ARI
8 LIB	9 SAG	7 SAG	8 AQU	8 PIS	9 TAU
10 SCO	11 CAP	9 CAP	11 PIS	10 ARI	12 GEM
12 SAG	14 AQU	12 AQU	13 ARI	13 TAU	14 CAN
15 CAP	16 PIS	14 PIS	16 TAU	15 GEM	16 LEO
17 AQU	19 ARI	17 ARI	18 GEM	17 CAN	18 VIR
20 PIS	21 TAU	19 TAU	20 CAN	20 LOE	20 LIB
22 ARI	23 GEM	22 GEM	22 LEO	22 VIR	22 SCO
25 TAU	26 CAN	24 CAN	24 VIR	24 LIB	24 SIG
27 GEM	28 LEO	26 LEO	27 LIG	26 SCO	27 CAP
29 CAN		28 VIR	29SCO	28 SAG	29 ARI
31 LEO		30 LIB		30 CAP	

JUL	AUG	SEP	OCT	NOV	DEC
2 PIS	3 TAU	2 GEM	1 CAN	2 VIR	1 LIB
4 ARI	5 GEM	4 CAN	3 LEO	4 LIB	3 SCO
7 TAU	8 CAN	6 LEO	5 VIR	6 SCO	5 SAG
9 GEM	10 LOE	8 VIR	7 LIB	8 SAG	8 CAP
11 CAN	12 VIR	10 LIB	9 SCO	10 CAP	10 AQU
13 LEO	14 LIB	13 SCO	11 SAG	12 AQU	12 PIS
15 VIR	16 SCO	14 SAG	14 CAP	15 PIS	15 ARI
17 LIB	18 SAG	16 CAP	16 AQU	18 ARI	17 TAU
19 SCO	20 CAP	19 AQU	19 PIS	20 TAU	20 GEM
22 SAG	23 AQU	21 PIS	21 ARI	22 GEM	22 CAN
24 CAP	25 PIS	24 RAI	24 TAU	25 CAN	24 LEO
26 AQU	28ARI	26 TAU	26 GEM	27 LEO	26 VIR
29 PIS	30 TAU	29 GEM	28 CAN	29 VIR	28 LIB
31 ARI			31 LEO		31 SCO

1905 MOON SIGNS

JAN	FEB	MAR	APR	MAY	JUN
2 SAG	3 AQU	2 AQU	1 PIS	3 TAU	2 GEM
4 CAP	4 PIS	4 PIS	3 ARI	4 GEM	4 CAN
6 AQU	8 ARI	7 ARI	6 TAU	8 CAN	6 LEO
9 PIS	10 TAU	9 TAU	8 GEM	10 LEO	8 VIR
11 ARI	13 GEM	12 GEM	10 CAN	12 VIR	11 LIB
14 TAU	15 CAN	14 CAN	13 LEO	14 LIB	13 SAG
16 GEM	17 LEO	16 LEO	15 VIR	16 CSO	15 SAG
18 CAN	19 VIR	19 VIR	17 LIB	18 SAG	17 CAP
21 LEO	21 LIB	21 LIB	19 SCO	20 CAP	19 AQU
23 VIR	23 SCO	23 SCO	21 SAG	23 AQU	21 PIS
25 LIB	25 SAG	25 SAG	23 CAP	25 PIS	24 ARI
27 SCO	27 CAP	27 CAP	25 AQU	28 ARI	26 TAU
29 SAG		29 AQU	28 PIS	30 TAU	29 GEM
31 CAP			30 ARI		

JUL	AUG	SEP	OCT	NOV	DEC
1 CAN	2 VIR	2 SCO	2 SAG	3 AQU	2 PIS
4 LEO	4 LIB	5 SAG	4 CAP	4 PIS	4 ARI
6 VIR	6 SCO	7 CAP	6 AQU	7 ARI	7 TAU
8 LIB	8 SAG	9 AQU	9 PIS	10 TAU	10 GEM
10 SCO	10 CAP	11 PIS	11 ARI	12 GEM	12 CAN
12 SAG	13 AQU	14 RAI	14 TAU	15 CAN	15 LEO
14 CAP	15 PIS	16 TAU	16 GEM	17 LEO	17 VIR
16 AQU	18 ARI	19 GEM	19 CAN	20 VIR	19 LIB
19 PIS	20 TAU	21 CAN	21 LEO	22 LIB	21 SCO
21 ARI	23 GEM	24 LEO	23 VIR	24 SCO	23 SAG
24 TAU	25 CAN	26 VIR	25 LIB	26 SAG	25 CAP
26 GEM	27 LEO	28 LIB	27 SCO	28 CAP	27 AQU
29 CAN	29 VIR	30 SCO	29 SAG	30 AQU	30 PIS
31 LEO	31 LIB		31 CAP		

1906 MOON SIGNS

JAN	FEB	MAR	APR	MAY	JUN
1 ARI	2 GEM	2 GEM	1 CAN	3 VIR	1 LIB
3 TAU	5 CAN	4 CAN	3 LEO	4 LIB	3 SCO
6 GEM	7 LEO	7 LEO	5 VIR	7 SCO	5 SAG
8 CAN	9 VIR	9 VIR	7 LIB	9 SAG	7 CAP
11 LEO	12 LIB	11 LIB	9 SCO	11 CAP	9 AQU
13 VIR	14 SCO	13 SCO	11 SAG	13 AQU	11 PIS
15 LIB	16 SAG	15 SAG	13 CAP	15 PIS	14 ARI
17 SCO	18 CAP	17 CAP	16 AQU	17 ARI	16 TAU
19 SAG	20 AQU	19 AQU	18 PIS	20 TAU	19 GEM
22 CAP	22 PIS	22 PIS	20 ARI	23 GEM	21 CAN
24 AQU	25 ARI	24 ARI	23 TAU	25 CAN	24 LEO
26 PIS	27 TAU	26 TAU	25 GEM	28 LEO	26 VIR
28 ARI		29 GEM	28 CAN	30 VIR	28 LIB
31 TAU			30 LEO		

JUL	AUG	SEP	OCT	NOV	DEC
1 SCO	1 CAP	2 PIS	1 ARI	2 GEM	2 CAN
3 SAG	3 AQU	4 ARI	4 TAU	5 CAN	5 LEO
5 CAP	5 PIS	6 TAU	6 GEM	7 LEO	7 VIR
7 AQU	7 ARI	9 GEM	9 CAN	10 VIR	9 LIB
9 PIS	10 TAU	11 CAN	11 LEO	12 LIB	12 SCO
11 ARI	12 GEM	14 LEO	13 VIR	14 SCO	14 SAG
14 TAU	15 CAN	16 VIR	16 LIB	16 SAG	16 CAP
16 GEM	17 LEO	18 LIB	18 SCO	18 CAP	18 AQU
19 CAN	20 VIR	20 SCO	20 SAG	20 AQU	20 PIS
21 LOE	22 LIB	22SAG	22CAP	22PIS	22ARI
23 VIR	24 SCO	24 CAP	24 AQU	25 ARI	24 TAU
26 LIB	26 SAG	27 AQU	26 PIS	27 TAU	27 GEM
28 SCO	28 CAP	29 PIS	28 ARI	30 GEM	29 CAN
30 SAG	30 AQU		31 TAU		

1907 MOON SIGNS

JAN	FEB	MAR	APR	MAY	JUN
1 LEO	2 LIB	1 LIB	2 SAG	1 CAP	2 PIS
3 VIR	4 SCO	3 CSO	4 CAP	3 AQU	4 ARI
6 LIB	6 SAG	6 SAG	6 AQU	5 PIS	6 TAU
8 SCO	8 CAP	8 CAP	8 PIS	8 ARI	9 GEM
10 SAG	10 AQU	10 AQU	10 ARI	10 TAU	11 CAN
12 CAP	13 PIS	12 PIS	13 TAU	12 GEM	14 LEO
14 AQU	15 ARI	14 ARI	15 GEM	15 CAN	16 VIR
16 PIS	17 TAU	16 TAU	18 CAN	17 LEO	19 LIB
18 ARI	19 GEM	19 GEM	20 LEO	20 VIR	21 SCO
21 TAU	22 CAN	21 CAN	23 VIR	22 LIB	23 SAG
23 GEM	24 LEO	24 LEO	25 LIB	25 SCO	25 CAP
26 CAN	27 VIR	26 VIR	27 SCO	27 SAG	27 AQU
28 LEO		29 LIB	29 SAG	29 CAP	29 PIS
31 VIR		31 SCO		31 AQU	

JUL	AUG	SEP	OCT	NOV	DEC
1 ARI	2 GEM	1 CAN	1 LEO	2 LIB	2 SCO
3 TAU	5 CAN	4 LEO	3 VIR	4 SCO	4 SAG
6 GEM	7 LOE	6 VIR	6 LIB	6 SAG	6 CAP
8 CAN	10 VIR	8 LIB	8 SCO	9 CAP	8 AQU
11 LEO	12 LIB	11 SCO	10 SAG	11 AQU	10 PIS
14 VIR	14 SCO	13 SAG	12 CAP	13 PIS	12 ARI
16 LIB	17 SAG	15 CAP	14 AQU	15 ARI	14 TAU
18 SCO	19 CAP	17 AQU	17 PIS	17 TAU	17 GEM
20 SAG	21 AQU	19 PIS	19 ARI	20 GEM	19 CAN
22 CAP	23 PIS	21 ARI	21 TAU	22 CAN	22 LEO
24 AQU	25 ARI	24 TAU	23 GEM	25 LEO	24 VIR
25 PIS	27 TAU	26 GEM	26 CAN	27 VIR	27 LIB
29 ARI	30 GEM	28 CAN	28 LEO	30 LIB	29 SCO
31 TAU			31 VIR		31 SAG

1908 MOON SIGNS

JAN	FEB	MAR	APR	MAY	JUN
2 CAP	1 AQU	1 PIS	2 TAU	1 GEM	3 LEO
4 AQU	3 PIS	3 ARI	4 GEM	4 CAN	5 VIR
6 PIS	5 ARI	5 TAU	6 CAN	6 LEO	8 LIB
8 ARI	7 TAU	8 GEM	9 LOE	9 VIR	10 SCO
11 TAU	9 GEM	10 CAN	11 VIR	11LIB	12 SAG
13 GEM	12 CAN	13 LEO	14 LUIB	14 SCO	14 CAP
16 CAN	13 LEO	15 VIR	16 SCO	16 SAG	16 AQU
18 LEO	17 VIR	18 LIB	19 SAG	18 CAP	18 PIS
21 VIR	19 LIB	20 SCO	21 CAP	20 AQU	21 ARI
23 LIB	22 CSO	22 SAG	23 AQU	22 PIS	23 TAU
26 SCO	24 SAG	24 CAP	25 PIS	24 ARI	25 GEM
28 SAG	26 CAP	27 AQU	27 ARI	27 TAU	26 CAN
30 CAP	28 AQU	29 PIS	29 TAU	29 GEM	30 LEO
31 VIR		31 ARI		31 CAN	

JUL	AUG	SEP	OCT	NOV	DEC
2 VIR	1 LIB	2 SAG	2 CAP	2 PIS	2 ARI
5 LIB	4 SCO	5 CAP	4 AQU	4 ARI	4 TAU
7 SCO	6 SAG	7 AQU	6 PIS	7 TAU	6 GEM
10 SAG	8 CAP	9 PIS	8 ARI	9 GEM	8 CAN
12 CAP	10 AQU	11ARI	10 TAU	11 CAN	11 LEO
14 AQU	12 PIS	13 TAU	12 GEM	13 LEO	13 VIR
16 PIS	14 ARI	15 GEM	15 CAN	16 VIR	16 LIB
18 ARI	16 TAU	17 CAN	17 LEO	18 LIB	18 SCO
20 TAU	19 GEM	20 LEO	20 VIR	21 SCO	21 SAG
22 GEM	21 CAN	22 VIR	22 LIB	23 SAG	23 CAP
25 CAN	23 LEO	25 LIB	24 SCO	25 CAP	25 AQU
27 LOE	26 VIR	27 SCO	27 SAG	27 AQU	27 PIS
30 VIR	28 LIB	30 SAG	29 CAP	30 PIS	29 ARI
	31 SCO		31 AQU		31 TAU

1909 MOON SIGNS

JAN	FEB	MAR	APR	MAY	JUN
2 GEM	1 CAN	3 LEO	1 VIR	1 LIB	2 SAG
5 CAN	3 LOE	5 VIR	4 LIB	4 SCO	5 CAP
7 LEO	6 VIR	8 LIB	6 SCO	6 SAG	7 AQU
10 VIR	8 LIB	10 SCO	9 SAG	8 CAP	9 PIS
12 LIB	11 SCO	13 SAG	11 CAP	11 AQU	11 ARI
15 SCO	13 SAG	15 CAP	13 AQU	13 PIS	13 TAU
17 SAG	16 CAP	17 AQU	16 PIS	15 ARI	15 GEM
19 CAP	18 AQU	19 PIS	18 ARI	17 TAU	18 CAN
21 AQU	20 PIS	21 ARI	20 TAU	19 GEM	20 LEO
23 PIS	22 ARI	23 TAU	22 GEM	21 CAN	22 VIR
25 ARI	24 TAU	25 GEM	24 CAN	23 LEO	25 LIB
27 TAU	26 GEM	27 CAN	26 LEO	26 VIR	27 SCO
30 GEM	28 CAN	30 LEO	29 VIR	28 LIB	30 SAG
				31 SCO	

JUL	AUG	SEP	OCT	NOV	DEC
2 CAP	1 AQU	1 ARI	2 GEM	1 CAN	1 LEO
4 AQU	3 PIS	3 TAU	2 CAN	3 LEO	3 VIR
6 PIS	5 ARI	5 GEM	7 LEO	6 VIR	5 LIB
8 ARI	7 TAU	7 CAN	9 VIR	8 LIB	8 SCO
10 TAU	9 GEM	10 LEO	12 LIB	11 SCO	11 SAG
13 GEM	11 CAN	12 VIR	14 SCO	13 SAG	13 CAP
15 CAN	13 LEO	15 LIB	17 SAG	16 CAP	15 AQU
17 LEO	16 VIR	17 SCO	19 CAP	18 AQU	17 PIS
20 VIR	18 LIB	20 SAG	22 AQU	20 PIS	20 ARI
22 LIB	21 SCO	22 CAP	24 PIS	22 ARI	22 TAU
25 SCO	23 SAG	24 AQU	26 ARI	24 TAU	24 GEM
27 SAG	26 CAP	26 PIS	28 TAU	26 GEM	26 CAN
29 CAP	28 AQU	28 ARI	30 GEM	28 CAN	28 LEO
	30 PIS	30 TAU			30 VIR

1910 MOON SIGNS

JAN	FEB	MAR	APR	MAY	JUN
2 LIB	1 SCO	3 SAG	1 CAP	1 AQU	2 ARI
4 SCO	3 SAG	5 CAP	4 AQU	3 PIS	4 TAU
7 SAG	6 CAP	7 AQU	6 PIS	5 ARI	6 GEM
9 CAP	8 AQU	9 PIS	8 ARI	7 TAU	8 CAN
11 AQU	10 PIS	11 ARI	10 TAU	9 GEM	10 LEO
14 PIS	12 ARI	13 TAU	12 GEM	11 CAN	12 VIR
16 ARI	14 TAU	16 GEM	14 CAN	14 LEO	15 VIR
18 TAU	16 GEM	18 CAN	16 LEO	16 VIR	17 SCO
20 GEM	18 CAN	20 LEO	19 VIR	18 LIB	20 SAG
22 CAN	21 LEO	22 VIR	21 LIB	21 SCO	22 CAP
24 LEO	23 VIR	25 LIB	24 SCO	23 SAG	24 AQU
27 VIR	25 LIB	27 SCO	26 SAG	26 CAP	27 PIS
29 LIB	28 SCO	30 SAG	29 CAP	28 AQU	29 ARI
				31 PIS	

JUL	AUG	SEP	OCT	NOV	DEC
1 TAU	2 CAN	2 VIR	2 LIB	1 SCO	3 CAP
3 GEM	4 LEO	5 LIB	4 SCO	3 SAG	5 AQU
5 CAN	6 VIR	7 SCO	7 SAG	6 CAP	8 PIS
7 LEO	8 LIB	10 SAG	9 CAP	8 AQU	10 ARI
10 VIR	11 SCO	12 CAP	12 AQU	11 PIS	12 TAU
12 LIB	13 SAG	14 AQU	14 PIS	13 ARI	14 GEM
14 SCO	16 CAP	17 PIS	16ARI	15 TAU	16 CAN
17 SAG	18 AQU	19 ARI	18 TAU	17 GEM	18 LEO
19 CAP	20 PIS	21 TAU	20 GEM	19 CAN	20 VIR
22AQU	22ARI	23 GEM	22 CAN	21 LEO	23 LIB
24 PIS	25 TAU	25 CAN	24 LEO	23 VIR	25 SCO
26 ARI	27 GEM	27 LEO	27 VIR	25 LIB	28 SAG
28 TAU	29 CAN	30 VIR	29 LIB	28 SCO	30 CAP
30 GEM	31 LEO			30 SAG	

1911 MOON SIGNS

JAN	FEB	MAR	APR	MAY	JUN
2 AQU	3 ARI	2 ARI	2 GEM	2 CAN	2 VIR
4 PIS	5 TAU	4 TAU	4 CAN	4 LEO	5 LIB
6 ARI	7 GEM	6 GEM	7 LEO	6 VIR	7 SCO
9 TAU	9 CAN	8 CAN	9 VRI	8 LIB	10 SAG
11 GEM	11 LEO	10 LEO	11 LIB	11 SCO	12 CAP
13 CAN	13 VIR	13 VIR	14 SCO	13 SAG	15 AQU
15 LEO	15 LIB	15 LIB	16 SAG	16 CAP	17 PIS
17 VIR	18 SCO	18 SAG	19 CAP	18 AQU	19 ARI
19 LIB	20 SAG	20 SAG	21 AQU	21 PIS	22 TAU
21 SCO	23 CAP	22 CAP	23 PIS	23 ARI	24 GEM
24 SAG	25 AQU	25 AQU	26 ARI	25 TAU	26 CAN
27 CAP	28 PIS	27 PIS	28 TAU	27 GEM	28 LEO
29 AQU		29 ARI	30 GEM	29 CAN	30 VIR
31 PIS		31 TAU		31 LEO	

JUL	AUG	SEP	OCT	NOV	DEC
2 LIB	1 SCO	2 CAP	2 AQU	1 PIS	2 TAU
4 SCO	3 SAG	4 AQU	4 PIS	3 ARI	4 GEM
7 SAG	6 CAP	7 PIS	6 ARI	5 TAU	7 CAN
9 CAP	8 AQU	9 ARI	9 TAU	7 GEM	9 LEO
12 AQU	11 PIS	11 TAU	11 GEM	9 CAN	11 VIR
14 PIS	13 ARI	13 GEM	13 CAN	11 LEO	13 LIB
17 ARI	15 TAU	16 CAN	15 LEO	13 VIR	15 SCO
19 TAU	17 GEM	18 LEO	17 VIR	16 LIB	18 SAG
21 GEM	19 CAN	20 VIR	19 LIB	18 SCO	20 CAP
23 CAN	21 LEO	22 LIB	22 SCO	20 SAG	23 AQU
25 LEO	23 VIR	24 SCO	24 SAG	23 CAP	25 PIS
27 VIR	26 LIB	27 SAG	27 CAP	26 AQU	28 ARI
29 LIB	28 SCO	29 CAP	29 AQU	28 PIS	30 TAU
	30 SAG			30 ARI	

1912 MOON SIGNS

JAN	FEB	MAR	APR	MAY	JUN
1 GEM	1 LEO	2 VIR	3 SCO	2 SAG	1 CAP
3 CAN	3 VIR	4 LIB	5 SAG	5 CAP	3 AQU
5 LEO	6 LIB	6 SCO	7 CAP	7 AQU	6 PIS
7 VIR	8 SCO	8 SAG	10 AQU	10 PIS	8 ARI
9 LIB	10 SAG	11 CAP	12 PIS	12 ARI	11 TAU
11 SCO	13 CAP	13 AQU	15 ARI	14 TAU	13 GEM
14 SAG	15 AQU	16 PIS	17 TAU	17 GEM	15 CAN
16 CAP	18 PIS	18 ARI	19 GEM	19 CAN	19 LEO
19 AQU	20 ARI	21 TAU	21 CAN	21 LEO	19 VIR
21 PIS	22 TAU	23 GEM	23 LOE	23 VIR	21 LIB
24 ARI	25 GEM	25 CAN	26 VIR	25 LIB	23 SCO
26 TAU	27 CAN	27 LEO	28 LIB	27 SCO	26 SAG
28 GEM	29 LEO	28 VIR	30 SCO	30 SAG	28 CAP
30 CAN		31 LIB			

JUL	AUG	SEP	OCT	NOV	DEC
1 AQU	2 ARI	1 TAU	2 CAN	1 LEO	2 LIB
3 PIS	4 TAU	3 GEM	5 LEO	3 VIR	4 SCO
6 ARI	7 GEM	5 CAN	7 VIR	5 LIB	7 SAG
8 TAU	9 CAN	7 LEO	9 LIB	7 SCO	9 CAP
10 GEM	11 LEO	9 VIR	11 SCO	9 SAG	12 AQU
12 CAN	13 VIR	11 LIB	13 SAG	12 CAP	14 PIS
14 LEO	15 LIB	13 SCO	15 CAP	14 AQU	17 ARI
16 VIR	17 SCO	16 SAG	18 AQU	17 PIS	19 TAU
18 LIB	19 SAG	18 CAP	20 PIS	19 ARI	21 GEM
21 SCO	22 CAP	21 AQU	23 ARI	22 TAU	23 CAN
23 SAG	24 AQU	23 PIS	25 TAU	24 GEM	25 LEO
25 CAP	27 PIS	26 ARI	27 GEM	26 CAN	27 VIR
28 AQU	29 ARI	29 TAU	30 CAN	28 LEO	29 LIB
31 PIS		30 GEM		30 VIR	

1914 MOON SIGNS

JAN	FEB	MAR	APR	MAY	JUN
3 ARI	2 TAU	1 TAU	2 CAN	2 LEO	2 LIB
5 TAU	4 GEM	3 GEM	4 LEO	4 VIR	4 SCO
8 GEM	6 CAN	6 CAN	7 VIR	6 LIB	6 SAG
10 CAN	9 LEO	8 LEO	9 LIB	8 SCO	8 CAP
12 LEO	11 VIR	10 VIR	11 SCO	10 SAG	11 AQU
14 VIR	13 LIB	12 LIB	13 SAG	12 CAP	13 PIS
16 LIB	15 SCO	14 SCO	15 CAP	14 AQU	15 ARI
18 SCO	17 SAG	16 SAG	17 AQU	17 PIS	18 TAU
21 SAG	19 CAP	18 CAP	19 PIS	19 ARI	20 GEM
23 CAP	21 AQU	21 AQU	22 ARI	22 TAU	23 CAN
25 AQU	24 PIS	23 PIS	23 TAU	24 GEM	25 LEO
28 PIS	26 ARI	26 ARI	27 GEM	27 CAN	27 VIR
30 ARI		28 TAU	29 CAN	29 LEO	30 LIB
		31 GEM		31 VIR	

JUL	AUG	SEP	OCT	NOV	DEC
2 SCO	2 CAP	1 AQU	3 ARI	1 TAU	1 GEM
4 SAG	4 AQU	3 PIS	5 TAU	4 GEM	4 CAN
6 CAP	7 PIS	5 ARI	8 GEM	6 CAN	6 LEO
8 AQU	9 ARI	8 TAU	10 CAN	9 LEO	8 VIR
10 PIS	12 TAU	10 GEM	13 LEO	11 VIR	11 LIB
13 ARI	14 GEM	13 CAN	15 VIR	13 LIB	13 SCO
15 TAU	17 CAN	15 LEO	17 LIB	15 SCO	15 SAG
18 GEM	19 LEO	17 VIR	19 SCO	17 SAG	17 CAP
20 CAN	21 VIR	19 LIB	21 SAG	19 CAP	19 AQU
22 LEO	23 LIB	21 SCO	23 CAP	21 AQU	21 PIS
25 VIR	25 SCO	23 SAG	25 AQU	24 PIS	25 ARI
27 LIB	27 SAG	26 CAP	27 PIS	26 ARI	26 TAU
29 SCO	29 CAP	28 AQU	30 ARI	29 TAU	29 GEM
31 SAG		30 PIS			31 CAN

1913 MOON SIGNS

JAN	FEB	MAR	APR	MAY	JUN
1 SCO	2 CAP	1 CAP	2 PIS	2 ARI	1 TAU
3 SAG	4 AQU	3 AQU	5 ARI	4 TAU	3 GEM
5 CAP	7 PIS	6 PIS	7 TAU	7 GEM	5 CAN
8 AQU	9 ARI	8 ARI	10 GEM	8 CAN	7 LEO
10 PIS	12 TAU	11 TAU	12 CAN	11 LEO	10 VIR
13 ARI	14 GEM	13 GEM	14 LEO	13 VIR	12 LIB
15 TAU	16 CAN	16 CAN	16 VIR	15 LIB	14 SCO
18 GEM	18 LEO	18 LEO	18 LIB	18 SCO	16 SAG
20 CAN	20 VIR	20 VIR	20 SCO	20 SAG	18 CAP
22 LEO	22 LIB	22 LIB	22 SAG	22 CAP	21 AQU
24 VIR	24 SCO	24 SCO	25 CAP	24 AQU	23 PIS
26 LIB	27 SAG	26 SAG	27 AQU	27 PIS	26 ARI
28 SCO		28 CAP	29 PIS	29 ARI	28 TAU
30 PIS		31 AQU			30 GEM

JUL	AUG	SEP	OCT	NOV	DEC
3 CAN	1 LEO	2 LIB	1 SCO	1 CAP	1 AQU
5 LEO	3 LIB	4 SCO	3 SAG	4 AQU	4 PIS
7 VIR	5 SCO	6 SAG	5 CAP	7 PIS	6 ARI
9 LIB	7 SCO	8 CAP	8 AQU	9 ARI	9 TAU
11 SCO	10 SAG	10 AQU	10 PIS	12 TAU	11 GEM
13 SAG	12 CAP	13 PIS	13 ARI	14 GEM	14 CAM
16 CAP	14 AQU	15 ARI	15 TAU	16 CAN	16 LEO
18 AQU	17 PIS	18 TAU	18 GEM	19 LEO	18 VIR
20 PIS	19 ARI	20 GEM	20 CAN	21 VIR	20 LIB
23 ARI	22 TAU	23 CAN	22 LEO	24 LIB	23 SCO
25 TAU	24 GEM	25 LEO	25 VIR	25 SCO	24 SAG
28 GEM	26 CAN	27 VIR	27 LIB	27 SAG	27 CAP
30 CAN	29 LEO	29 LIB	29 SCO	29 CAP	29 AQU
	31 VIR		31 SAG		31 PIS

1915 MOON SIGNS

JAN	FEB	MAR	APR	MAY	JUN
2 LEO	1 VIR	2 LIB	1 SCO	2 CAP	1 ARI
5 VIR	3 LIB	5 SCO	3 SAG	4 AQU	3 PIS
7 LIB	5 SCO	7 SAG	5 CAP	7 PIS	5 ARI
9 SCO	7 SAG	9 CAP	7 AQU	9 ARI	8 TAU
11 SAG	10 CAP	11 AQU	9 PIS	12 TAU	10 GEM
13 CAP	12 AQU	13 PIS	12 ARI	14 GEM	13 CAN
15 AQU	14 PIS	14 ARI	14 TAU	17 CAN	15 LEO
18 PIS	16 ARI	18 TAU	17 GEM	19 LEO	18 VIR
20 ARI	19 TAU	21 GEM	19 CAN	21 VIR	20 LIB
22 TAU	21 GEM	23 CAN	22 LEO	24 LIB	22 SCO
25 GEM	24 CAN	25 LEO	24 VIR	26 SCO	24 SAG
27 CAN	26 LEO	28 VIR	26 LIB	28 SAG	28 CAP
30 LEO	29 VIR	30 LIB	28 SCO	30 CAP	28 AQU
			30 SAG		30 PIS

JUL	AUG	SEP	OCT	NOV	DEC
3 ARI	1 TAU	3 CAN	3 LEO	1 VIR	1 LIB
5 TAU	4 GEM	5 LEO	5 VRI	4 LIB	3 SCO
8 GEM	6 CAN	8 VIR	7 LIB	6 SCO	5 SAG
10 CAN	9 LEO	10 LIB	9 SCO	8 SAG	7 CAP
13 LEO	11 VIR	13 SCO	11 SAG	10 CAP	9 AQU
15 VIR	13 LIB	14 SAG	13 CAP	12 AQU	11 PIS
17 LIB	16 SCO	16 CAP	16 AQU	14 PIS	14 ARI
19 SCO	18 SAG	18 AQU	18 PIS	19 ARI	16 TAU
22 SAG	20 CAP	20 PIS	20 ARI	19 TAU	18 GEM
24 CAP	22 AQU	23 ARI	22 TAU	21 GEM	21 CAN
26 AQU	24 PIS	25 TAU	25 GEM	24 CAN	23 LEO
28 PIS	26 ARI	28 GEM	27 CAN	26 LEO	26 VIR
30 ARI	29 TAU	30 CAN	30 LEO	29 VIR	28 LIB
	31 GEM				31 SCO

1916 MOON SIGNS

JAN	FEB	MAR	APR	MAY	JUN
2 SAG	2 AQU	3 PIS	1 ARI	1 TAU	2 CAN
4 CAP	4 PIS	5 ARI	3 TAU	3 GEM	4 LEO
6 AQU	6 ARI	7 TAU	6 GEM	5 CAN	7 VIR
8 PIS	9 TAU	9 GEM	8 CAN	8 LEO	9 LIB
10 ARI	11 GEM	12 CAN	11 LEO	11 VIR	11 SCO
12 TAU	14 CAN	14 LEO	13 VIR	13 LIB	14 SAG
15 GEM	16 LEO	17 VIR	15 LIB	15 SCO	16 CAP
17 CAN	18 VIR	19 LIB	18 SCO	17 SAG	18 AQU
20 LEO	21 LIB	21 SCO	20 SAG	19 CAP	20 PIS
22 VIR	23 SCO	24 SAG	22 CAP	21 AQU	22 ARI
25 LIB	25 SAG	26 CAP	24 AQU	23 PIS	24 TAU
27 SCO	27 CAP	28 AQU	26 PIS	26 ARI	27 GEM
29 SAG	29 AQU	30 PIS	28 ARI	28 TAU	29 CAN
31 CAP				30 GEM	

JUL	AUG	SEP	OCT	NOV	DEC
2 LEO	3 LIB	1 SCO	1 SAG	1 AQU	1 PIS
4 VIR	5 SCO	4 SAG	3 CAP	3 PIS	3 ARI
7 LIB	7 SAG	6 CAP	5 AQU	6 ARI	5 TAU
9 SCO	9 CAP	8 AQU	7 PIS	8 TAU	7 GEM
11 SAG	11 AQU	10 PIS	9 ARI	10 GEM	10 CAN
13 CAP	13 PIS	12 ARI	11 TAU	13 CAN	12 LEO
15 AQU	16 ARI	14 TAU	14 GEM	15 LEO	15 VIR
17 PIS	18 TAU	16 GEM	16 CAN	18 VIR	17 LIB
19 ARI	20 GEM	19 CAN	19 LEO	20 LIB	20 SCO
21 TAU	23 CAN	21 LEO	21 VIR	22 SCO	22 SAG
24 GEM	25 LEO	24 VIR	24 LIB	24 SAG	24 CAP
26 CAN	28 VIR	26 LIB	26 SCO	27 CAP	26 AQU
29 LEO	30 LIB	29 SCO	28 SAG	29 AQU	28 PIS
31 VIR			30 CAP		30 ARI

1917 MOON SIGNS

JAN	FEB	MAR	APR	MAY	JUN
1 TAU	2 CAN	2 CAN	3 VIR	3 LIB	2 SCO
4 GEM	5 LEO	4 LEO	5 LIB	5 SCO	4 SAG
6 CAN	7 VIR	7 VIR	8 SCO	7 SAG	6 CAP
9 LEO	10 LIB	9 LIB	10 SAG	10 CAP	8 AQU
11 VIR	12 SCO	12 SCO	12 CAP	12 AQU	10 PIS
14 LIB	15 SAG	14 SAG	15 AQU	14 PIS	12 ARI
16 SCO	17 CAP	16 CAP	17 PIS	16 ARI	14 TAU
18 SAG	19 AQU	18 AQU	19 ARI	18 TAU	17 GEM
20 CAP	21 PIS	20 PIS	21 TAU	20 GEM	19 CAN
22 AQU	23 ARI	22 ARI	23 GEM	23 CAN	21 LEO
24 PIS	25 TAU	24 TAU	25 CAN	25 LEO	24 VIR
26 ARI	27 GEM	27 GEM	28 LEO	28 VIR	27 LIB
29 TAU		29 CAN	30 VIR	30 LIB	29 SCO
31 GEM		31 LEO			

JUL	AUG	SEP	OCT	NOV	DEC
1 SAG	2 AQU	2 ARI	2 TAU	2 CAN	2 LEO
3 CAP	4 PIS	4 TAU	4 GEM	5 LEO	5 VIR
5 AQU	6 ARI	6 GEM	6 CAN	7 VIR	7 LIB
7 PIS	8 TAU	9 CAN	9 LEO	10 LIB	10 SCO
9 ARI	10 GEM	11 LEO	11 VIR	12 SCO	12 SAG
12 TAU	13 CAN	14 VIR	14 LIB	15 SAG	14 CAP
14 GEM	15 LEO	16 LIB	16 SCO	17 CAP	16 AQU
16 CAN	18 VIR	19 SCO	18 SAG	19 AQU	19 PIS
19 LEO	20 LIB	21 SAG	21 CAP	21 PIS	21 ARI
21 VIR	23 SCO	23 CAP	23 AQU	23 ARI	23 TAU
24 LIB	25 SAG	26 AQU	25 PIS	25 TAU	25 GEM
26 SCO	27 CAP	28 PIS	27 ARI	28 GEM	27 CAN
29 SAG	29 AQU	30 ARI	29 TAU	30 CAN	30 LEO
31 CAP	31 PIS		31 GEM		

1918 MOON SIGNS

JAN	FEB	MAR	APR	MAY	JUN
1 VIR	2 SCO	2 SCO	3 CAP	2 AQU	1 PIS
4 LIB	5 SAG	4 SAG	5 AQU	4 PIS	3 ARI
6 SCO	7 CAP	7 CAP	7 PIS	7 ARI	5 TAU
8 SAG	9 AQU	9 AQU	9 ARI	9 TAU	7 GEM
11 CAP	11 PIS	11 PIS	11 TAU	11 GEM	9 CAN
13 AQU	13 ARI	13 ARI	13 GEM	13 CAN	11 LEO
15 PIS	15 TAU	15 TAU	15 CAN	15 LEO	14 VIR
17 ARI	17 GEM	17 GEM	18 LEO	17 VIR	16 LIB
19 TAU	20 CAN	19 CAN	20 VIR	20 LIB	19 SCO
21 GEM	22 LEO	21 LEO	23 LIB	22 SCO	21 SAG
24 CAN	25 VIR	24 VIR	25 SCO	25 SAG	24 CAP
26 LEO	27 LIB	26 LIB	28 SAG	27 CAP	26 AQU
28 VIR		29 SCO	30 CAP	30 AQU	28 PIS
31 LIB		31 SAG			30 ARI

JUL	AUG	SEP	OCT	NOV	DEC
2 TAU	1 GEM	1 LEO	1 VIR	2 SCO	2 SAG
4 GEM	3 CAN	4 VIR	3 LIB	5 SAG	4 CAP
6 CAN	5 LEO	6 LIB	6 SCO	7 CAP	7 AQU
9 LEO	7 VIR	9 SCO	9 SAG	10 AQU	9 PIS
11 VIR	10 LIB	11 SAG	11 CAP	12 PIS	11 ARI
14 LIB	12 SCO	13 CAP	13 AQU	14 ARI	13 TAU
16 SCO	15 SAG	16 AQU	15 PIS	16 TAU	15 GEM
19 SAG	17 CAP	18 PIS	18 ARI	18 GEM	17 CAN
21 CAP	20 AQU	20 ARI	20 TAU	20 CAN	20 LEO
23 AQU	22 PIS	22 TAU	22 GEM	22 LEO	22 VIR
25 PIS	24 ARI	24 GEM	24 CAN	25 VIR	24 LIB
27 ARI	26 TAU	26 CAN	26 LEO	27 LIB	27 SCO
29 TAU	28 GEM	29 LEO	28 VIR	30 SCO	29 SAG
	30 CAN		31 LIB		

1919 MOON SIGNS

JAN	FEB	MAR	APR	MAY	JUN
1 CAP	2 PIS	1 PIS	1 TAU	1 GEM	1 LEO
3 AQU	4 ARI	3 ARI	3 GEM	3 CAN	4 VIR
5 PIS	6 TAU	5 TAU	6 CAN	5 LEO	6 LIB
7 ARI	8 GEM	7 GEM	8 LEO	7 VIR	9 SCO
10 TAU	10 CAN	9 CAN	10 VIR	10 LIB	11 SAG
12 GEM	12 LEO	12 LEO	13 LIB	12 SCO	14 CAP
14 CAN	15 VIR	14 VIR	15 SCO	15 SAG	16 AQU
16 LEO	17 LIB	16 LIB	18 SAG	17 CAP	18 PIS
18 VIR	20 SCO	19 SCO	20 CAP	20 AQU	21 ARI
21 LIB	22 SAG	21 SAG	23 AQU	22 PIS	23 TAU
23 SCO	25 CAP	24 CAP	25 PIS	24 ARI	25 GEM
26 SAG	27 AQU	26 AQU	27 ARI	26 TAU	27 CAN
28 CAP		28 PIS	29 TAU	28 GEM	29 LEO
30 AQU		30 ARI		30 CAN	

JUL	AUG	SEP	OCT	NOV	DEC
1 VIR	2 SCO	1 SAG	1 CAP	2 PIS	2 ARI
3 LIB	4 SAG	4 CAP	3 AQU	4 ARI	4 TAU
6 SCO	7 CAP	6 AQU	6 PIS	6 TAU	6 GEM
8 SAG	10 AQU	8 PIS	8 ARI	8 GEM	8 CAN
11 CAP	12 PIS	10 ARI	10 TAU	10 CAN	10 LEO
13 AQU	14 ARI	12 TAU	12 GEM	12 LEO	12 VIR
16 PIS	16 ATU	15 GEM	14 CAN	15 VIR	14 LIB
18 ARI	18 GEM	17 CAN	16 LEO	17 LIB	17 SCO
20 TAU	20 CAN	19 LEO	18 VIR	19 SCO	19 SAG
22 GEM	23 LEO	21 VIR	21 LIB	22 SAG	22 CAP
24 CAN	25 VIR	23 LIB	23 SCO	25 CAP	24 AQU
26 LEO	27 LIB	26 SCO	26 SAG	27 AQU	27 PIS
28 VIR	30 SCO	28 SAG	28 CAP	29 PIS	29 ARI
31 LIB			31 AQU		31 TAU

1920 MOON SIGNS

JAN	FEB	MAR	APR	MAY	JUN
2 GEM	1 CAN	1 LEO	2 LIB	1 SCO	3 CAP
4 CAN	3 LEO	3 VIR	4 SCO	4 SAG	5 AQU
6 LEO	5 VIR	5 LIB	7 SAG	6 CAP	8 PIS
8 VIR	7 LIB	8 SAG	9 CAP	9 AQU	10 ARI
11 LIB	9 SCO	10 SAG	12 AQU	11 PIS	12 TAU
13 SCO	12 SAG	13 CAP	14 PIS	14 ARI	14 GEM
16 SAG	14 CAP	15 AQU	16 ARI	16 TAU	16 CAN
18 CAP	17 AQU	18 PIS	18 TAU	18 GEM	18 LEO
21 AQU	19 PIS	20 ARI	20 GEM	20 CAN	20 VIR
23 PIS	21 ARI	22 TAU	22 CAN	22 LEO	22 LIB
25 ARI	24 TAU	24 GEM	24 LEO	24 VIR	25 SCO
27 TAU	26 GEM	26 CAN	27 VIR	26 LIB	27 SAG
30 GEM	28 CAN	28 LEO	29 LIB	29 SCO	30 CAP
		30 VIR		31 SAG	

JUL	AUG	SEP	OCT	NOV	DEC
2 AQU	1 PIS	2 TAU	1 GEM	2 LEO	1 VIR
5 PIS	3 ARI	4 GEM	4 CAN	4 VIR	3 LIB
7 ARI	6 TAU	6 CAN	6 LEO	6 LIB	6 CSO
9 TAU	8 GEM	8 LEO	8 VIR	9 SCO	8 SAG
12 GEM	10 CAN	10 VIR	10 LIB	11 SAG	11 CAP
14 CAN	12 LEO	13 LIB	12 SCO	13 CAP	13 AQU
16 LEO	14 VIR	15 SCO	15 SAG	16 AQU	16 PIS
18 VIR	16 LIB	17 SAG	17 CAP	18 PIS	18 ARI
20 LIB	18 SCO	20 CAP	20 ARI	21 ARI	20 TAU
22 SCO	21 SAG	22 AQU	22 PIS	23 TAU	23 GEM
25 SAG	23 CAP	25 PIS	24 ARI	25 GEM	25 CAN
27 CAP	26 AQU	27 ARI	27 TAU	27 CAN	27 LEO
30 AQU	28 PIS	29 TAU	29 GEM	29 LEO	29 VIR
	31 ARI		31 CAN		31 LIB

1921 MOON SIGNS

JAN	FEB	MAR	APR	MAY	JUN
2 SCO	1 SAG	2 CAP	1 AQU	1 PIS	2 TAU
4 SAG	3 CAP	5 AQU	4 PIS	4 ARI	5 GEM
7 CAP	6 AQU	8 PIS	6 ARI	6 TAU	7 CAN
9 AQU	8 PIS	10 ARI	9 TAU	8 GEM	9 LEO
12 PIS	11 ARI	12 TAU	11 GEM	11 CAN	11 VIR
14 ARI	13 TAU	15 GEM	13 CAN	12 LEO	13 LIB
17 TAU	15 GEM	17 CAN	15 LEO	14 VIR	15 SCO
19 GEM	17 CAN	19 LEO	17 VIR	17 LIB	17 SCO
21 CAN	19 LEO	21 VIR	19 LIB	19 SCO	20 CAP
23 LEO	21 VIR	23 LIB	21 SCO	21 SAG	22 AQU
25 VIR	24 LIB	25 SCO	24 SAG	24 CAP	25 PIS
27 LIB	26 SCO	27 SAG	26 CAP	26 AQU	27 ARI
29 SCO	28 SAG	30 CAP	29 AQU	29 PIS	30 TAU
				31 ARI	

JUL	AUG	SEP	OCT	NOV	DEC
2 GEM	2 LEO	1 VIR	2 SCO	1 SAG	1 CAP
4 CAN	4 VIR	3 LIB	5 SAG	3 CAP	3 AQU
6 LEO	6 LIB	5 SCO	7 CAP	6 AQU	6 PIS
8 VIR	9 SCO	7 SAG	9 AQU	8 PIS	8 ARI
10 LIB	11 SAG	10 CAP	12 PIS	11 ARI	11 TAU
12 SCO	13 CAP	12 AQU	14 ARI	13 TAU	13 GEM
15 SAG	16 AQU	15 PIS	17 TAU	15 GEM	15 CAN
17 CAP	18 PIS	17 ARI	19 GEM	18 CAN	17 LEO
20 AQU	21 ARI	20 TAU	21 CAN	10 LEO	19 VIR
22 PIS	23 TAU	22 GEM	23 LEO	22 VIR	21 LIB
25 ARI	26 GEM	24 CAN	26 VIR	24 LIB	23 SCO
27 TAU	28 CAN	26 LEO	28 LIB	26 SCO	26 SAG
29 GEM	30 LEO	28 VIR	30 SCO	28 SAG	28 CAP
31 CAN		30 LIB			30 AQU

1922 MOON SIGNS

JAN	FEB	MAR	APR	MAY	JUN
2 PIS	1 ARI	2 TAU	1 GEM	1 CAN	1 VIR
4 ARI	4 TAU	5 GEM	3 CAN	3 LEO	3 LIB
7 TAU	6 GEM	7 CAN	6 LEO	5 VIR	5 SCO
9 GEM	8 CAN	9 LEO	8 VIR	7 LIB	8 SAG
11 CAN	10 LEO	11 VIR	10 LIB	9 SCO	10 CAP
13 LEO	12 VIR	13 LIB	12 SCO	11 SAG	12 AQU
15 VIR	14 LIB	15 SCO	14AG	13 CAP	15 PIS
17 LIB	16 SCO	17 SAG	16 CAP	16 AQU	17 ARI
20 SCO	18 SAG	20 CAP	18 AQU	18 PIS	20 TAU
22 SAG	20 CAP	22 AQU	21 PIS	21 ARI	22 GEM
24 CAP	23 AQU	25 PIS	24 ARI	23 TAU	24 CAN
27 AQU	25 PIS	27 ARI	26 TAU	26 GEM	26 LEO
29 PIS	28 ARI	30 TAU	28 GEM	28 CAN	28 VIR
				30 LEO	

JUL	AUG	SEP	OCT	NOV	DEC
1 LIB	1 SAG	2 AQU	2 PIS	1 ARI	3 GEM
3 SCO	3 CAP	5 PIS	4 ARI	3 TAU	5 CAN
5 SAG	6 AQU	7 ARI	7 TAU	6 GEM	7 LEO
7 CAP	8 PIS	10 TAU	9 GEM	8 CAN	10 VIR
10 AQU	11 ARI	12 GEM	12 CAN	10 LEO	12 LIB
12 PIS	13 TAU	14 CAN	14 LEO	12 VIR	14 SCO
14 ARI	16 GEM	17 LEO	16 VIR	15 LIB	16 SAG
17 TAU	18 CAN	19 VIR	18 LIB	17 SCO	18 CAP
19 GEM	20 LEO	21 LIB	20 SCO	19 SAG	20 AQU
22 CAN	22 VIR	23 SCO	22 SGA	21 CAP	23 PIS
24 LEO	24 LIB	25 SAG	24 CAP	23 AQU	25 ARI
26 VRI	26 SCO	27 CAP	27 AQU	25 PIS	28 TAU
28 LIB	28 SAG	29 AQU	29 PIS	28ARI	30 GEM
30 SCO	31 CAP			30 TAU	

1923 MOON SIGNS

JAN	FEB	MAR	APR	MAY	JUN
2 CAN	2 VIR	2 VIR	2 SCO	2 SAG	2 AQU
4 LEO	4 LIB	4 LIB	4 SAG	4 CAP	4 PIS
6 VIR	6 SCO	6 SCO	6 CAP	6 AQU	7 ARI
8 LIB	9 SAG	8 SAG	8 AQU	8 PIS	9 TAU
10 SCO	11 CAP	10 CAP	11 PIS	11 ARI	12 GEM
12 SAG	13 AQU	12 AQU	13 ARI	13 TAU	14 CAN
14 CAP	15 PIS	15 PIS	16 TAU	16 GEM	17 LEO
17 AQU	18 ARI	17 ARI	18 GEM	18 CAN	19 VIR
19 PIS	20 TAU	20 TAU	21 CAN	21 LEO	21 LIB
22 ARI	23 GEM	22 GEM	23 LEO	23 VIR	23 SCO
24 TAU	25 CAN	25 CAN	25 VIR	25 LIB	25 SAG
27 GEM	28 LEO	27 LEO	28 LIB	27 SCO	27 CAP
29 CAN		29 VIR	30 SCO	29 SAG	30 AQU
31 LEO		31 LIB		31 CAP	

JUL	AUG	SEP	OCT	NOV	DEC
2 PIS	1 ARI	2 GEM	2 CAN	1 LEO	2 LIB
4 ARI	3 TAU	4 CAN	4 LEO	3 VIR	4 SCO
7 TAU	6 GEM	7 LEO	6 VIR	5 LIB	6 SCO
9 GEM	8 CAN	9 VIR	8 LIB	7 SCO	8 CAP
12 CAN	10 LEO	11 LIB	10 SCO	9 SAG	10 AQU
14 LEO	13 VIR	13 SCO	12 SAG	11 CAP	13 PIS
16 VIR	15 LIB	15 SAG	15 CAP	13 AQU	15 ARI
18 LIB	17 SCO	17 CAP	17 AQU	15 PIS	18 TAU
21 SCO	19 SAG	19 AQU	19 PIS	18 ARI	20 GEM
23 SAG	21 CAP	22 PIS	21 SCO	20 TAU	23 CAN
25 CAP	23 AQU	24 ARI	24 TAU	23 GEM	25 LEO
27 AQU	26 PIS	27 TAU	27 GEM	25 CAN	27 VIR
29 PIS	28 ARI	29 GEM	29 CAN	28 LEO	30 LIB
	30 TAU			30 VIR	

1924 MOON SIGNS

JAN	FEB	MAR	APR	MAY	JUN
1 SCO	1 CAP	2 AQU	2 ARI	2 TAU	1 GEM
3 SAG	3 AQU	4 PIS	5 TAU	5 GEM	3 CAN
5 CAP	5 PIS	6 ARI	7 GEM	7 CAN	6 LEO
7 AQU	8 ARI	9 TAU	10 CAN	10 LEO	8 VIR
9 PIS	10 TAU	11 GEM	12 LEO	12 VIR	11 LIB
11 ARI	13 GEM	14 CAN	15 VIR	14 LIB	13 SCO
14 TAU	15 CAN	16 LEO	17 LIB	16 SCO	15 SAG
16 GEM	18 LEO	18 VIR	19 SCO	18 SAG	17 CAP
19 CAN	20 VIR	20 LIB	21 SAG	20 CAP	19 AQU
21 LEO	22 LIB	23 SCO	23 CAP	23 AQU	21 PIS
24 VIR	24 SCO	25 SAG	25 AQU	25 PIS	23 ARI
26 LIB	26 SAG	27 CAP	27 PIS	27 ARI	26 TAU
28 SCO	28 CAP	29 AQU	30 ARI	29 TAU	28 GEM
30 SAG		31 PIS			

JUL	AUG	SEP	OCT	NOV	DEC
1 CAN	2 VIR	3 SCO	2 SAG	2 AQU	2 PIS
3 LEO	4 LIB	5 SAG	4 CAP	5 PIS	4 ARI
6 VIR	6 SCO	7 CAP	6 AQU	7 ARI	7 TAU
8 LIB	8 SAG	9 AQU	8 PIS	9 TAU	9 GEM
10 SCO	11 CAP	11 PIS	11 ARI	12 GEM	12 CAN
12 SAG	13 AQU	13 ARI	13 TAU	14 CAN	14 LEO
14 CAP	15 PIS	16 TAU	15 GEM	17 LEO	17 VIR
16 AQU	17 ARI	18 GEM	18 CAN	19 VIR	19 LIB
18 PIS	19 TAU	21 CAN	20 LEO	22 LIB	21 SCO
21 ARI	22 GEM	23 LEO	23 VIR	24 SCO	24 SAG
23 TAU	24 CAN	26 VIR	25 LIB	26 SAG	25 CAP
25 GEM	27 LEO	28 LIB	27 SCO	27 CAP	27 AQU
28 CAN	29 VIR	30 SCO	29 SAG	30 AQU	29 PIS
30 LEO	31 LIB		31 CAP		31 ARI

1925 MOON SIGNS

JAN	FEB	MAR	APR	MAY	JUN
3 TAU	2 GEM	1 GEM	2 LEO	2 VIR	1 LIB
5 GEM	4 CAN	3 CAN	5 VIR	4 LIB	3 SCO
8 CAN	7 LEO	6 LEO	7 LIB	7 SCO	5 SAG
10 LEO	9 VIR	8 VIR	9 SCO	9 SAG	7 CAP
13 VIR	11 LIB	11 LIB	11 SAG	11 CAP	9 AQU
15 LIB	14 SCO	13 SCO	13 CAP	13 AQU	11 PIS
17 SCO	16 SAG	15 SAG	16 AQU	15 PIS	13 ARI
20 SAG	18 CAP	17 CAP	18 PIS	17 ARI	16 TAU
22 CAP	20 AQU	19 AQU	20 ARI	19 TAU	18 GEM
24 AQU	22 PIS	21 PIS	22 TAU	22 GEM	21 CAN
26 PIS	24 ARI	24 ARI	25 GEM	24 CAN	23 LEO
28 ARI	26 TAU	26 TAU	27 CAN	27 LEO	26 VIR
30 TAU		28 GEM	30 LEO	29 VIR	28 LIB
		31 CAN			31 SCO

JUL	AUG	SEP	OCT	NOV	DEC
3 SAG	1 CAP	1 PIS	1 ARI	2 GEM	1 CAN
5 CAP	3 AQU	4 ARI	3 TAU	4 CAN	4 LEO
7 AQU	5 PIS	6 TAU	5 GEM	7 LEO	6 VIR
9 PIS	7 ARI	8 GEM	8 CAN	9 VIR	9 LIB
11 ARI	9 TAU	10 CAN	10 LEO	12 LIB	11 SCO
13 TAU	12 GEM	13 LEO	11 VIR	14 SCO	13 SAG
15 GEM	14 CAN	15 VIR	15 LIB	16 SAG	16 CAP
18 CAN	17 LEO	18 LIB	18 SCO	18 CAP	18 AQU
20 LEO	19 VIR	20 SCO	20 SAG	20 AQU	20 PIS
23 VIR	22 LIB	22 SAG	22 CAP	22 PIS	22 ARI
25 LIB	24 SCO	25 CAP	24 AQU	25 ARI	24 TAU
28 SCO	26 SAG	27 AQU	26 PIS	27 TAU	26 GEM
30 SAG	28 CAP	29 PIS	28 ARI	29 GEM	29 CAN
	30 AQU		30 TAU		31 LEO

1926 MOON SIGNS

JAN	FEB	MAR	APR	MAY	JUN
3 VIR	2 LIB	1 LIB	2 SAG	1 CAP	2 PIS
5 LIB	4 SCO	3 SCO	4 CAP	3 AQU	4 ARI
8 SCO	6 SAG	6 SAG	6 AQU	6 PIS	6 TAU
10 SAG	8 CAP	8 CAP	8 PIS	8 ARI	8 GEM
12 CAP	10 AQU	10 AQU	10 ARI	10 TAU	11 CAN
14 AQU	12 PIS	12 PIS	12 TAU	12 GEM	13 LEO
16 PIS	14 ARI	14 ARI	15 GEM	14 CAN	16 VIR
18 ARI	17 TAU	16 TAU	17 CAN	17 LEO	18 LIB
20 TAU	19 GEM	18 GEM	19 LEO	19 VIR	20 SCO
23 GEM	21 CAN	21 CAN	22 VIR	22 LIB	23 SAG
25 CAN	24 LEO	23 LEO	24 LIB	24 SCO	25 CAP
27 LEO	26 VIR	26 VIR	27 SCO	26 SAG	27 AQU
30 VIR		28 LIB	29 SAG	29 CAP	29 PIS
		30 SCO		31 AQU	

JUL	AUG	SEP	OCT	NOV	DEC
1 ARI	2 GEM	3 LEO	3 VIR	1 LIB	1 SCO
3 TAU	4 CAN	5 VIR	5 LIB	4 SCO	4 SAG
6 GEM	7 LEO	8 LIB	8 SCO	6 SCO	6 CAP
8 CAN	9 VIR	10 SCO	10 SAG	8 CAP	8 AQU
10 LEO	12 LIB	13 SAG	12 CAP	11 AQU	10 PIS
13 VIR	14 SCO	15 CAP	15 AQU	13 PIS	12 ARI
15 LIB	17 SAG	17 AQU	17 PIS	15 ARI	14 TAU
18 SCO	19 CAP	19 PIS	19 ARI	17 TAU	17 GEM
20 SAG	21 AQU	21 ARI	21 TAU	19 GEM	19 CAN
22 CAP	23 PIS	23 TAU	23 GEM	21 CAN	21 LEO
24 AQU	25 ARI	25 GEM	25 CAN	24 LEO	24 VIR
26 PIS	27 TAU	28 CAN	27 LEO	26 VIR	26 LIB
28 AQU	29 GEM	30 LEO	30 VIR	29 LIB	29 SCO
31 TAU	31 CAN				31 SAG

1927 MOON SIGNS

JAN	FEB	MAR	APR	MAY	JUN
2 CAP	1 AQU	2 PIS	1 ARI	2 GEM	1 CAN
4 AQU	3 PIS	4 ARI	3 TAU	4 CAN	3 LEO
6 PIS	5 ARI	6 TAU	5 GEM	7 LEO	5 VIR
9 ARI	7 TAU	8 GEM	7 CAN	9 VIR	8 LIB
11 TAU	9 GEM	11 CAN	9 LEO	13 LIB	10 CSO
13 GEM	11 CAN	13 LEO	12 VIR	14 SCO	14 SAG
15 CAN	14 LEO	15 VIR	14 LIB	16 SAG	15 CAP
17 LEO	16 VIR	18 LIB	17 SCO	18 CAP	17 ARI
20 VIR	19 LIB	21 SCO	19 SAG	21 AQU	10 PIS
22 LIB	21 SCO	23 SAG	22 CAP	23 PIS	22 ARI
25 SCO	24 SAG	25 CAP	24 AQU	25 ARI	24 TAU
27 SAG	26 CAP	28 AQU	26 PIS	28 TAU	26 GEM
30 CAP	28 AQU	30 PIS	28 ARI	30 GEM	28 CAN
			30 TAU		30 LEO

JUL	AUG	SEP	OCT	NOV	DEC
3 VIR	1 LIB	3 SAG	3 CAP	1 AQU	1 PIS
5 LIB	4 SCO	5 CAP	5 AQU	3 PIS	3 ARI
8 SCO	7 SAG	8 AQU	7 PIS	6 ARI	5 TAU
10 SAG	9 CAP	10 PIS	9 ARI	8 TAU	7 GEM
13 CAP	11 AQU	12 ARI	11 TAU	10 GEM	9 CAN
15 AQU	13 PIS	14 TAU	13 GEM	12 CAN	11 LEO
17 PIS	15 ARI	16 GEM	15 CAN	14 LEO	13 VIR
19 ARI	17 TAU	18 CAN	17 LEO	16 VIR	16 LIB
21 TAU	19 GEM	20 LEO	20 VIR	19 LIB	18 SCO
23 GEM	22 CAN	23 VIR	22 LIB	21 SCO	21 SAG
21 CAN	24 LEO	25 LIB	25 SCO	24 SAG	23 CAP
28 LEO	26 VIR	28 SCO	27 SAG	26 CAP	26 AQU
30 VIR	29 LIB	30 SAG	30 CAP	28 AQU	28 PIS
	31 SCO				30 ARI

1928 MOON SIGNS

JAN	FEB	MAR	APR	MAY	JUN
1 TAU	2 CAN	2 LEO	1 VIR	3 SCO	2 SAG
2 GEM	4 LEO	5 VIR	3 LIB	5 SAG	4 CAP
5 CAN	6 VIR	7 LIB	6 SCO	8 CAP	7 AQU
8 LEO	9 LIB	9 SCO	8 SAG	10 AQU	9 PIS
10 VIR	11 SCO	12 SCO	11 CAP	13 PIS	11 ARI
12 LIB	14 SAG	14 CAP	13 AQU	15 ARI	13 TAU
15 SCO	16 CAP	17 AQU	15 PIS	17 TAU	15 GEM
17 SAG	18 AQU	19 PIS	18 ARI	19 GEM	17 CAN
20 CAP	21 PIS	21 ARI	20 TAU	21 CAN	19 LEO
22 AQU	23 ARI	23 TAU	22 GEM	23 LEO	22 VIR
24 PIS	25 TAU	25 GEM	24 CAN	25 VIR	24 LIB
26 ARI	27 GEM	27 CAN	26 LEO	28 LIB	26 SCO
28 TAU	29 CAN	29 LEO	28 VIR	30 SCO	29 SAG
31 GEM			30 LIB		

JUL	AUG	SEP	OCT	NOV	DEC
2 CAP	3 PIS	1 ARI	3 GEM	1 CAN	3 VIR
4 AQU	5 ARI	3 TAU	5 CAN	3 LEO	5 LIB
6 PIS	7 TAU	5 GEM	7 LEO	5 VIR	7 SCO
9 ARI	9 GEM	7 CAN	9 VIR	8 LIB	10 SAG
11 TAU	11 CAN	10 LEO	11 LIB	10 SCO	12 CAP
13 LEO	13 LEO	12 VIR	13 SCO	13 SAG	15 AQU
15 CAN	15 VIR	14 LIB	16 SAG	15 CAP	17 PIS
17 LEO	18 LIB	16 SCO	19 CAP	18 AQU	20 ARI
19 VIR	20 SCO	19 SAG	21 AQU	20 PIS	22 TAU
21 LIB	23 SAG	22 CAP	24 PIS	22 ARI	24 GEM
24 SCO	25 CAP	24 AQU	26 ARI	24 TAU	26 CAN
26 SAG	28 AQU	26 PIS	28 TAU	26 GEM	28 LEO
29 CAN	30 PIS	28 ARI	30 GEM	28 CAN	30 VIR
31 AQU		30 TAU		30 LEO	

1930 MOON SIGNS

JAN	FEB	MAR	APR	MAY	JUN
1 AQU	2 ARI	2 ARI	2 GEM	2 CAN	2 VIR
4 PIS	5 TAU	4 TAU	5 CAN	4 LEO	4 LIB
6 ARI	7 GEM	6 GEM	7 LEO	6 VIR	7 SCO
8 ATU	9 CAN	8 CAN	9 VIR	8 LIB	9 SAG
11 GEM	11 LEO	10 LEO	11 LIB	10 SCO	11 CAP
13 CAN	13 VIR	12 VIR	13 SCO	13 SAG	14 AQU
15 LEO	15 LIB	15 LIB	15 SAG	15 CAP	16 PIS
17 VIR	17 SCO	17 SCO	18 CAP	18 AQU	19 ARI
19 LIB	20 SAG	19 SAG	20 AQU	20 PIS	21 TAU
21 SCO	22 CAP	21 CAP	23 PIS	23 ARI	24 GEM
23 SAG	25 AQU	24 AQU	25 ARI	25 TAU	26 CAN
26 CAP	27 PIS	26 PIS	28 TAU	27 GEM	28 LEO
28 AQU		29 ARI	30 GEM	29 CAN	30 VIR
31 PIS		31 TAU		31 LEO	

JUL	AUG	SEP	OCT	NOV	DEC
2 LIB	2 SAG	1 CAP	1 AQU	2 ARI	2 TAU
4 SCO	5 CAP	4 AQU	3 PIS	5 TAU	4 GEM
6 SAG	7 AQU	6 PIS	6 ARI	7 GEM	7 CAN
8 CAP	10 PIS	8 ARI	8 TAU	8 CAN	9 LEO
11 AQU	12 ARI	11 TAU	11 GEM	11 LEO	11 VIR
14 PIS	15 TAU	14 GEM	13 CAN	13 VIR	13 LIB
16 ARI	17 GEM	16 CAN	15 LEO	16 LIB	15 SCO
19 TAU	19 CAN	18 LEO	17 VIR	18 SCO	17 SAG
21 GEM	21 LEO	20 VIR	19 LIB	20 SAG	20 CAP
23 CAN	23 VIR	22 LIB	21 SCO	22 CAP	22 AQU
25 LEO	25 LIB	24 SCO	24 SAG	25 AQU	24 PIS
27 VIR	28 SCO	26 SAG	26 CAP	27 PIS	27 ARI
29 LIB	30 SAG	28 CAP	28 AQU	30 ARI	29 TAU
31 SCO			31 PIS		

1929 MOON SIGNS

JAN	FEB	MAR	APR	MAY	JUN
1 LIB	2 SAG	2 SAG	1 CAP	3 PIS	2 ARI
4 SCO	5 CAP	4 CAP	3 AQU	5 ARI	4 TAU
6 SAG	7 AQU	7 AQU	5 PIS	7 TAU	6 GEM
9 CAP	10 PIS	9 PIS	8 ARI	9 GEM	8 CAN
11 AQU	12 ARI	11 ARI	10 TAU	11 CAN	10 LEO
14 PIS	14 TAU	14 TAU	12 GEM	13 LEO	12 VIR
16 ARI	16 GEM	16 GEM	14 CAN	16 VIR	14 LIB
18 TAU	19 CAN	18 CAN	16 LEO	18 LIB	16 SCO
20 GEM	21 LEO	20 LEO	18 VIR	20 SCO	19 SAG
22 CAN	23 VIR	22 VIR	21 LIB	23 SAG	21 CAP
24 LEO	25 LIB	24 LIB	23 SCO	25 CAP	24 AQU
26 VIR	27 SCO	27 SCO	25 SAG	28 AQU	26 PIS
29 LIB		29 SAG	28 CAP	30 PIS	29 ARI
31 SCO			30 AQU		

JUL	AUG	SEP	OCT	NOV	DEC
1 TAU	2 CAN	2 VIR	2 LIB	2 SAG	2 CAP
3 GEM	4 LEO	4 LIB	4 SCO	5 CAP	5 AQU
5 CAN	6 VIR	6 SCO	6 SAG	7 AQU	7 PIS
7 LEO	8 LIB	9 SAG	9 CAP	10 PIS	10 ARI
9 VIR	10 SCO	11 CAP	11 AQU	12 ARI	12 TAU
11 LIB	12 SAG	14 AQU	14 PIS	15 TAU	14 GEM
14 SCO	15 CAP	16 PIS	16 ARI	17 GEM	16 CAN
16 SAG	17 AQU	19 ARI	18 TAU	19 CAN	18 LEO
19 CAP	20 PIS	21 TAU	20 GEM	21 LEO	20 VIR
21 AQU	22 ARI	23 GEM	22 CAN	23 VIR	22 LIB
24 PIS	25 TAU	25 CAN	25 LEO	25 LIB	25 SCO
26 ARI	27 GEM	27 LEO	27 VIR	27 SCO	27 SAG
28 TAU	29 CAN	29 VIR	29 LIB	30 SAG	30 CAP
31 GEM	31 LEO		31 SCO		

1931 MOON SIGNS

JAN	FEB	MAR	APR	MAY	JUN
1 GEM	1 LEO	1 LEO	1 LIB	1 SCO	1 CAP
3 CAN	3 VIR	3 VIR	3 SCO	3 SAG	4 AQU
5 LEO	5 LIB	5 LIB	5 SAG	5 CAP	6 PIS
7 VIR	8 SCO	7 SCO	8 CAP	7 AQU	9 ARI
9 LIB	10 SAG	9 SAG	10 AQU	10 PIS	11 TAU
11 SCO	12 CAP	11 CAP	13 PIS	12 ARI	14 GEM
13 SAG	15 AQU	14 AQU	15 ARI	15 TAU	16 CAN
16 CAP	17 PIS	16 PIS	18 TAU	17 GEM	18 LEO
18 AQU	20 ARI	19 ARI	20 GEM	20 CAN	20 VIR
21 PIS	22 TAU	21 TAU	22 CAN	22 LEO	22 LIB
23 ARI	25 GEM	24 GEM	25 LEO	24 VIR	24 SCO
26 TAU	27 CAN	26 CAN	27 VIR	26 LIB	27 SAG
28 GEM		28 LEO	29 LIB	28 SCO	29 CAP
30 CAN		30 VIR		30 SAG	

JUL	AUG	SEP	OCT	NOV	DEC
1 AQU	2 ARI	1 TAU	1 GEM	2 LEO	1 VIR
4 PIS	5 TAU	4 GEM	3 CAN	4 VIR	3 LIB
6 ARI	7 GEM	6 CAN	6 LEO	6 LIB	6 SCO
9 TAU	10 CAN	8 LEO	8 VIR	8 SCO	8 SAG
11 GEM	12 LEO	10 VIR	10 LIB	10 SAG	10 CAP
13 CAN	14 VIR	12 LIB	12 SCO	12 CAP	12 AQU
15 LEO	16 LIB	14 SCO	14 SCO	14 AQU	14 PIS
17 VIR	18 SCO	16 SAG	16 CAP	17 PIS	17 ARI
20 LIB	20 SAG	19 SAG	18 AQU	19 ARI	19 TAU
22 SCO	22 CAP	21 AQU	21 PIS	22 TAU	22 GEM
24 SAG	25 AQU	23 PIS	23 ARI	24 GEM	24 CAN
26 CAP	27 PIS	26 ARI	26 TAU	27 CAN	26 LEO
28 AQU	30 ARI	28 TAU	28 GEM	29 LEO	29 VIR
31 PIS			31 CAN		31 LIB

1932 MOON SIGNS

JAN	FEB	MAR	APR	MAY	JUN
2 SCO	2 CAP	1 CAP	2 PIS	1 ARI	2 GEM
4 SAG	5 AQU	3 AQU	4 ARI	4 TAU	5 CAN
6 CAP	7 PIS	5 PIS	7 TAU	6 GEM	7 LEO
8 AQU	9 ARI	8 ARI	9 GEM	9 CAN	10 VIR
11 PIS	12 TAU	10 TAU	12 CAN	11 LEO	12 LIB
13 ARI	14 GEM	13 GEM	14 LEO	13 VIR	14 SCO
16 TAU	17 CAN	15 CAN	16 VIR	16 LIB	16 SAG
18 GEM	19 LEO	18 LEO	18 LIB	18 SCO	18 CAP
20 CAN	21 VIR	20 VIR	20 SCO	20 SAG	20 AQU
23 LEO	23 LIB	22 LIB	22 SAG	22 CAP	22 PIS
25 VIR	25 SCO	24 SCO	24 CAP	24 AQU	25 ARI
27 LIB	27 SAG	26 SAG	26 AQU	26 PIS	27 TAU
29 SCO		28 CAP	29 PIS	29 ARI	30 GEM
31 SAG		30 AQU		31 TAU	

JUL	AUG	SEP	OCT	NOV	DEC
2 CAN	1 LEO	2 LIB	1 SCO	2 CAP	1 AQU
5 LEO	3 VIR	4 SCO	3 SAG	4 AQU	3 PIS
7 VIR	5 LIB	6 SAG	5 CAP	6 PIS	6 ARI
9 LIB	7 SCO	8 CAP	7 AQU	8 ARI	8 TAU
11 SCO	10 SAG	10 AQU	10 PIS	11 TAU	11 GEM
13 SAG	12 CAP	12 PIS	12 ARI	13 GEM	13 CAN
15 CAP	14 AQU	15 ARI	15 TAU	16 CAN	16 LEO
18 AQU	16 PIS	17 TAU	17 GEM	18 LEO	18 VIR
20 PIS	19 ARI	20 GEM	20 CAN	21 VIR	20 LIB
22 ARI	21 TAU	22 CAN	22 LOE	23 LIB	22 SCO
25 TAU	24 GEM	25 LEO	24 VIR	25 SCO	24 SAG
27 GEM	26 CAN	27 VIR	27 LIB	27 SAG	26 CAN
30 CAN	28 LEO	29 LIB	29 SCO	29 CAP	28 AQU
	31 VIR		31 SAG		31 PIS

1934 MOON SIGNS

JAN	FEB	MAR	APR	MAY	JUN
2 LEO	1 VIR	2 LIB	1 SCO	2 CAP	1 AQU
4 VIR	3 LIB	5 SCO	3 SAG	5 AQU	3 PIS
7 LIB	5 SCO	7 SAG	5 CAP	7 PIS	5 ARI
9 SCO	8 SAG	8 CAP	7 AQU	9 ARI	7 TAU
11 SAG	10 CAP	11 AQU	9 PIS	11 TAU	10 GEM
13 CAP	12 AQU	13 PIS	12 ARI	13 GEM	12 CAN
15 AQU	14 PIS	15 ARI	14 TAU	16 CAN	15 LEO
17 PIS	16 ARI	17 TAU	16 GEM	18 LEO	17 VIR
19 ARI	18 TAU	20 GEM	19 CAN	21 VIR	20 LIB
22 TAU	20 GEM	22 CAN	21 LEO	23 LIB	22 SCO
24 GEM	23 CAN	25 LEO	24 VIR	26 SCO	24 SAG
27 CAN	25 LEO	27 VIR	26 LIB	28 SAG	26 CAP
29 LEO	28 VIR	30 LIB	28 SCO	30 CAP	28 AQU
			30 SAG		30 PIS

JUL	AUG	SEP	OCT	NOV	DEC
2 ARI	1 TAU	2 CAN	2 LEO	1 VRI	3 SCO
5 TAU	3 GEM	4 LEO	4 VIR	3 LIB	5 SAG
7 GEM	6 CAN	7 VIR	7 LIB	5 SCO	7 CAP
9 CAN	8 LEO	9 LIB	9 SCO	8 SAG	8 AQU
12 LEO	11 VIR	12 SCO	11 SAG	10 CAP	11 PIS
15 VIR	13 LIB	14 SAG	14 CAP	12 AQU	13 ARI
17 LIB	16 SCO	16 CAP	16 AQU	14 PIS	16 TAU
19 SCO	18 SAG	18 AQU	18 PIS	16 ARI	18 GEM
22 SAG	20 CAP	20 PIS	20 ARI	18 TAU	20 CAN
24 CAP	22 AQU	23 ARI	22 TAU	21 GEM	23 LEO
26 AQU	24 PIS	25 TAU	24 GEM	23 CAN	25 VIR
28 PIS	26 ARI	27 GEM	27 CAN	25 LEO	28 LIB
30 ARI	28 TAU	29 CAN	29 LEO	28 VIR	30 SCO
	30 GEM			30 LIB	

1933 MOON SIGNS

JAN	FEB	MAR	APR	MAY	JUN
2 ARI	1 TAU	3 GEM	1 CAN	1 LEO	2 LIB
4 TAU	3 GEM	5 CAN	4 LEO	4 VIR	4 SCO
7 GEM	6 CAN	8 LEO	6 VIR	6 LIB	6 SAG
9 CAN	8 LEO	10 VIR	8 LIB	8 SCO	8 CAP
12 LEO	10 VIR	12 LIB	11 SCO	10 SAG	10 AQU
14 VIR	13 LIB	14 SCO	13 SAG	14 CAP	14 PIS
16 LIB	15 SCO	16 SAG	15 CAP	14 AQU	15 ARI
19 SCO	17 SAG	18 CAP	17 AQU	16 PIS	17 TAU
21 SAG	19 CAP	20 AQU	19 PIS	19 ARI	20 GEM
23 CAP	21 AQU	23 PIS	21 ARI	21 TAU	22 CAN
25 AQU	23 PIS	25 ARI	24 TAU	23 GEM	25 LEO
27 PIS	26 ARI	27 TAU	26 GEM	26 CAN	27 VIR
29 ARI	28 TAU	30 GEM	29 CAN	29 LEO	30 LIB
				31 VIR	

JUL	AUG	SEP	OCT	NOV	DEC
2 SCO	2 CAP	1 AQU	2 ARI	1 TAU	1 GEM
4 SAG	4 AQU	3 PIS	5 TAU	3 GEM	3 CAN
6 CAP	6 PIS	5 ARI	7 GEM	6 CAN	6 LEO
8 AQU	8 ARI	7 TAU	9 CAN	8 LEO	8 VIR
10 PIS	11 TAU	10 GEM	12 LEO	11 VIR	11 LIB
12 ARI	13 GEM	12 CAN	14 VIR	13 LIB	13 SCO
14 TAU	16 CAN	15 LEO	17 LIB	15 SCO	15 SAG
17 GEM	18 LEO	17 VIR	19 SCO	17 SAG	17 CAP
20 CAN	21 VIR	19 LIB	21 SAG	19 CAP	19 AQU
22 LEO	23 LIB	21 SCO	23 CAP	21 AQU	21 PIS
24 VIR	25 SCO	24 SAG	25 AQU	23 PIS	23 ARI
27 LIB	27 SAG	26 CAP	27 PIS	26 ARI	25 TAU
29 SCO	29 CAP	28 AQU	30 ARI	28 TAU	28 GEM
31 SAG		30 PIS			

1935 MOON SIGNS

JAN	FEB	MAR	APR	MAY	JUN
1 SAG	2 AQU	2 AQU	2 ARI	1 TAU	2 CAN
4 CAP	4 PIS	4 PIS	4 TAU	4 GEM	5 LEO
6 AQU	6 ARI	6 ARI	6 GEM	6 CAN	7 VIR
8 PIS	8 TAU	8 TAU	8 CAN	8 LEO	10 LIB
10 ARI	10 GEM	10 GEM	11 LEO	11 VIR	12 SCO
12 TAU	13 CAN	12 CAN	13 VIR	13 LIB	14 SAG
14 GEM	15 LEO	15 LEO	16 LIB	16 SCO	17 CAP
17 CAN	18 VIR	17 VIR	18 SCO	18 SCO	19 AQU
19 LEO	20 LIB	29 LIB	21 SAG	20 CAP	21 PIS
22 VIR	23 SCO	22 SCO	23 CAP	22 AQU	23 ARI
24 LEO	25 SAG	24 SAG	25 AQU	24 PIS	25 TAU
27 SCO	27 CAP	27 CAP	27 PIS	27 ARI	27 GEM
29 SAG		29 AQU	29 ARI	29 TAU	29 CAN
31 CAP		31 PIS		31 VIR	

JUL	AUG	SEP	OCT	NOV	DEC
2 LEO	1 VIR	2 SCO	2 SAG	2 AQU	2 PIS
4 VIR	3 LIB	4 SAG	4 CAP	5 PIS	4 ARI
7 LIB	6 SCO	7 CAP	6 AQU	7 ARI	6 TAU
9 SCO	8 SAG	9 AQU	8 PIS	9 TAU	8 GEM
12 SAG	10 CAP	11 PIS	10 ARI	11 GEM	10 CAN
14 CAP	12 AQU	13 ARI	12 TAU	13 CAN	13 LOE
16 AQU	14 PIS	15 TAU	14 GEM	15 LEO	15 VIR
18 PIS	16 ARI	17 GEM	17 CAN	18 VIR	18 LIB
20 ARI	18 TAU	19 CAN	19 LEO	20 LIB	20 SCO
22 TAU	21 GEM	22 LEO	21 VIR	23 SCO	23 SAG
24 GEM	23 CAN	24 VIR	24 LIB	25 SAG	25 CAP
27 CAN	25 LEO	27 LIB	26 SCO	27 CAP	26 AQU
29 LEO	28 VIR	29 SCO	29 SAG	30 AQU	29 PIS
	30 LIB		31 CAP		31 ARI

1936 MOON SIGNS

JAN	FEB	MAR	APR	MAY	JUN
2 TAU	1 GEM	1 CAN	2 VIR	2 LIB	1 SCO
5 GEM	3 CAN	4 LEO	5 LIB	5 SCO	3 SAG
7 CAN	5 LEO	6 VIR	7 SCO	7 SAG	6 CAP
9 LEO	8 VIR	9 LIB	10 SAG	9 CAP	8 AQU
11 VIR	10 LIB	11 SCO	12 CAP	12 AQU	10 PIS
14 LIB	13 SCO	14 SAG	15 AQU	14 PIS	12 ARI
16 SCO	15 SAG	16 CAP	17 PIS	16 ARI	14 TAU
19 SAG	18 CAP	18 AQU	19 ARI	18 TAU	17 GEM
21 CAP	20 AQU	20 PIS	21 TAU	20 GEM	19 CAN
23 AQU	22 PIS	22 ARI	23 GEM	22 CAN	21 LEO
25 PIS	24 ARI	24 TAU	25 CAN	24 LEO	23 VIR
27 ARI	26 TAU	26 GEM	27 LEO	27 VIR	26 LIB
30 TAU	28 GEM	28 CAN	30 VIR	29 LIB	28 SCO
		31 LEO			

JUL	AUG	SEP	OCT	NOV	DEC
1 SCO	2 AQU	2 ARI	2 TAU	2 CAN	2 LEO
3 CAP	4 PIS	4 TAU	4 GEM	4 LEO	4 VIR
5 AQU	6 ARI	6 GEM	6 CAN	7 VIR	6 LIB
8 PIS	8 TAU	9 CAN	8 LEO	9 LIB	9 SCO
10 GEM	10 GEM	11 LEO	10 VIR	12 SCO	11 SAG
12 TAU	12 CAN	13 VIR	13 LIB	14 SAG	14 CAP
14 GEM	15 LEO	15 LIB	15 SCO	17 CAP	16 AQU
16 CAN	17 VIR	18 SCO	18 SAG	19 AQU	19 PIS
18 LEO	19 LIB	21 SAG	20 CAP	21 PIS	21 ARI
21 VIR	22 SCO	23 CAP	23 AQU	24 ARI	23 TAU
23 LIB	24 SAG	25 AQU	25 PIS	26 TAU	25 GEM
26 SCO	27 CAP	28 PIS	27 ARI	28 GEM	27 CAN
28 SAG	29 AQU	30 ARI	29 TAU	30 CAN	29 LEO
30 CAP	31 PIS		31 GEM		31 VIR

1937 MOON SIGNS

JAN	FEB	MAR	APR	MAY	JUN
3 LIB	2 SCO	1 SCO	2 CAP	2 AQU	1 PIS
5 SCO	4 SAG	3 SAG	5 AQU	4 PIS	3 ARI
8 SAG	7 CAP	6 CAP	7 PIS	7 ARI	5 TAU
10 CAP	9 AQU	8 AQU	9 ARI	9 TAU	7 GEM
13 AQU	11 PIS	11 PIS	11 TAU	11 GEM	9 CAN
15 PIS	13 ARI	13 ARI	13 GEM	13 CAN	11 LEO
17 ARI	15 TAU	15 TAU	15 CAN	15 LEO	13 VIR
19 TAU	18 GEM	17 GEM	17 LEO	17 VIR	16 LIB
21 CAN	20 CAN	19 CAN	20 VIR	19 LIB	18 SCO
23 CAN	22 LEO	21 LEO	22 LIB	22 SCO	21 SAG
26 LEO	24 VIR	23 VIR	25 SCO	24 SAG	23 CAP
28 VIR	26 LIB	26 LIB	27 SAG	27 CAP	26 AQU
28 LIB		28 SCO	30 CAP	29 AQU	28 PIS
		31 SAG			30 ARI

JUL	AUG	SEP	OCT	NOV	DEC
2 TAU	1 GEM	1 LEO	1 VIR	2 SCO	1 SAG
4 GEM	3 CAN	3 VIR	3 LIB	4 SAG	4 CAP
6 CAN	5 LEO	6 LIB	5 SCO	7 CAP	6 AQU
8 LEO	7 VIR	8 SCO	8 SAG	9 AQU	9 PIS
11 VIR	9 LIB	11 SAG	10 CAP	12 PIS	11 ARI
13 LIB	12 SCO	13 CAP	13 AQU	14 ARI	13 TAU
15 SCO	14 SAG	15 AQU	15 PIS	16 TAU	15 GEM
18 SAG	17 CAP	18 PIS	17 ARI	18 GEM	17 CAN
20 AQU	19 AQU	20 ARI	19 TAU	20 CAN	19 LEO
23 AQU	21 PIS	22 TAU	21 GEM	22 LEO	21 VIR
25 LIS	24 ARI	24 GEM	24 CAN	24 VIR	24 LIB
27 ARI	26 TAU	26 CAN	26 LEO	26 LIB	26 SCO
30 TAU	28 GEM	28 LEO	28 VIR	29 SCO	29 SAG
	30 CAN		30 LIB		31 CAP

1938 MOON SIGNS

JAN	FEB	MAR	APR	MAY	JUN
3 AQU	1 PIS	1 PIS	1 TAU	1 GEM	1 LEO
5 PIS	4 ARI	3 ARI	4 GEM	3 CAN	3 VIR
7 ARI	6 TAU	5 TAU	6 CAN	5 LEO	6 LIB
10 TAU	8 GEM	7 GEM	8 LEO	7 VIR	8 SCO
12 GEM	10 CAN	9 CAN	10 VIR	9 LIB	10 SAG
14 CAN	12 LOE	12 LEO	12 LIB	12 SCO	13 CAP
16 LEO	14 VIR	14 VIR	15 SCO	14 SAG	16 AQU
18 VIR	16 LIB	16 LIB	17 SAG	17 CAP	18 PIS
20 LIB	19 SCO	18 SCO	19 CAP	19 AQU	20 ARI
22 SCO	21 SAG	21 SAG	22 AQU	22 PIS	23 TAU
25 SAG	24 CAP	23 CAP	24 PIS	24 ARI	25 GEM
27 CAP	26 AQU	26 AQU	27 ARI	26 TAU	27 CAN
30 AQU		28 PIS	29 TAU	28 GEM	29 LEO
		30 ARI		30 CAN	

JUL	AUG	SEP	OCT	NOV	DEC
1 VIR	2 SCO	3 CAP	3 AQU	2 PIS	1 ARI
3 LIB	4 SAG	5 AQU	5 PIS	4 ARI	4 TAU
5 SCO	7 CAP	8 PIS	8 ARI	6 TAU	6 GEM
8 SAG	9 AQU	10 ARI	10 TAU	8 GEM	8 CAN
10 CAP	12 PIS	12 TAU	12 GEM	10 CAN	10 LEO
13 AQU	14 ARI	15 GEM	14 CAN	12 LEO	12 VIR
15 PIS	16 TAU	17 CAN	16 LEO	15VIR	14 LIB
18 ARI	18 GEM	19 LEO	18 VIR	17 LIB	16 SCO
20 TAU	21 CAN	21 VIR	20 LIB	19 SCO	19 SAG
22 GEM	23 LEO	23 LIB	23 SCO	21 SAG	21 CAP
24 CAN	25 VIR	25 SCO	25 SAG	24 CAP	24 AQU
26 LEO	27 LIB	28 SAG	27 CAP	26 AQU	26 PIS
28 VIR	29 SCO	30 CAP	30 AQU	29 PIS	29 ARI
30 LIB	31 SAG				31 TAU

1939 MOON SIGNS

JAN	FEB	MAR	APR	MAY	JUN
2 GEM	1 CAN	2 LEO	3 LIB	2 SCO	1 SAG
4 CAN	3 LEO	4 VIR	5 SCO	4 SAG	3 CAP
6 LEO	5 VIR	6 LIB	7 SAG	7 CAP	7 AQU
8 VIR	7 LIB	8 SCO	9 CAP	9 AQU	8 PIS
10 LIB	9 SCO	10 SAG	12 AQU	12 PIS	10 ARI
12 SCO	11 SAG	13 CAP	14 PIS	14 ARI	13 TAU
15 SAG	14 CAP	15 AQU	17 ARI	16 TAU	15 GEM
17 CAP	16 AQU	18 PIS	19 TAU	19 GEM	17 CAN
20 AQU	19 PIS	20 ARI	21 GEM	21 CAN	19 LEO
22 PIS	21 ARI	23 TAU	24 CAN	23 LEO	21 VIR
25 ARI	24 TAU	25 GEM	26 LEO	25 VIR	23 LIB
27 TAU	26 GEM	27 CAN	28 VIR	27 LIB	26 SCO
30 GEM	28 CAN	29 LEO	30 LIB	29 SCO	28 SAG
		31 VIR			30 CAP

JUL	AUG	SEP	OCT	NOV	DEC
3 AQU	1 PIS	2 TAU	2 GEM	1 CAN	2 VIR
5 PIS	4 ARI	5 GEM	5 CAN	3 LEO	5 LIB
8 ARI	6 TAU	7 CAN	7 LEO	5 VIR	7 SCO
10 TAU	9 GEM	9 LEO	9 VIR	7 LIB	9 SAG
12 GEM	11 CAN	11 VIR	11 LIB	9 SCO	11 CAP
15 CAN	13 LEO	13 LIB	13 SCO	11 SAG	13 AQU
17 LEO	15 VIR	15 SCO	15 SAG	14 CAP	16 PIS
19 VIR	17 LIB	18 SAG	17 CAP	16 AQU	19 ARI
21 LIB	19 SCO	20 CAP	20 AQU	19 PIS	21 TAU
23 SCO	21 SAG	22 AQU	22 PIS	21 ARI	23 GEM
25 SAG	24 CAP	25 PIS	25 ARI	24 TAU	26 CAN
27 CAP	26 AQU	28 ARI	27 TAU	26 GEM	28 LEO
30 AQU	29 PIS	30 TAU	30 GEM	28 CAN	30 VIR
	31 ARI			30 LEO	

1940 MOON SIGNS

JAN	FEB	MAR	APR	MAY	JUN
1 LIB	1 SAG	2 CAP	1 AQU	3 ARI	2 TAU
3 SCO	4 CAP	4 AQU	3 PIS	5 TAU	4 GEM
5 SAG	6 AQU	7 PIS	6 ARI	8 GEM	6 CAN
7 CAP	9 PIS	9 ARI	8 TAU	10 CAN	9 LEO
10 AQU	11 ARI	12 TAU	11 GEM	12 LEO	12 VIR
12 LIS	14 TAU	14 GEM	13 CAN	15 VIR	13 LIB
15 ARI	16 GEM	17 CAN	15 LEO	17 LIB	15 SCO
17 TAU	18 CAN	19 LEO	17 VIR	19 SCO	17 SAG
20 GEM	20 LEO	21 VIR	19 LIB	21 SAG	19 CAP
22 CAN	22 VIR	23 LIB	21 SCO	23 CAP	22 AQU
24 LEO	24 LIB	25 SCO	23 SAG	25 AQU	24 PIS
26 VIR	26 SCO	27 SAG	26 CAP	28 PIS	27 ARI
28 LIB	29 SAG	29 CAP	28 AQU	30 ARI	28 TAU
30 SCO			30 PIS		

JUL	AUG	SEP	OCT	NOV	DEC
2 GEM	2 LEO	1 VIR	2 SCO	1 SAG	2 AQU
4 CAN	4 VIR	3 LIB	4 SAG	3 CAP	5 PIS
6 LEO	6 LIB	5 SCO	6 CAP	5 AQU	7 ARI
8 VIR	9 SCO	7 SAG	9 AQU	7 PIS	10 TAU
10 LIB	11 SAG	9 CAP	11 PIS	10 ARI	12 GEM
12 SCO	13 CAP	11 AQU	14 ARI	13 TAU	15 CAN
14 SAG	15 AQU	14 PIS	16 TAU	15 GEM	17 LEO
17 CAP	18 PIS	16 ARI	19 GEM	17 CAN	19 VIR
19 AQU	20 ARI	19 TAU	21 CAN	20 LEO	21 LIB
21 PIS	23 TAU	22 GEM	23 LEO	22 VIR	23 SCO
24 ARI	25 GEM	24 CAN	26 VIR	24 LIB	26 SAG
26 TAU	28 CAN	26 LEO	28 LIB	26 SCO	28 CAP
29 GEM	30 LEO	28 VIR	30 SCO	28 SAG	30 AQU
31 CAN		30 LIB		30 CAP	

1941 MOON SIGNS

JAN	FEB	MAR	APR	MAY	JUN
1 PIS	2 TAU	2 TAU	1 GEM	3 LEO	1 VIR
4 ARI	5 GEM	4 GEM	3 CAN	5 VIR	4 LIB
6 TAU	7 CAN	7 CAN	5 LEO	7 LIB	6 SCO
9 GEM	10 LEO	9 LEO	8 VIR	9 SCO	8 SAG
11 CAN	12 VIR	11 VIR	10 LIB	11 SAG	10 CAP
13 LEO	14 LIB	13 LIB	12 SCO	13 CAP	12 AQU
15 VIR	16 SCO	15 SCO	14 SAG	15 AQU	14 PIS
18 LIB	18 SAG	17 SAG	16 CAP	18 PIS	16 ARI
20 SCO	20 CAP	19 CAP	18 AQU	20 ARI	19 TAU
22 SAG	23 AQU	22 AQU	20 PIS	23 TAU	21 GEM
24 CAP	25 PIS	24 PIS	23 ARI	25 GEM	24 CAN
26 AQU	27 ARI	27 ARI	25 TAU	28 CAN	26 LEO
29 PIS		29 TAU	28 GEM	30 LEO	29 VIR
31 ARI			30 CAN		

JUL	AUG	SEP	OCT	NOV	DEC
1 LIB	1 SAG	2 AQU	1 PIS	2 TAU	2 GEM
3 SCO	3 CAP	4 PIS	4 ARI	5 GEM	5 CAN
5 SAG	5 AQU	6 ARI	6 TAU	7 CAN	7 LEO
7 CAP	8 PIS	9 TAU	9 GEM	10 LEO	10 VIR
9 AQU	10 ARI	11 GEM	11 CAN	12 VIR	12 LIB
11 PIS	13 TAU	14 CAN	14 LEO	15 LIB	14 SCO
14 ARI	15 GEM	16 LEO	16 VIR	17 SCO	16 SAG
16 TAU	18 CAN	19 VIR	18 LIB	19 SAG	18 CAP
19 GEM	20 LEO	21 LIB	20 SCO	21 CAP	20 AQU
21 CAN	22 VIR	23 SCO	22 SAG	23 AQU	22 PIS
24 LEO	24 LIB	25 SAG	24 CAP	25 PIS	24 ARI
26 VIR	26 SCO	27 CAP	26 AQU	27 ARI	27 TAU
28 LIB	28 SAG	29 AQU	29 PIS	30 TAU	29 GEM
30 SCO	31 CAP		31 ARI		

1942 MOON SIGNS

JAN	FEB	MAR	APR	MAY	JUN
1 CAN	2 VIR	1 VIR	2 SCO	2 SAG	2 AQU
3 LEO	4 LIB	4 LIB	4 SAG	4 CAP	4 PIS
6 VIR	7 SCO	6 SCO	6 CAP	6 AQU	6 ARI
8 LIB	9 SAG	8 SAG	8 AQU	8 PIS	9 TAU
10 SCO	11 CAP	10 CAP	11 PIS	10 ARI	11 GEM
12 SAG	13 AQU	12 AQU	13 ARI	13 TAU	14 CAN
14 CAP	15 PIS	14 PIS	15 TAU	15 GEM	16 LEO
16 AQU	17 ARI	17 ARI	18 GEM	18 CAN	19 VIR
18 PIS	20 TAU	19 TAU	20 CAN	20 LEO	21 LIB
21 ARI	22 GEM	21 GEM	23 LEO	23 VIR	23 SCO
23 TAU	25 CAN	24 CAN	25 VIR	25 LIB	25 SAG
26 GEM	27 LEO	26 LEO	27 LIB	27 SCO	27 CAP
28 CAN		29 VIR	30 SCO	29 SAG	29 AQU
31 LEO		31 LIB		31 CAP	

JUL	AUG	SEP	OCT	NOV	DEC
1 PIS	2 TAU	1 GEM	1 CAN	2 VIR	2 LIB
4 ARI	5 GEM	4 CAN	4 LEO	5 LIB	4 SCO
6 TAU	7 CAN	6 LEO	6 VIR	7 SCO	6 SAG
9 GEM	10 LEO	9 VIR	8 LIB	9 SAG	8 CAP
11 CAN	12 VIR	11 LIB	10 SCO	11 CAP	10 AQU
14 LEO	15 LIB	13 SCO	13 SAG	13 AQU	12 PIS
16 VIR	17 SCO	15 SAG	15 CAP	15 PIS	15 ARI
18 LIB	19 SAG	17 CAP	17 AQU	17 ARI	17 TAU
21 SCO	21 CAP	20 AQU	19 PIS	20 TAU	19 GEM
23 SAG	23 AQU	22 PIS	21 ARI	22 GEM	22 CAN
25 CAP	25 PIS	24 ARI	23 TAU	25 CAN	24 LEO
27 AQU	27 ARI	26 TAU	26 GEM	27 LEO	27 VIR
29 PIS	30 TAU	29 GEM	28 CAN	30 VIR	29 LIB
31 ARI			31 LEO		

1943 MOON SIGNS

JAN	FEB	MAR	APR	MAY	JUN
1 SCO	1 CAP	1 CAP	1 PIS	3 TAU	1 GEM
3 SAG	3 AQU	3 AQU	3 ARI	5 GEM	4 CAN
5 CAP	5 PIS	5 PIS	5 TAU	7 CAN	6 LEO
7 AQU	7 ARI	7 ARI	8 GEM	10 LEO	9 VIR
9 PIS	10 TAU	9 TAU	10 CAN	12 VIR	11 LIB
11 ARI	12 GEM	11 GEM	13 LEO	15 LIB	14 SCO
13 TAU	14 CAN	14 CAN	15 VIR	17 SCO	16 SAG
16 GEM	17 LEO	16 LEO	18 LIB	19 SAG	18 CAP
18 CAN	19 VIR	19 VIR	20 SCO	21 CAP	20 AQU
21 LEO	22 LIB	21 LIB	22 SAG	23 AQU	22 PIS
23 VIR	24 SCO	23 SCO	24 CAP	26 PIS	24 ARI
26 LIB	26 SAG	26 SAG	26 AQU	28 ARI	26 TAU
28 SCO		28 CAP	28 PIS	30 TAU	29 GEM
30 SAG		30 AQU	30 ARI		

JUL	AUG	SEP	OCT	NOV	DEC
1 CAN	2 VIR	1 LIB	1 SCO	1 CAP	1 AQU
4 LEO	5 LIB	3 SCO	3 SAG	4 AQU	3 PIS
6 VIR	7 SCO	6 SAG	5 CAP	6 PIS	5 ARI
9 LIB	10 SAG	8 CAP	7 AQU	8 ARI	7 TAU
11 SCO	12 CAP	10 AQU	9 PIS	10 TAU	10 GEM
13 SAG	14 AQU	12 PIS	12 ARI	12 GEM	12 CAN
15 CAP	16 PIS	14 ARI	14 TAU	15 CAN	14 LEO
17 AQU	18 ARI	16 TAU	16 GEM	17 LEO	17 VIR
19 PIS	20 TAU	18 GEM	18 CAN	20 VIR	19 LIB
21 ARI	22 GEM	21 CAN	21 LEO	22 LIB	22 SCO
23 TAU	25 CAN	23 LEO	23 VIR	24 SCO	24 SAG
26 GEM	27 LEO	26 VIR	25 LIB	27 SAG	27 CAP
28 CAN	30 VIR	28 LIB	28 SCO	28 CAP	29 AQU
31 LEO			30 SAG		30 PIS

1944 MOON SIGNS

JAN	FEB	MAR	APR	MAY	JUN
1 ARI	2 GEM	3 CAN	1 LEO	1 VIR	3 SCO
3 TAU	4 CAN	5 LEO	4 VIR	4 LIB	5 SAG
6 GEM	7 LEO	8 VIR	6 LIB	6 SCO	7 CAP
8 CAN	9 VIR	10 LIB	9 SCO	9 SAG	9 AQU
11 LEO	12 LIB	13 SCO	11 SAG	11 CAP	11 PIS
13 VIR	14 SCO	15 SAG	14 CAP	13 AQU	13 ARI
16 LIB	17 SAG	17 CAP	16 AQU	15 PIS	16 TAU
18 SCO	19 CAP	19 AQU	18 PIS	17 ARI	18 GEM
20 SAG	21 PIS	22 PIS	20 ARI	19 TAU	20 CAN
23 CAP	23 PIS	24 ARI	22 TAU	21 GEM	22 LEO
25 AQU	25 ARI	26 TAU	24 GEM	24 CAN	25 VIR
27 PIS	27 TAU	28 GEM	26 CAN	26 LEO	27 LIB
29 ARI	29 GEM	30 CAN	29 LEO	29 VIR	30 SCO
31 TAU				31 LIB	

JUL	AUG	SEP	OCT	NOV	DEC
2 SAG	1 CAP	1 PIS	1 ARI	1 GEM	1 CAN
4 CAP	3 ARI	3 ARI	3 TAU	4 CAN	3 LEO
7 AQU	5 PIS	5 TAU	5 GEM	6 LEO	6 VIR
9 PIS	7 ARI	8 GEM	7 CAN	8 VIR	8 LIB
11 ARI	9 TAU	10 CAN	10 LEO	11 LIB	11 SCO
13 TAU	11 GEM	12 LEO	12 VIR	13 SCO	13 SAG
15 GEM	14 CAN	15 VIR	15 LIB	16 SAG	15 CAP
17 CAN	16 LEO	17 LIB	17 SCO	18 CAP	18 AQU
20 LEO	18 VI	20 SCO	19 SAG	20 AQU	20 PIS
22 VIR	21 LIB	22 SAG	22 CAP	23 PIS	22 ARI
25 LIB	24 SCO	25 CAP	24 AQU	25 ARI	24 TAU
27 SCO	26 SAG	27 AQU	26 PIS	27 TAU	26 GEM
30 SAG	28 CAP	29 PIS	28 ARI	29 GEM	28 CAN
	30 AQU		30 TAU		31 LEO

1945 MOON SIGNS

JAN	FEB	MAR	APR	MAY	JUN
2 VIR	1 LIB	3 SCO	1 SAG	1 CAP	2 PIS
4 LIB	3 SCO	5 SAG	4 CAP	3 AQU	4 ARI
7 SCO	6 SAG	8 CAP	6 AQU	6 PIS	6 TAU
9 SAG	8 CAP	10 AQU	8 PIS	8 ARI	8 GEM
12 CAP	10 AQU	12 PIS	10 ARI	10 TAU	10 CAN
14 AQU	12 PIS	14 ARI	12 ATU	12 GEM	12 LEO
16 PIS	14 ARI	16 TAU	14 GEM	14 CAN	15 VIR
18 ARI	17 TAU	18 GEM	16 CAN	16 LEO	17 LIB
20 TAU	19 GEM	20 CAN	19 LEO	18 VIR	20 SCO
22 GEM	21 CAN	22 LEO	21 VIR	21 LIB	22 SAG
25 CAN	23 LEO	25 VIR	24 LIB	23 SCO	25 CAP
27 LEO	26 VIR	27 LIB	26 SCO	26 SAG	27 AQU
29 VIR	28 LIB	30 SCO	29 SAG	29 CAP	29 PIS
				31 AQU	

JUL	AUG	SEP	OCT	NOV	DEC
1 ARI	2 GEM	2 LEO	2 VIR	1 LIB	3 SAG
3 TAU	4 CAN	5 VIR	4 LIB	3 SCO	6 CAP
6 GEM	6 LEO	7 LIB	7 SCO	6 SAG	8 AQU
8 CAN	8 VIR	10 SCO	10 SAG	8 CAP	10 PIS
10 LEO	11 LIB	12 SAG	12 CAP	11 AQU	12 ARI
12 VIR	13 SCO	15 CAP	14 AQU	13 PIS	15 TAU
15 LIB	16 SAG	17 AQU	17 PIS	15 ARI	17 GEM
17 SCO	18 CAP	19 PIS	19 ARI	17 TAU	19 CAN
20 SAG	21 AQU	21 ARI	21 TAU	19 GEM	21 LEO
22 CAP	23 PIS	23 TAU	23 GEM	21 CAN	23 VIR
24 AQU	25 ARI	25 GEM	25 CAN	23 LEO	25 LIB
26 PIS	27 TAU	27 CAN	27 LEO	26 VIR	28 SCO
29 ARI	29 GEM	30 LEO	29 VIR	28 LIB	30 SAG
31 TAU	31 CAN			30 SCO	

1946 MOON SIGNS

JAN	FEB	MAR	APR	MAY	JUN
2 CAP	1 AQU	2 PIS	1 ARI	2 GEM	1 CAN
4 AQU	3 PIS	4 ARI	3 TAU	4 CAN	3 LEO
6 PIS	5 ARI	6 TAU	5 GEM	6 LEO	5 VIR
9 ARI	7 TAU	8 GEM	7 CAN	6 VIR	7 LIB
11 TAU	9 GEM	11 CAN	9 LEO	11 LIB	10 SCO
13 GEM	11 CAN	13 LEO	11 VIR	13 SCO	12 SAG
15 CAN	13 LEO	15 VIR	14 LIB	16 SAG	15 CAP
17 LEO	16 VIR	17 LIB	16 SCO	18 CAP	17 AQU
19 VIR	18 LIB	20 SCO	19 SAG	21 AQU	20 PIS
22 LIB	20 SCO	22 SAG	21 CAP	23 PIS	22 ARI
24 SCO	23 SAG	25 CAP	24 AQU	26 ARI	24 TAU
27 SAG	26 CAP	27 AQU	26 PIS	28 TAU	26 GEM
29 CAP	28 AQU	30 PIS	28 ARI	30 GEM	28 CAN
			30 TAU		30 LEO

JUL	AUG	SEP	OCT	NOV	DEC
2 VIR	1 LIB	2 SAG	2 CAP	1 AQU	3 ARI
4 LIB	3 SCO	5 CAP	4 AQU	3 PIS	5 TAU
7 SCO	6 SCO	7 AQU	7 PIS	5 ARI	7 GEM
9 SAG	8 CAP	9 PIS	9 ARI	7 TAU	9 CAP
12 CAP	11 AQU	12 ARI	11 TAU	10 GEM	11 LEO
14 AQU	13 PIS	14 TAU	13 GEM	12 CAN	13 VIR
17 PIS	15 ARI	16 GEM	15 CAN	14 LEO	15 LIB
19 ARI	17 TAU	18 CAN	17 LEO	16 VIR	18 SCO
21 TAU	20 GEM	20 LEO	20 VIR	18 LIB	20 SAG
23 GEM	22 CAN	22 VIR	22 LIB	20 SCO	23 CAP
25 CAN	24 LEO	25 LIB	24 SCO	23 SAG	25 AQU
27 LEO	26 VIR	27 SCO	27 SAG	25 CAP	28 PIS
30 VIR	28 LIB	29 SAG	29 CAP	28 AQU	30 ARI
	31 SCO			30 PIS	

1947 MOON SIGNS

JAN	FEB	MAR	APR	MAY	JUN
1 TAU	2 CAN	1 CAN	2 VIR	1 LIB	2 SAG
3 GEM	4 LEO	3 LEO	4 LIB	3 SCO	5 CAP
5 CAN	6 VIR	5 VIR	6 SCO	6 SAG	7 AQU
7 LEO	8 LIB	7 LIB	8 SAG	8 CAP	10 PIS
9 VIR	10 SCO	10 SCO	11 CAP	11 AQU	12 ARI
12 LIB	13 SAG	12 SAG	13 AQU	13 PIS	14 TAU
14 SCO	15 CAP	15 CAP	16 PIS	16 ARI	16 GEM
16 SAG	18 AQU	17 AQU	18 ARI	18 TAU	18 CAN
19 CAP	20 PIS	20 PIS	20 TAU	20 GEM	20 LEO
22 AQU	23 ARI	22 ARI	23 GEM	22 CAN	22 VIR
22 PIS	25 TAU	24 TAU	25 CAN	24 LEO	25 LIB
26 ARI	27 GEM	26 GEM	27 LEO	26 VIR	27 SCO
29 TAU		28 CAN	29 VIR	28 LIB	29 SAG
31 GEM		31 LEO		31 SCO	

JUL	AUG	SEP	OCT	NOV	DEC
2 CAP	1 AQU	2 ARI	1 TAU	2 CAN	1 LOE
4 AQU	3 PIS	4 TAU	4 GEM	4 LEO	3 VIR
7 PIS	6 ARI	6 GEM	6 CAN	6 VI	6 LIB
9 ARI	8 TAU	9 CAN	8 LEO	8 LIB	8 SCO
12 TAU	10 GEM	11 LEO	10 VIR	11 SCO	10 SAG
14 GEM	12 CAN	13 VIR	12 LIB	13 SAG	13 CAP
16 CAN	14 LEO	17 SCO	14 SCO	15 CAP	15 AQU
18 LEO	16 VIR	19 SAG	17 SAG	18 AQU	18 PIS
20 VIR	18 LIB	22 CAP	19 CAP	20 PIS	20 ARI
20 LIB	20 SCO	24 AQU	22 AQU	23 ARI	23 TAU
24 SCO	23 SAG	27 PIS	24 PIS	25 TAU	25 GEM
27 SAG	25 CAP	29 ARI	26 ARI	27 GEM	27 CAN
29 CAP	28 AQU		29 TAU	29 CAN	29 LEO
	30 PIS		31 GEM		31 VIR

1948 MOON SIGNS

JAN	FEB	MAR	APR	MAY	JUN
2 LIB	3 SAG	1 SAG	2 AQU	2 PIS	1 ARI
4 SCO	5 CAP	3 CAP	5 PIS	5 ARI	3 TAU
6 SAG	8 AQU	6 AQU	7 ARI	7 TAU	6 GEM
9 CAP	10 PIS	8 PIS	10 TAU	9 GEM	8 CAN
11 AQU	13 ARI	11 ARI	12 GEM	11 CAN	10 LEO
14 PIS	15 TAU	13 TAU	14 CAN	14 LEO	12 VIR
16 ARI	17 GEM	16 GEM	16 LEO	16 VIR	14 LIB
19 TAU	20 CAN	18 CAN	18 VIR	18 LIB	16 SCO
21 GEM	22 LEO	20 LEO	21 LIB	20 SCO	18 SAG
23 CAN	24 VIR	22 VIR	23 SCO	22 SAG	21 CAP
25 LEO	26 LIB	24 LIB	25 SAG	25 CAP	23 AQU
27 VIR	28 SCO	26 SCO	27 CAP	27 AQU	26 PIS
29 LIB		28 SAG	30 AQU	29 PIS	28 PRI
31 SCO		31 CAP			

JUL	AUG	SEP	OCT	NOV	DEC
1 TAU	2 CAN	2 VIR	1 LIB	2 SAG	2 CAP
3 GEM	4 LEO	4 LIB	3 SCO	4 CAP	4 AQU
5 CAN	6 VIR	6 SCO	6 SAG	7 AQU	6 PIS
7 LEO	8 LIB	8 SAG	8 CAP	9 PIS	9 ARI
9 VIR	10 SCO	11 CAP	10 AQU	12 AQU	12 ARI
11 LIB	12 SAG	13 AQU	13 PIS	14 TAU	14 GEM
13 SCO	14 CAP	16 PIS	15 ARI	17 GEM	16 CAN
16 SAG	17 AQU	18 ARI	18 TAU	19 CAN	18 LEO
18 CAP	19 PIS	21 TAU	20 GEM	21 LEO	20 VIR
21 AQU	22 ARI	23 GEM	23 CAN	23 VIR	22 LIB
23 PIS	24 TAU	25 CAN	25 LEO	25 LIB	25 SCO
26 ARI	27 GEM	27 LEO	27 VIR	27 SCO	27 CAP
28 TAU	29 CAN	29 VIR	29 LIB	29 SAG	29 CAP
30 GEM	31 LEO		31 SCO		31 AQU

1950 MOON SIGNS

JAN	FEB	MAR	APR	MAY	JUN
3 CAN	1 LEO	1 LEO	1 LIB	1 SCO	1 CAP
5 LEO	3 VIR	3 VIR	3 SCO	3 SAG	3 AQU
7 VIR	6 LIB	5 LIB	5 SAG	5 CAP	5 PIS
9 LIB	8 SCO	7 SCO	7 CAP	7 AQU	8 ARI
11 SCO	10 SAG	8 SAG	10 AQU	9 PIS	10 TAU
14 SAG	12 CAP	11 CAP	12 PIS	12 ARI	13 GEM
16 CAP	14 AQU	13 AQU	14 ARI	14 TAU	15 CAN
18 AQU	16 PIS	16 PIS	17 TAU	17 GEM	18 LEO
20 PIS	19 ARI	18 ARI	19 GEM	19 CAN	20 VIR
22 ARI	21 TAU	21 TAU	22 CAN	22 LEO	22 LIB
25 TAU	24 GEM	23 GEM	24 LEO	24 VIR	25 SCO
28 GEM	26 CAN	26 CAN	27 VIR	26 LIB	27 SAG
30 CAN		28 LEO	29 LIB	28 SCO	29 CAP
		30 VIR		30 SAG	

JUL	AUG	SEP	OCT	NOV	DEC
1 AQU	1 ARI	3 GEM	3 CAN	2 LEO	1 VIR
3 PIS	4 TAU	5 CAN	5 LEO	4 VIR	3 LIB
5 ARI	7 GEM	8 LEO	7 VIR	6 LIB	6 SCO
8 TAU	9 CAN	10 VIR	10 LIB	8 SCO	8 SAG
10 GEM	11 LEO	12 LIB	12 SCO	10 SAG	10 CAP
13 CAN	14 VIR	14 SCO	14 SAG	12 CAP	12 AQU
15 ELO	16 LIB	16 SAG	16 CAP	14 AQU	14 PIS
17 VIR	18 SCO	18 CAP	18 AQU	16 PIS	16 ARI
20 LIB	20 SAG	21 AQU	20 PIS	19 ARI	19 TAU
22 SCO	22 CAP	23 PIS	23 ARI	21 TAU	21 GEM
24 SAG	24 AQU	25 ARI	25 TAU	24 GEM	24 CAN
26 CAP	27 PIS	28 TAU	28 GEM	26 CAN	26 LEO
28 AQU	29 ARI	30 GEM	30 CAN	29 LEO	28 VIR
30 PIS	31 TAU				31 LIB

1949 MOON SIGNS

JAN	FEB	MAR	APR	MAY	JUN
3 PIS	2 ARI	1 ARI	2 GEM	2 CAN	2 VIR
5 ARI	4 TAU	3 TAU	5 CAN	4 LEO	5 LIB
8 TAU	7 GEM	6 GEM	7 LEO	6 VIR	7 SCO
10 GEM	9 CAN	8 CAN	9 VIR	8 LIB	9 SAG
12 CAN	11 LEO	10 LEO	11 LIB	10 SCO	11 CAP
15 LEO	13 VIR	13 VIR	13 SCO	12 SAG	13 AQU
17 VIR	15 LIB	15 LIB	15 SAG	15 CAP	16 PIS
19 LIB	17 SCO	17 SCO	17 CAP	17 AQU	18 ARI
21 SCO	19 SAG	19 SAG	19 AQU	20 PIS	21 TAU
23 SAG	22 CAP	21 CAP	22 PIS	22 ARI	23 GEM
25 CAP	24 AQU	23 AQU	24 ARI	24 TAU	25 CAN
28 AQU	26 PIS	26 PIS	27 TAU	27 GEM	28 LEO
30 PIS		28 ARI	29 GEM	29 CAN	30 VIR
		31 TAU		31 LEO	

JUL	AUG	SEP	OCT	NOV	DEC
2 LIB	2 SAG	1 CAP	3 PIS	2 ARI	1 TAU
4 SCO	5 CAP	3 AQU	5 ARI	4 TAU	4 GEM
6 SAG	7 AQU	6 PIS	8 TAU	7 GEM	6 CAN
8 CAP	9 PIS	8 ARI	10 GEM	9 CAN	9 LEO
11 AQU	12 ARI	11 TAU	13 CAN	11 LEO	11 VIR
13 PIS	14 TAU	13 GEM	15 LEO	14 VIR	13 LIB
15 ARI	17 GEM	15 CAN	17 VIR	16 LIB	15 SCO
18 TAU	19 CAN	18 LEO	19 LIB	18 SCO	17 SAG
20 GEM	21 LEO	20 VIR	21 SCO	20 SAG	19 CAP
23 CAN	23 VIR	22 LIB	23 SAG	22 CAP	21 AQU
25 LEO	25 LIB	24 SCO	25 CAP	24 AQU	24 PIS
27 VIR	27 SCO	26 SAG	28 AQU	26 PIS	27 ARI
29 LIB	30 SAG	28 CAP	30 PIS	29 ARI	29 TAU
31 SCO		30 AQU			31 GEM

1951 MOON SIGNS

JAN	FEB	MAR	APR	MAY	JUN
2 SCO	2 CAP	2 CAP	2 PIS	2 ARI	3 GEM
4 SAG	4 AQU	4 AQU	5 ARI	4 TAU	5 CAN
6 CAP	7 PIS	6 PIS	7 TAU	7 GEM	8 LEO
8 AQU	9 ARI	8 ARI	9 GEM	9 CAN	10 VIR
10 PIS	11 TAU	11 TAU	12 CAN	12 LEO	13 LIB
12 ARI	14 GEM	13 GEM	14 LEO	14 VIR	15 SCO
15 TAU	· 16 CAN	16 CAN	17 VIR	16 LIB	17 SAG
17 GEM	19 LEO	18 LEO	19 LIB	19 SCO	19 CAP
20 CAN	21 VIR	20 VIR	21 SCO	21 SAG	21 AQU
22 LEO	23 LIB	23 LIB	23 SAG	23 CAP	23 PIS
25 VIR	25 SCO	25 SCO	25 CAP	25 AQU	25 ARI
27 LIB	28 SAG	27 SAG	27 AQU	27 PIS	28 TAU
29 SCO		29 CAP	29 PIS	29 ARI	30 GEM
31 SAG		31 AQU		31 TAU	

JUL	AUG	SEP	OCT	NOV	DEC
3 CAN	1 LEO	3 LIB	2 SCO	1 SAG	2 AQU
5 LEO	4 VIR	5 SCO	4 SAG	3 CAP	4 PIS
8 VIR	6 LIB	7 SAG	6 CAP	5 AQU	6 ARI
10 LIB	9 SCO	9 CAP	8 AQU	7 PIS	9 TAU
12 SCO	11 SAG	11 AQU	11 PIS	9 ARI	11 GEM
14 SAG	13 CAP	13 PIS	13 ARI	11 TAU	13 CAN
16 CAP	15 AQU	15 ARI	15 TAU	14 GEM	16 LEO
18 AQU	17 PIS	18 TAU	17 GEM	16 CAN	19 VIR
20 PIS	19 ARI	20 GEM	20 CAN	19 LEO	21 LIB
23 ARI	21 TAU	23 CAN	22 LEO	21 VIR	23 SCO
25 TAU	24 GEM	25 LEO	25 VIR	24 LIB	25 CAP
27 GEM	26 CAN	28 VIR	27 LIB	26 SCO	27 CAP
30 CAN	29 LEO	30 LIB	29 SCO	28 SAG	29 AQU
	31 VIR			30 CAP	31 PIS

235

1952 MOON SIGNS

JAN	FEB	MAR	APR	MAY	JUN
3 ARI	1 TAU	2 GEM	1 CAN	3 VIR	2 LIB
5 TAU	3 GEM	4 CAN	3 LEO	5 LIB	4 SCO
7 GEM	6 CAN	7 LEO	6 VIR	8 SCO	6 SAG
10 CAN	9 LEO	9 VIR	8 LIB	10 SAG	8 CAP
12 LEO	11 VIR	12 LIB	10 SCO	12 CAP	10 AQU
15 VIR	14 LIB	14 SCO	13 SAG	14 AQU	12 PIS
17 LIB	16 SCO	16 SAG	15 CAP	16 PIS	15 ARI
20 SCO	18 SAG	19 CAP	17 AQU	18 ARI	17 TAU
22 SAG	20 CAP	21 AQU	19 PIS	21 TAU	19 GEM
24 CAP	22 AQU	23 PIS	21 ARI	23 GEM	22 CAN
26 AQU	24 PIS	25 ARI	23 TAU	25 CAN	24 LEO
28 PIS	26 ARI	27 TAU	26 GEM	28 LEO	27 VIR
30 ARI	29 TAU	29 GEM	28 CAN	30 VIR	29 LIB
			30 LEO		

JUL	AUG	SEP	OCT	NOV	DEC
2 SCO	2 CAP	1 AQU	2 ARI	1 TAU	2 CAN
4 SAG	4 AQU	3 PIS	4 TAU	3 GEM	5 LEO
6 CAP	6 PIS	5 ARI	6 GEM	5 CAN	7 VIR
8 AQU	8 ARI	7 TAU	9 CAN	7 LEO	10 LIB
10 PIS	10 TAU	9 GEM	11 LEO	10 VIR	12 SCO
12 ARI	13 GEM	11 CAN	14 VIR	13 LIB	15 SAG
14 TAU	15 CAN	14 LEO	16 LIB	15 SCO	17 CAP
16 GEM	18 LEO	16 VIR	19 SCO	17 SAG	19 AQU
19 CAN	20 VIR	19 LIB	21 SAG	19 CAP	21 PIS
21 LEO	23 LIB	21 SCO	23 CAP	21 AQU	23 ARI
24 VIR	25 SCO	24 SAG	25 AQU	24 PIS	25 TAU
26 LIB	27 SAG	26 CAP	26 PIS	26 ARI	27 GEM
29 SCO	30 CAP	28 AQU	29 ARI	28 TAU	30 CAN
31 SAG		30 PIS		30 GEM	

1954 MOON SIGNS

JAN	FEB	MAR	APR	MAY	JUN
1 SAG	2 AQU	1 AQU	2 ARI	1 TAU	2 CAN
3 CAP	4 PIS	3 PIS	4 TAU	3 GEM	4 LEO
6 AQU	6 ARI	5 ARI	6 GEM	5 CAN	6 VIR
8 PIS	8 TAU	7 TAU	8 CAN	8 LEO	9 LIB
10 ARI	10 GEM	10 GEM	10 LEO	10 VIR	11 SCO
12 TAU	13 CAN	12 CAN	13 VIR	12 LIB	14 SAG
14 GEM	15 LEO	14 LEO	15 LIB	15 SCO	16 CAP
16 CAN	17 VIR	16 VIR	18 SCO	17 SAG	19 AQU
19 LEO	20 LIB	19 LIB	20 SAG	20 CAP	21 PIS
21 VIR	22 SCO	21 SCO	23 CAP	22 AQU	23 ARI
23 LIB	25 SAG	24 SAG	25 AQU	25 PIS	25 TAU
26 SCO	27 CAP	26 CAP	27 PIS	27 ARI	27 GEM
28 SAG		29 AQU	29 ARI	29 TAU	29 CAN
31 CAP		31 PIS		31 GEM	

JUL	AUG	SEP	OCT	NOV	DEC
1 LEO	2 LIB	1 SCO	1 SAG	2 AQU	2 PIS
4 VIR	5 SCO	5 SAG	4 CAP	5 PIS	4 ARI
6 LIB	7 SAG	6 CAP	6 AQU	7 ARI	6 TAU
9 SCO	10 CAP	9 AQU	9 PIS	9 TAU	8 GEM
11 SAG	12 AQU	11 PIS	10 ARI	11 GEM	10 CAN
13 CAP	14 PIS	13 ARI	12 TAU	13 CAN	12 LEO
16 AQU	16 ARI	15 TAU	14 GEM	15 LEO	14 VIR
18 PIS	19 TAU	17 GEM	16 CAN	17 VIR	17 LIB
20 ARI	21 GEM	19 CAN	19 LEO	20 LIB	19 SCO
22 TAU	23 CAN	21 LEO	21 VIR	22 SCO	22 SAG
24 GEM	25 LEO	24 VIR	23 LIB	25 SAG	24 CAP
27 CAN	27 VIR	26 LIB	26 SCO	27 CAP	27 AQU
29 LEO	30 LIB	29 SCO	28 SAG	30 AQU	29 PIS
31 VIR			31 CAP		31 ARI

1953 MOON SIGNS

JAN	FEB	MAR	APR	MAY	JUN
1 LEO	3 LIB	2 LIB	1 SCO	2 CAP	1 AQU
4 VIR	5 SCO	4 SCO	3 SAG	5 AQU	3 PIS
6 LIB	7 SAG	7 SAG	5 CAP	7 PIS	5 ARI
9 SCO	10 CAP	9 CAP	7 AQU	9 ARI	7 TAU
11 SAG	12 AQU	11 AQU	10 PIS	11 TAU	9 GEM
13 CAP	14 PIS	13 PIS	12 ARI	13 GEM	12 CAN
15 AQU	16 ARI	15 ARI	14 TAU	15 CAN	14 LEO
17 PIS	18 TAU	17 TAU	16 GEM	18 LEO	16 VIR
19 ARI	20 GEM	19 GEM	18 CAN	20 VIR	19 LIB
21 TAU	22 CAN	22 CAN	20 LEO	23 LIB	21 SCO
24 GEM	25 LEO	24 LEO	23 VIR	25 SCO	24 SAG
26 CAN	27 VIR	27 VIR	25 LIB	27 SAG	26 CAP
28 LEO		29 LIB	26 SCO	30 CAP	28 AQU
31 VIR			30 SAG		30 PIS

JUL	AUG	SEP	OCT	NOV	DEC
2 ARI	1 TAU	1 CAN	1 LEO	2 LIB	2 SCO
5 TAU	3 GEM	4 LEO	4 VIR	5 SCO	5 SAG
7 GEM	5 CAN	6 VIR	6 LIB	7 SAG	7 CAP
9 CAN	8 LEO	9 LIB	9 SCO	10 CAP	9 AQU
11 LEO	10 VIR	11 SCO	11 SAG	12 AQU	11 PIS
14 VIR	13 LIB	14 SAG	13 CAP	14 PIS	14 ARI
16 LIB	15 SCO	16 CAP	16 AQU	16 ARI	16 TAU
19 SCO	18 SAG	18 AQU	18 PIS	18 TAU	18 GEM
21 SAG	20 CAP	21 PIS	20 ARI	20 GEM	20 CAN
23 CAP	22 AQU	23 ARI	22 TAU	22 CAN	22 LEO
26 AQU	24 PIS	25 TAU	24 GEM	25 LEO	25 VIR
28 PIS	26 ARI	27 GEM	26 CAN	27 VIR	27 LIB
30 ARI	28 TAU	29 CAN	28 LEO	30 LIB	30 SCO
	30 GEM		31 VIR		

1955 MOON SIGNS

JAN	FEB	MAR	APR	MAY	JUN
3 TAU	1 GEM	2 CAN	1 LEO	2 LIB	1 SCO
5 GEM	3 CAN	4 LEO	3 VIR	5 SCO	4 SAG
7 CAN	5 LEO	7 VIR	5 LIB	7 SAG	6 CAP
9 LEO	7 VIR	9 LIB	8 SCO	10 CAP	9 AQU
11 VIR	10 LIB	11 SCO	10 SAG	12 AQU	11 PIS
13 LIB	12 SCO	14 SAG	13 CAP	15 PIS	13 ARI
16 SCO	15 SAG	16 CAP	15 AQU	17 ARI	16 TAU
18 SAG	17 CAP	19 AQU	18 P/IS	19 TAU	18 GEM
21 CAP	19 AQU	21 PIS	20 ARI	21 GEM	20 CAN
23 AQU	22 PIS	23 ARI	22 TAU	23 CAN	22 LEO
25 PIS	24 ARI	25 TAU	24 GEM	25 LEO	24 VIR
28 ARI	26 TAU	27 GEM	26 CAN	27 VIR	26 LIB
30 TAU	28 GEM	29 CAN	28 LEO	30 LIB	28 SCO
			30 VIR		

JUL	AUG	SEP	OCT	NOV	DEC
1 SAG	2 AQU	1 PIS	1 ARI	1 GEM	1 CAN
3 CAP	5 PIS	3 ARI	3 TAU	3 CAN	3 LEO
6 AQU	7 ARI	5 TAU	5 GEM	5 LEO	5 VIR
8 PIS	9 TAU	8 GEM	7 CAN	7 VIR	7 LEB
11 ARI	11 GEM	10 CAN	9 LEO	10 LIB	9 SCO
13 TAU	13 CAN	12 LEO	11 VIR	12 SCO	12 SAG
15 GEM	15 LEO	14 VIR	13 LIB	15 SAG	14 CAP
17 CAN	18 VIRF	16 LIB	16 SCO	17 CAP	17 AQU
19 LEO	20 LIB	18 SCO	18 SAG	20 AQU	19 PIS
21 VIR	22 SCO	21 SAG	21 CAP	22 PIS	22 ARI
23 LIB	25 SAG	23 CAP	23 AQU	24 ARI	24 TAU
26 SCO	27 CAP	26 AQU	26 PUIS	27 TAU	26 GEM
28 SAG	30 AQU	28 PIS	28 ARI	29 GEM	28 CAN
31 CAP			30 TAU		30 LEO

1956 MOON SIGNS

AN	FEB	MAR	APR	MAY	JUN
1 VIR	2 SCO	3 SAG	1 CAP	1 AQU	3 ARI
3 LIB	4 SAG	5 CAP	4 AQU	4 PIS	5 TAU
6 SCO	7 CAP	8 AQU	6 PIS	6 ARI	7 GEM
8 SAG	9 AQU	10 PIS	8 ARI	8 TAU	9 CAP
11 CAP	12 PIS	12 ARI	11 TAU	11 GEM	11 LEO
13 AQU	14 ARI	15 TAU	13 GEM	13 CAN	13 VIR
16 PIS	16 TAU	17 GEM	15 CAN	15 LEO	15 LIB
18 ARI	19 GEM	19 CAN	17 LEO	17 VIR	18 SCO
20 TAU	21 CAN	21 LEO	20 VIR	19 LIB	20 SAG
22 GEM	23 LEO	23 VIR	22 LIB	21 SCO	22 CAP
24 CAN	25 VIR	25 LIB	24 SCO	24 SAG	25 AQU
26 LEO	27 LIB	28 SCO	26 SAG	26 CAP	27 PIS
28 VIR	29 SCO	30 SAG	29 CAP	29 AQU	30 ARI
31 LIB					31 PIS
JUL	**AUG**	**SEP**	**OCT**	**NOV**	**DEC**
2 TAU	1 GEM	1 LEO	1 VIR	1 SCO	1 SAG
4 GEM	3 CAN	3 VIR	3 LIB	3 SAG	3 CAP
6 CAN	5 LEO	5 LIB	5 SCO	6 CAP	6 AQU
8 LEO	7 VIR	7 SCO	7 SAG	8 AQU	8 PIS
10 VIR	9 LIB	10 SAG	10 CAP	11 PIS	11 ARI
12 LIB	11 SCO	12 CAP	12 AQU	13 ARI	13 TAU
15 SCO	13 SAG	15 AQU	15 PIS	16 TAU	15 GEM
17 SAG	16 CAP	17 PIS	17 ARI	18 GEM	17 CAN
20 CAP	18 AQU	20 ARI	19 TAU	20 CAN	19 LEO
22 AQU	21 PIS	22 TAU	22 GEM	22 LEO	21 VIR
25 PIS	23 ARI	24 GEM	24 CAN	24 VIR	24 LIB
27 ARI	26 TAU	26 CAN	26 LEO	26 LIB	26 SCO
30 TAU	28 GEM	29 LEO	28 VIR	29 SCO	28 SAG
	30 CAN		30 LIB		31 CAP

1957 MOON SIGNS

JAN	FEB	MAR	APR	MAY	JUN
2 AQU	1 PIS	3 ARI	1 TAU	1 GEM	1 LEO
5 PIS	3 ARI	5 TAU	4 GEM	3 CAN	4 VIR
7 ARI	6 TAU	7 GEM	6 CAN	5 LEO	6 LIB
9 TAU	8 GEM	10 CAN	8 LEO	7 VIR	8 SCO
12 GEM	10 CAN	12 LEO	10 VIR	9 LIB	10 SCO
14 CAN	12 LEO	14 VIR	12 LIB	12 SCO	12 CAP
16 LEO	14 VIR	16 LIB	14 SCO	14 SAG	15 AQU
18 VIR	16 LIB	18 SCO	16 SAG	16 CAP	17 PIS
20 LIB	18 SCO	20 SAG	19 CAP	18 AQU	20 ARI
22 SCO	21 SAG	22 CAP	21 AQU	21 PIS	22 TAU
24 SAG	23 CAP	25 AQU	24 PIS	23 ARI	25 GEM
27 CAP	26 AQU	27 PIS	26 ARI	26 TAU	27 CAN
29 AQU	28 PIS	30 ARI	29 TAU	28 GEM	29 LEO
				30 CAN	
JUL	**AUG**	**SEP**	**OCT**	**NOV**	**DEC**
1 VIR	1 SCO	2 CAP	2 AQU	1 PIS	1 ARI
LIB	4 SAG	5 AQU	4 PIS	3 ARI	3 TAU
5 SCO	6 CAP	7 PIS	7 ARI	6 TAU	5 GEM
7 SAG	8 AQU	10 ARI	9 TAU	8 GEM	8 CAN
10 CAP	11 PIS	12 TAU	12 GEM	10 CAN	10 LEO
12 AQU	13 ARI	15 GEM	14 CAN	13 LEO	12 VIR
15 PIS	16 TAU	17 CAN	16 LEO	15 VIR	14 LIB
17 ARI	18 GEM	19 LEO	18 VIR	17 LIB	16 SCO
20 TAU	21 CAN	21 VIR	21 LIB	19 SCO	18 SAG
22 GEM	23 LEO	23 LIB	23 SCO	21 SAG	21 CAP
24 CAN	25 VIR	25 SCO	25 SAG	23 CAP	23 AQU
26 LEO	27 LIB	27 SAG	27 CAP	26 AQU	25 PIS
28 VIR	29 SCO	29 CAP	29 AQU	28 PIS	28 ARI
30 LIB	31 SAG				30 TAU

1958 MOON SIGNS

JAN	FEB	MAR	APR	MAY	JUN
2 GEM	3 LEO	2 LEO	1 VIR	2 SCO	3 CAP
4 CAN	5 VIR	4 VIR	3 LIB	4 SAG	5 AQU
6 LEO	7 LIB	6 LIB	5 SCO	6 CAP	7 PIS
8 VIR	9 SCO	8 SCO	7 SAG	8 AQU	10 ARI
10 LIB	11 SAG	10 SAG	9 CAP	11 PIS	12 TAU
12 SCO	13 CAP	12 CAP	11 AQU	13 ARI	15 GEM
15 SAG	16 AQU	15 AQU	13 PIS	16 TAU	17 CAN
17 CAP	18 PIS	17 PIS	16 ARI	18 GEM	19 LEO
19 AQU	21 ARI	20 ARI	19 TAU	21 CAN	21 VIR
22 PIS	23 TAU	22 TAU	21 GEM	23 LEO	24 LIB
24 ARI	26 GEM	25 GEM	23 CAN	25 VIR	26 SCO
27 TAU	28 CAN	27 CAN	26 LEO	27 LIB	28 SAG
29 GEM		29 LEO	28 VIR	29 SCO	30 CAP
31 CAN			30 LIB	31 SAG	
JUL	**AUG**	**SEP**	**OCT**	**NOV**	**DEC**
2 AQU	1 PIS	2 TAU	2 GEM	1 CAN	3 VIR
4 PIS	3 ARI	5 GEM	4 CAN	3 LEO	5 LIB
7 ARI	6 TAU	7 CAN	7 LEO	5 VIR	7 SCO
9 TAU	8 GEM	9 LEO	9 VIR	7 LIB	9 SCO
12 GEM	11 CAN	11 VIR	11 LIB	9 SCO	11 CAP
14 CAN	13 LEO	13 LIB	13 SCO	11 SAG	13 AQU
17 LEO	15 VIR	15 SCO	15 SAG	13 CAP	15 PIS
19 VIR	17 LIB	18 SAG	17 CAP	16 AQU	18 ARI
21 LIB	19 SCO	20 CAP	19 AQU	18 PIS	20 TAU
23 SCO	21 SAG	22 AQU	22 PIS	20 ARI	23 GEM
25 SAG	23 CAP	24 PIS	24 ARI	23 TAU	25 CAN
27 CAP	26 AQU	27 ARI	27 TAU	25 GEM	27 LEO
29 AQU	28 PIS	29 TAU	29 GEM	28 CAN	30 VIR
	31 ARI			30 LEO	

1959 MOON SIGNS

JAN	FEB	MAR	APR	MAY	JUN
1 LIB	1 SAG	1 SAG	1 AQU	1 PIS	2 TAU
3 SCO	4 CAP	3 CAP	4 PIS	3 ARI	5 GEM
5 SAG	6 AQU	5 AQU	6 ARI	6 TAU	7 CAN
7 CAP	8 PIS	7 PIS	8 TAU	8 GEM	9 LEO
9 AQU	10 ARI	10 ARI	11 GEM	11 CAN	12 VIR
12 PIS	13 TAU	12 TAU	14 CAN	13 LEO	14 LIB
14 ARI	15 GEM	15 GEM	16 LEO	16 VIR	16 SCO
17 TAU	18 CAN	17 CAN	18 VIR	18 LIB	18 SAG
19 GEM	20 LEO	20 LEO	20 LIB	20 SCO	20 CAP
21 CAN	22 VIR	22 VIR	22 SCO	22 SAG	22 AQU
24 LEO	24 LIB	24 LIB	24 SAG	24 CAP	24 PIS
26 VIR	27 SCO	26 SCO	26 CAP	26 AQU	27 ARI
28 LIB		28 SAG	28 AQU	28 PIS	29 TAU
30 SCO		30 CAP		30 ARI	
JUL	**AUG**	**SEP**	**OCT**	**NOV**	**DEC**
2 GEM	1 CAN	2 VIR	1 LIB	2 SAG	1 CAP
4 CAN	3 LEO	4 LIB	3 SCO	4 CAP	3 AQU
7 LEO	5 VIR	6 SCO	5 SAG	6 AQU	5 PIS
9 VIR	8 LIB	8 SAG	7 CAP	8 PIS	8 ARI
11 LIB	10 SCO	10 CAP	10 AQU	10 ARI	10 TAU
13 SCO	12 SAG	12 AQU	12 PIS	13 TAU	13 GEM
16 SAG	14 CAP	15 PIS	14 ARI	15 GEM	15 CAN
18 CAP	16 AQU	17 ARI	17 TAU	18 CAN	18 LEO
20 AQU	18 PIS	19 TAU	19 GEM	20 LEO	20 VIR
22 PIS	20 ARI	22 GEM	22 CAN	23 VIR	22 LIB
24 ARI	23 TAU	24 CAN	24 LEO	25 LIB	25 SCO
27 TAU	25 GEM	27 LEO	26 VIR	27 SCO	27 SAG
29 GEM	28 CAN	29 VIR	29 LIB	29 SAG	29 CAP
	30 LEO		31 SCO		31 AQU

1960 MOON SIGNS

JAN	FEB	MAR	APR	MAY	JUN
2 PIS	3 TAU	1 TAU	2 CAN	2 LEO	1 VIR
4 ARI	5 GEM	4 GEM	5 LEO	5 VIR	3 LIB
6 TAU	8 CAN	6 CAN	7 VIR	7 LIB	6 SCO
9 GEM	10 LEO	9 LEO	10 LIB	9 SCO	8 SAG
11 CAN	13 VIR	11 VIR	12 SCO	11 SAG	10 CAP
14 LEO	15 LIB	13 LIB	14 SAG	13 CAP	12 AQU
16 VIR	17 SCO	15 SCO	16 CAP	15 AQU	14 PIS
19 LIB	19 SAG	17 SAG	18 AQU	17 PIS	16 ARI
21 SCO	21 CAP	20 CAP	20 PIS	20 ARI	18 TAU
23 SAG	23 AQU	22 AQU	22 ARI	22 TAU	21 GEM
25 CAP	26 PIS	24 PIS	25 TAU	24 GEM	23 CAN
27 AQU	28 ARI	26 ARI	27 GEM	27 CAN	26 LEO
29 PIS		28 TAU	30 CAN	29 LEO	28 VIR
31 ARI		31 GEM			

JUL	AUG	SEP	OCT	NOV	DEC
1 LIB	1 SAG	2 AQU	1 PIS	2 TAU	2 GEM
3 SCO	3 CAP	4 PIS	3 ARI	4 GEM	4 CAN
5 SAG	5 AQU	6 ARI	6 TAU	7 CAN	7 LEO
7 CAP	7 PIS	8 TAU	8 GEM	9 LEO	9 VIR
9 AQU	10 ARI	11 GEM	10 CAN	12 VIR	12 LIB
11 PIS	12 TAU	13 CAN	13 LEO	14 LIB	14 SCO
13 ARI	14 GEM	16 LEO	15 VIR	16 SCO	16 SAG
15 TAU	17 CAN	18 VIR	18 LIB	19 SAG	18 CAP
18 GEM	19 LEO	20 LIB	20 SCO	21 CAP	20 AQU
20 CAN	22 VIR	23 SCO	22 SAG	23 AQU	22 PIS
23 LEO	24 LIB	25 SAG	24 CAP	25 PIS	24 ARI
25 VIR	26 SCO	27 CAP	26 AQU	27 ARI	26 TAU
28 LIB	29 SAG	29 AQU	28 PIS	29 TAU	29 GEM
30 SCO	31 CAP		31 ARI		31 CAN

1962 MOON SIGNS

JAN	FEB	MAR	APR	MAY	JUN
3 SAG	1 CAP	1 CAP	1 PIS	1 ARI	1 GEM
5 CAP	3 AQU	3 AQU	3 ARI	3 TAU	3 CAN
7 AQU	5 PIS	5 PIS	5 TAU	5 GEM	6 LEO
9 PIS	7 ARI	7 ARI	7 GEM	7 CAN	8 VIR
11 ARI	9 TAU	9 TAU	9 CAN	9 LEO	10 LIB
13 TAU	12 GEM	11 GEM	12 LEO	12 VIR	13 SCO
15 GEM	14 CAN	13 CAN	14 VIR	14 LIB	15 SAG
18 CAN	16 LEO	16 LEO	17 LIB	17 SCO	18 CAP
20 LEO	19 VIR	18 VIR	19 SCO	19 SAG	20 AQU
23 VIR	21 LIB	21 LIB	22 SAG	21 CAP	22 PIS
25 LIB	24 SCO	23 SCO	24 CAP	24 AQU	24 ARI
28 SCO	26 SAG	26 SAG	26 AQU	26 PIS	26 TAU
30 SAG		28 CAP	29 PIS	28 ARI	28 GEM
		30 AQU		30 TAU	

JUL	AUG	SEP	OCT	NOV	DEC
1 CAN	2 VIR	3 SCO	3 SAG	1 CAP	1 AQU
3 LEO	4 LIB	5 SAG	5 CAP	4 AQU	3 PIS
5 VIR	7 SCO	8 CAP	7 AQU	6 PIS	5 ARI
8 LIB	9 SAG	10 AQU	10 PIS	8 ARI	7 TAU
10 SCO	11 CAP	12 PIS	12 ARI	10 TAU	9 GEM
13 SAG	14 AQU	14 ARI	14 TAU	12 GEM	11 CAN
15 CAP	16 PIS	16 TAU	16 GEM	14 CAN	14 LEO
17 AQU	18 ARI	18 GEM	18 CAN	16 LEO	16 VIR
19 PIS	20 TAU	20 CAN	20 LEO	19 VIR	19 LIB
21 ARI	22 GEM	23 LEO	22 VIR	21 LIB	21 SCO
23 TAU	24 CAN	25 VIR	25 LIB	24 SCO	24 SAG
26 GEM	26 LEO	28 LIB	27 SCO	26 SAG	26 CAP
28 CAN	29 VIR	30 SCO	30 SAG	29 CAP	28 AQU
30 LEO	31 LIB				30 PIS

1961 MOON SIGNS

JAN	FEB	MAR	APR	MAY	JUN
3 LEO	2 VIR	1 VIR	2 SCO	2 SAG	2 AQU
5 VIR	4 LIB	3 LIB	4 SAG	4 CAP	4 PIS
8 LIB	6 SCO	6 SCO	6 CAP	6 AQU	6 ARI
10 SCO	9 SAG	8 SAG	9 AQU	8 PIS	8 TAU
12 SAG	11 CAP	10 CAP	11 PIS	10 ARI	11 GEM
14 CAP	13 AQU	12 AQU	13 ARI	12 TAU	13 CAN
16 AQU	15 PIS	14 PIS	15 TAU	14 GEM	16 LEO
18 PIS	17 ARI	16 ARI	17 GEM	17 CAN	18 VIR
20 ARI	19 TAU	18 TAU	19 CAN	19 LEO	21 LIB
23 TAU	21 GEM	21 GEM	22 LEO	22 VIR	23 SCO
25 GEM	24 CAN	23 CAN	25 VIR	24 LIB	25 SAG
28 CAN	26 LEO	26 LEO	27 LIB	27 SCO	27 CAP
30 LEO		28 VIR	29 SCO	29 SAG	29 AQU
		31 LIB		31 CAP	

JUL	AUG	SEP	OCT	NOV	DEC
1 PIS	2 TAU	1 GEM	3 LEO	2 VIR	1 LIB
4 ARI	4 GEM	3 CAN	5 VIR	4 LIB	4 SCO
6 TAU	7 CAN	5 LEO	8 LIB	6 SCO	6 SAG
8 GAM	9 LEO	8 VIR	10 SCO	9 SAG	9 CAP
10 CAN	12 VIR	10 LIB	13 SAG	11 CAP	10 AQU
13 LEO	14 LIB	13 SCO	15 CAP	13 AQU	13 PIS
15 VIR	17 SCO	15 SAG	17 AQU	15 PIS	15 ARI
18 LIB	19 SAG	18 CAP	19 PIS	17 ARI	17 TAU
20 SCO	21 CAP	20 AQU	21 ARI	20 TAU	19 GEM
23 SAG	23 AQU	22 PIS	23 TAU	22 GEM	21 CAN
25 CAP	25 PIS	24 ARI	25 GEM	24 CAN	24 LEO
27 AQU	27 ARI	26 TAU	28 CAN	26 LEO	26 VIR
29 PIS	29 TAU	28 GEM	30 LEO	29 VIR	29 LIB
31 ARI		30 CAN			31 SCO

1963 MOON SIGNS

JAN	FEB	MAR	APR	MAY	JUN
1 ARI	2 GEM	1 GEM	2 LEO	2 VIR	3 SCO
4 TAU	4 CAN	3 CAN	4 VIR	4 LIB	5 SAG
6 GEM	6 LEO	6 LEO	7 LIB	7 SCO	8 CAP
8 CAN	9 VIR	8 VIR	9 SCO	9 SAG	10 AQU
10 LEO	11 LIB	11 LIB	12 SAG	12 CAP	12 PIS
12 VIR	14 SCO	13 SCO	14 CAP	14 AQU	15 ARI
15 LIB	16 SAG	16 SAG	17 AQU	16 PIS	17 TAU
17 SCO	19 CAP	18 CAP	19 PIS	18 ARI	19 GEM
20 SAG	21 AQU	20 AQU	21 ARI	20 TAU	21 CAN
22 CAP	23 PIS	23 PIS	23 TAU	22 GEM	23 LEO
25 AQU	25 ARI	25 ARI	25 GEM	24 CAN	25 VIR
27 PIS	27 TAU	27 TAU	27 CAN	27 LEO	28 LIB
29 ARI		29 GEM	29 LEO	29 VIR	30 SCO
31 TAU		31 CAN		31 LIB	

JUL	AUG	SEP	OCT	NOV	DEC
3 SCO	1 CAP	2 PIS	2 ARI	2 GEM	2 CAN
5 CAP	4 AQU	4 ARI	4 TAU	4 CAN	4 LEO
7 AQU	6 PIS	7 TAU	6 GEM	6 LEO	6 VIR
10 PIS	8 ARI	9 GEM	8 CAN	9 VIR	8 LIB
12 ARI	10 TAU	11 CAN	10 LEO	11 LIB	11 SCO
14 TAU	12 GEM	13 LEO	12 VIR	14 SCO	13 SAG
16 GEM	14 CAN	15 VIR	15 LIB	16 SAG	16 CAP
18 CAN	17 LEO	18 LIB	17 SCO	19 CAP	18 AQU
20 LEO	19 VIR	20 SCO	20 SAG	21 AQU	21 PIS
23 VIR	21 LIB	23 SAG	22 CAP	24 PIS	23 ARI
25 LIB	24 SCO	25 CAP	25 AQU	26 ARI	25 TAU
27 SCO	26 SAG	28 AQU	27 PIS	28 TAU	27 GEM
30 SAG	29 CAP	30 PIS	29 ARI	30 GEM	29 CAN
	31 AQU		31 TAU		31 LEO

1964 MOON SIGNS

JAN	FEB	MAR	APR	MAY	JUN
2 VIR	1 LIB	2 SCO	1 SAG	1 CAP	2 PIS
5 LIB	4 SCO	4 SAG	3 CAP	3 AQU	4 ARI
7 SCO	6 SAG	7 CAP	6 AQU	5 PIS	6 TAU
10 SAG	9 CAP	9 AQU	8 PIS	8 ARI	8 GEM
12 CAP	11 AQI	12 PIS	10 ARI	10 TAU	10 CAN
15 AQU	13 PIS	14 ARI	12 TAU	12 GEM	12 LEO
17 PIS	16 ARI	16 TAU	14 GEM	14 CAN	14 VIR
19 ARI	18 TAU	18 GEM	16 CAN	16 LEO	17 LIB
21 TAU	20 GEM	20 CAN	19 LEO	18 VIR	19 SAG
24 GEM	22 CAN	22 CAP	21 VIR	20 LIB	20 SAG
26 CAN	24 LEO	25 VIR	23 LIB	23 SCO	24 CAP
28 LEO	26 VIR	27 LIB	26 SCO	25 SAG	27 AQU
30 VIR	28 LIB	29 SCO	28 SAG	28 CAP	29 PIS
				30 AQU	

JUL	AUG	SEP	OCT	NOV	DEC
1 ARI	2 GEM	2 LEO	2 VIR	3 SCO	2 SAG
4 TAU	4 CAN	5 VIR	4 LIB	5 SAG	5 CAP
6 GEM	6 LEO	7 LIB	6 SCO	8 CAP	7 AQU
8 CAN	8 VIR	9 SCO	9 SAG	10 AQU	10 PIS
10 LEO	10 LIB	11 SAG	11 CAP	13 PIS	12 ARI
12 VIR	13 SCO	14 CAP	14 AQU	15 ARI	15 TAU
14 LIB	15 SAG	16 AQU	16 PIS	17 TAU	17 GEM
16 SCO	18 CAP	19 PIS	19 ARI	19 GEM	19 CAN
19 SAG	20 AQU	21 ARI	21 TAU	21 CAN	21 LEO
21 CAP	23 PIS	23 TAU	23 GEM	23 LEO	23 VIR
24 AQU	25 ARI	25 GEM	25 CAN	25 VIR	25 LIB
26 PIS	27 TAU	28 CAN	27 LEO	28 LIB	27 SCO
29 ARI	29 GEM	30 LEO	29 VIR	30 SCO	30 SAG
31 TAU	31 CAN		31 LIB		

1966 MOON SIGNS

JAN	FEB	MAR	APR	MAY	JUN
1 TAU	2 CAN	1 CAN	2 VIR	1 LIB	2 SAG
3 GEM	4 LEO	3 LEO	4 LIB	3 SCO	4 CAP
5 CAN	6 VIR	5 VIR	6 SCO	5 SAG	6 AQU
7 LEO	8 LIB	7 LIB	8 SAG	8 CAP	8 PIS
9 VIR	10 SCO	9 SCO	10 CAP	10 AQU	11 ARI
11 LIB	12 SAG	12 SAG	13 AQU	12 PIS	14 TAU
14 SCO	15 CAP	14 CAP	15 PIS	15 ARI	16 GEM
16 SAG	17 AQU	16 AQU	18 ARI	17 TAU	18 CAN
18 CAP	20 PIS	19 PIS	20 TAU	20 GEM	20 LEO
21 AQU	22 ARI	21 ARI	22 GEM	22 CAN	23 VIR
23 PIS	25 TAU	24 TAU	25 CAN	24 LEO	25 LIB
26 ARI	27 GEM	26 GEM	27 LEO	26 VIR	27 SCO
28 TAU		29 CAN	29 VIR	28 LIB	29 SAG
31 GEM		31 LEO		31 SCO	

JUL	AUG	SEP	OCT	NOV	DEC
1 CAP	2 PIS	1 ARI	1 TAU	2 CAN	2 LEO
4 AQU	5 ARI	4 TAU	3 GEM	4 LEO	4 VIR
6 PIS	7 TAU	6 GEM	6 CAN	6 VIR	6 LIB
9 ARI	10 GEM	9 CAN	8 LEO	8 LIB	8 SCO
11 TAU	12 CAN	11 LEO	10 VIR	11 SCO	10 SAG
14 GEM	14 LEO	13 VIR	12 LIB	13 SAG	12 CAP
14 CAN	16 VIR	15 LIB	14 SCO	15 CAP	14 AQU
18 LEO	18 LIB	17 SCO	16 SAG	17 AQU	17 PIS
20 VIR	20 SCO	19 SAG	18 CAP	20 PIS	19 ARI
22 LIB	22 SAG	21 CAP	21 AQU	22 ARI	22 TAU
24 SCO	25 CAP	23 AQU	23 PIS	25 TAU	24 GEM
26 SAG	27 AQU	26 PIS	26 ARI	28 GEM	27 CAN
29 CAP	30 PIS	28 ARI	28 TAU	29 CAN	29 LEO
31 AQU			31 GEM		31 VIR

1965 MOON SIGNS

JAN	FEB	MAR	APR	MAY	JUN
1 CAP	2 PIS	2 PIS	3 TAU	2 GEM	1 CAN
4 AQU	5 ARI	4 ARI	5 GEM	4 CAN	3 LEO
6 PIS	7 TAU	6 TAU	7 CAN	6 LEO	5 VIR
9 ARI	9 GEM	9 GEM	9 LEO	8 VIR	7 LIB
11 TAU	11 CAN	11 CAN	11 VIR	11 LIB	9 SCO
13 GEM	13 LEO	13 LEO	13 LIB	13 SCO	12 SAG
15 CAN	16 VIR	15 VIR	16 SCO	15 SAG	14 CAP
17 LEO	18 LIB	17 LIB	18 SAG	18 CAP	16 AQU
19 VIR	20 SCO	19 SCO	20 CAP	20 AQU	19 PIS
21 LIB	22 SAG	22 SAG	23 AQU	23 PIS	21 ARI
23 SCO	25 CAP	24 CAP	25 PIS	25 ARI	24 TAU
26 SAG	27 AQU	27 AQU	28 ARI	27 TAU	26 GEM
28 CAP		29 PIS	30 TAU	30 GEM	28 CAN
31 AQU		31 ARI			30 LEO

JUL	AUG	SEP	OCT	NOV	DEC
2 VIR	3 SCO	1 SAG	1 CAP	2 PIS	2 ARI
4 LIB	5 SAG	4 CAP	4 AQU	5 ARI	5 TAU
6 SCO	7 CAP	6 AQU	6 PIS	7 TAU	7 GEM
9 SAG	10 AQU	9 PIS	9 ARI	9 GEM	9 CAN
11 CAP	13 PIS	11 ARI	11 TAU	12 CAN	11 LEO
14 AQU	15 ARI	14 TAU	13 GEM	14 LEO	13 VIR
16 PIS	17 TAU	16 GEM	15 CAN	16 VIR	15 LIB
19 ARI	20 GEM	18 CAN	17 LEO	18 LIB	17 SCO
21 TAU	22 CAN	20 LEO	20 VIR	20 SCO	20 SAG
23 GEM	24 LEO	22 VIR	22 LIB	22 SAG	22 CAP
25 CAN	26 VIR	24 LIB	24 SCO	25 CAP	25 AQU
27 LEO	28 LIB	26 SCO	26 SAG	27 AQU	27 PIS
29 VIR	30 SCO	29 SAG	28 CAP	30 PIS	30 ARI
31 LIB			31 AQU		

1967 MOON SIGNS

JAN	FEB	MAR	APR	MAY	JUN
2 LIB	3 SAG	2 SAG	3 AQU	2 PIS	1 ARI
4 SCO	5 CAP	4 CAP	5 PIS	5 ARI	4 TAU
6 SAG	7 AQU	6 AQU	8 ARI	7 TAU	6 GEM
9 CAP	10 PIS	9 PIS	10 TAU	10 GEM	9 CAN
11 AQU	12 ARI	11 ARI	13 GEM	12 CAN	11 LEO
13 PIS	15 TAU	14 TAU	15 CAN	15 LEO	13 VIR
16 ARI	17 GEM	16 GEM	17 LEO	17 VIR	15 LIB
18 TAU	19 CAN	19 CAN	20 VIR	19 LIB	17 SCO
21 GEM	22 LEO	21 LEO	22 LIB	21 SCO	19 SAG
23 CAN	24 VIR	23 VIR	24 SAG	23 SAG	21 CAP
25 LEO	26 LIB	25 LIB	26 SCO	25 CAP	24 AQU
27 VIR	28 SCO	27 SCO	29 CAP	27 AQU	26 PIS
29 LIB		29 SAG	30 AQU	30 PIS	28 ARI
31 SCO		31 CAP			

JUL	AUG	SEP	OCT	NOV	DEC
1 TAU	2 CAN	1 LEO	2 LIB	1 SCO	2 CAP
3 GEM	4 LEO	3 VIR	4 SCO	3 SAG	4 AQU
6 CAN	7 VIR	5 LIB	6 SAG	5 CAP	7 PIS
8 LEO	9 LIB	7 SCO	9 CAP	7 AQU	9 ARI
10 VIR	11 SCO	9 SAG	11 AQU	9 PIS	12 TAU
12 LIB	13 SAG	11 CAP	13 PIS	12 ARI	14 GEM
15 SCO	15 CAP	14 AQU	16 ARI	14 TAU	17 CAN
17 SAG	17 AQU	16 PIS	18 TAU	17 GEM	19 LEO
19 CAP	20 PIS	18 ARI	21 GEM	19 CAN	21 VIR
21 AQU	22 ARI	21 TAU	23 CAB	22 LEO	24 LIB
23 PIS	25 TAU	23 GEM	26 LEO	2 VIR	26 SCO
26 ARI	27 GEM	26 CAN	28 VIR	26 LIB	28 SAG
28 TAU	30 CAN	28 LEO	30 LIB	28 SCO	30 CAP
31 GEM		30 VIR		30 SAG	

1968 MOON SIGNS

JAN	FEB	MAR	APR	MAY	JUN
1 AQU	2 ARI	3 TAU	2 GEM	1 CAN	2 VIR
3 PIS	4 TAU	5 GEM	4 CAN	4 LEO	5 LIB
6 ARI	7 GEM	8 CAN	7 LEO	6 VIR	7 SCO
8 TAU	9 CAN	10 LEO	9 VIR	8 LIB	9 SAG
11 GEM	12 LEO	12 VIR	11 LIB	10 SCO	11 CAP
13 CAN	14 VIR	14 LIB	13 SCO	12 SAG	13 AQU
15 LEO	16 LIB	17 SCO	15 SAG	14 CAP	15 PIS
18 VIR	18 SCO	19 SAG	17 CAP	16 AQU	17 ARI
20 LIB	20 SAG	21 CAP	19 AQU	19 PIS	20 TAU
22 SCO	22 CAP	23 AQU	21 PIS	21 ARI	22 GEM
24 SAG	25 AQU	25 PIS	24 ARI	24 TAU	24 CAN
26 CAP	27 PIS	28 ARI	26 TAU	26 GEM	27 LEO
28 AQU	29 ARI	30 TAU	29 GEM	29 CAN	30 VIR
31 PIS				31 LEO	

JUL	AUG	SEP	OCT	NOV	DEC
2 LIB	3 SAG	1 CAP	2 PIS	1 ARI	1 TAU
4 SCO	5 CAP	3 AQU	5 ARI	3 TAU	3 GEM
6 SAG	7 AQU	5 PIS	7 TAU	6 GEM	6 CAN
8 CAP	9 PIS	7 ARI	10 GEM	8 CAN	8 LEO
10 AQU	11 ARI	10 TAU	12 CAN	11 LEO	11 VIR
12 PIS	13 TAU	13 GEM	15 LEO	13 VIR	13 LIB
15 ARI	16 GEM	15 CAN	17 VIR	16 LIB	15 SCO
17 TAU	18 CAN	17 LEO	19 LIB	18 SCO	17 SAG
20 GEM	21 LEO	20 VIR	21 SCO	20 SAG	19 CAP
22 CAN	23 VIR	22 LIB	23 SAG	22 CAP	21 AQU
25 LEO	25 LIB	24 SCO	25 CAP	24 AQU	23 PIS
27 VIR	28 SCO	26 SAG	27 AQU	26 PIS	26 ARI
29 LIB	30 SAG	28 CAP	30 PIS	28 ARI	28 TAU
31 SCO		30 AQU			30 GEM

1970 MOON SIGNS

JAN	FEB	MAR	APR	MAY	JUN
2 SCO	2 CAP	2 CAP	2 PIS	2 ARI	2 GEM
4 SAG	4 AQU	4 AQU	4 ARI	4 TAU	5 CAN
6 CAP	6 PIS	6 PIS	6 TAU	6 GEM	7 LEO
8 AQU	8 ARI	8 ARI	9 GEM	8 CAN	10 VIR
10 PIS	11 TAU	10 TAU	11 CAN	11 LEO	12 LIB
12 ARI	13 GEM	12 GEM	14 LEO	13 VIR	15 SCO
14 TAU	15 CAN	15 CAN	16 VIR	16 LIB	17 SAG
17 GEM	18 LEO	17 LEO	19 LIB	18 SCO	19 CAP
19 CAN	20 VIR	20 VIR	21 SCO	20 SCG	21 AQU
22 LEO	23 LIB	22 LIB	23 SAG	23 CAP	23 PIS
24 VIR	25 SCO	25 SCO	25 CAP	25 AQU	25 ARI
27 LIB	28 SAG	27 SAG	27 AQU	27 PIS	27 TAU
29 SCO		29 CAP	30 PIS	29 ARI	30 GEM
31 SAG		31 AQU		31 TAU	

JUL	AUG	SEP	OCT	NOV	DEC
2 CAN	1 LEO	2 LIB	2 SCO	3 CAP	2 AQU
4 LEO	3 VIR	5 SCO	4 SAG	5 AQU	4 PIS
7 VIR	6 LIB	7 SAG	6 CAP	6 PIS	6 ARI
10 LIB	8 SCO	9 CAP	9 AQU	9 ARI	8 TAU
12 SCO	11 SAG	11 AQU	11 PIS	11 TAU	11 GEM
14 SAG	13 CAP	13 PIS	13 ARI	13 GEM	13 CAN
16 CAP	15 AQU	15 ARI	15 TAU	16 CAN	15 LEO
18 AQU	17 PIS	17 TAU	17 GEM	18 LEO	18 VIR
20 PIS	19 ARI	19 GEM	19 CAN	20 VIR	20 LIB
22 ARI	21 TAU	22 CAN	22 LEO	23 LIB	23 SCO
25 TAU	23 GEM	24 LEO	24 VIR	25 SCO	25 SAG
27 GEM	26 CAN	27 VIR	26 LIB	28 SAG	27 CAP
29 CAN	28 LEO	28 LIB	29 SCO	30 CAP	29 AQU
31 VIR			31 SAG		31 PIS

1969 MOON SIGNS

JAN	FEB	MAR	APR	MAY	JUN
2 CAN	1 LEO	2 VIR	1 LIB	1 SCO	1 CAP
4 LEO	3 VIR	5 LIB	3 SCO	3 SAG	3 AQU
7 VIR	6 LIB	7 SCO	5 SAG	5 CAP	5 PIS
9 LIB	8 SCO	9 SAG	8 CAP	7 AQU	7 ARI
12 SCO	10 SAG	11 CAP	10 AQU	9 PIS	10 TAU
14 CAP	12 CAP	13 ARU	12 PIS	11 ARI	12 GEM
16 CAP	14 AQU	16 PIS	14 ARI	14 TAU	15 CAN
18 AQU	16 PIS	18 ARI	16 TAU	16 GEM	17 LEO
20 PIS	18 ARI	20 TAU	19 GEM	19 CAN	20 VIR
22 ARI	21 TAU	22 GEM	21 CAN	21 LEO	22 LIB
24 TAU	23 GEM	25 CAN	24 LEO	23 VIR	25 SCO
27 GEM	26 CAN	27 LEO	26 VIR	26 LIB	27 SAG
29 CAN	28 LEO	30 VIR	29 LIB	28 SCO	29 CAP
				30 SAG	

JUL	AUG	SEP	OCT	NOV	DEC
1 AQU	1 ARI	2 GEM	2 CAN	1 LEO	1 VIR
3 PIS	3 TAU	5 CAN	4 LEO	3 VIR	3 LIB
5 ARI	6 GEM	7 LEO	7 VIR	6 LIB	5 SCO
7 TAU	8 CAN	10 VIR	9 LIB	8 SCO	8 SAG
10 GEM	11 LEO	12 LIB	12 SCO	10 SAG	10 CAP
12 CAN	13 VIR	14 SCO	14 SAG	12 CAP	12 AQU
15 LEO	16 LIB	17 SAG	16 CAP	14 AQU	14 PIS
17 VIR	18 SCO	19 CAP	18 AQU	16 PIS	16 ARI
20 LIB	20 SAG	21 AQU	20 PIS	19 ARI	18 TAU
22 SCO	22 CAP	23 PIS	22 ARI	21 TAU	20 GEM
24 SAG	24 AQU	25 ARI	25 TAU	23 GEM	23 CAN
26 CAP	26 PIS	27 TAU	27 GEM	26 CAN	25 LEO
28 AQU	29 ARI	28 GEM	29 CAN	28 LEO	28 VIR
30 PIS	31 TAU				30 LIB

1971 MOON SIGNS

JAN	FEB	MAR	APR	MAY	JUN
3 ARI	1 TAU	2 GEM	1 CAN	1 LEO	2 LIB
5 TAU	3 GEM	5 CAN	3 LEO	3 VIR	5 SCO
7 GEM	5 CAN	7 LEO	6 VIR	6 LIB	7 SAG
9 CAN	8 LEO	10 VIR	8 LIB	8 SCO	9 CAP
12 LEO	10 VIR	12 LIB	11 SCO	11 SAG	11 AQU
14 VIR	13 LIB	15 SCO	13 SAG	13 CAP	14 PIS
17 LIB	15 SCO	17 SAG	16 CAP	15 AQU	16 ARI
19 SCO	18 SAG	19 CAP	18 AQU	17 PIS	18 TAU
22 SAG	20 CAP	22 AQU	20 PIS	20 ARI	20 GEM
24 CAP	22 AQU	24 PIS	22 ARI	22 TAU	22 CAN
26 AQU	24 PIS	26 ARI	24 TAU	24 GEM	24 LEO
28 PIS	26 ARI	28 TAU	26 GEM	26 CAN	27 VIR
30 ARI	28 TAU	30 GEM	28 CAN	28 LEO	29 LIB
				30 VIR	

JUL	AUG	SEP	OCT	NOV	DEC
2 SCO	1 SAG	2 AQU	1 PIS	2 TAU	1 GEM
4 SAG	3 CAP	4 PIS	3 ARI	4 GEM	3 CAN
7 CAP	5 AQU	6 ARI	5 TAU	6 CAN	5 LEO
9 AQU	7 PIS	8 TAU	7 GEM	8 LEO	8 VIR
11 PIS	9 ARI	10 GEM	9 CAN	10 VIR	10 LIB
13 ARI	11 TAU	12 CAN	12 L;EO	13 LIB	13 SCO
15 TAU	13 GEM	14 LEO	14 VIR	15 SCO	15 SAG
17 GEM	16 CAN	17 VIR	16 LIB	18 SAG	17 CAP
19 CAN	18 LEO	19 LIB	19 SCO	20 CAP	20 AQU
22 LEO	20 VIR	22 SCO	22 SAG	23 AQU	22 PIS
24 VIR	23 LIB	24 SAG	24 CAP	25 PIS	24 ARI
27 L;IB	26 SCO	27 CAP	26 AQU	27 ARI	25 TAU
29 SCO	28 SAG	29 AQU	29 PIS	29 TAU	28 GEM
	30 CAP		31 ARI		30 CAN

1972 MOON SIGNS

JAN	FEB	MAR	APR	MAY	JUN
2 LEO	3 LIB	1 LIB	2 SAG	2 CAP	1 AQU
4 VIR	5 SCO	5 SCO	5 CAP	6 AQU	3 PIS
6 LIB	8 SAG	6 SAG	7 AQU	7 PIS	5 ARI
9 SCO	10 CAP	9 CAP	10 PIS	9 ARI	7 TAU
11 AQU	12 AQU	11 AQU	12 ARI	11 TAU	9 GEM
14 CAP	15 ARI	13 PIS	14 ATU	13 GEM	11 CAN
16 AQU	17 ARI	15 ARI	16 GEM	15 CAN	14 LEO
18 PIS	19 TAU	17 TAU	18 CAN	17 LEO	16 VIR
20 ARI	21 GEM	19 GEM	20 LEO	19 VIR	18 LIB
23 TAU	23 CAN	21 CAN	22 VIR	22 LIB	21 SCO
25 GEM	25 LEO	24 LEO	25 LIB	24 SCO	23 SAG
27 CAN	28 VIR	26 VIR	27 SCO	27 SAG	26 CAP
27 LEO		28 LIB	30 SAG	29 CAP	28 AQU
31 VIR		31 SCO			30 PIS

JUL	AUG	SEP	OCT	NOV	DEC
3 ARI	1 TAU	1 CAN	1 LEO	2 LIB	1 CAO
5 TAU	3 GEM	4 LEO	3 VIR	4 SCO	4 SAG
7 GEM	5 CAN	6 VIR	5 LIB	7 SAG	7 CAP
9 CAN	7 LEO	8 LIB	8 SCO	9 CAP	9 AQU
11 LEO	10 VIR	11 SCO	10 SAG	12 AQU	11 PIS
13 VIR	12 LIB	13 SAG	13 CAP	14 PIS	14 ARI
15 LIB	14 SCO	16 CAP	15 AQU	16 ARI	16 TAU
18 SCO	17 SAG	18 AQU	18 PIS	18 TAU	18 GEM
20 SAG	19 CAP	20 PIS	20 ARI	20 GEM	20 CAN
23 CAP	22 AQU	22 ARI	22 TAU	22 CAN	22 LEO
25 AQU	24 PIS	24 TAU	24 GEM	24 LEO	24 VIR
28 PIS	26 ARI	27 GEM	26 CAN	27 VIR	26 LIB
30 ARI	28 TAU	29 CAN	28 LEO	29 LIB	29 SCO
	30 GEM		30 VIR		31 SAG

1974 MOON SIGNS

JAN	FEB	MAR	APR	MAY	JUN
2 TAU	1 GEM	2 CAN	1 LEO	2 LIB	1 SCO
4 GEM	3 CAN	4 LEO	3 VIR	4 SCO	3 SAG
7 CAN	5 LEO	7 VIR	5 LIB	7 SAG	6 CAP
9 LEO	7 VIR	9 LIB	7 SCO	9 CAP	8 AQU
11 VIR	9 LIB	11 SCO	9 SAG	12 AQU	11 PIS
13 LIB	11 SCO	13 SAG	12 CAP	14 PIS	13 ARI
15 SCO	14 AG	16 CAP	14 AQU	17 ARI	15 TAU
17 SAG	16 CAP	18 AQU	17 PIS	19 TAU	18 GEM
20 CAP	19 AQU	21 PIS	19 ARI	21 GEM	20 CAN
22 AQU	21 PIS	23 ARI	22 TAU	23 CAN	22 LEO
25 PIS	24 ARI	25 TAU	24 GEM	25 LEO	24 VIR
27 ARI	26 TAU	27 GEM	26 CAN	27 VIR	26 LIB
30 TAU	28 GEM	30 CAN	28 LEO	30 LIB	28 SCO
			30 VIR		30 SAG

JUL	AUG	SEP	OCT	NOV	DEC
3 CAP	2 AQU	3 ARI	2 TAU	1 GEM	1 CAN
5 ARI	4 PIS	5 TAU	5 GEM	3 CAN	3 LEO
8 PIS	7 ARI	8 GEM	7 CAN	5 LEO	5 VIR
10 ARI	9 TAU	10 CAN	9 LEO	8 VIR	7 LIB
13 TAU	11 GEM	12 LEO	11 VIR	10 LIB	9 SCO
15 GEM	13 CAN	14 VIR	13 LIB	12 SCO	11 SAG
17 CAN	15 LEO	16 LIB	15 SCO	14 SAG	14 CAP
19 LEO	17 VIR	18 SCO	18 SAG	16 CAP	16 ARI
21 VIR	20 LIB	20 SAG	20 CAP	19 AQU	19 PIS
23 LIB	22 SCO	23 CAP	22 AQU	21 PIS	21 ARI
25 SCO	24 SAG	25 AQU	25 PIS	24 ARI	24 TAU
28 SAG	26 CAP	28 PIS	28 ARI	26 TAU	26 GEM
30 CAP	29 AQU	30 ARI	30 TAU	28 GEM	28 CAN
	31 PIS				30 LEO

1973 MOON SIGNS

JAN	FEB	MAR	APR	MAY	JUN
3 CAP	2 AQU	1 AQU	2 ARI	1 TAU	2 CAN
5 AQU	4 PIS	3 PIS	4 TAU	3 GEM	4 LEO
8 PIS	6 ARI	5 ARI	6 GEM	5 CAN	6 VIR
10 ARI	8 TAU	8 TAU	8 CAN	7 LEO	8 LIB
12 TAU	10 GEM	10 GEM	10 LEO	10 VIR	11 SCO
14 GEM	13 CAN	12 CAN	12 VIR	12 LIB	13 SAG
16 CAN	15 LEO	14 LEO	15 LIB	14 SCO	16 CAP
18 LEO	17 VIR	16 VIR	17 SCO	17 SAG	18 AQU
20 VIR	19 LIB	19 LIB	20 SAG	19 CAP	21 PIS
23 LIB	21 SCO	21 SCO	22 CAP	22 AQU	23 ARI
25 SCO	24 SAG	23 SAG	25 AQU	24 PIS	25 TAU
28 SAG	26 CAP	26 CAP	27 PIS	27 ARI	27 GEM
30 CAP		28 AQU	29 ARI	29 TAU	29 CAN
		31 PIS		31 GEM	

JUL	AUG	SEP	OCT	NOV	DEC
1 LEO	2 LIB	1 SCO	3 CAP	2 AQU	1 PIS
3 VIR	4 SCO	3 SAG	5 AQU	4 PIS	4 ARI
5 LIB	7 SAG	5 CAP	8 PIS	6 ARI	6 TAU
8 SCO	9 CAP	8 AQU	10 ARI	9 TAU	8 GEM
10 SAG	12 AQU	10 PIS	12 TAU	11 GEM	10 CAN
13 CAP	14 PIS	13 ARI	14 GEM	13 CAN	12 LEO
15 AQU	16 ARI	15 TAU	16 CAN	15 LEO	14 VIR
18 PIS	19 TAU	17 GEM	19 LEO	17 VIR	16 LIB
20 ARI	21 GEM	19 CAN	21 VIR	19 LIB	19 SCO
22 TAU	23 CAN	21 LEO	23 LIB	22 SCO	21 SAG
25 GEM	25 LEO	23 VIR	25 SCO	24 SAG	24 CAP
27 CAN	27 VIR	26 LIB	28 SAG	26 CAP	26 AQU
29 LEO	29 LIB	28 SCO	30 CAP	29 AQU	29 PIS
31 VIR		30 SAG			31 ARI

1975 MOON SIGNS

JAN	FEB	MAR	APR	MAY	JUN
1 VIR	2 SCO	1 SCO	2 CAP	2 AQU	3 ARI
3 LIB	4 SAG	3 SAG	4 AQU	4 PIS	5 TAU
5 SCO	6 CAP	5 CSP	7 PIS	7 ARI	8 GEM
8 SAG	9 AQU	8 AQU	9 ARI	9 TAU	10 CAN
10 CAP	11 PIS	10 PIS	12 TAU	11 GEM	12 LEO
12 AQU	14 ARI	13 ARI	14 GEM	14 CAN	14 VIR
15 PIS	16 TAU	15 TAU	16 CAN	16 LEO	16 LIB
17 ARI	19 GEM	18 GEM	19 LEO	18 VIR	18 SCO
20 TAU	21 CAN	20 CAN	21 VIR	20 LIB	21 SAG
22 GEM	23 LEO	22 LEO	21 LIB	22 SCO	23 CAP
24 CAN	25 VIR	24 VIR	SCO	24 SAG	25 AQU
26 LEO	27 LIB	26 LIB	27 SAG	27 CAP	28 PIS
28 VIR		28 SCO	29 CAP	29 AQU	30 ARI
30 LIB		30 SAG		31 PIS	

JUL	AUG	SEP	OCT	NOV	DEC
3 TAU	1 GEM	2 LEO	2 VIR	2 SCO	2 SAG
5 GEM	4 CAN	4 VIR	4 LIB	4 SAG	4 CAP
7 CAN	6 LEO	6 LIB	6 SCO	6 CAP	6 AQU
9 LEO	8 VIR	8 SCO	8 SAG	9 AQU	8 PIS
11 VIR	10 LIB	10 SAG	10 CAP	11 PIS	11 ARI
14 LIB	12 SCO	13 CAP	12 AQU	14 ARI	13 TAU
16 SCO	14 SAG	15 AQU	15 PIS	16 TAU	16 GEM
18 SAG	16 CAP	18 PIS	17 ARI	19 GEM	18 CAN
20 CAP	19 AQU	20 ARI	20 TAU	21 CAN	20 LEO
23 AQU	21 PIS	23 TAU	22 GEM	23 LEO	23 VIR
25 PIS	24 ARI	25 GEM	25 CAN	25 VIR	25 LIB
28 ARI	26 TAU	27 CAN	27 LEO	27 LIB	27 SCO
30 TAU	29 GEM	30 LEO	29 VIR	30 SCO	29 SAG
	31 CAN		31 LIB		31 CAP

241

1976 MOON SIGNS

JAN	FEB	MAR	APR	MAY	JUN
2 AQU	1 PIS	2 ARI	1 TAU	3 CAN	1 LEO
5 PIS	4 ARI	4 TAU	3 GEM	5 LEO	4 VIR
7 ARI	6 TAU	7 GEM	6 CAN	7 VIR	6 LIB
10 TAU	9 GEM	9 CAN	8 LEO	10 LIB	8 SCO
12 GEM	11 CAN	12 LEO	10 VIR	12 SCO	10 SAG
15 CAN	13 LEO	14 VIR	12 LIB	14 SAG	12 CAP
17 LEO	15 VIR	16 LIB	14 SCO	16 CAP	14 AQU
19 VIR	17 LIB	18 SCO	16 SAG	18 AQU	17 PIS
21 LIB	19 SCO	20 SAG	18 CAP	20 PIS	19 ARI
23 SCO	21 SAG	22 CAP	20 AQU	23 ARI	22 TAU
25 SAG	24 CAP	24 AQU	23 PIS	25 TAU	24 GEM
27 CAP	26 AQU	27 PIS	25 ARI	28 GEM	26 CAN
30 AQU	28 PIS	29 ARI	28 TAU	30 CAN	29 LEO
			30 GEM		
JUL	**AUG**	**SEP**	**OCT**	**NOV**	**DEC**
1 VIR	1 SCO	2 CAP	1 AQU	2 ARI	2 TAU
3 LIB	4 SAG	4 AQU	4 PIS	5 TAU	5 GEM
5 SCO	6 CAP	7 PIS	6 ARI	8 GEM	7 CAN
7 SAG	8 AQU	9 ARI	9 TAU	10 CAN	10 LEO
9 CAP	10 PIS	11 TAU	11 GEM	12 LEO	12 VIR
12 AQU	13 ARI	14 GEM	14 CAN	15 VIR	14 LIB
14 PIS	15 TAU	17 CAN	16 LEO	17 LIB	16 SCO
16 ARI	18 GEM	19 LEO	18 VIR	19 SCO	18 SAG
19 TAU	20 CAN	21 VIR	21 LIB	21 SAG	20 CAP
21 GEM	22 LEO	23 LIB	23 SCO	23 CAP	22 AQU
24 CAN	25 VIR	25 SCO	25 SAG	25 AQU	25 PIS
26 LEO	27 LIB	27 SAG	27 CAP	27 PIS	27 ARI
28 VIR	29 SCO	29 CAP	29 AQU	30 ARI	30 TAU
30 LIB	31 SAG		31 PIS		

1977 MOON SIGNS

JAN	FEB	MAR	APR	MAY	JUN
1 GEM	2 LEO	2 LEO	2 LIB	2 SCO	2 CAP
4 CAN	5 VIR	4 VIR	5 SCO	4 SAG	4 AQU
6 LEO	7 LIB	6 LIB	7 SAG	6 CAP	7 PIS
8 VIR	9 SCO	8 SCO	8 CAP	8 AQU	9 ARI
10 LIB	11 SAG	10 SAG	11 AQU	10 PIS	11 TAU
13 SCO	13 CAP	12 CAP	13 PIS	13 ARI	14 GEM
15 SAG	15 AQU	15 AQU	15 ARI	15 TAU	16 CAN
17 CAP	17 PIS	17 PIS	18 TAU	18 GEM	19 LEO
19 AQU	20 ARI	19 ARI	20 GEM	20 CAN	21 VIR
21 PIS	22 TAU	22 TAU	23 CAN	23 LEO	24 LIB
23 ARI	25 GEM	24 GEM	25 LEO	25 VIR	26 SCO
26 TAU	27 CAN	27 CAN	28 VIR	27 LIB	28 SAG
28 GEM		29 LEO	30 LIB	29 SCO	30 CAP
31 CAN		31 VIR		31 SAG	
JUL	**AUG**	**SEP**	**OCT**	**NOV**	**DEC**
2 AQU	3 ARI	1 TAU	1 GEM	3 LEO	2 VIR
4 PIS	5 TAU	4 GEM	4 CAN	5 VIR	5 LIB
6 ARI	7 GEM	6 CAN	6 LEO	7 LIB	7 SCO
9 TAU	10 CAN	9 LEO	9 VIR	9 SCO	9 SAG
11 GEM	12 LEO	11 VIR	11 LIB	11 SAG	11 CAP
14 CAN	15 VIR	13 LIB	13 SCO	13 CAP	13 AQU
16 LEO	17 LIB	16 SCO	15 SAG	15 AQU	15 PIS
19 VIR	19 SCO	18 SAG	17 CAP	18 PIS	17 ARI
21 LIB	21 SAG	20 CAP	19 AQU	20 ARI	19 TAU
23 SCO	24 CAP	22 AQU	21 PIS	22 TAU	22 CAM
25 SAG	26 AQU	24 PIS	24 ARI	25 GEM	25 CAN
27 CAP	28 PIS	26 ARI	26 TAU	27 CAN	27 LEO
29 AQU	30 ARI	28 TAU	28 GEM	30 LEO	30 VIR
31 PIS			31 CAN		

1978 MOON SIGNS

JAN	FEB	MAR	APR	MAY	JUN
1 LIB	2 SAG	1 SAG	1 AQU	1 PIS	1 TAU
3 SCO	4 CAP	3 CAP	3 PIS	3 ARI	4 GEM
5 SAG	6 AQU	5 AQU	6 ARI	5 TAU	6 CAN
7 CAP	8 PIS	7 PIS	8 TAU	8 GEM	9 LEO
9 AQU	10 ARI	9 ARI	10 GEM	10 CAN	11 VIR
11 PIS	12 TAU	12 TAU	13 CAN	13 LEO	14 LIB
13 ARI	15 GEM	14 GEM	15 LEO	15 VIR	16 SCO
16 TAU	17 CAN	16 CAN	18 VIR	17 LIB	18 SAG
18 GEM	20 LEO	19 LEO	20 LIB	20 SCO	20 CAP
21 CAN	22 VIR	21 VIR	22 SCO	22 SAG	22 AQU
23 LEO	24 LIB	24 LIB	24 SAG	24 CAP	24 PIS
26 VIR	27 SCO	26 SCO	26 CAP	26 AQU	26 ARI
28 LIB		28 SAG	29 AQU	28 PIS	29 TAU
30 SCO		30 CAP		30 ARI	
JUL	**AUG**	**SEP**	**OCT**	**NOV**	**DEC**
1 GEM	2 LEO	1 VIR	1 LIB	2 SAG	1 CAP
4 CAN	5 VIR	4 LIB	3 SCO	4 CAP	3 AQU
6 LEO	7 LIB	6 SCO	5 SAG	6 AQU	5 PIS
9 VIR	10 SCO	8 SAG	8 CAP	8 PIS	7 ARI
11 LIB	12 SAG	10 CAP	10 AQU	10 ARI	10 TAU
13 SCO	14 CAP	12 AQU	12 PIS	12 TAU	12 GEM
16 SAG	16 AQU	14 PIS	14 ARI	15 GEM	14 CAN
18 CAP	18 PIS	17 ARI	16 TAU	17 CAN	17 LEO
20 AQU	20 ARI	19 TAU	18 GEM	20 LEO	19 VIR
22 PIS	22 TAU	21 GEM	21 CAN	22 VIR	22 LIB
24 ARI	25 GEM	23 CAN	23 LEO	25 LIB	24 SCO
26 TAU	27 CAN	26 LEO	26 VIR	27 SCO	27 SAG
28 GEM	30 LEO	28 VIR	28 LIB	29 SAG	29 CAP
31 CAN			31 SCO		31 AQU

1979 MOON SIGNS

JAN	FEB	MAR	APR	MAY	JUN
2 PIS	2 TAU	2 TAU	3 CAN	2 LEO	1 VIR
4 ARI	5 GEM	4 GEM	5 LEO	5 VIR	4 LIB
6 TAU	7 CAN	6 CAN	8 VIR	7 LIB	6 SCO
8 GEM	9 LEO	9 LEO	10 LIB	10 SCO	8 SAG
11 CAN	12 VIR	11 VIR	12 SCO	12 SAG	11 CAP
13 LEO	15 LIB	14 LIB	15 SAG	14 CAP	13 AQU
16 VIR	17 SCO	16 SCO	17 CAP	16 AQU	15 PIS
18 LIB	19 SAG	19 SAG	19 AQU	18 PIS	17 ARI
21 SCO	21 CAP	21 CAP	21 PIS	21 ARI	19 TAU
23 SAG	24 AQU	23 AQU	23 ARI	23 TAU	21 GEM
25 CAP	26 PIS	25 PIS	25 TAU	25 GEM	24 CAN
27 AQU	28 ARI	27 ARI	28 GEM	27 CAN	26 LEO
29 PIS		29 TAU	30 CAN	30 LEO	29 VIR
31 ARI		31 GEM			
JUL	**AUG**	**SEP**	**OCT**	**NOV**	**DEC**
1 LIB	2 SAG	1 CAP	2 PIS	1 ARI	2 GEM
4 SCO	4 CAP	3 AQU	4 ARI	3 TAU	4 CAN
6 SAG	6 AQU	5 PIS	6 TAU	5 GEM	7 LEO
8 CAP	8 PIS	7 ARI	8 GEM	7 CAN	9 VIR
10 AQU	10 ARI	9 TAU	11 CAN	9 LEO	12 LIB
12 PIS	13 TAU	11 GEM	13 LEO	12 VIR	14 SCO
14 ARI	15 GEM	13 CAN	16 VIR	14 LIB	17 SAG
16 TAU	17 CAN	16 LEO	18 LIB	17 SCO	19 CAP
19 GEM	20 LEO	18 VIR	21 SCO	19 SAG	21 AQU
21 CAN	22 VIR	21 LIB	23 SAG	22 CAP	23 PIS
23 LEO	25 LIB	23 SCO	25 CAP	24 AQU	25 ARI
26 VIR	27 SCO	26 SAG	28 AQU	26 PIS	27 TAU
28 LIB	30 SAG	28 CAP	30 PIS	28 ARI	30 GEM
31 SCO		30 AQU		30 TAU	

1980 MOON SIGNS

JAN	FEB	MAR	APR	MAY	JUN
1 CAN	2 VRI	3 LIB	2 SCO	1 SAG	2 AQU
3 LEO	4 LIB	5 SCO	4 SAG	4 CAP	4 PIS
6 VIR	7 SCO	8 SAG	6 CAP	6 AQU	6 ARI
8 LIB	9 SAG	10 CAP	9 AQU	8 PIS	9 TAU
11 SCO	12 CAP	12 AQU	11 PIS	10 ARI	11 GEM
13 AQU	14 AQU	14 PIS	13 ARI	12 TAU	13 CAN
15 CAP	16 PIS	16 ARI	15 TAU	14 GEM	15 LEO
17 AQU	18 ARI	18 TAU	17 GEM	16 CAN	17 VIR
19 PIS	20 TAU	20 GEM	19 CAN	19 LEO	20 LIB
21 ARI	22 GEM	23 CAN	21 LEO	21 VIR	22 SCO
24 TAU	24 CAN	25 LEO	24 VIR	24 LIB	25 SAG
26 GEM	27 LEO	27 VIR	26 LIB	26 SCO	27 CAP
28 CAN	29 VIR	30 LIB	29 SCO	29 SAG	29 AQU
30 LEO				31 CAP	

JUL	AUG	SEP	OCT	NOV	DEC
2 PIS	2 TAU	3 CAN	2 LEO	1 VIR	1 LIB
4 ARI	4 GEM	5 LEO	5 VIR	3 LIB	3 SCO
6 TAU	6 CAN	7 VIR	7 LIB	6 SCO	6 SAG
8 GEM	9 LEO	10 LIB	10 SCO	8 SAG	8 CAP
10 CAN	11 VIR	12 SCO	12 SAG	11 CAP	10 AQU
12 LEO	14 LIB	15 SAG	15 CAP	13 AQU	13 PIS
15 VIR	16 SCO	17 CAP	17 AQU	15 PIS	15 ARI
17 LIB	19 SAG	20 AQU	19 PIS	18 ARI	17 TAU
20 SCO	21 CAP	22 PIS	21 ARI	20 TAU	19 GEM
22 AQU	23 AQU	24 ARI	23 TAU	22 GEM	21 CAN
25 CAP	25 PIS	26 TAU	25 GEM	24 CAN	23 LEO
27 AQU	27 ARI	28 GEM	27 CAN	26 LEO	25 VIR
29 PIS	29 TAU	30 CAN	29 LEO	28 VIR	28 LIB
31 ARI	31 GEM				30 SCO

1981 MOON SIGNS

JAN	FEB	MAR	APR	MAY	JUN
2 SAG	1 CAP	2 AQU	1 PIS	1 ARI	1 GEM
4 CAP	3 AQU	5 PIS	3 ARI	3 TAU	3 CAN
7 AQU	5 PIS	7 ARI	5 TAU	5 GEM	5 LEO
9 PIS	7 ARI	9 TAU	7 GEM	7 CAN	7 VIR
11 ARI	9 TAU	11 GEM	9 CAN	9 LEO	10 LIB
13 TAU	12 GEM	13 CAN	11 LEO	11 VIR	12 SCO
15 GEM	14 CAN	15 LEO	14 VIR	13 LIB	15 SAG
17 CAN	16 LEO	18 VIR	16 LIB	16 SCO	17 CAP
20 LEO	18 VIR	20 LIB	19 SCO	18 SAG	20 AQU
22 VIR	21 LIB	22 SCO	21 SAG	21 CAP	22 PIS
24 LIB	23 SCO	25 SAG	24 CAP	23 AQU	24 ARI
27 SCO	26 SAG	27 CAP	26 AQU	26 PIS	26 TAU
29 SAG	28 CAP	30 AQU	28 PIS	28 ARI	28 GEM
				30 TAU	30 CAN

JUL	AUG	SEP	OCT	NOV	DEC
2 LEO	1 VIR	2 SCO	2 SAG	1 CAP	1 AQU
5 VIR	3 LIB	5 SAG	5 CAP	3 AQU	3 PIS
7 LIB	6 SCO	7 CAP	7 AQU	6 PIS	5 ARI
10 SCO	8 SAG	10 AQU	9 PIS	8 ARI	7 TAU
12 SAG	11 CAP	12 PIS	11 ARI	10 TAU	9 GEM
15 CAP	13 AQU	14 ARI	13 TAU	12 GEM	11 CAN
17 AQU	16 PIS	16 TAU	15 GEM	14 CAN	13 LEO
19 PIS	18 ARI	18 GEM	18 CAN	16 LEO	16 VIR
21 ARI	20 TAU	20 CAN	20 LEO	18 VIR	18 LIB
24 TAU	22 GEM	22 LEO	22 VIR	21 LIB	20 SCO
26 GEM	24 CAN	25 VIR	24 LIB	23 SCO	23 SAG
28 CAN	26 LEO	27 LIB	27 SCO	26 SAG	25 CAP
30 LEO	28 VIR	29 SCO	29 SAG	28 CAP	28 AQU
	31 LIB				30 PIS

1982 MOON SIGNS

JAN	FEB	MAR	APR	MAY	JUN
2 ARI	3 GEM	1 GEM	2 LEO	1 VIR	2 SCO
4 TAU	4 CAN	3 CAN	4 VIR	4 LIB	5 SAG
6 GEM	6 LEO	6 LEO	6 LIB	6 SCO	7 CAP
8 CAN	8 VIR	8 VIR	9 SCO	8 SAG	10 AQU
10 LEO	11 LIB	10 LIB	11 SAG	11 CAP	12 PIS
12 VIR	13 SCO	12 SCO	14 CAP	13 AQU	15 ARI
14 LIB	15 SAG	15 SAG	16 AQU	16 PIS	17 TAU
17 SCO	18 CAP	17 CAP	19 PIS	18 ARI	19 GEM
19 SAG	20 AQU	20 AQU	21 ARI	20 TAU	21 CAN
22 CAP	23 PIS	22 PIS	23 TAU	22 GEM	23 LEO
24 AQU	25 ARI	24 ARI	25 GEM	24 CAN	25 VIR
26 PIS	27 TAU	27 TAU	27 CAN	26 LEO	27 LIB
29 ARI		29 GEM	29 LEO	29 VIR	29 SCO
31 TAU		31 CAN		31 LIB	

JUL	AUG	SEP	OCT	NOV	DEC
2 SAG	1 CAP	2 PIS	2 ARI	2 GEM	2 CAN
4 CAP	3 AQU	4 ARI	4 TAU	4 CAN	4 LEO
7 AQU	6 PIS	7 TAU	6 GEM	6 LEO	6 VIR
9 PIS	8 ARI	9 GEM	8 CAN	9 VIR	8 PIB
12 ARI	10 TAU	11 CAN	10 LEO	11 LIB	10 SCO
14 TAU	13 GEM	13 LEO	12 VIR	13 SCO	13 SAG
16 GEM	16 CAN	15 VIR	15 LIB	15 SAG	15 CAN
18 CAN	17 LEO	17 LIB	17 SCO	18 CAP	18 AQU
20 LEO	19 VIR	19 SCO	19 SAG	21 AQU	20 PIS
22 VIR	21 LIB	22 SAG	22 CAP	23 PIS	23 ARI
24 LIB	23 SCO	24 CAP	24 AQU	24 ARI	25 TAU
27 SCO	25 SAG	27 AQU	27 PIS	28 TAU	27 GEM
29 SAG	28 CAP	29 PIS	29 ARI	30 GEM	29 CAN
	31 AQU	31 TAU			31 LEO

1983 MOON SIGNS

JAN	FEB	MAR	APR	MAY	JUN
2 VIR	1 LIB	2 SCO	1 SAG	1 CAP	2 PIS
4 LIB	3 SCO	5 SAG	3 CAP	3 AQU	5 ARI
7 SCO	5 SAG	7 CAP	6 AQU	6 PIS	7 TAU
9 SAG	8 CAP	10 AQU	8 PIS	8 ARI	9 GEM
12 CAP	12 AQU	12 PIS	11 ARI	11 TAU	11 CAN
14 AQU	13 PIS	15 ARI	13 TAU	13 GEM	13 LEO
17 PIS	15 ARI	15 TAU	15 GEM	15 CAN	15 VIR
19 ARI	18 TAU	19 GEM	18 CAN	17 LEO	17 LIB
21 TAU	20 GEM	21 CAN	20 LEO	19 VIR	20 SCO
24 GEM	22 CAN	23 LEO	22 VIR	21 LIB	22 SAG
26 CAN	24 LEO	26 VIR	24 LIB	23 SCO	24 CAP
28 LEO	26 VIR	28 LIB	26 SCO	26 SAG	26 AQU
30 VIR	28 LIB	30 SCO	28 SAG	28 CAP	29 PIS
				31 AQU	

JUL	AUG	SEP	OCT	NOV	DEC
2 ARI	1 TAU	1 CAN	1 LEO	1 LIB	1 SCO
4 TAU	3 GEM	3 LEO	3 VIR	3 SCO	3 SAG
7 GEM	5 CAN	5 VIR	5 LIB	6 SAG	5 CAP
9 CAN	7 LEO	7 LIUB	7 SCO	8 CAP	8 AQU
11 LEO	9 LOIB	10 SCO	9 SAG	10 AQU	10 PIS
13 VIR	11 LIB	12 SAG	11 CAP	`13 PIS	13 ARI
15 LIB	13 SCO	14 CAP	14 AQU	15 ARI	15 TAU
17 SCO	15 SAG	17 AQU	16 PIS	18 TAU	17 GEM
19 SAG	18 CAP	19 PIS	19 ARI	20 GEM	20 CAN
22 CAP	20 AQU	22 ARI	21 TAU	22 CAN	22 LEO
24 AQU	23 PIS	24 TAU	24 GEM	24 LEO	24 VIR
27 PIS	27 ARI	26 GEM	26 CAN	26 VIR	26 LIB
29 ARI	28 TAU	29 CAN	28 LEO	29 LIB	28 SCO
	30 GEM		30 VIR		30 SAG

1984 MOON SIGNS

JAN	FEB	MAR	APR	MAY	JUN
2 CAP	3 PIS	1 PIS	2 TAU	2 GEM	1 CAN
4 AQU	5 ARI	4 ARI	5 GEM	4 CAN	3 LEO
6 PIS	8 TAU	6 TAU	7 CAN	6 LEO	5 VIR
9 ARI	10 GEM	8 GEM	9 LEO	9 VIR	7 LIB
11 TAU	12 CAN	11 CAN	11 VIR	11 LIB	9 SCO
14 GEM	15 LEO	13 LEO	13 LIB	13 SCO	13 SAG
16 CAN	17 VIR	15 VIR	15 SCO	15 SAG	13 CAP
18 LEO	19 LEO	17 LIB	17 SAG	17 CAP	16 AQU
20 SCO	21 LIB	19 SCO	20 CAP	19 AQU	18 PIS
22 LIB	23 SAG	21 SAG	22 AQU	22 PIS	21 ARI
24 SCO	25 CAP	23 CAP	25 PIS	24 ARI	23 TAU
26 SAG	28 AQU	26 AQU	27 ARI	27 TAU	26 GEM
29 CAP		28 PIS	30 TAU	29 GEM	28 CAN
31 AQU		31 ARI			30 LEO

JUL	AUG	SEP	OCT	NOV	DEC
2 VIR	3 SCO	1 SAG	1 CAP	2 PIS	1 ARI
4 LIB	5 SAG	3 CAP	3 AQU	4 ARI	4 TAU
6 SCO	7 CAP	6 AQU	5 PIS	7 TAU	6 GEM
9 SAG	9 AQU	9 PIS	8 ARI	9 GEM	9 CAN
11 CAP	12 PIS	11 ARI	10 TAU	12 CAN	11 LEO
13 AQU	14 ARI	13 TAU	13 GEM	14 LEO	13 VIR
16 PIS	17 TAU	16 GEM	15 CAN	16 VIR	15 LIB
18 ARI	19 GEM	18 CAN	18 LEO	18 LIB	17 SCO
21 TAU	22 CAN	20 LEO	20 VIR	20 SCO	20 SCO
23 GEM	24 LEO	22 VIR	22 LIB	22 SAG	22 CAP
25 CAN	26 VIR	24 LIB	24 SCO	24 CAP	24 AQU
27 LEO	28 LIB	26 SCO	26 SAG	27 AQU	26 PIS
29 VIR	30 SCO	28 SAG	28 CAP	29 PIS	29 ARI
31 LIB			30 AQU	31 ARI	31 TAU

1985 MOON SIGNS

JAN	FEB	MAR	APR	MAY	JUN
3 GEM	2 CAN	1 CAN	2 VIR	1 LIB	2 SAG
5 CAN	4 LEO	3 LIB	4 LIB	3 SCO	4 CAP
7 LOE	6 VIR	5 LIB	6 SCO	5 SAG	6 AQU
9 VIR	8 LIB	7 LIB	8 SAG	7 CAP	8 PIS
12 LIB	10 SCO	9 SCO	10 CAP	9 AQU	11 ARI
14 SCO	12 SAG	11 SAG	12 AQU	12 PIS	13 TAU
16 SAG	14 CAP	14 CAP	14 PIS	14 ARI	16 GEM
18 CAP	17 AQU	16 AQU	17 ARI	17 TAU	18 CAN
20 AQU	19 PIS	18 PIS	20 TAU	19 GEM	20 LEO
22 PIS	21 ARI	21 ARI	22 GEM	22 CAN	23 VIR
25 ARI	24 TAU	23 TAU	25 CAN	24 LEO	25 LIB
28 TAU	27 GEM	26 GEM	27 LEO	26 VIR	27 SCO
30 GEM		28 CAN	29 VIR	29 LIB	29 SAG
		31 LEO		31 SCO	

JUL	AUG	SEP	OCT	NOV	DEC
1 CAP	2 PIS	1 ARI	3 GEM	2 CAN	1 LEO
3 AQU	4 ARI	3 TAU	5 CAN	4 LEO	4 VIR
5 PIS	7 TAU	6 GEM	8 LEO	6 VIR	6 LIB
8 ARI	9 GEM	8 CAN	10 VIR	9 LIB	8 SCO
10 TAU	12 CAN	10 LEO	12 LIB	11 SCO	10 SAG
13 GEM	14 LEO	13 VIR	14 SCO	13 SAG	12 CAP
15 CAN	16 VIR	15 LIB	16 SAG	15 CAP	14 AQU
18 LEO	18 LIB	17 SCO	18 CAP	17 AQU	16 PIS
20 VIR	20 SCO	19 SAG	20 AQU	19 PIS	19 ARI
22 LIB	22 SAG	21 CAP	23 PIS	21 ARI	21 TAU
24 SCO	25 CAP	23 AQU	25 ARI	24 TAU	24 GEM
26 SAG	27 AQU	25 PIS	28 TAU	26 GEM	26 CAN
28 CAP	29 PIS	28 ARI	30 GEM	29 CAN	29 LEO
31 AQU		30 TAU			31 VIR

1986 MOON SIGNS

JAN	FEB	MAR	APR	MAY	JUN
2 LIB	1 SCO	2 SAG	2 AQU	2 PIS	3 TAU
4 SCO	3 SAG	4 CAP	4 PIS	4 ARI	5 GEM
6 SAG	5 CAP	6 AQU	7 ARI	7 TAU	8 CAN
8 CAP	7 AQU	8 PIS	9 TAU	9 GEM	11 LEO
11 AQU	9 PIS	11 ARI	12 GEM	12 CAN	13 VIR
13 PIS	11 ARI	13 TAU	14 CAN	14 LEO	15 LIB
15 ARI	14 TAU	16 GEM	17 LEO	17 VIR	17 SCO
17 TAU	16 GEM	18 CAN	19 VIR	19 LIB	19 SAG
20 GEM	19 CAN	21 LEO	21 LIB	21 SCO	21 CAP
22 CAN	21 LEO	23 VIR	24 SCO	23 SAG	23 AQU
25 LEO	243 VIR	25 SAG	26 SAG	25 CAP	26 PIS
27 VIR	26 LIB	27 SCO	28 CAP	27 AQU	28 ARI
29 LIB	28 SCO	29 SAG	30 AQU	29 PIS	30 TAU
		31 CAP		31 ARI	

JUL	AUG	SEP	OCT	NOV	DEC
3 GEM	2 CAN	3 VIR	2 LIB	1 SCO	2 CAP
5 CAN	4 LEO	5 LIB	4 SCO	3 SAG	4 AQU
8 LEO	6 VIR	7 SCO	7 SAG	5 CAP	6 PIS
10 VIR	9 LIB	9 SAG	9 CAP	7 AQU	9 ARI
12 LIB	11 SCO	11 CAP	11 AQU	9 PIS	11 TAU
15 SCO	13 SAG	14 AQU	13 PIS	11 ARI	14 GEM
17 SAG	15 CAP	16 PIS	15 ARI	14 TAU	16 CAN
19 CAP	17 AQU	18 ARI	18 TAU	16 GEM	19 LEO
21 AQU	19 PIS	20 TAU	20 GEM	19 CAN	21 VIR
23 PIS	22 ARI	23 GEM	23 CAN	21 LEO	24 LIB
25 ARI	24 TAU	25 CAN	25 LEO	24 VIR	26 SCO
28 TAU	26 GEM	28 LEO	27 VIR	26 LIB	28 SCO
30 GEM	29 CAN	30 VIR	30 LIB	28 SCO	30 CAP
	30 LEO			30 SAG	

1987 MOON SIGNS

JAN	FEB	MAR	APR	MAY	JUN
1 AQU	1 ARI	1 ARI	2 GEM	2 CAN	3 VIR
3 PIS	4 TAU	3 TAU	4 CAN	4 LEO	5 LIB
5 ARU	6 GEM	5 GEM	7 LEO	7 VIR	8 SCO
7 TAU	9 CAN	8 CAN	9 VIR	9 LIB	10 SAG
10 GEM	11 LEO	10 LEO	12 LIB	11 SCO	12 CAP
12 CAN	14 VIR	13 VIR	14 SCO	13 SAG	14 AQU
15 LEO	16 LIB	15 LIB	16 SAG	15 CAP	16 PIS
17 VIR	18 SCO	18 SCO	18 CAP	17 AQU	18 ARI
20 LIB	21 SAG	20 SAG	20 AQU	20 PIS	20 TAU
22 SCO	23 CAP	22 CAP	22 PIS	22 ARI	23 GEM
24 SAG	25 AQU	24 AQU	25 ARI	24 TAU	25 CAN
26 CAP	27 PIS	26 PIS	27 TAU	26 GEM	28 LEO
28 AQU		28 ARI	29 GEM	29 CAN	30 VIR
30 PIS		30 TAU		31 LEO	

JUL	AUG	SEP	OCT	NOV	DEC
3 LIB	1 SCO	1 CAP	1 AQU	2 ARI	1 TAU
5 SCO	4 SAG	4 AQU	3 PIS	4 TAU	4 GEM
7 SAG	6 CAP	6 PUIS	6 ARI	6 GEM	6 CAN
9 CAP	8 AQU	8 ARI	8 TAU	9 CAN	8 LEO
11 AQU	10 PIS	10 TAU	10 GEM	11 LEO	11 VIR
13 PIS	12 ARI	13 GEM	12 CAN	14 VIR	14 L;IB
15 ARI	14 TAU	15 CAN	15 LEO	16 LIB	16 SCO
18 TAU	16 GEM	17 LEO	17 VIR	18 SCO	18 SAG
20 GEM	19 CAN	20 VIR	20 LIB	21 SAG	20 CAP
22 CAN	21 LEO	22 LIB	22 SCO	23 CAP	22 AQU
25 LEO	24 VIR	25 SCO	24 SAG	25 AQU	24 PIS
27 VIR	26 LIB	27 SAG	26 CAP	27 PIS	26 ARI
30 LIB	29 SCO	29 CAP	29 AQU	29 ARI	29 TAU
	31 SAG		31 PIS		31 GEM

244

1988 MOON SIGNS

JAN	FEB	MAR	APR	MAY	JUN
2 CAN	1 LEO	2 VIR	1 LIB	3 SAG	1 CAP
5 LEO	4 VIR	4 LIB	3 SCO	5 CAP	3 AQU
7 VIR	6 LIB	7 SCO	5 SAG	7 AQU	5 PIS
10 LIB	9 SCO	9 SAG	8 CAP	9 PIS	8 ARI
12 SCO	11 SAG	11 CAP	10 AQU	11 ARI	10 TAU
15 SAG	13 CAP	14 AQU	12 PIS	13 TAU	12 GEM
17 CAP	14 AQU	16 PIS	14 ARI	16 GEM	14 CAN
19 AQU	17 PIS	18 ARI	16 TAU	18 CAN	17 LEO
21 PIS	19 ARI	20 TAU	18 GEM	20 LEO	19 VIR
23 ARI	21 TAU	22 GEM	20 CAN	23 VIR	22 LIB
25 TAU	23 GEM	24 CAN	23 LEO	25 LIB	24 SCO
27 GEM	26 CAN	27 LEO	25 VIR	28 SCO	26 SAG
30 CAN	28 LEO	29 VIR	28 LIB	30 SAG	29 CAP
			30 SCO		

JUL	AUG	SEP	OCT	NOV	DEC
1 AQU	1 ARI	2 GEM	1 CAN	2 VIR	2 LIB
3 PIS	3 TAU	4 CAN	4 LEO	5 LIB	5 SCO
5 ARI	5 GEM	6 LEO	6 VIR	7 SCO	7 SAG
7 TAU	8 CAN	9 VIR	9 LIB	10 SAG	9 CAP
9 GEM	10 LEO	11 LIB	11 SCO	12 CAP	12 AQU
11 CAN	13 VIR	14 SCO	14 SAG	14 AQU	14 PIS
14 LEO	15 LIB	16 SAG	16 CAP	17 PIS	16 ARI
16 VIR	18 SCO	19 CAP	18 AQU	19 ARI	18 TAU
19 LIB	20 SAG	21 AQU	20 PIS	21 TAU	20 GEM
21 SCO	22 CAP	23 PIS	22 ARI	23 GEM	22 CAN
24 SAG	24 AQU	25 ARI	24 TAU	25 CAN	25 LEO
26 CAP	26 PIS	27 TAU	26 GEM	27 LEO	27 VIR
28 AQU	28 ARI	29 GEM	29 CAN	30 VIR	30 LIB
30 PIS	30 TAU		31 LEO		

1989 MOON SIGNS

JAN	FEB	MAR	APR	MAY	JUN
1 SCO	2 CAP	2 CAP	2 PIS	2 ARI	2 GEM
4 SAG	4 AQU	4 AQU	4 ARI	4 TAU	4 CAN
6 CAP	6 PIS	6 PIS	6 TAU	6 GEM	7 LEO
8 AQU	8 ARI	8 ARI	8 GEM	8 CAN	9 VIR
10 PIS	11 TAU	10 TAU	11 CAN	10 LEO	11 LIB
12 ARI	13 GEM	12 GEM	13 LEO	13 VIR	14 SCO
14 TAU	15 CAN	14 CAN	15 VIR	15 LIB	16 SAG
16 GEM	17 LEO	17 LEO	18 LIB	18 SCO	19 CAP
19 CAN	20 VIR`	19 VIR	20 SCO	20 SAG	21 AQU
21 LEO	22 LIB	22 LIB	23 SAG	22 CAP	23 PIS
23 VIR	25 SCO	24 SCO	25 CAP	23 AQU	25 ARI
26 LIB	27 SAG	27 SAG	28 AQU	27 PIS	27 TAU
29 SCO		29 CAP	30 PIS	29 ARI	30 GEM
31 VIR		31 AQU		31 TAU	

JUL	AUG	SEP	OCT	NOV	DEC
2 CAN	3 VIR	1 LIB	1 SCO	2 CAP	2 AQU
4 LEO	5 LIB	4 SCO	4 SAG	5 AQU	4 PIS
6 VIR	8 SCO	6 SAG	6 CAP	7 PIS	7 ARI
9 LIB	10 SAG	9 CAP	9 AQU	9 ARI	9 TAU
11 SCO	12 CAP	11 AQU	11 PIS	11 TAU	11 GEM
14 SAG	15 AQU	13 PIS	13 ARI	13 GEM	13 CAN
16 CAP	17 PIS	15 ARI	15 TAU	15 CAN	15 LEO
18 AQU	19 ARI	17 TAU	17 GEM	17 LEO	17 VIR
20 PIS	21 TAU	19 GEM	19 CAN	20 VIR	19 LIB
23 ARI	23 GEM	21 CAN	21 LEO	22 LIB	22 SCO
25 TAU	25 CAN	25 LEO	23 VIR	25 SCO	24 SAG
27 GEM	28 LEO	26 VIR	26 LIB	27 SAG	27 CAP
29 CAN	30 VIR	29 LIB	28 SCO	30 CAP	29 AQU
31 LEO			31 SAG		

1990 MOON SIGNS

JAN	FEB	MAR	APR	MAY	JUN
1 PIS	1 TAU	2 GEM	1 CAN	3 VIR	1 LIB
3 ARI	3 GEM	5 CAN	3 LEO	5 LIB	4 SCO
5 TAU	5 CAN	7 LEO	5 VIR	8 SCO	6 SAG
7 GEM	8 LEO	9 VIR	8 LIB	10 SAG	9 CAP
9 CAN	10 VIR	12 LIB	10 SCO	13 CAP	11 AQU
11 LEO	12 LIB	14 SCO	13 SAG	15 AQU	14 PIS
13 VIR	15 SCO	16 SAG	15 CAP	17 PIS	16 ARI
16 LIB	17 SAG	19 CAP	18 AQU	20 ARI	18 TAU
18 SCO	20 CAP	21 AQU	20 PIS	22 TAU	20 GEM
21 SAG	22 AQU	24 PIS	22 ARI	24 GEM	22 CAN
23 CAP	24 PIS	26 ARI	24 TAU	26 CAN	24 LEO
26 AQU	26 ARI	28 TAU	26 GEM	28 LEO	26 VIR
28 PIS	28 TAU	30 GEM	28 CAN	30 VIR	29 LIB
30 ARI			30 LEO		

JUL	AUG	SEP	OCT	NOV	DEC
1 SCO	2 CAP	1 AQU	1 PIS	2 TAU	1 GEM
4 SAG	5 AQU	3 PIS	3 ARI	4 GEM	3 CAN
6 CAP	7 PIS	6 ARI	5 TAU	6 CAN	5 LEO
9 ARQ	9 ARI	8 TAU	7 GEM	8 LEO	7 VIR
11 PIS	12 TAU	10 GEM	9 CAN	10 VIR	9 LIB
13 ARI	14 GEM	12 CAN	11 LEO	12 LIB	12 SAG
15 TAU	16 CAN	14 LEO	14 VIR	15 SCO	14 SAG
17 GEM	18 LEO	16 VIR	16 LIB	17 SAG	17 CAP
19 CAN	20 VIR	19 LIB	18 SCO	20 CAP	19 AQU
21 LEO	22 LIB	21 SCO	21 SAG	22 AQU	22 PIS
24 VIR	25 SCO	24 SAG	23 CAP	25 PIS	24 ARI
26 LIB	27 SAG	26 CAP	26 AQU	27 ARI	26 TAU
28 SCO	30 CAP	29 AQU	28 PIS	29 TAU	28 GEM
31 SAG			30 ARI		30 CAN

1991 MOON SIGNS

JAN	FEB	MAR	APR	MAY	JUN
1 LEO	2 LIB	2 LIB	3 SAG	2 CAP	1 AQU
4 VIR	4 SCO	4 SCO	5 CAP	5 AQU	4 PIS
6 LIB	7 SAG	6 SAG	8 AQU	7 PIS	6 ARI
8 SCO	9 CAP	9 CAP	10 PIS	10 ARI	8 TAU
11 SAG	12 AQU	11 AQU	12 ARI	12 TAU	10 GEM
13 CAP	14 PIS	14 PIS	15 TAU	14 GEM	12 CAN
16 AQU	17 ARI	16 ARI	17 GAM	16 CAN	14 LEO
18 PIS	19 TAU	18 TAU	19 CAN	18 LEO	16 VIR
20 ARI	21 GEM	20 GEM	21 LEO	20 VIR	19 LIB
23 TAU	23 CAN	22 CAN	23 VIR	22 LIB	21 SCO
25 GEM	25 LEO	25 LEO	25 LIB	25 SCO	23 SAG
27 CAN	27 VIR	27 VIR	28 SCO	27 SAG	26 CAP
29 LEO		29 LIB	30 SAG	30 CAP	29 AQU
31 VIR		31 SCO			

JUL	AUG	SEP	OCT	NOV	DEC
1 PIS	2 TAU	3 CAN	2 LEO	2 LIB	2 SCO
3 ARI	4 GEM	5 LEO	4 VIR	5 SCO	4 SAG
6 TAU	6 CAN	7 VIR	6 LIB	7 SAG	7 CAN
8 GEM	8 LEO	9 LIB	8 SCO	10 CAP	9 AQU
10 CAN	10 VIR	11 SCO	11 SAG	12 AQU	12 PIS
12 LEO	12 LIB	13 SAG	13 CAP	15 PIS	14 ARI
14 VIR	15 SCO	16 CAP	16 AQU	17 ARI	17 TAU
16 LIB	17 SAG	18 AQU	18 PIS	19 TAU	19 GEM
18 SCO	20 CAP	21 PIS	21 ARI	21 GEM	21 CAN
21 SAG	22 AQU	23 ARI	23 TAU	23 CAN	23 LEO
23 CAP	25 PIS	25 TAU	25 GEM	25 LEO	25 VIR
26 AQU	27 ARI	28 GEM	27 CAN	28 VIR	27 LIB
28 PIS	29 TAU	30 CAN	29 LEO	30 LIB	29 SCO
31 ARI	31 GEM		31 VIR		

1992 MOON SIGNS

JAN	FEB	MAR	APR	MAY	JUN
1 SAG	2 AQU	3 PIS	1 ARI	1 TAU	2 CAN
3 CAP	4 PIS	5 ARI	4 TAU	3 GEM	4 LEO
6 AQU	7 ARI	8 TAU	6 GEM	5 CAN	6 VIR
8 PIS	9 TAU	10 GEM	8 CAN	8 LEO	8 LIB
11 ARI	12 GEM	12 CAN	10 LEO	10 VIR	10 SCO
13 TAU	14 CAN	14 CAN	12 VIR	12 LIB	13 SAG
15 GEM	16 LEO	16 VIR	15 LIB	14 SCO	15 CAP
17 CAN	18VIR	18 LIB	17 SCO	16 SAG	18 AQU
19 LEO	20 LIB	20 SCO	19 SAG	19 CAP	20 PIS
21 VIR	22 SCO	23 SAG	21 CAP	21 AQU	22 ARI
23 LIB	24 SAG	25 CAP	24 AQU	24 PIS	24 TAU
25 SCO	27 CAP	27 AQU	26 PIS	26 ARI	27 GEM
28 SAG	29 AQU	30 PIS	29 ARI		29 CAN
30 CAP				31 GEM	

JUL	AUG	SEP	OCT	NOV	DEC
1 LEO	2 LOB	2 SAG	2 CAP	1 AQU	1 PIS
3 VIR	4 SCO	5 CAP	5 AQU	3 PIS	3 ARI
5 LIB	6 SAG	7 AQU	7 PIS	6 ARI	6 TAU
7 SCO	8 CAP	10 PIS	10 ARI	8 TAU	8 GEM
10 SAG	11 AQU	12 ARI	12 TAU	11 GEM	10 CAN
12 CAP	13 PIS	15 TAU	14 GEM	13 CAN	12 LEO
15 AQU	16 ARI	17 GEM	17 CAN	15 LEO	14 VIR
17 PIS	18 TAU	19 CAN	19 LEO	17 VIR	16 LIB
20 ARI	21 GEM	21 LEO	21 VIR	19 LIB	19 SCO
22 TAU	23 CAN	24 VIR	23 LIB	21 SCO	21 SAG
24 GEM	25 LEO	26 LIB	25 SCO	24 SAG	23 CAP
27 CAN	27 VIR	28 SCO	27 SAG	26 CAP	26 AQU
29 LEO	29 LIB	30 SAG	29 CAP	28 AQU	28 PIS
31 VIR	31 SCO				31 ARI

1994 MOON SIGNS

JAN	FEB	MAR	APR	MAY	JUN
1 VIR	2 SCO	1 SCO	1 CAP	1 AQU	2 ARI
3 LIB	4 SAG	3 SAG	4 AQU	3 PIS	5 TAU
5 SCO	6 CAP	5 CAP	6 PIS	6 ARI	7 GEM
8 SAG	8 AQU	7 AQU	9 ARI	8 TAU	10 CAN
10 CAP	11 PIS	10 PIS	11 TAU	11 GEM	12 LEO
12 AQU	13 ARI	12 ARI	14 GEM	13 CAN	14 VIR
14 PIS	16 TAU	15 TAU	16 CAN	16 LEO	16 LIB
17 ARI	18 GEM	17 GEM	18 LEO	18 VIR	19 SCO
19 TAU	20 CAN	20 CAN	21 VIR	20 LIB	21 SAG
22 GEM	23 LEO	22 LEO	23 LIB	22 SCO	23 CAP
24 CAN	25 VIR	24 VIR	25 SCO	24 SAG	25 AQU
26 LEO	27 LIB	26 LIB	27 SAG	26 CAP	27 PIS
28 VIR		28 SCO	29 CAP	28 AQU	29 ARI
31 LIB		30 SAG		31 PIS	

JUL	AUG	SEP	OCT	NOV	DEC
2 TAU	1 GEM	2 LEO	2 VIR	2 SCO	2 SAG
4 GEM	3 CAN	4 VIR	4 LIB	4 SAG	4 CAP
7 CAN	6 LEO	6 LIB	6 SCO	6 CAP	6 AQU
9 LEO	8 VIR	8 SCO	8 SAG	8 AQU	8 PIS
11 VIR	10 LIB	10 SAG	10 CAP	11 PIS	10 ARI
14 LIB	12 SCO	13 CAP	12 AQU	13 ARI	13 TAU
16 SCO	14 SAG	15 AQU	14 PIS	15 TAU	15 GEM
18 SAG	16 CAP	17 PIS	17 ARI	18 GEM	18 CAN
20 CAP	18 AQU	19 ARI	19 TAU	20 CAN	20 LEO
22 AQU	21 PIS	22 TAU	22 GEM	23 LEO	23 VIR
24 PIS	23 ARI	24 GEM	24 CAN	25 VIR	25 LIB
27 ARI	26 TAU	27 CAN	27 LEO	28 LIB	27 SCO
29 TAU	28 GEM	29 LEO	29 VIR	30 SCO	29 SAG
	31 CAN		31 LIB		31 CAP

1993 MOON SIGNS

JAN	FEB	MAR	APR	MAY	JUN
2 TAU	1 GEM	2 CAN	1 LEO	2 LIB	1 SCO
4 GEM	3 CAN	5 LEO	1 VIR	4 SCO	3 SAG
7 CAN	5 LEO	7 VIR	5 LIB	6 SAG	5 CAP
9 LEO	7 VIR	9 LIB	7 SCO	9 CAP	7 AQU
11 VIR	9 LIB	11 SCO	9 SAG	11 AQU	10 PIS
13 LIB	11 SCO	13 SAG	13 CAP	13 PIS	12 ARI
15 SCO	13 SAG	15 CAP	14 AQU	16 ARI	15 TAU
17 SAG	16 CAP	17 AQU	16 PIS	18 TAU	17 GEM
19 CAP	18 AQU	20 PIOS	19 ARI	21 GEM	19 CAN
22 AQU	21 PIS	22 ARI	21 TAU	23 CAN	22 LEO
24 PIS	23 ARI	25 TAU	24 GEM	25 LEO	24 VIR
27 ARI	26 TAU	27 GEM	26 CAN	28 VIR	26 LIB
29 TAU	28 GEM	30 CAN	28 LEO	30 LIB	28 SCO
			30 VIR		30 SAG

JUL	AUG	SEP	OCT	NOV	DEC
2 CAP	1 AQU	2 ARI	2 TAU	1 GEM	3 LEO
5 AQU	3 PIS	5 TAU	4 GEM	3 CAN	5 VIR
7 PIS	6 ARI	7 GEM	7 CAN	5 LEO	7 LIB
10 ARI	8 TAU	10 CAN	9 LEO	8 VIR	9 SCO
12 TAU	11 GEM	12 LEO	11 VIR	10 LIB	11 SAG
15 GEM	13 CAN	14 VIR	13 LIB	12 SCO	13 CAP
17 CAN	15 LEO	16 LIB	15 SCO	14 SAG	15 AQU
19 LEO	17 VIR	18 SCO	17 SAG	16 CAP	18 PIS
21 VIR	19 LIB	20 SAG	19 CAP	18 AQU	20 ARI
23 LIB	21 SCO	22 CAP	22 AQU	20 PIS	23 TAU
25 SCO	24 SAG	24 AQU	24 PIS	23 ARI	25 GEM
27 SAG	26 CAP	27 PIS	27 ARI	26 TAU	28 CAN
30 CAP	28 AQU	29 ARI	29 TAU	28 GEM	30 LEO
	31 PIS			30 CAN	

1995 MOON SIGNS

JAN	FEB	MAR	APR	MAY	JUN
2 AQU	1 PIS	2 ARI	1 TAI	1 GEM	2 LEO
4 PIS	3 ARI	5 TAU	3 GEM	3 CAN	5 VIR
7 ARI	5 TAU	7 GEM	6 CAN	6 LEO	7 LIB
9 TAU	8 GEM	10 CAN	9 LEO	8 VIR	9 SCO
12 GEM	10 CAN	12 LEO	11 VIR	10 LIB	11 SAG
14 CAN	13 LEO	14 VIR	13 LIB	13 SCO	13 CAP
16 LEO	15 VIR	17 LIB	15 SCO	15 SAG	15 AQU
19 VIR	17 LIB	19 SCO	17 SAG	17 CAP	17 PIS
21 LIB	19 SCO	21 SAG	19 CAP	19 AQU	19 ARI
23 SCO	22 SAG	23 CAP	21 AQU	21 PIS	22 TAU
25 SAG	24 CAP	25 AQU	24 PIS	23 ARI	24 GEM
27 CAP	26 AQU	27 PIS	26 ARI	26 TAU	27 CAN
30 AQU	28 PIS	30 ARI	28 TAU	28 GEM	29 LEO
				31 CAN	

JUL	AUG	SEP	OCT	NOV	DEC
2 VIR	3 SCO	1 SAG	2 AQU	1 PIS	3 TAU
4 LIB	5 SAG	3 CAP	5 PIS	3 ARI	5 GEM
6 SCO	7 CAP	5 AQU	7 ARI	5 TAU	8 CAP
8 SAG	9 AQU	7 PIS	9 TAU	8 GEM	10 LEO
10 CAP	11 PIS	9 ARI	12 GEM	10 CAN	13 VIR
12 AQU	13 ARI	12 TAU	14 CAN	13 LEO	15 LIB
14 PIS	15 TAU	14 GEM	17 LEO	15 VIR	17 SCO
17 ARI	18 GEM	17 CAN	19 VIR	18 LIB	19 SAG
19 TAU	20 CAN	19 LEO	21 LIB	20 SCO	21 CAP
22 CAN	23 LEO	22 VIR	23 SCO	22 SAG	23 AQU
24 CAN	25 VIR	24 LIB	26 SAG	24 CAP	25 PIS
27 LEO	28 LIB	26 SCO	28 CAP	26 AQU	28 ARI
29 VIR	30 SCO	28 SAG	30 AQU	28 PIS	30 TAU
31 LIB		30 CAP		30 ARI	

1996 MOON SIGNS

JAN	FEB	MAR	APR	MAY	JUN
1 GEM	3 LEO	1 LEO	2 LIB	2 SCO	2 CAP
4 CAN	5 VIR	3 VIR	4 SCO	4 SAG	4 AQU
6 LEO	8 LIB	6 LIB	7 SAG	6 CAP	6 PIS
9 VIR	10 SCO	8 SCO	9 CAP	8 AQU	9 ARI
11 LIB	12 SAG	10 SAG	11 AQU	10 PIS	11 TAU
14 SCO	14 CAP	13 CAP	13 PIS	12 ARI	13 GEM
16 SAG	16 AQU	15 AQU	15 ARI	15 TAU	16 CAN
18 CAP	18 PIS	17 PIS	17 TAU	17 GEM	18 LEO
20 AQU	20 ARI	19 ARI	20 GEM	19 CAN	21 VIR
22 PIS	23 TAU	21 TAU	22 CAN	22 LEO	23 LIB
24 ARI	25 GEM	23 GEM	25 LEO	25 VIR	26 SCO
26 TAU	27 CAN	26 CAN	27 VIR	27 LIB	28 SAG
29 PIS		28 LEO	30 LIB	29 SCO	30 CAP
31 CAN		31 VIR		31 SAG	
JUL	**AUG**	**SEP**	**OCT**	**NOV**	**DEC**
2 AQU	2 ARI	1 TAU	3 CAN	2 LEO	2 VIR
4 PIS	2 TAU	3 GEM	5 LEO	4 VIR	4 LIB
6 ARI	7 GEM	6 CAN	8 VIR	7 LIB	6 SCO
8 TAU	9 CAN	8 LEO	10 LIB	9 SCO	9 SAG
11 GEM	12 LEO	11 VIR	13 SCO	11 SAG	11 CAP
13 CAN	14 VIR	13 LIB	15 SAG	13 CAP	13 AQU
16 LEO	17 LIB	15 SCO	17 CAP	16 AQU	15 PIS
18 VIR	19 SCO	18 SAG	19 AQU	18 PIS	17 ARI
21 LIB	21 SAG	20 CAP	21 PIS	20 ARI	19 TAU
23 SCO	24 CAP	22 AQU	23 ARI	22 TAU	22 GEM
25 SAG	26 AQU	24 PIS	26 TAU	24 GEM	24 CAN
27 CAP	28 PIS	26 ARI	28 GEM	27 CAN	26 LEO
29 AQU	30 ARI	28 TAU	30 CAN	29 LEO	29 VIR
31 PIS		30 GEM			31 LIB

1998 MOON SIGNS

JAN	FEB	MAR	APR	MAY	JUN
2 PIS	2 TAU	2 CAN	2 CAN	2 LEO	3 LIB
4 ARI	4 GEM	4 GEM	4 LEO	4 VIR	5 SCO
6 TAU	7 CAN	6 CAN	7 VIR	7 LIB	8 SAG
8 GEM	9 LEO	8 LEO	9 LIB	9 SCO	10 CAP
10 CAN	11 VIR	11 VIR	12 SCO	12 SAG	13 AQU
13 LEO	14 LIB	13 LIB	14 SAG	14 CAP	15 PIS
15 VIR	16 SCO	16 SCO	17 CAP	16 AQU	17 ARI
18 LIB	19 SAG	18 SAG	19 AQU	19 PIS	19 TAU
20 SCO	21 CAP	21 CAP	21 PIS	21 ARI	21 GEM
23 SAG	23 AQU	23 AQU	23 ARI	23 TAU	23 CAN
25 CAP	25 PIS	25 PIS	25 TAU	25 GEM	25 LEO
27 AQU	27 ARI	27 ARI	27 GEM	27 CAN	28 VIR
29 PIS		29 TAU	29 CAN	29 LEO	30 LIB
31 ARI		31 GEM		31 VIR	
JUL	**AUG**	**SEP**	**OCT**	**NOV**	**DEC**
3 SCO	2 SAG	3 AQU	2 PIS	1 ARI	2 GEM
5 SAG	4 CAP	5 PIS	4 ARI	3 TAU	4 CAN
8 CAP	6 AQU	7 ARI	6 TAU	5 GEM	6 LEO
10 AQU	8 PIS	9 TAU	8 GEM	7 CAN	9 VIR
12 PIS	11 ARI	11 GEM	10 CAN	9 LEO	11 LIB
14 ARI	13 TAU	13 CAN	13 LEO	11 VIR	14 SCO
16 TAU	15 GEM	15 LEO	15 VIR	14 LIB	16 SAG
18 GEM	17 CAN	18 VIR	17 LIB	16 SCO	19 CAP
21 CAN	19 LEO	20 LIB	20 SCO	19 SAG	21 AQU
23 LEO	21 VIR	23 SCO	23 SAG	21 CAP	23 PIS
25 VIR	24 LIB	25 SAG	25 CAP	24 AQU	25 ARI
28 LIB	26 SCO	28 CAP	27 AQU	26 PIS	28 TAU
30 SCO	29 SAG	30 AQU	30 PIS	28 ARI	30 GEM
	31 CAP			30 TAU	

1997 MOON SIGNS

JAN	FEB	MAR	APR	MAY	JUN
3 SCO	1 SAG	1 SAG	1 AQU	1 PIS	1 TAU
5 SAG	4 CAP	3 CAP	4 PIS	3 ARI	4 GEM
7 CAP	6 AQU	5 AQU	6 ARI	5 TAU	6 CAN
9 AQU	8 PIS	7 PIS	8 TAU	7 GEM	8 LEO
11 PIS	10 ARI	9 ARI	10 GEM	9 CAN	11 VIR
13 ARI	12 TAU	11 TAU	12 CAN	12 LEO	13 LIB
15 TAU	14 GEM	13 GEM	14 LEO	14 VIR	16 SCO
18 GEM	16 CAN	16 CAN	17 VIR	17 LIB	18 SAG
20 CAN	19 LEO	18 LEO	19 LIB	19 SCO	20 CAP
23 LEO	21 VIR	21 VIR	22 SCO	22 SAG	22 AQU
25 VIR	24 LIB	23 LIB	24 SAG	24 CAP	24 PIS
28 LIB	26 SCO	26 SCO	27 CAP	26 AQU	26 ARI
30 SCO		28 SAG	29 AQU	28 PIS	29 TAU
		30 CAP		30 ARI	
JUL	**AUG**	**SEP**	**OCT**	**NOV**	**DEC**
1 GEM	2 LEO	3 LIB	3 SCO	1 SAG	1 CAP
3 CAN	4 VIR	6 SCO	5 SAG	4 CAP	3 AQU
5 LEO	7 LIB	8 SAG	8 CAP	6 AQU	5 PIS
8 VIR	9 SCO	10 CAP	10 AQU	8 PIS	8 ARI
10 LIB	12 SAG	12 AQU	12 PIS	10 ARI	10 TAU
13 SCO	14 CAP	15 PIS	14 ARI	12 TAU	12 GEM
15 SCG	16 AQU	17 ARI	16 TAI	14 GEM	14 CAN
18 CAP	18 PIS	19 TAU	18 GEM	17 CAN	16 LEO
20 AQU	20 ARI	21 GEM	20 CAN	19 LEO	19 VIR
22 PIS	22 TAU	23 CAN	23 LEO	21 VIR	21 LIB
24 ARI	24 GEM	25 LEO	25 VIR	24 LIB	26 SCO
26 TAU	27 CAN	28 VIR	28 LIB	26 SCO	26 SAG
28 GEM	29 LEO	30 LIB	30 SCO	29 SAG	28 CAP
30 CAN	31 VIR				31 AQU

1999 MOON SIGNS

JAN	FEB	MAR	APR	MAY	JUN
1 CAN	1 VIR	1 VIR	2 SCO	2 SAG	3 AQU
3 LEO	4 LIB	3 LIB	4 SAG	4 CAP	5 PIS
5 VIR	6 SCO	6 SCO	7 CAP	7 AQU	8 ARI
7 LIB	9 SAG	8 SAG	9 AQU	9 PIS	10 TAU
10 SCO	11 CAP	11 CAP	12 PIS	11 ARI	12 GEM
12 SAG	14 AQU	13 AQU	14 ARI	13 TAU	14 CAN
15 CAP	16 PIS	15 PIS	16 TAU	15 GEM	16 LEO
17 LEO	18 ARI	17 ARI	18 GEM	17 CAN	18 VIR
19 PIS	20 TAU	19 TAU	20 CAN	19 LEO	20 LIB
22 ARI	22 GEM	21 GEM	22 LEO	21 VIR	23 SCO
24 TAU	24 CAN	23 CAN	24 VIR	24 LIB	25 SAG
26 GEM	26 LEO	26 LEO	27 LIB	26 SCO	28 CAP
28 CAN		28 VIR	29 SCO	29 SAG	30 AQU
30 LEO		30 LIB	31 CAP		
JUL	**AUG**	**SEP**	**OCT**	**NOV**	**DEC**
2 PIS	1 ARI	2 GEM	1 CAN	1 VIR	1 LIB
5 ARI	3 TAU	4 CAN	3 LEO	4 LIB	3 SCO
7 TAU	5 GEM	6 LEO	5 VIR	6 SCO	6 SAG
9 GEM	7 CAN	8 VIR	8 LIB	9 SAG	8 CAP
11 CAN	9 LEO	10 LIB	10 SCO	11 CAP	11 AQU
13 LEO	12 VIR	13 SCO	12 SAG	14 AQU	13 PIS
15 VIR	14 LIB	15 SAG	15 CAP	16 PIS	16 ARI
17 LIB	16 SCO	18 CAP	17 CAP	18 ARI	18 TAU
20 SCO	19 SAG	20 AQU	20 PIS	21 TAU	20 GEM
22 SAG	21 CAP	22 PIS	22 ARI	23 GEM	22 CAN
25 CAP	24 AQU	25 ARI	24 TAU	25 CAN	24 LEO
27 AQU	26 PIS	27 TAU	26 GEM	27 LEO	26 VIR
30 PIS	28 ARI	29 GEM	28 CAN	29 VIR	28 LIB
	30 TAU		30 LEO		31 SCO

2000 MOON SIGNS

JAN	FEB	MAR	APR	MAY	JUN
3 SAG	1 CAP	2 AQU	1 PIS	3 TAU	1 GEM
3 CAP	4 AQU	4 PIS	3 ARI	5 GEM	3 CAN
7 AQU	6 PIS	7 ARI	5 TAU	7 CAN	5 LEI
10 PIS	8 ARI	9 TAU	7 GEM	9 LEO	7 VIR
12 ARI	11 TAU	11 GEM	9 CAN	11 VIR	9 LIB
14 TAU	13 GEM	13 CAN	11 LEO	13 LIB	12 SCO
16 GEM	15 CAN	15 LEO	14 VIR	15 SCO	14 SAG
18 CAN	17 LEO	17 VIR	16 LIB	18 SAG	17 CAN
20 LEO	19 VIR	20 LIB	18 SCO	20 CAP	19 AQU
23 VIR	21 LIB	22 SCO	21 SAG	23 AQU	22 PIS
25 LIB	23 SCO	24 SAG	23 CAP	25 PIS	24 ARI
27 SCO	26 SAG	27 CAP	26 AQU	28 ARI	26 TAU
29 TAU	28 CAP	29 AQU	28 PIS	30 TAU	28 GEM
			30 ARI		30 CAN

JUL	AUG	SEP	OCT	NOV	DEC
2 LEO	1 LIB	2 SCO	1 SAG	3 AQU	2 PIS
4 VIR	3 LIB	4 SAG	4 CAP	5 PIS	5 ARI
7 LIB	5 SCO	6 CAP	6 AQU	8 ARI	7 TAU
9 SCO	8 SAG	9 AQU	9 PIS	10 TAU	9 GEM
11 SAG	10 CAP	11 PIS	11 ARI	12 GEM	11 CAN
14 CAP	13 AQU	14 ARI	13 TAU	14 CAN	13 LEO
16 AQU	15 PIS	16 TAU	16 GEM	16 LEO	15 VIR
19 PIS	18 ARI	18 GEM	18 CAN	18 VIR	18 LIB
21 ARI	20 TAU	20 CAN	20 LEO	20 LIB	20 SCO
24 TAU	22 GEM	23 LEO	22 VIR	23 SCO	22 SAG
26 GEM	24 CAN	25 VIR	24 LIB	25 SAG	25 CAP
28 CAN	26 LEO	27 LIB	26 SCO	27 CAP	27 AQU
30 LEO	28 VIR	29 SCO	29 SAG	30 AQU	30 PIS
	30 LIB		31 CAP		

2001 MOON SIGNS

JAN	FEB	MAR	APR	MAY	JUN
1 ARI	2 GEM	1 GEM	2 LEO	1 VIR	2 SCO
4 TAU	4 CAN	4 CAN	4 VIR	3 LIB	4 SAG
6 GEM	6 LEO	6 LEO	6 LIB	6 SCO	7 CAP
8 CAN	8 VIR	8 VIR	8 SCO	8 SAG	9 AQO
10 LEO	10 LIB	10 LIB	10 SAG	10 CAP	10 PIS
12 VIR	12 SCO	12 SCO	13 CAP	13 AQU	14 ARI
14 LIB	15 SAG	14 SAG	15 AQU	15 PIS	16 TAU
16 SCO	17 CAP	16 CAP	18 PIS	18 ARI	19 GEM
19 SAG	20 AQU	19 AQU	20 ARI	20 TAU	21 CAN
21 CAP	22 PIS	22 PIS	23 TAU	22 GEM	23 LEO
23 AQU	25 ARI	24 ARI	25 GEM	24 CAN	25 VIR
26 PIS	27 TAU	26 TAU	27 CAN	27 LEO	27 LIB
28 ARI		29 GEM	29 LEO	29 VIR	29 SCO
31 TAU		31 CAN		31 LIB	

JUL	AUG	SEP	OCT	NOV	DEC
1 SAG	3 AQU	1 PIS	1 ARI	2 GEM	2 CAN
4 CAP	5 PIS	4 ARI	4 TAU	4 CAN	4 LEO
6 AQU	8 ARI	6 TAU	6 GEM	7 LEO	6 VIR
8 PIS	10 TAU	9 GEM	8 CAN	9 VIR	8 LIB
11 ARI	12 GEM	11 CAN	10 LEO	11 LIB	10 SCO
14 TAU	15 CAN	13 LEO	13 VIR	13 SCO	12 SAG
16 GEM	17 LEO	15 VIR	15 LIB	15 SAG	15 CAP
18 CAN	19 VIR	17 LIB	17 SCO	17 CAP	17 AQU
20 LEO	21 LIB	19 SCO	19 SAG	20 AQU	20 PIS
22 VIR	23 SCO	21 SAG	21 CAP	22 PIS	22 ARI
24 LIB	25 SAG	24 CAP	23 AQU	25 ARI	25 TAU
26 SCO	27 CAP	26 AQU	26 PIS	27 TAU	27 GEM
29 SAG	30 AQU	29 PIS	28 ARI	30 GEM	29 CAN
31 CAP			31 TAU		31 LEO

2002 MOON SIGNS

JAN	FEB	MAR	APR	MAY	JUN
2 VIR	1 LIB	2 SCO	1 SAG	2 AQU	1 PIS
4 LIB	3 SCO	4 SAG	3 CAP	5 PIS	4 ARI
6 SCO	5 SAG	6 CAP	5 AQU	7 ARI	6 TAU
9 SAG	7 CAP	9 ARI	8 PIS	7 TAU	9 GEM
11 CAP	10 AQU	11 PIS	10 ARI	12 GEM	11 CAN
13 AQU	12 PIS	14 ARI	13 TAU	15 CAN	13 CAN
16 PIS	15 ARI	16 TAU	15 GEM	17 LEO	15 VIR
18 ARI	17 TAU	19 GEM	18 CAN	19 VIR	18 LIB
21 TAU	20 GEM	21 CAN	20 LEO	21 LIB	20 SCO
23 GEM	22 CAN	24 LEO	22 VIR	23 SCO	23 SAG
26 CAN	24 LEO	26 VIR	24 LIB	25 SAG	24 CAP
28 LEO	26 VIR	28 LIB	26 SCO	28 CAP	26 AQU
30 VIR	28 LIB	30 SCO	28 SAG	30 AQU	29 PIS
			30 CAP		

JUL	AUG	SEP	OCT	NOV	DEC
1 ARI	2 GEM	1 CAN	1 LEO	1 LIB	1 SCO
4 TAU	5 CAN	3 LEO	3 VIR	3 SCO	3 SAG
6 GEM	7 LEO	5 VIR	5 LIB	5 SAG	5 CAP
8 CAN	9 VIR	7 LIB	7 SCO	7 CAP	7 AQU
11 LEO	11 LIB	9 SCO	9 SCO	10 AQU	9 PIS
13 VIR	13 SCO	12 SAG	11 CAP	12 PIS	12 ARI
15 LIB	15 SAG	14 CAP	13 AQU	15 ARI	14 TAU
17 SCO	18 CAP	16 AQU	16 P[IS	17 TAU	17 GEM
19 SAG	20 AQU	19 PIS	18 ARI	20 GEM	19 CAN
21 CAP	22 PIS	21 ARI	21 TAU	22 CAN	22 LEO
24 AQU	25 ARI	24 TAU	23 GEM	24 LEO	24 VIR
26 PIS	27 TAU	26 GEM	26 CAN	27 VIR	26 LIB
28 ARI	30 GEM	29 CAN	28 LEO	29 LIB	28 SCO
31 TAU			30 VIR		30 SAG

2003 MOON SIGNS

JAN	FEB	MAR	APR	MAY	JUN
1 CAP	2 PIS	1 PIS	3 TAU	2 GEM	1 CAN
3 AQU	5 ARI	4 ARI	5 GEM	5 CAN	4 LEO
6 PIS	7 TAU	6 TAU	8 CAN	7 LEO	6 VIR
8 ARI	10 GEM	9 GEM	10 LEO	10 VIR	8 LIB
11 TAU	11 CAN	11 CAN	12 VIR	12 LIB	10 SCO
13 GEM	14 LEO	14 LEO	14 LIB	14 SCO	12 SAG
16 CAN	16 VIR	16 VIR	16 SCO	16 SAG	14 CAP
18 LEO	18 LIB	18 LIB	18 SAG	18 CAP	16 AQU
20 VIR	21 SCO	20 SCO	20 CAP	20 AQU	19 PIS
22 LIB	23 SAG	22 SAG	23 AQU	22 PIS	21 ARI
24 SCO	25 CAP	24 CAP	25 PIS	25 ARI	23 TAU
26 SAG	27 AQU	26 AQU	27 ARI	27 TAU	26 GEM
29 CAP		29 PIS	30 TAU	30 GEM	28 CAN
31 AQU		31 ARI			

JUL	AUG	SEP	OCT	NOV	DEC
1 LEO	2 LIB	2 SAG	1 CAP	2 PIS	2 ARI
3 VIR	4 SCO	4 CAP	4 AQU	5 ARI	4 TAU
5 LIB	6 SAG	6 AQU	6 PIS	7 TAU	7 GEM
7 SCO	8 CAP	9 PIS	8 ARI	10 GEM	9 CAN
10 SAG	10 AQU	11 ARI	11 TAU	12 CAN	12 LEO
12 CAP	12 PIS	13 TAU	13 GEM	15 LEO	14 VIR
14 AQU	15 ARI	16 GEM	16 CAN	17 VIR	16 LIB
16 PIS	17 TAU	18 CAN	18 LEO	19 LIB	19 SCO
18 ARI	20 GEM	21 LEO	21 VIR	21 SCO	21 SAG
21 TAU	22 CAN	23 VIR	23 LIB	23 SAG	23 CAP
23 GEM	24 LEO	25 LIB	25 SCO	25 CAP	25 AQU
26 CAN	27 VIR	27 SCO	27 SAG	27 AQU	27 PIS
28 LEO	29 LIB	29 SAG	29 CAP	29 PIS	29 ARI
30 VIR	31 SCO		31 AQU		

2004 MOON SIGNS

JAN	FEB	MAR	APR	MAY	JUN
1 TAU	2 CAN	3 LEO	1 VIR	1 LIB	2 SAG
3 GEM	4 LEO	4 VIR	4 LIB	3 SCO	4 CAP
6 CAN	76 VIR	7 LIB	6 SCO	5 SAG	6 AQU
8 LEO	9 LIB	9 SCO	8 SAG	7 CAP	8 PIS
10 VIR	11 SCO	12 SAG	10 CAP	9 AQU	10 ARI
13 LIB	13 SAG	14 CAP	12 AQU	11 PIS	12 TAU
15 SCO	15 CAP	16 AQU	14 PIS	14 ARI	15 GEM
17 SAG	17 AQU	18 PIS	16 ARI	16 TAU	17 CAN
19 CAP	20 PIS	20 ARI	19 TAU	19 GEM	20 LEO
21 ARI	22 ARI	23 TAU	21 GEM	21 CAN	22 VIR
23 PIS	24 TAU	25 GEM	24 CAN	24 LEO	24 LIB
25 ARI	27 GEM	28 CAN	26 LEO	26 VIR	27 SCO
28 TAU	29 CAN	30 LEO	29 VIR	28 LIB	29 SAG
30 GEM				31 SCO	

JUL	AUG	SEP	OCT	NOV	DEC
1 CAP	1 PIS	2 TAU	2 GEM	1 CAN	1 LEO
3 AQU	4 ARI	5 GEM	5 CAN	3 LEO	3 VIR
5 PIS	6 TAU	7 CAN	7 LEO	6 VIR	6 LIB
7 ARI	8 GEM	10 LEO	10 VIR	8 LIB	8 SCO
10 TAU	11 CAN	12 VIR	12 LIB	10 SCO	10 SAG
12 GEM	13 LEO	14 LIB	14 SCO	13 SAG	12 CAP
15 CAN	16 VIR	17 SCO	16 SAG	15 CAP	14 AQU
17 LEO	18 LIB	19 SAG	18 CAP	17 AQU	16 PIS
20 VIR	20 SCO	21 CAP	20 AQU	19 PIS	18 ARI
22 LIB	23 SAG	23 AQU	23 PIS	21 ARI	21 TAU
24 SAO	25 CAP	25 PIS	25 ARI	23 TAU	23 GEM
26 SAG	27 AQU	27 ARI	27 TAU	26 GEM	25 CAN
28 CAP	29 PIS	30 TAU	29 GEM	28 CAN	28 LEO
30 AQU	31 ARI				31 VIR

2005 MOON SIGNS

JAN	FEB	MAR	APR	MAY	JUN
2 LIB	1 SCO	2 SAG	3 AQU	2 PIS	3 TAU
4 SCO	3 SAG	4 CAP	5 PIS	4 ARI	5 GEM
6 SAG	5 CAP	6 AQU	7 ARI	6 TAU	7 CAN
8 CAP	7 AQU	8 PIS	9 TAU	9 GEM	10 LEO
10 AQU	10 PIS	10 ARI	11 GEM	11 CAN	12 VIR
12 PIS	11 ARI	13 TAU	14 CAN	14 LEO	15 LIB
15 ARI	13 TAU	15 GEM	16 LEO	16 VIR	17 SCO
17 TAU	16 GEM	17 CAN	19 VIR	18 LIB	19 SAG
19 GEM	18 CAN	20 LEO	21 LIB	21 SCO	21 CAP
22 CAN	21 LEO	22 VIR	23 SCO	23 SCO	23 AQU
24 LEO	23 VIR	25 LIB	26 SAG	25 CAP	25 PIS
27 VIR	25 LIB	27 SCO	28 CAP	27 AQU	28 ARI
29 LIB	28 SCO	29 SAG	30 AQU	29 PIS	30 TAU
		31 CAP		31 ARI	

JUL	AUG	SEP	OCT	NOV	DEC
2 GEM	1 CAN	2 VIR	2 LIB	1 SCO	2 CAP
5 CAN	3 LEO	5 LIB	4 SCO	3 SCG	4 AQU
7 LEO	6 VIR	7 SCO	7 SAG	5 CAP	7 PIS
10 VIR	8 LIB	9 SAG	9 CAP	7 AQU	9 ARI
12 LIB	11 SCO	12 CAP	11 AQU	9 PIS	11 TAU
15 SCO	13 SAG	14 AQU	13 PIS	11 ARI	13 GEM
17 SAG	15 CAP	16 PIS	15 ARI	14 TAU	15 CAN
19 CAP	17 AQU	18 ARI	17 TAU	16 GEM	18 LEO
21 AQU	19 PIS	20 TAU	19 GEM	18 CAN	20 VIR
23 PIS	21 ARI	22 GEM	22 CAN	21 LEO	23 LIB
25 ARI	23 TAU	24 CAN	24 LEO	24 LEO	23 VIR
27 TAU	26 GEM	27 LEO	27 VIR	26 LIB	28 SAG
29 GEM	28 CAN	29 VIR	29 LIB	28 SCO	30 CAP
	31 LEO			30 SAG	

2006 MOON SIGNS

JAN	FEB	MAR	APR	MAY	JUN
1 AQU	1 ARI	1 ARI	1 GEM	1 CAN	2 VIR
3 PIS	3 TAU	3 TAU	4 CAN	3 LEO	5 LIB
5 ARI	6 GEM	5 GEM	6 LEO	6 VIR	7 SCO
7 TAU	8 CAN	7 CAN	9 VIR	8 LIB	10 SAG
9 GEM	10 LEO	10 LEO	11 LIB	11 SCO	12 CAP
12 CAN	13 VIR	12 VIR	14 SCO	13 SAG	14 AQU
14 LEO	16 LIB	15 LIB	16 SAG	15 CAP	16 PIS
17 VIR	18 SCO	17 SCO	18 CAP	18 AQU	18 ARI
19 LIB	20 SAG	20 SAG	20 AQU	20 PIS	20 TAU
22 SCO	23 CAP	22 CAP	22 PIS	22 ARI	22 GEM
24 SAG	25 AQU	24 AQU	25 ARI	24 TAU	25 CAN
26 CAP	27 PIS	26 PIS	27 TAU	26 GEM	27 LEO
28 AQU		28 ARI	29 GEM	28 CAN	29 VIR
30 PIS		30 TAU		31 LEO	

JUL	AUG	SEP	OCT	NOV	DEC
2 LIB	1 SCO	2 CAP	1 AQU	2 ARI	1 TAU
5 SCO	3 SAG	4 AQU	3 PIS	4 TAU	3 GEM
7 SAG	6 CAP	6 PIS	6 ARI	6 GEM	6 CAN
9 CAP	8 AQU	8 ARI	8 TAU	8 CAN	8 LEO
11 AQU	10 PIS	10 TAU	10 GEM	10 LEO	10 VIR
13 PIS	12 ARI	12 GEM	12 CAN	13 VIR	13 LIB
15 ARI	14 TAU	14 CAN	14 LEO	15 LIB	15 SCO
17 TAU	16 GEM	17 LEO	17 VIR	18 SCO	18 SAG
20 GEM	18 CAN	19 VIR	19 LIB	20 SAG	20 CAP
22 CAN	21 LEO	22 LIB	22 SCO	23 CAP	23 AQU
24 LEO	23 VIR	24 SCO	24 SAG	25 AQU	24 PIS
27 VIR	26 LIB	27 SAG	26 CAP	27 PIS	27 ARI
29 LIB	28 SCO	29 CAP	29 AQU	29 ARI	29 TAU
	31 SAG		31 PIS		31 GEM

2007 MOON SIGNS

JAN	FEB	MAR	APR	MAY	JUN
2 CAN	1 LEO	2 VIR	1 LIB	1 SCO	1 CAP
4 LEO	3 VIR	5 LIB	3 SCO	3 SAG	4 AQU
7 VIR	5 LIB	7 SCO	6 SAG	6 CAP	6 PIS
7 LIB	8 SCO	10 SAG	8 CAP	8 AQU	9 ARI
12 SCO	10 SAG	12 CAP	11 AQU	10 PIS	11 TAU
14 SAG	13 CAP	14 AQU	13 PIS	12 ARI	13 GEM
16 CAP	15 AQU	17 PIS	15 ARI	14 TAU	13 CAN
19 AQU	17 PIS	19 ARI	17 TAU	16 GEM	17 LEO
21 PIS	19 ARI	21 TAU	19 GEM	18 CAN	19 VIR
23 ARI	21 TAU	23 GEM	21 CAN	21 LEO	22 LIB
25 TAU	23 GEM	25 CAN	23 LEO	23 VIR	24 SCO
27 GEM	25 CAN	27 LEO	26 VIR	25 LIB	27 SAG
29 CAN	28 LEO	29 VIR	28 LIB	28 SCO	29 CAP
				31 SAG	

JUL	AUG	SEP	OCT	NOV	DEC
2 AQU	2 ARI	1 TAU	2 CAN	3 VIR	3 LIB
4 PIS	4 TAU	3 GEM	4 LEO	5 LIB	5 SCO
6 ARI	6 GEM	5 CAN	7 VIR	8 SCO	8 SAG
8 TAU	9 CAN	7 LEO	9 LIB	10 SAG	10 CAP
10 GEM	11 LEO	9 VIR	12 SCO	13 CAP	13 AQU
12 CAN	13 VIR	12 LIB	14 SAG	15 AQU	15 PIS
14 LEO	15 LIB	14 SCO	17 CAP	18 PIS	17 ARI
17 VIR	18 SCO	17 SAG	19 AQU	20 ARI	19 TAU
17 LIB	20 SAG	19 CAP	21 PIS	22 TAU	21 GEM
22 SCO	23 CAP	22 AQU	23 ARI	24 GEM	23 CAN
24 SAG	25 AQU	24 PIS	25 TAU	26 CAN	25 LEO
27 CAP	27 PIS	26 ARI	27 GEM	28 LEO	27 VIR
29 AQU	29 ARI	28 TAU	29 CAN	30 VIR	30 LIB
31 PIS		30 GEM	31 LEO		

2008 MOON SIGNS

JAN	FEB	MAR	APR	MAY	JUN
1 SCO	3 CAP	1 CAP	2 PIS	2 ARI	2 GEM
4 SAG	5 AQU	3 AQU	4 ARI	4 TAU	4 CAN
6 CAP	7 PIS	6 PIS	6 TAU	6 GEM	6 LEO
9 AQU	10 ARI	8 ARI	8 GEM	8 CAN	8 VIR
11 PIS	12 TAU	10 TAU	10 CAN	10 LEO	11 LIB
13 ARI	14 GEM	12 GEM	13 LEO	12 VIR	13 SCO
15 TAU	16 CAN	14 CAN	15 VIR	14 LIB	16 SAG
18 GEM	18 LEO	16 LEO	17 LIB	17 SCO	18 CAP
20 CAN	20 VIR	19 VIR	20 SCO	19 SAG	21 AQU
22 LEO	23 LIB	21 LIB	22 SAG	22 CAP	23 PIS
24 VIR	25 SCO	23 SCO	25 CAP	24 AQU	25 ARI
26 LIB	28 SCO	26 SAG	27 AQU	27 PIS	28 TAU
29 SCO		28 CAP	30 PIS	29 ARI	30 GEM
31 SAG		31 AQU		31 TAU	

JUL	AUG	SEP	OCT	NOV	DEC
2 CAN	2 VIR	1 LIB	3 SAG	2 CAP	2 AQU
4 LEO	4 LIB	3 SCO	5 CAP	4 AQU	4 PIS
6 VIR	7 SCO	6 SAG	8 AQU	7 PIS	6 ARI
8 LIB	9 SAG	8 CAP	10 PIS	9 ARI	9 TAU
10 SCO	12 CAP	11 AQU	13 ARI	11 TAU	11 GEM
13 SAG	14 AQU	13 PIS	15 TAU	13 GEM	13 CAN
15 CAP	17 PIS	15 ARI	17 GEM	15 CAN	15 LEO
18 AQU	19 ARI	17 TAU	19 CAN	17 LEO	17 VIR
20 PIS	21 TAU	19 GEM	21 LEO	19 VIR	19 LIB
23 ARI	23 GEM	22 CAN	23 VIR	22 LIB	21 SCO
25 TAU	25 CAN	24 LEO	25 LIB	24 SCO	24 SAG
27 GEM	27 LEO	26 VIR	28 SCO	27 SAG	26 CAP
29 CAN	30 VIR	28 LIB	30 SAG	29 CAP	29 AQU
31 LEO		30 SCO			31 PIS

2009 MOON SIGNS

JAN	FEB	MAR	APR	MAY	JUN
3 ARI	1 TAU	3 GEM	1 CAN	2 VIR	1 LIB
5 TAU	3 GEM	5 CAN	3 LEO	5 LIB	3 SCO
7 GEM	5 CAN	7 LEO	5 VIR	7 SCO	6 SAG
9 CAN	7 LEO	9 VIR	7 LIB	9 SAG	8 CAP
11 LEO	10 VIR	11 LIB	10 SCO	12 CAP	11 AQU
13 VIR	12 LIB	13 CSO	12 SAG	14 AQU	13 PIS
15 LIB	14 SCO	16 SAG	15 CAP	17 PIS	16 ARI
18 SCO	16 SAG	18 CAP	17 AQU	19 ARI	18 TAU
20 SAG	19 CAP	21 AQU	20 PIS	21 TAU	20 GEM
23 CAP	21 AQU	23 PIS	22 ARI	24 GEM	22 CAN
25 AQU	24 PIS	26 ARI	24 TAU	26 CAN	24 LEO
28 PIS	26 ARI	28 TAU	26 GEM	28 LEO	26 VIR
30 ARI	28 TAU	30 GEM	28 CAN	30 VIR	28 LIB
			30 LEO		30 SCO

JUL	AUG	SEP	OCT	NOV	DEC
3 SAG	2 CAP	3 PIS	3 ARI	1 TAU	1 GEM
5 CAP	4 AQU	5 ARI	5 TAU	4 GEM	3 CAN
8 AQU	7 PIS	8 TAU	7 GEM	6 CAN	5 LEO
10 PIS	9 ARI	10 GEM	9 CAN	8 LEO	7 VIR
13 ARI	11 TAU	12 CAN	12 LEO	10 VIR	9 LIB
15 TAU	14 GEM	14 LEO	14 VIR	12 LIB	11 SCO
17 GEM	16 CAN	16 VIR	16 LIB	14 SCO	14 SAG
19 CAN	18 LEO	18 LIB	18 SCO	17 SAG	16 CAP
21 LEO	20 VIR	20 SCO	20 SAG	19 CAP	19 AQU
23 VIR	22 LIB	23 SAG	23 CAP	21 AQU	21 PIS
26 LIB	24 SCO	25 CAP	25 AQU	24 PIS	24 ARI
28 SCO	26 SAG	28 AQU	28 PIS	26 ARI	26 TAU
30 SAG	29 CAP	30 PIS	30 ARI	29 TAU	29 GEM
	31 AQU				30 CAN

2010 MOON SIGNS

JAN	FEB	MAR	APR	MAY	JUN
1 LEO	2 LIB	1 LIB	2 SAG	2 CAP	1 AQU
3 VIR	4 SCO	3 SCO	4 CAP	4 AQU	3 PIS
6 LIB	6 SAG	6 SAG	7 AQU	7 PIS	6 ARI
8 SCO	9 CAP	8 CAP	9 PIS	9 ARI	8 TAU
10 SAG	11 AQU	11 AQU	12 ARI	12 TAU	10 GEM
13 CAP	14 PIS	13 PIS	14 TAU	14 GEM	12 CAN
15 AQU	16 ARI	16 ARI	17 GEM	16 CAN	14 LEO
18 PIS	19 TAU	18 TAU	19 CAN	18 LEO	17 VIR
20 ARI	21 GEM	20 GEM	21 LEO	20 VIR	19 LIB
22 TAU	23 CAN	23 CAN	23 VIR	22 LIB	21 SCO
25 GEM	25 LEO	25 LEO	25 LIB	25 SCO	23 SAG
27 CAN	27 VIR	27 VIR	27 SCO	27 SAG	25 CAP
29 LEO		28 LIB	29 SCO	29 CAP	28 AQU
31 VIR		31 SCO			30 PIS

JUL	AUG	SEP	OCT	NOV	DEC
3 ARI	2 TAU	3 CAN	2 LEO	3 LIB	2 SCO
5 TAU	4 GEM	5 LEO	4 VIR	5 SCO	4 SAG
8 GEM	6 CAN	7 VIR	6 LIB	7 SAG	6 CAP
10 CAN	8 LEO	9 LIB	8 SCO	9 CAP	9 AQU
12 LEO	10 VIR	11 SCO	10 SAG	11 AQU	11 PIS
14 VIR	12 LIB	13 SAG	12 CAP	14 PIS	14 ARI
16 LIB	14 SCO	15 CAP	15 AQU	16 ARI	16 TAU
18 SCO	17 SAG	18 AQU	17 PIS	19 TAU	18 GEM
20 SAG	19 CAP	20 PIS	20 ARI	21 GEM	21 CAN
23 CAP	21 AQU	23 ARI	22 TAU	23 CAN	23 LEO
25 AQU	24 PIS	25 TAU	25 GEM	26 LEO	25 VIR
28 PIS	26 ARI	28 GEM	27 CAN	28 VIR	27 LIB
30 ARI	29 TAU	30 CAN	29 LEO	30 LIB	29 SCO
	31 GEM		31 VIR		31 SAG

2011 MOON SIGNS

JAN	FEB	MAR	APR	MAY	JUN
3 CAP	1 AQU	1 AQU	2 ARI	2 TAU	3 CAN
5 AQU	4 PIS	3 PIS	4 TAU	4 GEM	5 LEO
7 PIS	6 ARI	6 ARI	7 GEM	6 CAN	7 VIR
10 ARI	9 TAU	8 TAU	9 CAN	9 LEO	9 LIB
12 TAU	11 GEM	11 GEM	11 LEO	11 VIR	11 SCO
15 GEM	14 CAN	13 CAN	14 VIR	13 LIB	13 SAG
17 CAN	16 LEO	15 LEO	16 LIB	15 SCO	16 CAP
19 LEO	18 VIR	17 VIR	18 SCO	17 SAG	18 AQU
21 VIR	20 LIB	19 LIB	20 SAG	19 CAP	20 PIS
23 LIB	22 SCO	21 SCO	22 CAP	21 AQU	23 ARI
25 SCO	24 SAG	23 SAG	24 AQU	24 PIS	25 TAU
28 SAG	26 CAP	25 CAP	27 PIS	26 ARI	28 GEM
30 CAP		28 AQU	29 ARI	29 TAU	30 CAN
		30 PIS		31 GEM	

JUL	AUG	SEP	OCT	NOV	DEC
2 LEO	1 VIR	1 SCO	1 SAG	1 AQU	1 PIS
4 VIR	3 LIB	3 SAG	3 CAP	4 PIS	3 ARI
6 LIB	5 SCO	5 CAP	5 AQU	6 ARI	6 TAU
9 SCO	7 SAG	8 AQU	7 PIS	9 TAU	8 GEM
11 SAG	9 CAP	10 PIS	10 ARI	11 GEM	11 CAN
13 CAP	11 AQU	13 ARI	12 TAU	11 CAN	13 LEO
15 AQU	14 PIS	15 TAU	15 GEM	16 LEO	15 VIR
18 PIS	16 ARI	18 GEM	17 CAN	18 VIR	18 LIB
20 ARI	19 TAU	20 CAN	20 LEO	20 LIB	20 SCO
23 TAU	21 GEM	22 LEO	22 VIR	22 SCO	22 SAG
25 GEM	24 CAN	24 VIR	24 LIB	24 SAG	24 CAP
27 CAN	26 LEO	27 LIB	26 SCO	26 CAP	26 AQU
30 LEO	28 VIR	29 SCO	28 SAG	29 AQU	28 PIS
	30 LIB		30 CAP		31 ARI

2012 MOON SIGNS

JAN	FEB	MAR	APR	MAY	JUN
2 TAU	1 GEM	2 CAN	1 LEO	2 LIB	1 SCO
5 GEM	4 CAN	4 LEO	3 VIR	4 SCO	3 SAG
7 CAN	6 LEO	6 VIR	5 LIB	6 SAG	5 CAP
9 LEO	8 VIR	8 LIB	7 SCO	8 CAP	7 AQU
12 VIR	10 LIB	11 SCO	9 SAG	11 AQU	9 PIS
14 LIB	12 SCO	13 SAG	11 CAP	13 PIS	11 ARI
16 SCO	14 SAG	15 CAP	13 AQU	15 ARI	14 TAU
18 SAG	17 CAP	17 AQU	16 PIS	18 TAU	17 GEM
20 CAP	19 AQU	19 PIS	18 ARI	20 GEM	19 CAN
22 AQU	21 PIS	22 ARI	20 TAU	23 CAN	21 LEO
25 PIS	23 ARI	24 TAU	23 GEM	25 LEO	24 VIR
27 ARI	26 TAU	27 GEM	26 CAN	28 VIR	26 LIB
30 TAU	28 GEM	29 CAN	28 LEO	30 LIB	28 SCO
			30 VIR		30 SAG

JUL	AUG	SEP	OCT	NOV	DEC
2 CAP	1 AQU	2 ARI	1 TAU	1 CAN	2 LEO
4 AQU	3 PIS	4 TAU	4 GEM	5 LEO	5 VIR
6 PIS	5 ARI	6 GEM	6 CAN	7 VIR	7 LIB
9 ARI	8 TAU	9 CAN	9 LEO	10 LIB	9 SCO
11 TAU	10 GEM	11 LEO	11 VIR	12 SCO	11 SAG
14 GEM	13 CAN	14 VIR	13 LIB	14 SAG	13 CAP
16 CAN	15 LEO	16 LIB	15 SCO	16 CAP	15 AQU
19 LEO	17 VIR	18 SCO	17 SAG	18 AQU	17 PIS
21 VIR	19 LIB	20 SAG	19 CAP	20 PIS	20 ARI
23 LIB	22 SCO	22 CAP	22 AQU	22 ARI	22 TAU
25 SCO	24 SAG	24 AQU	24 PIS	25 TAU	25 GEM
28 SAG	26 CAP	27 PIS	26 ARI	27 GEM	27 CAN
30 CAP	28 AQU	29 ARI	28 TAU	30 TAU	30 LEO
	30 PIS		31 GEM		

2013 MOON SIGNS

JAN	FEB	MAR	APR	MAY	JUN
1 VIR	2 SCO	1 SCO	2 CAP	1 AQU	2 ARI
3 LIB	4 SAG	3 SAG	4 AQU	3 PIS	4 TAU
6 SCO	6 CAP	5 CAP	6 PIS	5 ARI	6 GAM
8 SAG	8 AQU	7 AQU	8 ARI	8 TAU	9 CAN
10 CAP	10 PIS	10 PIS	10 TAU	10 GEM	12 LEO
12 AQU	12 ARI	12 ARI	13 GEM	13 CAN	13 VIR
14 PIS	15 TAU	14 TAU	15 CAN	15 LEO	16 LIB
16 ARI	17 GEM	17 GEM	18 LEO	18 VIR	18 SCO
18 TAU	20 CAN	19 CAN	20 VIR	20 LIB	21 SAG
21 GEM	22 LEO	22 LEO	23 LIB	22 SCO	21 CAP
23 CAN	25 VIR	24 VIR	25 SCO	24 SAG	25 AQU
26 LEO	27 LIB	26 LIB	27 SAG	26 CAP	27 PIS
28 VIR		28 SCO	29 CAP	28 AQU	29 ARI
31 LIB		30 SAG		30 PIS	

JUL	AUG	SEP	OCT	NOV	DEC
1 TAU	2 CAN	1 LEO	1 VIR	2 SCO	2 SAG
4 GEM	5 LEO	4 VIR	3 LIB	4 SAG	4 CAP
6 CAN	7 VIR	6 LIB	6 SCO	6 CAP	6 AQU
9 LEO	10 LIB	8 SCO	8 SAG	8 AQU	8 PIS
11 VIR	12 SCO	11 SAG	10 CAP	10 PIS	10 ARI
14 LIB	14 SAG	13 CAP	12 AQU	12 ARI	12 TAU
16 SCO	16 CAP	15 AQU	14 PIS	15 TAU	15 GEM
18 SAG	18 AQU	17 PIS	16 ARI	18 GEM	17 CAN
20 CAP	20 PIS	19 ARI	19 TAU	20 CAN	20 LEO
22 AQU	23 ARI	21 TAU	21 GEM	22 LEO	22 VIR
24 PIS	25 TAU	24 GEM	23 CAN	25 VIR	25 LIB
26 ARI	27 GEM	26 CAN	26 LEO	27 LIB	27 SCO
28 TAU	30 CAN	29 LEO	28 VIR	29 SCO	29 SAG
31 GEM			31 LIB		31 CAP

2014 MOON SIGNS

JAN	FEB	MAR	APR	MAY	JUN
2 AQU	1 PIS	2 ARI	1 TAU	3 CAN	1 LEO
4 PIS	3 ARI	4 TAU	3 GEM	5 LEO	4 VIR
6 ARI	5 TAU	6 GEM	5 CAN	8 VIR	6 LIB
8 TAU	7 GEM	9 CAN	8 LEO	10 LIB	9 SCO
11 GEM	10 CAN	11 LEO	10 VIR	12 SCO	11 SAG
13 CAN	12 LEO	14 VIR	13 LIB	12 SAG	13 CAP
16 LEO	15 VIR	16 LIB	15 SCO	17 CAP	15 AQU
18 VIR	17 LIB	19 SCO	17 SAG	19 AQU	17 PIS
21 LIB	19 SCO	21 SAG	19 CAP	21 PIS	19 ARI
23 SCO	22 SAG	23 CAP	21 AQU	23 ARI	21 TAU
25 SAG	24 CAP	25 AQU	24 PIS	25 TAU	24 GEM
28 CAP	26 AQU	27 PIS	26 ARI	27 GEM	26 CAN
30 AQU	28 PIS	29 ARI	29 TAU	30 CAN	29 LEO
			30 GEM		

JUL	AUG	SEP	OCT	NOV	DEC
1 VIR	2 SCO	1 SAG	3 AQU	1 PIS	3 TAU
4 LIB	5 SAG	3 CAP	5 PIS	3 ARI	5 GEM
6 SCO	7 CAP	3 AQU	7 ARI	5 TAU	7 CAN
8 SAG	9 AQU	7 PIS	9 TAU	7 GEM	9 LEO
10 CAP	11 PIS	9 ARI	11 GEM	10 CAN	12 VIR
12 AQU	13 ARI	11 TAU	13 CAN	12 LEO	14 LIB
14 PIS	15 TAU	14 GEM	16 LEO	15 VIR	17 SCO
16 ARI	17 GEM	16 CAN	18 VIR	17 LIB	19 SAG
19 TAU	20 CAN	18 LEO	21 LIB	20 SCO	21 CAP
21 GEM	22 LEO	21 VIR	23 SCO	22 SAG	23 AQU
23 CAN	25 VIR	23 LIB	25 SAG	24 CAP	25 PIS
26 LEO	27 LIB	26 SCO	28 CAP	26 AQU	28 ARI
28 VIR	30 SCO	28 SAG	30 AQU	28 PIS	30 TAU
31 LIB		30 CAP		30 ARI	

2015 MOON SIGNS

JAN	FEB	MAR	APR	MAY	JUN
1 GEM	2 LEO	1 LEO	2 LIB	2 SCO	1 SAG
3 CAN	5 VIR	4 VIR	5 SCO	5 SAG	3 CAP
6 LEO	7 LIB	6 LIB	8 SAG	7 CAP	6 AQU
8 VIR	10 SCO	9 SCO	10 CAP	9 AQU	8 PIS
11 LIB	12 SAG	11 SAG	12 AQU	11 PIS	11 ARI
13 SCO	14 CAP	14 CAP	14 PIS	14 ARI	12 TAU
16 SAG	16 AQU	16 AQU	16 ARI	16 TAU	14 GEM
18 CAP	18 PIS	18 PIS	18 TAU	18 GEM	16 CAN
20 AQU	20 ARI	20 ARI	20 GEM	20 CAN	19 LEO
22 PIS	22 TAU	22 TAU	22 CAN	22 LEO	21 VIR
24 ARI	25 GEM	24 GEM	25 LEO	25 VIR	24 LIB
26 TAU	27 CAN	26 CAN	27 VIR	27 LIB	26 SCO
28 GEM		29 LEO	30 LIB	30 SCO	28 SAG
31 CAN		31 VIR			

JUL	AUG	SEP	OCT	NOV	DEC
1 CAP	1 PIS	2 TAU	1 GEM	2 LEO	2 VIR
3 AQU	3 ARI	4 GEM	3 CAN	4 VIR	4 LIB
5 PIS	5 TAU	6 CAN	6 LEO	7 LIB	7 SCO
7 ARI	8 GEM	8 LEO	8 VIR	9 SCO	9 SAG
9 TAU	10 CAN	11 VIR	11 LIB	12 SAG	12 CAP
11 GEM	12 LEO	13 LIB	13 SCO	14 CAP	14 AQU
14 CAN	15 VIR	16 SCO	16 SAG	17 AQU	16 PIS
16 LEO	16 LIB	18 SAG	18 CAP	19 PIS	18 ARI
18 VIR	20 SCO	21 CAP	20 AQU	21 ARI	20 TAU
21 LIB	22 SAG	23 AQU	23 PIS	23 TAU	22 GEM
23 SCO	24 CAP	25 PIS	25 ARI	25 GEM	25 CAN
26 SAG	27 AQU	27 ARI	27 TAU	27 CAN	27 LEO
28 CAP	29 PIS	29 TAU	29 GEM	28 LEO	29 VIR
30 AQU	31 ARI		31 CAN		

2016 MOON SIGNS

JAN	FEB	MAR	APR	MAY	JUN
1 LIB	2 SAG	3 CAP	1 AQU	1 PIS	1 TAU
3 SCO	4 CAP	5 AQU	4 PIS	3 ARI	3 GEM
6 SAG	7 AQU	7 PIS	6 ARI	5 TAU	5 CAN
8 CAP	9 PIS	9 ARI	8 TAU	7 GEM	8 LEO
10 AQU	11 ARI	11 TAU	10 GEM	9 CAN	10 VIR
12 PIS	13 TAU	13 GEM	12 CAN	11 LEO	12 LIB
14 ARI	15 GEM	15 CAN	14 LEO	14 VIR	15 SCO
17 TAU	17 CAN	18 LEO	16 VIR	16 LIB	17 SAG
19 GEM	19 LEO	20 VIR	19 LIB	19 SCO	20 CAP
21 CAN	22 VIR	23 LIB	21 SCO	21 SAG	22 AQU
23 LEO	24 LIB	25 SCO	24 SAG	24 CAP	24 PIS
25 VIR	27 SCO	28 SAG	26 CAP	26 AQU	27 ARI
28 LIB	29 SAG	30 CAP	29 AQU	28 PIS	29 TAU
30 SCO				30 ARI	

JUL	AUG	SEP	OCT	NOV	DEC
1 GEM	1 LEO	2 LEO	2 SCO	1 SAG	1 CAP
3 CAN	4 VIR	5 SCO	5 SAG	3 CAP	3 AQU
5 LEO	6 LIB	7 SAG	7 CAP	6 AQU	5 PIS
7 VIR	9 SCO	10 CAP	10 AQU	8 PIS	8 ARI
10 LIB	11 SAG	12 AQU	12 PIS	10 ARI	10 TAU
12 SCO	13 CAP	14 PIS	14 ARI	12 TAU	12 GEM
15 SAG	16 AQU	16 ARI	16 TAU	14 GEM	14 CAN
17 CAP	18 PIS	19 TAU	18 GEM	16 CAN	16 LEO
19 AQU	20 ARI	21 GEM	20 CAN	18 LEO	18 VIR
22 PIS	22 TAU	23 CAN	22 LEO	21 VIR	20 LIB
24 ARI	24 GEM	25 LEO	24 VIR	23 LIB	23 SCO
26 TAU	26 CAN	27 VIR	27 LIB	26 SCO	25 SAG
28 GEM	29 LEO	30 PIB	29 SCO	28 SAG	29 CAP
30 CAN	31 VIR				30 AQU

2017 MOON SIGNS

JAN	FEB	MAR	APR	MAY	JUN
2 PIS	2 TAU	2 TAU	2 CAN	1 LEO	2 LIB
4 ARI	4 GEM	4 GEM	4 LEO	4 VIR	5 SCO
6 TAU	7 CAN	6 CAN	6 VIR	6 LIB	7 SAG
8 GEM	9 LEO	8 LEO	9 LIB	9 SCO	10 CAP
10 CAN	11 VIR	10 VIR	11 SCO	11 SAG	12 AQU
12 LIB	13 LIB	13 LIB	14 SAG	14 CAP	15 PIS
14 VIR	16 SCO	15 SCO	16 CAP	16 AQU	17 ARI
17 LIB	18 SAG	17 SAG	19 AQU	18 PIS	19 TAU
19 SCO	21 CAP	20 CAP	21 PIS	21 ARI	21 GEM
22 SAG	23 AQU	22 AQU	23 ARI	23 TAU	23 GEM
24 CAP	25 PIS	25 PIS	25 TAU	25 GEM	25 LEO
27 AQU	28 ARI	27 ARI	27 GEM	27 CAN	27 VIR
29 PIS		29 TAU	29 CAN	29 LEO	30 LIB
31 ARI		31 GEM		31 VIR	

JUL	AUG	SEP	OCT	NOV	DEC
2 SCO	1 SAG	2 AQU	2 PIS	1 ARI	2 GEM
5 SAG	3 CAP	5 PIS	4 ARI	3 TAU	4 CAN
7 CAP	6 AQU	7 ARI	6 TAU	5 GEM	6 LEO
10 AQU	8 PIS	9 TAU	8 GEM	7 CAN	8 VIR
12 PIS	11 ARI	11 GEM	10 CAN	9 LEO	11 LIB
14 ARI	13 TAU	13 CAN	13 LEO	11 VIR	13 SCO
17 TAU	15 GEM	15 LEO	15 VIR	13 LIB	15 SAG
19 GEM	17 CAN	18 VIR	17 LIB	16 SAC	18 CAP
21 CAN	19 LEO	20 LIB	19 SCO	18 SAG	20 AQU
23 LEO	21 VIR	22 SCO	22 SAG	21 CAP	23 PIS
25 VIR	23 LIB	24 SAG	24 CAP	23 AQU	25 ARI
27 LIB	26 SCO	27 CAP	27 AQU	26 PIS	28 TAU
29 SCO	28 SAG	29 AQU	29 PIS	28 ARI	30 GEM
	31 CAP			30 TAU	

2018 MOON SIGNS

JAN	FEB	MAR	APR	MAY	JUN
1 CAN	1 VIR	1 VIR	1 SCO	1 SAG	2 AQU
3 LEO	3 LIB	3 LIB	3 SAG	3 CAP	5 PIS
5 VIR	5 SCO	5 SCO	6 CAP	6 AQU	7 ARI
7 LIB	8 SAG	7 SAG	9 AQU	8 PIS	9 TAI
9 SCO	10 CAP	10 CAP	11 PIS	11 ARI	12 GEM
12 SAG	13 AQU	12 AQU	13 ARI	13 TAU	14 CAN
14 CAP	15 PIS	15 PIS	16 TAU	15 GEM	16 LEO
17 AQU	18 ARI	17 ARI	18 GEM	17 CAN	18 VIR
19 PIS	20 TAU	19 TAU	20 CAN	19 LEO	20 LIB
22 ARI	22 GEM	22 GEM	22 LEO	21 VIR	22 SCO
24 TAU	24 CAN	24 CAN	24 VIR	24 LIB	24 SAG
26 GEM	26 LEO	26 LEO	26 LIB	26 SCO	27 CAP
28 CAN		28 VIR	28 SCO	28 SAG	29 AQU
30 LEO		30 LIB		31 CAP	

JUL	AUG	SEP	OCT	NOV	DEC
2 PIS	1 ARI	2 GEM	1 CAN	2 VIR	1 LIB
4 ARI	3 TAU	4 CAN	3 LEO	4 LIB	3 SCO
7 TAU	5 GEM	6 LEO	5 VIR	6 SCO	5 SAG
9 GEM	7 CAN	8 VIR	7 LIB	8 SAG	8 CAP
11 CAN	9 LEO	10 LIB	9 SCO	10 CAP	10 AQU
13 LEO	11 VIR	12 SCO	12 SAG	13 AQU	13 PIS
15 VIR	14 LIB	14 SAG	14 CAP	15 PIS	15 ARI
17 LIB	16 SCO	17 CAP	17 AQU	18 ARI	18 TAU
19 SCO	18 SAG	19 AQU	19 PIS	20 TAU	20 GEM
22 SAG	20 CAP	22 PIS	22 ARI	22 GEM	22 CAN
24 CAN	23 AQU	24 ARI	24 TAU	25 CAN	24 LEO
27 AQU	25 PIS	27 TAU	26 GEM	27 LEO	26 VIR
29 PIS	28 ARI	29 GEM	28 CAN	29 VIR	28 LIB
	30 TAU		30 LEO		30 SCO

2019 MOON SIGNS

JAN	FEB	MAR	APR	MAY	JUN
2 SAG	3 AQU	2 AQU	1 PIS	1 ARI	2 GEM
4 CAP	5 PIS	5 PIS	3 ARI	3 TAU	4 CAN
7 AQU	8 ARI	7 ARI	6 TAU	5 GEM	6 LEO
9 PIS	10 TAU	10 TAU	8 GEM	8 CAN	8 VIR
12 ARI	13 GEM	12 GEM	10 CAN	10 LEO	10 LIB
14 TAU	15 CAN	14 CAN	13 LEO	12 VIR	12 VIR
16 GEM	17 LEO	16 LEO	15 VIR	14 LIB	15 SAG
18 CAN	19 VIR	18 VIR	17 LIB	16 SCO	17 CAP
20 LEO	21 LIB	20 LIB	19 SCO	18 SAG	19 AQU
22 VIR	23 SCO	22 SCO	21 SAG	21 CAP	22 PIS
24 LIB	25 SAG	25 SAG	23 CAP	23 AQU	24 ARI
27 SCO	28 CAP	27 CAP	26 AQU	26 PIS	27 TAU
29 SAG		29 AQU	28 PIS	28 ARI	29 GEM
31 CAP				30 TAU	

JUL	AUG	SEP	OCT	NOV	DEC
1 CAN	2 VIR	2 SCO	3 SAG	3 AQU	3 PIS
3 LEO	4 LIB	4 SAG	4 CAP	5 PIS	5 ARI
5 VIR	6 SCO	7 CAP	6 AQU	8 ARI	8 TAU
8 LIB	8 SAG	9 AQU	9 PIS	10 TAU	10 GEM
10 SCO	11 CAP	12 PIS	11 ARI	13 GEM	12 CAN
12 SAG	13 AQU	14 ARI	14 TAU	15 CAN	14 LEO
14 CAP	15 PIS	17 TAU	16 GEM	17 LEO	17 VIR
17 AQU	18 ARI	19 GEM	19 CAN	19 VIR	19 LIB
19 PIS	20 TAU	22 CAN	21 LEO	21 LIB	21 SCO
22 ARI	23 GEM	24 LEO	23 VIR	24 SCO	23 SAG
24 TAU	25 CAN	26 VIR	25 LIB	26 SAG	25 CAP
27 GEM	27 LEO	28 LIB	27 SCO	28 CAP	28 AQU
29 CAN	29 VIR	30 SCO	29 SAG	30 AQU	30 PIS
31 LEO	31 LIB		31 CAP		

2020 MOON SIGNS

JAN	FEB	MAR	APR	MAY	JUN
1 ARI	3 GEM	1 GEM	2 LEO	2 VIR	2 SCO
4 TAU	5 CAN	3 CAN	4 VIR	4 LIB	4 SAG
6 GEM	7 LEO	6 LEO	6 LIB	6 SCO	6 CAP
9 CAN	9 VIR	8 VIR	8 SCO	8 SAG	8 AQU
11 LEO	11 LIB	10 LIB	10 SAG	10 CAP	11 PIS
13 VIR	13 SCO	12 SCO	12 CAP	12 AQU	13 ARI
15 LIB	15 SAG	14 SAG	15 AQU	14 PIS	16 TAU
17 SCO	18 CAP	16 CAP	17 PIS	17 ARI	18 GEM
19 SAG	20 AQU	18 AQU	20 ARI	19 TAU	21 CAN
22 CAP	23 PIS	21 PIS	22 TAU	22 GEM	23 LEO
24 AQU	25 ARI	23 ARI	25 GEM	24 CAN	25 VIR
26 PIS	28 TAU	26 TAU	27 CAN	27 LEO	27 LIB
29 ARI		28 GEM	29 LEO	29 VIR	29 SCO
31 TAU		31 CAN		31 LIB	

JUL	AUG	SEP	OCT	NOV	DEC
1 SAG	2 AQU	1 PIS	3 TAU	2 GEM	1 CAN
3 CAP	4 PIS	3 ARI	5 GEM	4 CAN	4 LEO
6 AQU	7 ARI	6 TAU	8 CAN	7 LEO	6 VIR
8 PIS	9 TAU	8 GEM	10 LEO	9 VIR	8 LIB
11 ARI	12 GEM	11 CAN	13 VIR	11 LIB	10 SCO
13 TAU	14 CAN	13 LEO	15 LIB	13 SCO	12 SAG
16 GEM	17 LEO	15 VIR	17 SCO	15 SAG	14 CAP
18 CAN	19 VIR	17 LIB	19 SAG	17 CAP	17 AQU
20 LEO	21 LIB	19 SCO	21 CAP	19 AQU	19 PIS
22 VIR	23 SCO	21 SAG	23 AQU	21 PIS	21 ARI
24 LIB	25 SAG	23 CAP	25 PIS	24 ARI	24 TAU
26 SCO	27 CAP	26 AQU	28 ARI	26 TAU	26 GEM
29 SAG	29 AQU	28 PIS	30 TAU	29 GEM	29 CAN
31 CAP		30 ARI			31 LEO

2021 MOON SIGNS

JAN	FEB	MAR	APR	MAY	JUN
2 VIR	1 LIB	2 SCO	1 SAG	2 AQU	1 PIS
5 LIB	3 SCO	4 SAG	3 CAP	4 PIS	3 ARI
7 SCO	5 SAG	6 CAP	5 AQU	7 ARI	6 TAU
9 SAG	7 CAP	9 AQU	7 PIS	9 TAU	8 GEM
11 CAP	9 AQU	11 PIS	10 ARI	12 GEM	11 CAN
13 AQU	12 PIS	13 ARI	12 TAU	14 CAN	13 LEO
15 PIS	14 ARI	16 TAU	15 GEM	17 LEO	15 VIR
18 ARI	16 TAU	18 GEM	17 CAN	19 VIR	18 LIB
20 TAU	19 GEM	21 CAN	20 LEO	21 LIB	20 SCO
23 GEM	21 CAN	23 LEO	22 VIR	23 SCO	22 SAG
25 CAN	24 LEO	25 VIR	24 LIB	25 SAG	24 CAP
27 LEO	26 VIR	28 LIB	26 SCO	27 CAP	26 AQU
30 VIR	28 LIB	30 SCO	28 SAG	29 AQU	28 PIS
			30 CAP		30 ARI

JUL	AUG	SEP	OCT	NOV	DEC
3 TAU	2 GEM	1 CAN	3 VIR	1 LIB	1 SCO
5 GEM	4 CAN	3 LEO	5 LIB	3 SCO	3 SAG
8 CAN	7 LEO	5 VIR	7 SCO	5 SAG	5 CAP
10 LEO	9 VIR	7 LIB	9 SAG	7 CAP	7 AQU
13 VIR	11 LIB	10 SCO	11 CAP	9 AQU	9 PIS
15 LIB	13 SCO	12 SAG	13 AQU	12 PIS	11 ARI
17 SCO	15 SAG	14 CAP	15 PIS	14 ARI	14 TAU
19 SAG	18 CAP	16 AQU	18 ARI	16 TAU	16 GEM
21 CAP	20 AQU	18 PIS	20 TAU	19 GEM	19 CAN
23 AQU	22 PIS	20 ARI	23 GEM	21 CAN	21 LEO
25 PIS	24 ARI	23 TAU	25 CAN	24 LEO	24 VIR
28 ARI	26 TAU	25 GEM	28 LEO	26 VIR	26 LIB
30 TAU	29 GEM	28 CAN	30 VIR	29 LIB	28 SCO
		30 LEO			30 SAG

2022 MOON SIGNS

JAN	FEB	MAR	APR	MAY	JUN
1 CAP	2 PIS	1 PIS	2 TAU	2 GEM	1 CAN
3 AQU	4 ARI	3 ARI	4 GEM	4 CAN	3 LEO
5 PIS	6 TAU	6 TAU	7 CAN	7 LEO	6 VIR
8 ARI	9 GEM	8 GEM	9 LEO	9 VIR	8 LIB
10 TAU	11 CAN	11 CAN	12 VIR	12 LIB	10 SCO
12 GEM	14 LEO	13 LEO	14 LIB	14 SCO	12 SAG
15 CAN	16 VIR	16 VIR	16 SCO	16 SAG	14 CAP
17 LEO	18 LIB	18 LIB	18 SAG	18 CAP	16 AQU
20 VIR	21 SCO	20 SCO	20 CAP	20 AQU	18 PIS
22 LIB	23 SAG	22 SAG	23 AQU	22 PIS	20 ARI
24 SCO	25 CAP	24 CAP	25 PIS	24 ARI	23 TAU
27 SAG	27 AQU	27 AQU	27 ARI	27 TAU	25 GEM
29 CAP		28 PIS	29 TAU	29 GEM	27 CAN
31 AQU		31 ARI			30 LEO

JUL	AUG	SEP	OCT	NOV	DEC
3 VIR	1 LIB	2 SAG	2 CAP	2 PIS	1 ARI
5 LIB	4 SCO	4 CAP	4 AQU	4 ARI	4 TAU
8 SCO	6 SAG	6 AQU	6 PIS	7 TAU	6 GEM
10 SAG	8 CAP	8 PIS	8 ARI	9 GEM	9 CAN
12 CAP	10 AQU	11 ARI	10 TAU	11 CAN	11 LEO
14 AQU	12 PIS	13 TAU	13 GEM	14 LEO	14 VIR
16 PIS	14 ARI	15 GEM	15 CAN	16 VIR	16 LIB
18 ARI	16 TAU	18 CAN	17 LEO	19 LIB	18 SCO
20 TAU	19 GEM	20 LEO	20 VIR	21 SCO	21 SAG
23 GEM	21 CAN	23 VIR	22 LIB	23 SAG	23 CAP
25 CAN	24 LEO	25 LIB	25 SCO	25 CAP	25 AQU
28 LEO	26 VIR	27 SCO	27 SAG	27 AQU	27 PIS
30 VIR	29 LIB	29 SAG	29 CAP	29 PIS	29 ARI
	31 SCO		31 AQU		31 TAU

2023 MOON SIGNS

JAN	FEB	MAR	APR	MAY	JUN
2 GEM	1 CAN	3 LEO	2 VIR	2 LIB	3 SAG
5 CAN	4 LEO	5 VIR	4 LIB	4 SCO	5 CAP
7 LEO	6 VIR	8 LIB	7 SCO	6 SAG	7 AQU
10 VIR	9 LIB	10 SCO	9 SAG	8 CAP	9 PIS
12 LIB	11 SCO	13 SAG	11 CAP	10 AQU	11 ARI
15 SCO	13 SAG	15 CAP	13 AQU	12 PIS	13 TAU
17 SAG	16 CAP	17 AQU	15 PIS	15 ARI	15 GEM
19 CAP	18 AQU	19 PIS	17 ARI	17 TAU	18 CAN
21 AQU	20 PIS	21 ARI	19 TAU	19 GEM	20 LEO
23 PIS	22 ARI	23 TAU	21 GEM	21 CAN	23 VIR
25 ARI	24 TAU	25 GEM	24 CAN	24 LEO	25 LIB
27 TAU	26 GEM	28 CAN	27 LEO	26 VIR	28 SCO
30 GEM	28 CAN	30 LEO	29 VIR	29 LIB	30 SAG
				31 SCO	

JUL	AUG	SEP	OCT	NOV	DEC
2 CAP	2 PIS	1 ARI	3 GEM	1 CAN	1 LEO
4 AQU	4 ARI	3 TAU	5 CAN	4 LEO	3 VIR
6 PIS	7 TAU	5 GEM	7 LEO	6 VIR	6 LIB
8 ARI	9 GEM	8 CAN	10 VIR	9 LIB	8 SCO
10 TAU	11 CAN	10 LEO	12 LIB	11 SCO	11 SAG
13 GEM	14 LEO	13 VIR	15 SCO	13 SAG	13 CAP
15 CAN	16 VIR	15 LIB	17 SAG	16 CAP	15 AQU
17 LEO	19 LIB	18 SCO	19 CAP	18 AQU	17 PIS
20 VIR	21 SCO	20 SAG	22 AQU	20 PIS	19 ARI
23 LIB	24 SAG	22 CAP	24 PIS	22 ARI	21 TAU
25 SCO	26 CAP	24 AQU	26 ARI	24 TAU	24 GEM
27 SAG	28 AQU	26 PIS	28 TAU	26 GEM	26 CAN
29 CAP	30 PIS	28 ARI	30 GEM	29 CAN	28 LEO
31 AQU		30 TAU			31 VIR

2024 MOON SIGNS

JAN	FEB	MAR	APR	MAY	JUN
2 LIB	1 SCO	2 SAG	3 AQU	2 PIS	3 TAU
5 SCO	4 SAG	4 CAP	5 PIS	4 ARI	5 GEM
7 SAG	6 CAP	6 AQU	7 ARI	6 TAU	7 CAN
9 CAP	8 AQU	8 PIS	9 TAU	8 GEM	9 LEO
11 AQU	10 PIS	10 ARI	11 GEM	10 CAN	12 VIR
13 PIS	12 AQU	12 TAU	13 CAN	13 LEO	14 LIB
16 ARI	14 TAU	14 GEM	15 LEO	15 VIR	17 SCO
18 TAU	16 GEM	17 CAN	18 VIR	18 LIB	19 SAG
20 GEM	18 CAN	19 LEO	20 LIB	20 SCO	21 CAP
22 CAN	21 LEO	22 VIR	23 SCO	23 SAG	23 AQU
25 LEO	23 VIR	24 LIB	25 SAG	25 CAP	26 PIS
27 VIR	26 LIB	27 SCO	28 CAP	27 AQU	27 ARI
30 LIB	28 SCO	29 SAG	30 AQU	29 PIS	30 TAU
		31 CAP		31 ARI	

JUL	AUG	SEP	OCT	NOV	DEC
2 GEM	3 LEO	1 VIR	1 LIB	3 SAG	2 CAP
2 CAN	5 VIR	4 LIB	4 SCO	5 CAP	4 AQU
6 LEO	8 LIB	7 SCO	6 SAG	7 AQU	7 PIS
9 VIR	10 SCO	9 SAG	9 CAP	9 PIS	9 ARI
11 LIB	13 SAG	11 CAP	11 AQU	12 ARI	11 TAU
14 SCO	15 CAP	14 AQU	13 PIS	14 TAU	13 GEM
16 SAG	17 AQU	16 PIS	15 ARI	16 GEM	16 CAN
19 CAP	19 PIS	18 ARI	17 TAU	18 CAN	17 LEO
21 AQU	21 ARI	20 TAU	19 GEM	20 LEO	20 VIR
23 PIS	23 TAU	22 GEM	21 CAN	22 VIR	22 LIB
25 ARI	25 GEM	24 CAN	24 LEO	25 LIB	25 SCO
27 TAU	28 CAN	26 LEO	26 VIR	27 SCO	27 SAG
29 GEM	30 LEO	29 VIR	28 LIB	30 SAG	29 CAP
31 CAN			31 SCO		

2025 MOON SIGNS

JAN	FEB	MAR	APR	MAY	JUN
1 AQU	1 ARI	1 ARI	1 GEM	1 CAN	1 VIR
3 PIS	3 TAU	3 TAU	3 CAN	3 LEO	4 VIR
5 ARI	6 GEM	5 GEM	5 LEO	5 VIR	6 SCO
7 TAU	8 CAN	7 CAN	8 VIR	8 LIB	9 SAG
9 GEM	10 LEO	9 LEO	10 LIB	10 SCO	11 CAP
11 CAN	12 VIR	12 VIR	13 SCO	13 SAG	14 AQU
14 LEO	15 LIB	14 LIB	15 SAG	15 CAP	16 PIS
16 VIR	17 SCO	17 SCO	18 CAP	18 AQU	18 ARI
18 LIB	20 SAG	19 SAG	20 AQU	20 PIS	20 TAU
21 SCO	22 CAP	22 CAP	23 PIS	22 ARI	22 GEM
23 SAG	25 AQU	24 AQU	25 ARI	24 TAU	24 CAN
26 CAP	27 PIS	26 PIS	27 TAU	26 GEM	27 LEO
28 AQU		28 ARI	29 GEM	28 CAN	29 VIR
30 PIS		30 TAU		30 LEO	

JUL	AUG	SEP	OCT	NOV	DEC
1 LEO	3 SAG	1 CAP	1 AQU	2 ARI	1 TAU
4 SCO	5 CAP	4 AQU	3 PIS	4 TAU	3 GEM
6 SAG	7 AQU	6 PIS	6 ARI	6 GEM	5 CAN
9 CAP	10 PIS	8 ARI	8 TAU	8 CAN	7 LEO
11 AQU	12 ARI	10 TAU	10 GEM	10 LEO	10 VIR
13 PIS	14 TAU	12 GEM	12 CAN	12 VIR	12 LIB
15 ARI	16 GEM	14 CAN	14 LEO	15 LIB	14 SCO
18 TAU	18 CAN	17 LEO	16 VIR	17 SCO	17 SAG
20 GEM	20 LEO	19 VIR	18 LIB	20 SAG	20 CAP
22 CAN	23 VIR	21 LIB	21 SCO	22 CAP	22 AQU
24 LEO	25 LIB	24 SCO	23 SAG	25 AQU	24 PIS
26 VIR	27 SCO	26 SAG	26 CAP	27 PIS	27 ARI
29 LIB	30 SAG	29 CAP	28 AQU	29 ARI	29 TAU
31 SCO			31 PIS		31 GEM

Aspects

Modern astrologers spend a significant amount of time analyzing and studying the birth chart. In this analysis of a person's birth chart, one of the considerations is the aspects of the planets which produce an influence on the individuals personal characteristics, traits, decision making process, and response to events. The aspects are determined by the number of degrees apart the planets are on the birth chart. That number varies among astrologers, but the closer the aspects are the stronger their influence. Well-aspected planets produce beneficial or harmonious influences while poorly aspected planets present the individual with obstacles and challenges. Also, the influence of one aspect may weaken or strengthen the effect of another. In interpreting a birth chart, what becomes apparent is that certain traits begin to repeat themselves and become more prevalent. Other characteristics may appear less often in the individuals chart, and it is these traits that can be overlooked or unrealized by the person.

A ephemeris is used to determine the exact degree of the planets, or a faster method for researching planetary locations on a birth chart is to use a computer program. These influences are divided into major aspects and minor aspects.

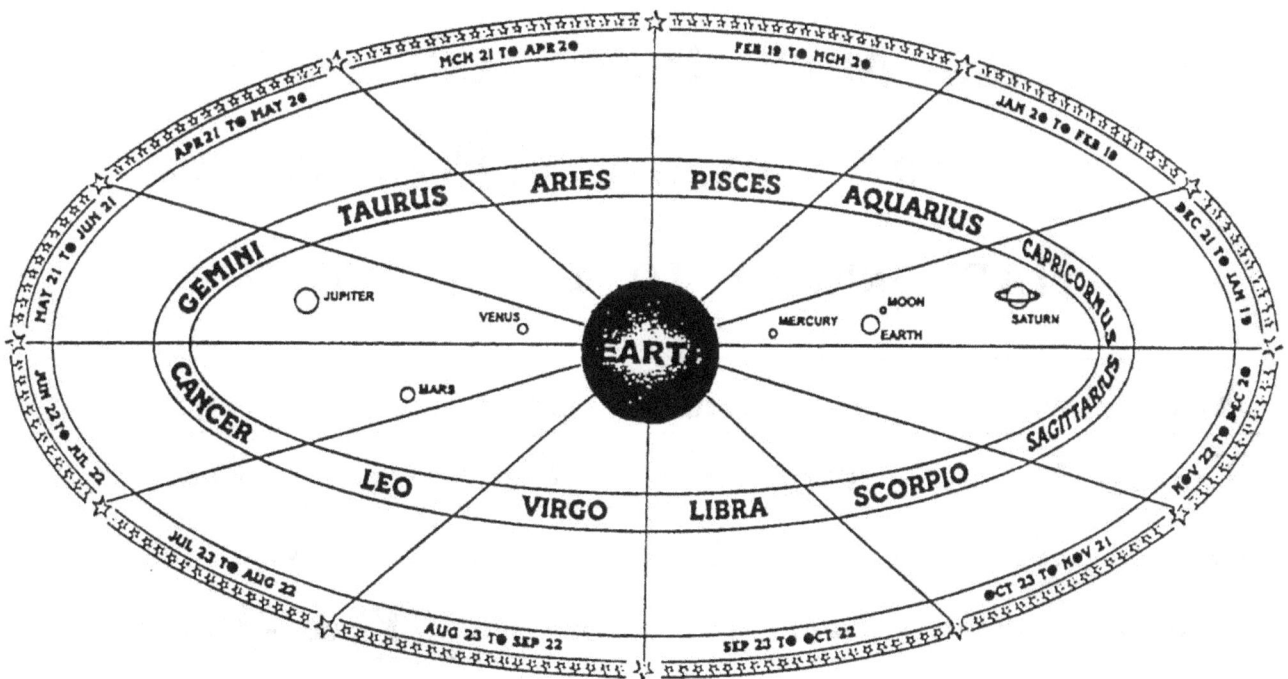

Major Aspects
Conjunction

Name: *Conjunction*

Angle in Degrees: 0 plus or minus 8 degrees in the same sign

Symbol:

The strongest aspect in a chart, a conjunction is a powerful influence intensifying and combining the positive traits of the planets. A harmonious influence, it is usually beneficial but not always, depending on the manner in which the individual utilizes the benefits.

Sun Signs, Moon Signs and Ascendents that are Conjunct (Positive or Harmonious) to Each Other:

Aries - Aries	Leo - Leo	Sagittarius - Sagittarius
Taurus - Taurus	Virgo - Virgo	Capricorn - Capricorn
Gemini - Gemini	Libra - Libra	Aquarius - Aquarius
Cancer - Cancer	Scorpio - Scorpio	Pisces - Pisces

Sextile

Name: *Sextile*

Angle in Degrees: 60 plus or minus 8 degrees and two   signs apart

Symbol:

$$\text{\textasteriskcentered}$$

The sextile is a harmonious aspect representing opportunities and luck but requiring the efforts and actions of the individual. Not as powerful as a trine, the sextile indicates the planets are cooperating to produce positive and harmonious results.

Sun Sign, Moon Signs and Ascendents that are Sextile (Positive or Harmonious) to Each Other:

Aries - Gemini	**Gemini - Leo**	**Libra- Sagittarius**
Aries - Aquarius	**Cancer - Virgo**	**Scorpio - Capricorn**
Taurus - Cancer	**Leo - Libra**	**Sagittarius - Aquarius**
Taurus - Pisces	**Virgo - Scorpion**	**Capricorn - Pisces**

Square

Name: *Square*

Angle in Degrees: 90 plus or minus 8 degrees and three   signs apart

Symbol:

This is a powerful aspect representing obstacles and stressful challenges which require action and decision on the part of the individual. The square indicates areas of one life where growth and development are needed.

Sun Signs, Moon Signs and Ascendents that are Square (Negative or Inharmonious) with Each Other:

Aries - Cancer	Gemini - Virgo	Virgo - Sagittarius
Aries - Capricorn	Gemini - Pisces	Libra - Capricorn
Taurus - Leo	Cancer - Libra	Scorpio - Aquarius
Taurus - Aquarius	Leo - Scorpio	Sagittarius - Pisces

Trine

Name: *Trine*

Angle in Degrees: 120 plus or minus 8 degrees and four signs apart

Symbol:

△

A trine is a powerful and strong aspect which is harmonious and advantageous. It indicates ease and good luck, but too many trines can influence the individual to be lazy with too few obstacles to challenge personal growth.

Sun Sign, Moon Signs and Ascendents that are Trine (Positive or Harmonious) to Each Other:

Aries - Leo	Gemini - Libra	Leo - Sagittarius
Aries - Sagittarius	Gemini - Aquarius	Virgo - Capricorn
Taurus - Virgo	Cancer - Scorpio	Libra - Aquarius
Taurus - Capricorn	Cancer - Pisces	Scorpio - Pisces

Opposition

Name: *Opposition*

Angle in Degrees: 180 plus or minus 8 degrees and six signs apart

Symbol:

The opposition is a difficult and inharmonious aspect which brings stress, opposing forces and confrontations and which present the individual with obstacles to overcome and adjustments to be made to balance the situation.

Sun Signs, Moon Signs and Ascendents that are in Opposition (Negative or Disharmonious) or to Each Other:

Aries - Libra	Cancer - Capricorn
Taurus - Scorpio	Leo - Aquarius
Gemini - Sagittarius	Virgo - Pisces

MINOR ASPECTS

SEMISEXTILE

NAME: *SEMISEXTILE*

ANGLE IN DEGREES: 30 plus or minus 3 degrees

SYMBOL:

The semisextile is similar to the sextile, but it is less powerful. It is somewhat favorable and brings opportunities, but the individual must learn to recognize these opportunities.

SEMISQUARE

NAME: *SEMISQUARE*

ANGLE IN DEGREES: 45 plus or minus 3 degrees

SYMBOL:

Less powerful than the square, the semisquare is mildly unfavorable and indicates areas producing friction and stress which require attention.

Sesquisquare

NAME: *Sesquisquare*

ANGLE IN DEGREES: 135 plus or minus 3 degrees

SYMBOL:

The sesquisquare is somewhat adverse and similar to the semisquare. It produces tension and mild stress in the life of the individual which require time and thought and effort to produce positive results.

Quincunx

NAME: *Quincunx*

ANGLE IN DEGREES: 150 plus or minus 3 degrees

SYMBOL:

The quincunx is somewhat adverse with an unpredictable influence which presents the individual with negative traits requiring growth and development or with situations requiring learning. In some situations, it relates to health problems.

Zodiacs Planets

ARIES	♈	SUN	☉
TAURUS	♉	MOON	☽
GEMINI	♊	MERCURY	☿
CANCER	♋	VENUS	♀
LEO	♌	MARS	♂
VIRGO	♍	JUPITER	♃
LIBRA	♎	SATURN	♄
SCORPIO	♏	URANUS	♅
SAGITTARIUS	♐	NEPTUNE	♆
CAPRICORN	♑	PLUTO	♇ or ⯓
AQUARIUS	♒		
PISCES	♓		

SUN

The Sun and the Moon are luminaries, but both hold a powerful influence in aspects, and are, therefore, considered along with planets in analyzing the birth chart. The only major aspect the Sun can have with Mercury of Venus is a conjunction. The Sun and Mercury are always within twenty-eight degrees of each other, and the locations of the Sun and Venus are always within forty-eight degrees.

SUN CONJUNCTIONS

SUN CONJUNCT MOON

The Sun influences the ego and personality while the Moon enhances the emotions. With this aspect, the individual possesses extremes in emotions and may be stubborn and willful, impersonal, and opinionated with a tendency to forceful and somewhat impulsive.

SUN CONJUNCT MERCURY

This aspect emphasizes the traits of forceful thinking granting mental energies and power, ambition, and creativity. The individual may be either highly introspective but with a tendency toward mental stress or a person who is opinionated and inflexible.

SUN CONJUNCT VENUS

This aspect brings optimism, cheerfulness, charm, affection, and sociable traits with a love of life, romance, and eroticism. Self-expression or talents in art and music may be enhanced. A close conjunction is indicative of egotism, selfishness and conceit.

Sun Conjunct Mars

This beneficial aspect combining the power of the Sun in conjunction with the energy of Mars is indicative of an industrious and enterprising person with a strong will power who is hard working, assertive, competitive, daring, dramatic and courageous. There may be quick, temperamental outbursts but the person is prone to forgive and forget. This person likes to lead.

Sun Conjunct Jupiter

A beneficial aspect granting a pleasant personality, generosity, success, optimism, enthusiasm, good luck, a love for life, and an indication for accomplishments and recognition when ambition is applied. A tendency toward extravagance, conceit, or laziness must be avoided.

Sun Conjunct Saturn

Saturn's influence toward self-discipline and being industrious is strengthened in order to overcome obstacles, limitations, disappointments, tensions and frustrations. This person gains through hard work and perseverance on the merits of their own efforts.

Sun Conjunct Uranus

This aspect grants strong perceptive and intuitive abilities resulting in inspiration, imagination, initiative, quick decisions, and leadership ability. This individual may be impulsive, unpredictable, or eccentric with a need for personal freedom. This person can also be stubborn, erratic, obstinate and opinionated. This aspect is associated with the qualities of genius. There is an indication for misunderstandings and problems with romantic relationships. There may also be strong psychic abilities.

Sun Conjunct Neptune

Imagination and strong creative self-expression enhances artistic abilities. Psychic abilities are indicated. The individual must guard against self-delusion, vagueness, a lack of self-confidence, and escaping into fantasies.

Sun Conjunct Pluto

A forceful conjunction, this aspect produces energy and power with strong.desires for physical, mental, and spiritual development. Strong psychic abilities are noted. A self-confident, strong character is indicated but there is a tendency to be egotistical or intolerant of weaknesses in others.

Sun Sextiles

Sun Sextile Moon

This aspect endows popularity and a happy personality with well-balanced emotions in a person who is not overly ambitious but who finds success without much effort. Harmonious relationships and a pleasant marriage are indicated.

Sun Sextile Mars

Mars brings energy, will power, determination, daring and an adventurous nature to the force of the Sun. Courage and endurance to overcome obstacles are enhanced.

Sun Sextile Jupiter

Generous, self-confident, and optimistic, this individual receives a protective influences that prevents harm and misfortune. There is a strong desire to travel and a tendency toward over indulgence.

Sun Sextile Saturn

Saturn brings self-discipline and patience to a methodical and practical person with strong concentration and organizational abilities. Efforts produce gains. This aspect is indicative of longevity and good health.

Sun Sextile Uranus

This aspect inspires imagination, inventiveness, creativity, and a cheerful and pleasant nature that attracts the attention of other people. The person may be outspoken.

Sun Sextile Neptune

Imagination, creativity, and a visionary outlook are emphasized perhaps producing a talent in writing or art. Sensitive and romantic, this person is responsive to the needs of others.

Sun Sextile Pluto

Leadership abilities, motivation, will power, resourcefulness, creativity, concentration and application are inherent in this aspect. Gains, inheritance, or legacies are indicated.

Sun Squares

Sun Square Moon

Emotional conflicts may produce obstacles to creative expressions. The conflict is generated through insecurities and problems associated with the family, associates, friends or partners. These insecurities can produce the will to endure and succeed.

Sun Square Mars

The Sun and Mars square to produce a forceful nature prone to argumentation and aggressive behavior which may result in temperamental outbursts. This is a risk taker who is outspoken but enthusiastic.

Sun Square Jupiter

An arrogant and egotistical person who can be extravagant with a "do it my way" approach who can unrealistic and overly optimistic. There is a tendency toward taking risks, speculation, looking for the quick deal, or even scheming and conning others. Pleasures and over indulgence are indicated.

Sun Square Saturn

Saturn combined with the strong influence of the Sun brings obstacles, limitations and hardships which either produces a strong person or one who is overwhelmed. Good judgment and developing self-assurance brings results, but determination and hard work are required.

Sun Square Uranus

This individual exhibits erratic tendencies, and impulsive, rash, and impractical behavior and ideas. There is a lack of consideration of consequences or the feelings of others. However, others find this individual original and magnetically appealing.

Sun Square Neptune

This aspect produces an emotional person with a lack of confidence in abilities, self-delusions and an avoidance of responsibility. This person is sensitive and caring of others but is susceptible to fraud and deceptions and being taken advantage of.

Sun Square Pluto

Pluto in this aspect brings traits that include forceful, domineering, willful, arrogant, temperamental, uncompro-mising, and rebellious with a lack of positive direction. This person can be overly aggressive in relationships.

Sun Trines

Sun Trine Moon

Harmony and well-balanced emotions, a pleasant personality, good health, and strong recuperative abilities are indicated. This person may be somewhat nonassertive, but enjoys an undemanding life, good home and support from friends and family.

Sun Trine Mars

Honor and integrity enhance a person who is enthusiastic, courageous, ambitious, decisive and adventurous. Physically strong, good health is indicated.

Sun Trine Jupiter

Similar to the influence of Mars, Jupiter trine the Sun also brings good luck, optimism, and success in endeavors. The individual enjoys happiness and positive interactions with others but must strive not to over indulge.

Sun Trine Saturn

Practical, self-disciplined, methodical and well organized traits produce a person who is conservative, finding success through efforts. This person can be frugal if necessary.

Sun Trine Uranus

This person possesses leadership ability, will power, creativity, enthusiasm and optimism. A magnetic personality, strong intuitions, and psychic abilities are indicated.

SUN TRINE NEPTUNE

Insightful, intuitive, and spiritual, this individual is a visionary who inspires others. Leadership ability and a good imagination may also be indicated.

SUN TRINE PLUTO

A motivated leader whose will power and determination produce opportunities for gain. Energy, concentration, and recuperative abilities are enhanced.

SUN OPPOSITIONS

SUN OPPOSITION MOON

Difficulties in interpersonal relationships may exist. Emotions and worries may produce nervous disorders. Other planetary aspects may weaken this opposition.

SUN OPPOSITION MARS

A combative and aggressive person who is outspoken and at times overly forceful with others. This is an energetic and physical person who may be prone to accidents.

SUN OPPOSITION JUPITER

Overly optimistic, arrogant, and egotistical, this person may produce more promises than results. There is indication for a tendency to over indulge in pleasures.

Sun Opposition Saturn

A reserved, aloof and conservative attitude restricts interpersonal relationship. There is a tendency for this person to marry later in life or not at all with the marriage being to someone of a different age bracket.

Sun Opposition Uranus

This person insists on having his or her own way and can be impulsive and rash with a tendency to be unprincipled and inconsiderate of others. Original but impractical, this person may be irritable, eccentric and nervous and accident prone. There is a personal magnetism about this individual that draws others to the individual.

Sun Opposition Neptune

This person may have good motives but low self-esteem may reduce achievements. There is a tendency toward fantasy and disceitfulness.

Sun Opposition Pluto

This aspect can influence a person to be willful, arrogant, temperamental, overly domineering, demanding, forceful and overbearing. This person builds a wall between himself and others.

Moon

Aspects to the Moon relate to emotions. These aspects may enhance and add positive traits or they may produce negative traits which the individual must deal with in order to achieve positive interactions with others.

Moon Conjuncts

Moon Conjunct Mercury

This person is imaginative, expressive, perceptive and intuitive and also sensitive to the reactions and remarks of others. A concern for family, home, and health is indicated. The intellect is enhanced along with creativity, voice and speech.

Moon Conjunct Venus

Charming, diplomatic, kind and sociable, this individual appreciates beauty and is usually successful in romance. There is indication for artistic creativity.

Moon Conjunct Mars

Temperamental, angry, and impatient, this person is easily offended but is insensitive to other people's feelings; may be overly concerned with home and family.

Moon Conjunct Jupiter

This aspect influences the individual to be generous, sympathetic, and practical with good intentions and spiritual inclinations. opportunities and success are noted but there is a restless quality that produces changeability and a desire to travel. Vanity may be a problem.

Moon Conjunct Saturn

Practical, hard workers with good common sense, this individual is also secretive, apprehensive and somewhat overly serious. A tendency to feel unappreciated by family or parents and to dwell on the past.

Moon Conjunct Uranus

Changeable, moody and unpredictable, this individual is involved in unusual experiences. This person is an independent thinker who is unconventional. Strong intuitions and a good imagination combine with a resourceful and determined nature. Family life is unusual in some way.

Moon Conjunct Neptune

This aspect influences a person to be moody, restless, and sensitive with a tendency to be idealistic, talented, sympathetic and easily influenced. Spiritualism, mystical experiences, and psychic abilities are enhanced.

Moon Conjunct Pluto

Intense, overbearing, strong willed, and overly emotional, this individual faces family problems which he or she may cause. There is a desire for love and emotional and sexual fulfillment.

KEY WORDS TO MOON ASPECTS

PLANETS CONJUNCTIONS	POSITIVE INFLUENCES	NEGATIVE
MOON/ MERCURY		
Awareness	*Retentive*	*Forgetful*
Restless	*Perceptive*	*Inconstant*
Sensitive	*Versatile*	*Nervous Tension*
MOON/ VENUS		
Epicurean	*Beautiful*	*Hedonistic*
Poised	*Graceful*	*Sybaritic*
Attractive	*Refined*	*Uncouth*
MOON/MARS		
Passionate	*Enthusiastic*	*Belligerent*
Forceful	*Vivacious*	*Sentimental*
Intense	*Zealous*	*Quarrelsome*
MOON/ JUPITER		
Expansive	*Exuberant*	*Careless*
Fertile	*Happy*	*Exaggerative*
Jovial	*Generous*	*Willful*
MOON/ SATURN		
Restrictive	*Cohesive*	*Moody*
Depressive	*Restrained*	*Clannish*
Limited	*Tenacious*	*Miserable*
MOON/ URANUS		
Electrifying	*Fascinating*	*Devious*
Impetuous	*Magnetic*	*Capricious*
Quixotic	*Unique*	*Eccentric*
MOON/ NEPTUNE		
Fanciful	*Appreciative*	*Diffuse*
Impressionable	*Perceptive*	*Fantasizes*
Receptive	*Sympathetic*	*Lazy*
MOON/ PLUTO		
Recessive	*Discreet*	*Sarcastic*
Receding	*Penetrating*	*Superficial*

Moon Sextiles

Moon Sextile Mercury

Good self-expression and communication skills combined with a careful, logical and at times shrewd ability with few emotional anxieties.

Moon Sextile Venus

Sociable and charming as well as affectionate, caring and kind, this individual finds marriage and home life rewarding. A calm optimism and cheerfulness are indicated with some artistic ability.

Moon Sextile Mars

This individual devotes his or her energies to hard work in an ambitious manner. Creative, imaginative and intensely emotional, this person is highly protective of home and family.

Moon Sextile Jupiter

Cheerful, optimistic and honest, this individual likes to help others. Good judgment and reasoning abilities are enhanced. Gains realized through associations with women.

Moon Sextile Saturn

Serious, conservative, practical, sensible, and with a strong sense of integrity, this individual strives to be thoughtful and cautious.

Moon Sextile Uranus

This aspect produces mental alertness, self-expression, and quick reactions to opportunities. An intuitive nature prone to change and advantageous actions are indicated. This aspect brings numerous female friends and psychic abilities.

Moon Sextile Neptune

The emotions are enhanced and unoriginal desires are experienced. An aspect that brings psychic sensitivities, the individual must also guard against an over active imagination. Talents are indicated and a pleasing nature.

Moon Sextile Pluto

This aspect leads an individual to establish selfcontrol over emotions and to use the ability to dream to create what they want in life. This person is intuitive and psychic but may withdraw into a reality of his or her own making.

Moon Sextile Mercury

This persons allow emotions to rule judgment and common sense. There is a tendency to be restless, overly sensitive, shrewd and critical of others. However, loyalty is indicated. A nervous nature may lead to digestive disorders and health problems.

Moon Sextile Venus

This aspect brings feelings of insecurity, moodiness, and overindulgence. This person is gullible and susceptible to being taken advantage of. Disharmony in marriage, interference by family, and unhappiness with romantic involvements are indicated.

Moon Sextile Mars

An emotional and temperamental nature prone to indiscretion is indicated. There is a tendency to be demanding, overly indulgent, and critical of others which results in bitterness and loneliness. when energies are focused, this person achieves changes and direction. Difficulties with interpersonal relationships with women are indicated.

Moon Sextile Jupiter

This square indicates obstacles brought about by an overly generous, gullible and susceptible nature. There are problems associated with laziness, apathy, morals, overindulgence, travel, and financial speculation.

Moon Sextile Saturn

This square produces a lack of self-confidence, insecurities, frustrations, moodiness and depression, pessimism, discontentment and feelings of martyrdom. This person must work hard to realize accomplishments.

Moon Sextile Uranus

Ingenuity, intellectual ability, and talent are indicated with a nature that is prone to be restless, changeable, stubborn, eccentric, and emotional. There is a tendency toward unpredictable changes, relocations, indecision, infidelity, and unusual romantic involvements. obstacles occur through sudden misfortunes, accidents, poor health, or problems with the opposite sex.

Moon Sextile Neptune

There is a tendency for escapism and to live in a fantasy world of the person's own making. There is also a tendency toward problems with romantic relationships, emotional disorders, and drug and alcohol addiction. Inner conflicts must be overcome in order to develop talents.

Moon Sextile Pluto

This individual does not want his or her personal freedoms limited and may be a loner with a tendency to be harshly outspoken and critical of others. A strong will power and emotions are indicated along with jealousy and possessiveness. Creative abilities may be used in a destructive manner.

Moon Trines

Moon Trine Mercury

This aspect influences a person to have well-balanced emotions with a pleasant nature and to be shrewd, sensible, and logical in handling business affairs. Communications and self-expression are enhanced.

Moon Trine Venus

This person is sociable, cheerful, diplomatic and tactful and interacts well with others. Good values and a refined and charming magnetism are indicated. Grace, beauty and artistic appreciation are indicated.

Moon Trine Mars

Energy produces an intensely emotional person who is ambitious and quick in actions. Self-control allows this person to be constructive and to focus energies on imaginative and creative pursuits.

Moon Trine Jupiter

This individual is pleasant, optimistic, sensible, and judicious with a strong interest in home and family. There are indications for an expansive imagination and good luck through inheritance, associations with women, or business.

MOON TRINE SATURN

Saturn trine the Moon brings a reserved dignity to a person who is serious and responsible. Good business acumen and management abilities are combined with a conservative nature. Family involvement is important and may bring gains or benefits perhaps through inheritance.

MOON TRINE URANUS

A creative imagination is combined with an unconventional nature to produce a mental alertness and desire for new and innovative ideas. An enterprising, inventive, and original outlook enhances benefits from change or travel. Psychic abilities are indicated.

MOON TRINE NEPTUNE

Creativity is enhanced by an expansive emotional nature producing a person who is genuinely warm and caring but may at times be overly idealistic. Psychic abilities are noted.

MOON TRINE PLUTO

Intense emotions are controlled to produce a strong will power, courage and determination. This person is personable but prefers serious relationships. A tendency to withdraw is indicated as are psychic abilities.

MOON OPPOSITIONS

MOON OPPOSITIONS MERCURY

An emotional, restless, and egotistical individual whose decisions and opinions are confused by feelings and self-interest. overly sensitive, clever and sarcastic, this person talks to hear his or her own voice and can't understand why others are uninterested in what he or she may have to say.

Moon Oppositions Venus

This person is gullible and lacks discretion and may suffer through romantic relationships, over eating, and over indulgence in pleasures and sexual affairs. There is an indication for domestic disharmony. This person seeks to be the center of attention desiring admiration and praise but can also be moody and changeable.

Moon Oppositions Mars

Irresponsible and indiscrete, this individual is prone to temperamental outbursts and some may become violent. This is a self-centered and demanding person who is intolerant and inconsiderate of others and who may be prone to alcoholism or drug addiction. In order to achieve success, energy must be properly focused.

Moon Oppositions Jupiter

This person is a big spender with a tendency to be extravagant, wasteful, and self-indulgent. Prone to exaggerations, this individual may promise more than is delivered. Also, there is a tendency to be gullible and taken advantage of. Loss through speculation is indicated by this aspect.

Moon Oppositions Saturn

This person is overly concerned with slights and injuries (either real or imagined) from the past, including childhood. Insecurities and a lack of self-confidence produce a person who is depressed and moody. Saturn's restrictions require hard work to realize achievement. Depression and discontentment with life and the personal situation are indicated.

Moon Oppositions Uranus

This aspect indicates a person lacking stability who is undependable and irresponsible. Changeable, moody, and irritable, this person neglects family and home. Restless and impulsive, there is an indication for unusual situations and relationships.

Moon Oppositions Neptune

Escapism from reality is indicated in a person who is prone to dreams, fantasy, or alcohol and drug addiction. Irresponsible and lazy, this person experiences disharmony in domestic affairs and complications in romantic relationships. A susceptibility to get-rich-quick schemes and questionable business dealings are indicated.

Moon Oppositions Pluto

Emotional growth is hampered by intense emotions, jealousy, possessiveness, and feelings of rejection producing problems in personal relationships with friends and family. This person may be a loner who is intolerant and fanatical and who seeks drastic actions to problems and situations. Some problems may develop through inheritance are indicated.

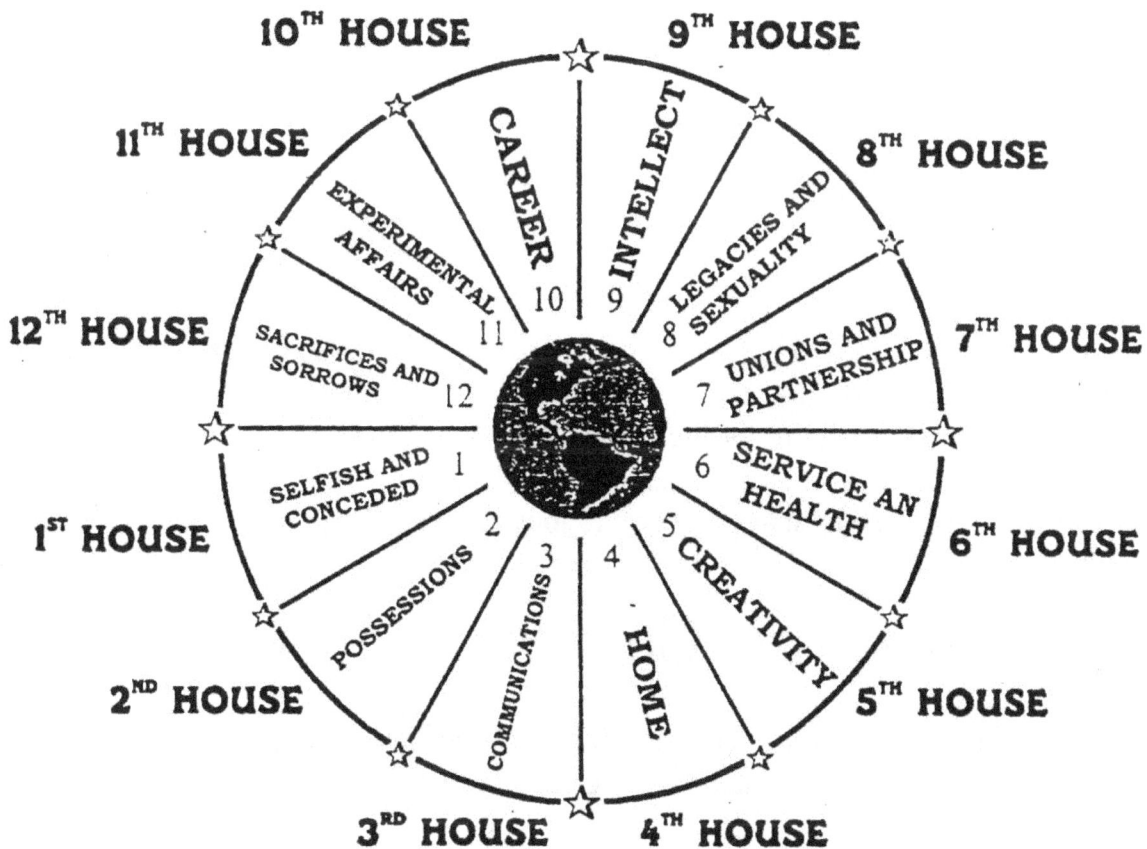

MERCURY

Aspects to Mercury produce an emphasis on mental alertness, awareness, tensions, or needed adjustments. These aspects also influence the ability or lack of ability to communicate with others.

MERCURY CONJUNCTIONS

MERCURY CONJUNCT VENUS

Artistic and expressive, this personable and charming individual possesses a love of romance and affections. Diplomacy, tact, and a concern for the feelings of others are enhanced along with the ability to translate such feelings into words and writing.

MERCURY CONJUNCT MARS

This aspects brings a desire for mental challenges which may be exhibited in a talent for debate, discussion, and well worded arguments. A restless mental energy produces a curious and impatient nature in a person who may be drawn to journalism, writing, law or politics.

MERCURY CONJUNCT JUPITER

This individual possesses an expansive intellect, curious nature, and self-confidence that produces enthusiasm and diverse interests. Personable, persuasive and inspiring, this person influences other.

MERCURY CONJUNCT SATURN

This is a well-disciplined, responsible, logical and methodical individual who speaks with an authority that others listen to. Hard working and diligent, this person is an organized planner and manager.

MERCURY CONJUNCT URANUS

This aspect grants an original and inventive creativity which may produce a genius. Quick thinking and independent, this person is also articulate and individualistic. Progressive thinking is highlighted, but a need for independence and a strong will may hamper relationships.

MERCURY CONJUNCT NEPTUNE

This aspect produces a dreamy and idealistic visionary who may at times prefer fantasy to reality. An imaginative person who can at times be deceitful and misleading in order to avoid unpleasantness. Creativity, artistic talent, and an interest in photography are indicated.

MERCURY CONJUNCT PLUTO

This individual is drawn to a mystery and loves to uncover secrets. A deep, penetrating mind and persuasive nature are indicated with a resourceful, probing and persistent nature. Research or detective work may appeal to this person. In some cases, deceit or questionable associates are indicated.

MERCURY SEXTILES

MERCURY SEXTILE VENUS

A calm, graceful person who is charming, sociable, and refined with an imaginative style. A skillful speaker and writer, this person enjoys the cooperation of others.

MERCURY SEXTILE MARS

An efficient and productive planner with a quick and practical mind. This person is courageous, perceptive and trustable with a tendency to perceive the intentions or weaknesses of others and to be a creative and convincing speaker or writer.

MERCURY SEXTILE JUPITER

This aspect brings a pleasant, easy going nature with an interest in intellectual pursuits. This person attracts good luck, opportunities and good fortune. Philosophical and optimistic, there is an enjoyment of travel, writing, and speaking.

MERCURY SEXTILE SATURN

The ability to concentrate and a good memory are highlighted in a person who is methodical, practical, logical, and resourceful. This person is responsible, dutiful, and sensible with the ability to use sound judgment in making decisions.

MERCURY SEXTILE URANUS

Talent and an unique originality are combined to produce an independent and brilliant thinker who possesses a dramatic flair and presentation abilities. Intuitive and enterprising, this person is progressive and innovative. There is an intolerance for ignorance in a person who is perceptive and who recognizes and sets trends with their brilliant insights.

Mercury Sextile Neptune

There is a tendency toward being secretive and intuitive with a flair for acting, writing, photography, and persuasive speech. Idealistic and imaginative, this person likes new ideas and innovative methods.

Mercury Sextile Pluto

Strong will power and effective, influential presentations and expressions are indicated in a person who is logical, analytical, diplomatic, and courageous. The ability to concentrate and comprehend enhances the intellect and mental processes.

Mercury Squares

Mercury Square Venus

The only major aspects that can exist between Mercury and Venus are the conjunction and the sextile.

Mercury Square Mars

A curious and active intellect in a person who is critical, opinionated and argumentative. A restless, impulsive, and distrustful nature leads this person to jump to conclusions.

Mercury Square Jupiter

A curious and impulsive individual who has good intentions but is overly optimistic, too eager to succeed, and attempts what is beyond capabilities. This person often looks for the easiest path or the route which offers the least resistance while bluffing or making promises that can't be kept. Confidences are broken.

Mercury Square Saturn

An ambitious, reserved and conservative individual who is hard working and responsible. Insecurities and inhibitions produce opinionated and intolerant tendencies in a person who can be defensive, suspicious, pessimistic and lacking imagination. In some individuals this aspect produces dishonesty and disceitfulness.

Mercury Square Uranus

An alert mind and ingenious intellect in a person who is eccentric and who defies authority in a rebellious manner. Impulsive and unpredictable, this person strives to assert his or her uniqueness and individualism often not heeding the advice of others. A nervous nature is indicated in a person who is arrogant, daring but discontent.

Mercury Square Neptune

Communication is hampered by an absent-minded nature and disorganized thought process in a person who fears competition. This person is creatively imaginative and insightful, but possesses unrealistic ideals in love and romance. Indiscretions and an inability to keep secrets are found in a person who is either well meaning but gullible or crafty and cunning.

Mercury Square Pluto

This aspect produces a strong will power in a person who can be diplomatic and tactful or brutally outspoken and blunt in their efforts to persuade and influence others. This person's moods swing from optimistic to pessimistic and from cheerfulness to complaining and fault finding. There is a tendency to take risks.

Mercury Trines

Mercury Trine Mars

Leadership abilities are enhanced by mental energy, concentration and alertness in a person who sizes up others easily. This is a practical, creative and perceptive individual who communications are delivered with a flair.

Mercury Trine Jupiter

A quick, alert mind is indicated in an easy going individual who is pleasant and sociable. Optimistic and fair with others, this person communicates well. Combined with other aspects in the chart, however, this one can produce laziness. There is interest in philosophy, literature, and travel.

Mercury Trine Saturn

A practical, logical, and disciplined mind combine with a good memory for facts and powers of concentration to produce good judgments and responsible behavior. This is a loyal and trustworthy friend or associate. Organization and manual dexterity result in diverse skills.

Mercury Trine Uranus

This is a progressive thinker who likes new ideas and methods. This person forms his or her own opinions and may intuitively and instinctively sense upcoming trends. A unique and creative individual who is original and talented.

Mercury Trine Neptune

Persuasiveness and talent marks the traits of this aspect. Intuition is enhanced in a logical and analytical intellect which can produce spiritual insights. A gift for words or artist ability are indicated.

Mercury Trine Pluto

An analytical intellect capable of grasping abstract theories is combined with a diplomatic and pleasant nature. Courageous and influential, this person has high expectations of themselves and others.

Mercury Oppositions

Mercury Opposition Mars

Argumentative and intolerant of the opinions of others, this person can be impulsive, critical and fault finding with a tendency to jump to conclusions. Too much mental energy leads to nervousness and may produce headaches.

Mercury Opposition Jupiter

An expansive nature leads this individual to make big plans, but he or she lacks the ability to produce the necessary results. Impulsive and impractical decisions and a tendency to seek the easy way are indicated.

Mercury Opposition Saturn

This aspect influences a conservative, reserved person who is prone to be pessimistic, overly critical, opinionated, and fault finding. A narrow-minded outlook in a distrustful, suspicious and defensive individual.

Mercury Opposition Uranus

This person is impulsive and changeable but also possesses strong preferences, opinions, and likes or dislikes. No amount of persuasion changes this person's mind. Blunt and outspoken, this person can be arrogant, conceited and rebellious.

Mercury Opposition Neptune

Insightful, creative and imaginative this person can also be unrealistic or overly idealistic. There is a tendency to be cunning, deceitful and scheming and easily distracted. Artistic talent may be present.

Mercury Opposition Pluto

There is an element of risk inherent in this aspect and the individual may be involved in secretive or dangerous endeavors which produces mental tension and stress. This person is tactful and diplomatic or sharply outspoken depending upon the situation. unnecessary risks and activities can lead to death.

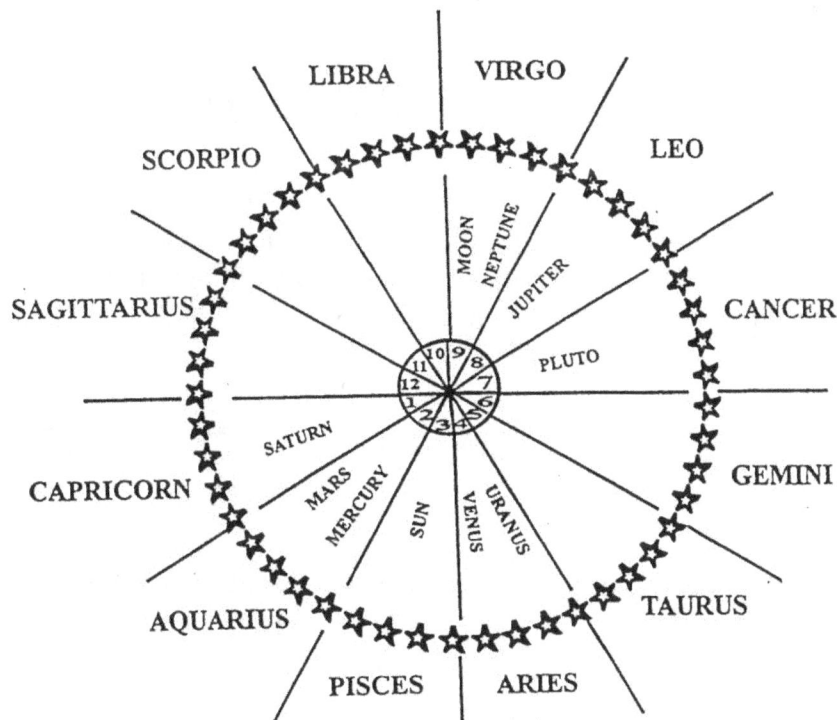

Venus

Aspects to Venus have an effect on the affections, love life, sensuality, social skills and values of the individual. The awareness that the person possesses of his or her own feelings is enhanced.

Venus Conjuncts

Venus Conjunct Mars

Mars brings energy to a charming, sensitive and sensuous nature bestowing passion, sexual desires and artistic self-expressions which is manifested in either a generous and loving concern for others or in sexual conquests. Romance is important to this person who easily becomes unhappy when demands are not met.

Venus Conjunct Jupiter

This aspect confers a generous and charming nature in a person who likes to make others happy. A refined and spiritual person who is honest, diplomatic and prefers to be a good role model. There are times when this person is easily taken advantage of and other times when life is easy and fortunate and the person becomes lazy, unappreciative, and pleasure seeking.

Venus Conjunct Saturn

Saturn adds a practical and sensible nature to the Venus generosity, sense of justice, harmony, and fairness. Creative talents are enhanced. This person finds happiness in being dutiful, responsible, and self-disciplined. A faithful, serious, hard working person who may prefer the security of love with an older person. When not appreciated or love is not reciprocated, this person can become depressed or moody.

VENUS CONJUNCT URANUS

Sudden attractions and affairs and changes in relationships and affections are indicated. Friendship and love mingle and this person may be in love with more than one person at a time or involved in unusual relationships. A personable and unique individual who is determined and passionate with a intellectual approach to life.

VENUS CONJUNCT NEPTUNE

An emotional and temperamental dreamer who can be impractical in romance or a person who inspires spiritual purity in love. This aspect may produce a love of beauty, artistic expression and music. In some it indicates an unreliable, deceitful manipulator who uses other people.

VENUS CONJUNCT PLUTO

A deeply emotional person who may fall in love at first sight. Passionate and possessive, they desire to give completely of themselves in love and may be disappointment when their intensity and total commitment is not reciprocated. This aspect intensifies talents in a person who is daring and willing to fight injustices. There is indication for either enlightenment or perversion.

VENUS SEXTILES

VENUS SEXTILE MARS

This person is generous, pleasant, energetic and seeks harmony in relationships. An affectionate, congenial and sociable nature is indicated in a person who is independent and fun loving. Good fortune is indicated for some but others live an easy going life with little ambition and fail to develop talents or abilities.

Venus Sextile Jupiter

Naturally popular and well liked, good fortune and a comfortable life are indicated for this person. This is a personable, graceful and refined person who is charitable, sympathetic, diplomatic, and outgoing. Partnerships are beneficial. Aspirations and goals are attainable if the individual doesn't become lazy, indolent, and extravagant.

Venus Sextile Saturn

Good judgment and economical sensibilities are combined with an artistic appreciation or talent. This person succeeds based on the merit of hard work and diligent effort, but gains are indicated through partnerships. Honest, serious and faithful, this person is dependable, reliable, fair, and dutiful.

Venus Sextile Uranus

Popular and surrounded by friends, this person possesses a magnetic personality and may suddenly fall in and out of love or marry just as suddenly. Marriage may or may not last for this unconventional person but opportunities come through friends.

Venus Sextile Neptune

A creative imagination and sensitivities combine to produce a person who is refined and appreciative of beauty in all forms. A romantic spiritualism draws this person to be sympathetic to the needs of others and capable of selfsacrifice if necessary. Laziness can lead this person to be pleasure seeking and dependent on others.

Venus Sextile Pluto

This aspect may join a person with their true soul mate. A creative ingenuity combines with leadership abilities in a person who is intuitive, honest and dedicated. You love to solve problems and mysteries and may be a bit secretive in your affairs.

Venus Squares

Venus Square Mars

Egotistical, impulsive and passionate, romantic affairs may produce problems in a person who seeks sexual gratification or is used for that reason. This person can be critical of others but is easily offended if others don't respond to his or her romantic overtures. This aspect, however, is indicative of the great romantic lover. In some, it signifies an over indulgence and preoccupation with romance and sex.

Venus Square Jupiter

A vain and overly emotional person who conceitedly uses their numerous love affairs to prove their self-worth. This person can be lazy, overindulgent, deceitful, and morally corrupt. A user who exaggerates, is overly optimistic, and resents being used by others. A hanger-on who is around during the good times but disappears during bad times. Other aspects in the chart may help this person to realize and accept better values in life.

Venus Square Saturn

This person faces obstacles and harsh realities and often experiences a loss love or financial burdens. This person can be discontent, easily depressed, and willing to place responsibilities and duties above personal happiness. A reserved, aloof or shy person who easily feels rejected or unaccepted. Relationships are often hampered by an inability to exprbss feelings and affection.

Venus Square Uranus

Self-centered and egotistical, this person uses personal charm and magnetism to win over others and to get his or her own way. A fickled person who falls in and out of love, is unwilling to make any personal sacrifices or to be inconvenienced in any way. Unpredictable in love, this person marries and divorces suddenly. Unconventional and changeable, this person seeks personal freedom over true love and responsibilities.

Venus Square Neptune

This aspect confers an artistic but overly emotional nature that depends heavily on intuition over intellect. There is a tendency to be easily taken advantage of or used physically or financially and then to blame the other person. This is an escapist who may be susceptible to fantasy, dreams, drugs and alcohol, sexual promiscuity, perversion, or deceit in romantic affairs. Other well aspected planets aid this person in focusing and redirecting sexual energy into positive and creative sources.

Venus Square Pluto

There is a tendency for strong sexual passions to be a dominate force in this person's life leading to numerous affairs and infidelities. This person is prone to marry for money rather than love. This is an uncompromising and demanding person who seeks perfection and an ideal partner, but uses that as an excuse to be unfaithful.

Venus Trines

Venus Trine Mars

Energy and passions are combined in happy, harmonious and lasting relationships and marriage. Affectionate, congenial, fun loving and faithful, this person enjoys a favorable family life and positive aspirations. Because life is easy, energies must be focused on ambitions in order to achieve success.

Venus Trine Jupiter

A tolerant, accepting, and fair person with a quick mind and curious intellect, this person knows how to be diplomatic and charming and to put others at ease. Successful partnerships and luck in love brings numerous romantic affairs. This is a refined, charitable and sympathetic person who likes people. Again, personal traits and opportunities must be utilized in order to overcome any tendencies to take life easy.

Venus Trine Saturn

A hard working and industrious person who is diligent in efforts and faithful in love. A good judge of character combined with good business sense helps this person to gain through partnerships, business and personal efforts and merits. A serious and responsible person who is a reliable and dependable marriage partner.

Venus Trine Uranus

An outgoing, fun loving person who attracts romance and love. This person experiences sudden relationships and sudden good fortune in finances as well. Expect the out of the ordinary and the unconventional with this person.

Venus Trine Neptune

This aspect brings creativity and imagination to life and love. There is a refined appreciation for beauty, art and music. Intuition and perceptions produce a romantic at heart who may be spiritual or sensitive to the feelings of others. This person is a sympathetic and sincere listener who anticipates the desires and wishes of others.

Venus Trine Pluto

Intense emotions, passions and a strong sexual drive produce a romantic nature in a person who may fall in love at first sight. Creative, ingenious, dedicated, and honest, this person wants to give fully and completely in love. Fairness and harmony produce a tranquility in the temperament.

Venus Oppositions

Venus Opposition Mars

A well-developed sexual awareness in a sensitive person who is easily offended by the unfairness of others. An impulsive person who may encounter some emotional problems in love and romance. A romantic lover, but one who is often dissatisfied with involvements.

Venus Opposition Jupiter

Sweet, overly optimistic people with spiritual inclinations who may experience difficulties in love and marriage. Vanity, conceit and hypocrisy in a person who may use others financially and romantically.

Venus Opposition Saturn

Limitations and restrictions bring financial hardships and difficulties in love and marriage. Frustrations, tensions and stress add to the obstacles and often result in moodiness or depression. Mental and physical exhaustion can develop. A reserved and serious person who may fear rejection and is overly sensitive.

Venus Opposition Uranus

A charming personality but a person who is selfcentered and has emotional problems. This person can be demanding and willful while also being changeable and impulsive. An unconventional lifestyle and sudden changes in affections lead to speedy marriage and divorce. The rebel instinct brings an inconsideration for the feelings of others and no regard for consequences.

Venus Opposition Neptune

Extremes in emotions and ideals are indicated in a person prone to sexual scandals and secret affairs. Trust is given to the wrong people with the individual often being taken advantage of in love and finances. Escapist tendencies may lead to alcohol or drug abuse. This person blames others for their problems.

Venus Opposition Pluto

This person is over sexed and prone to sexual addiction and numerous affairs. Promiscuity and questionable associations can lead to a life centered on sex perhaps for the purpose of financial gains. Depression and emotional problems can lead to thoughts of suicide. This person is never satisfied and is always looking for the perfect partner but is frequently disappointed.

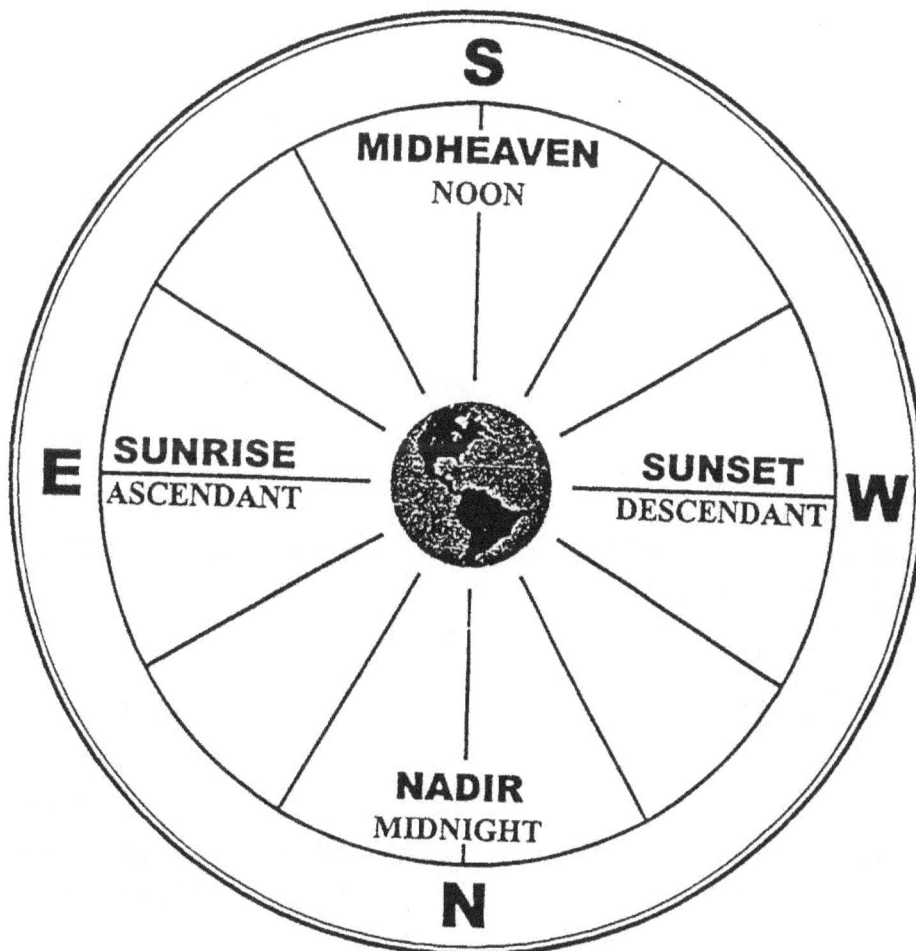

MARS

Mars is indicative of energy, action and aggressive tendencies. This planet represents the sexual drive and the focus of energy on motivations and expressions. It enhances the competitive urge, the ability to turn ideas into action, and tensions and stresses.

MARS CONJUNCTS

MARS CONJUNCT JUPITER

This aspect combines raw energy with the enthusiasm, optimism, and good luck of a confident, decisive, and generous person. Frank and outspoken, this is a fearless person who pushes right through or around obstacles in order to get where he or she is going. This is an idea person with a good mind who believes in their plans and ambitions.

MARS CONJUNCT SATURN

A well-organized and methodical person who applies logic and caution to decisions and actions. This is a restlessly energetic person who attempts to control his or her drives. Hard work and diligent effort are combined with courage. When poorly aspected with other planets this aspect can produce criminal tendencies, anger and violence.

MARS CONJUNCT URANUS

A courageous, energetic rebel who seeks change and revolutions. A strong will is found in a person who lives by their own rules. A restless, unpredictable nature in a person who is daring and fearless, but accident prone. This person is strongly opinionated and not easily persuaded.

Mars Conjunct Neptune

A highly energetic and emotional person who relies on intuition and perceptions. Psychic and healing abilities are indicated. This is an imaginative and enthusiastic individual who sometimes sets unrealistic goals. Care must be taken against a susceptibility to alcohol and drugs, reactions to drugs, poisoning, and contagious diseases.

Mars Conjunct Pluto

Energy is focused through will power and courage in an emotional intense person. This person is unafraid of hard work, possesses regenerative abilities, and is capable of enduring hardships or obstacles but can become obsessive. This person is fearless of death or danger and can be positive, daring and courageous or in some cases the individual is cruel and sadistic.

Mars Sextiles

Mars Sextile Jupiter

This person possesses an undefeatable enthusiasm and optimism. Expansive ideas add to direction, self-confidence and leadership abilities. This is an industrious person who strives to help others. A positive attitude results in a pleasant disposition and a person who is content with their accomplishments.

Mars Sextile Saturn

Energy combined with determination and self-discipline produces a person who is practical, efficient, hard working and ambitious. This person faces obstacles and hardships with strong endurance and fortitude and is responsible and dutiful to home and family.

Mars Sextile Uranus

An ambitious, self-directed person who displays creativity, originality, and an inventive streak. This is a forceful and dynamic person who is known for courage, action and a strong will power. Personal drive is focused on accomplishments.

Mars Sextile Neptune

An imaginative and artistic person who is intuitive, perceptive, and sympathetic to the needs of others. This person isn't easily fooled but often sees the best in other people. This person applies self-control over emotions and learns to focus energy well.

Mars Sextile Pluto

Energy combined with strong will power and courage enable this person to transform emotions into action in a positive manner. This aspect produces a person who is a self-confident leader who strives to fight wrongs in society and to be of service to others.

Mars Squares

Mars Square Jupiter

An undisciplined and impulsive person who lacks direction and whose energies are misdirected. Restlessness and impatience are displayed when results aren't forthcoming. In some people, this aspect represents a destructive tendency in a person who uses social issues to promote violence.

Mars Square Saturn

This aspect can produce a passive-aggressive nature in a person who can be selfish, self-centered, unsympathetic, callous, and egotistical. Obstacles are manifested in physical hardships, dangerous situations, violence, and temperamental and angry outbursts.

Mars Square Uranus

A willful person who makes his or her own rules and who is reckless, impulsive, and rash. This is a risk taker who makes sudden decisions and is unafraid to accept a dare. This person is prone to accidents, violence and over work. For some, there is an indication of accidental death.

Mars Square Neptune

An over active imagination and creativity leads this person into questionable or dangerous activities and situations. This is a casual person who is susceptible to deceit, drugs, alcohol, sexual abuse, food poisoning and infection. When energy is redirected, creativity brings success when focused in the proper direction.

Mars Square Pluto

When well aspected, this square is indicative of courageous feats. otherwise, it is indicative of an aggressive person who uses forceful tendencies to get what they want. Anger can lead to violence. There is an inherent interest in dangerous situations and in some this leads to a violent death. There is also a strong sexual drive with this aspect.

Mars Trines

Mars Trine Jupiter

An idealist whose optimism and energetic enthusiasm is propelled into constructive action. This person is confident and self-assured with an expansive, broad-minded outlook and loyalty to others. This aspect may lead to active spiritual beliefs.

Mars Trine Saturn

Neither responsibilities, hardships, nor danger deter this person from their ambitions and goals. An organized, methodical, and self-disciplined person who directs his or her energy and strong will power into overcoming any and all obstacles. This person is shrewd, skillful and daring when it is necessary.

Mars Trine Uranus

A unique individual with an energetic drive who seeks new methods and innovative ideas. This person is original and inventive offering creative and resourceful ideas. Freedom of movement, action, and ideas is important.

Mars Trine Neptune

Energy and drive are focused through a creative imagination in a person who is intuitive, perceptive, and sensitive. They instinctively sense the feelings and motivations of others and anticipate any dangerous situations. Psychic abilities are enhanced.

Mars Trine Pluto

This aspect brings a will power that is undeterred by obstacles. Emotionally intense and physically active, this person will fight for a cause. Leadership, courage, and self-confidence combine to produce a dynamic person.

Mars Oppositions

Mars Opposition Jupiter

Impulsive and impatient, this person is a self-promoter whose energies can be distracted in many directions. A mercenary or self-serving person who uses force to get his or her ideas across.

Mars Opposition Saturn

Tensions mount causing frustration and stress in a person who may resort to overly aggressive action, forcefulness, or violence. A self-centered and selfish person whose ideas may be sabotaged by others. Criminal activity is indicated in some.

Mars Opposition Uranus

An impulsive and changeable person with little concern for the consequences of his or her actions. There may be good intentions, but this person is on the fast track and recognizes no limitations. With this aspect there is an indication for a reckless risk taker who is prone to dangerous situations, numerous affairs, afflictions through friends, accidents, and a violent death.

Mars Opposition Neptune

A casual and apathetic person whose creativity and imagination are wasted on impractical ideas. This person can be well meaning but is often unreliable. Questionable activities can lead to problems. This person is susceptible to deceit, drugs, alcohol, poisons, and infections.

Mars Opposition Pluto

A forceful, aggressive, and selfish person who is dogmatic and driven to push through his personal ideas. This person can be verbally abusive and anger results in temperamental outbursts that can lead to violence. There is indication with this aspect for a violent death. The sexual drive is also strong.

THE PLANETS ARE THE DAUGHTERS OF THE SUN BY SCIENCE, PAIN, LOVE, OR BY DEATH

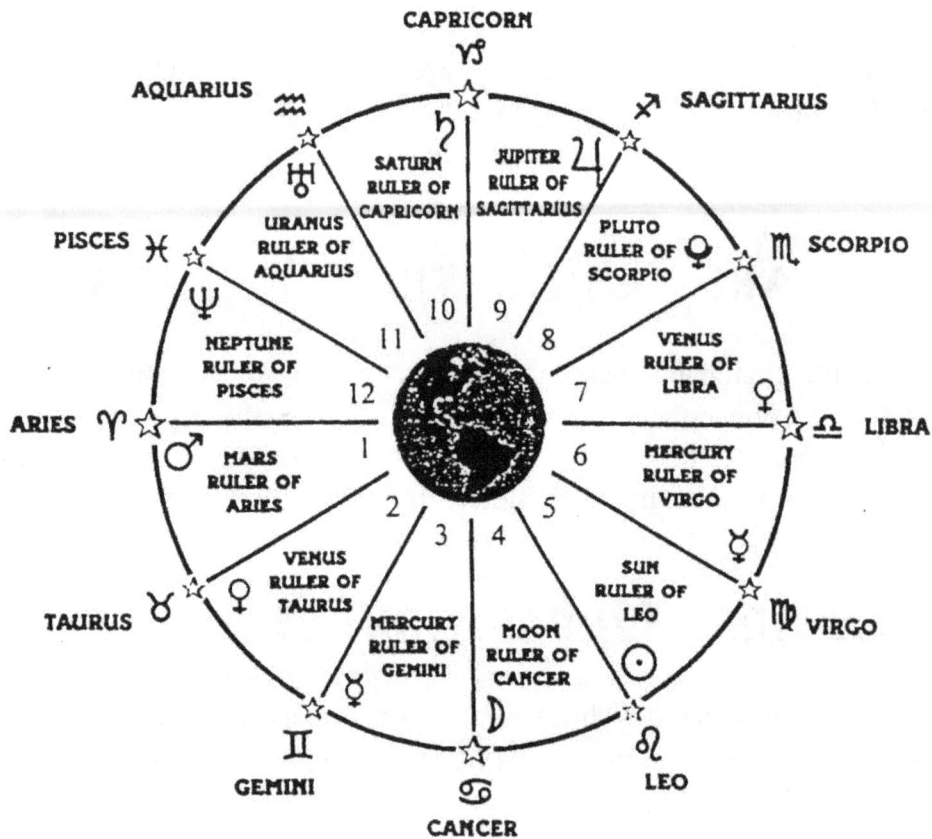

JUPITER

Jupiter is the planet of expansion, good luck, philosophy, generosity and protection. It is indicative of that area of a person's life where good things come easily and with little effort. Optimism and enthusiasm enhance the aspects of Jupiter.

JUPITER CONJUNCTS

JUPITER CONJUNCT SATURN

This aspect brings good fortune through a conservative lifestyle and orthodox business practices. This person is a diligent hard worker who is ambitious, responsible, and capable of accepting burdens. While life can bring struggles, this person feels the protection of Jupiter when times get rough.

JUPITER CONJUNCT URANUS

An innovative and progressive philosopher who experiences sudden benefits and opportunities. This is a restless person who dislikes limitations and restrictions to his or her expansive nature. A person who is generous to friends and lucky in an unconventional way.

JUPITER CONJUNCT NEPTUNE

This aspect confers an expansive imagination and creative talents. This is a sympathetic and generous person who is capable of great humility. Intuition and sensitivities are focused on opportunities, but this person must strive to be practical.

JUPITER CONJUNCT PLUTO

Strong ideals are combined with determination, dedication, will power and drive in a person who seeks to achieve goals for himself and others. Leadership abilities are enhanced as are the ability to influence others.

JUPITER SEXTILES

JUPITER SEXTILE SATURN

A fortunate aspect indicative of a person who sets practical and constructive goals and then applies his or her efforts to achieve them. Optimism and enthusiasm are combined with diligent effort, self-discipline, and a methodical, well organized approach. Saturn offers restrictions but Jupiter provides optimism.

JUPITER SEXTILE URANUS

This person discovers opportunities for creative and inventive abilities and applies enthusiasm and optimism to endeavors in order to achieve success. This is a kind and friendly person who finds protection in times of need and sudden good luck when least expected.

JUPITER SEXTILE NEPTUNE

An emotional and sensitive person with an expansive imagination and a desire to be of service to others. This person strives to make others happy through self-sacrifice if necessary. Spirituality is enhanced as is psychic abilities and awareness of the mystical.

Jupiter Sextile Pluto

Pluto adds insights to Jupiterts philosophical nature. This is an enthusiastic person who is determined and ambitious, but who likes to help others as well. Wisdom is sought through enlightenment.

Jupiter Squares

Jupiter Square Saturn

This person fails to recognize opportunities or is too cautious to act on them. In their determined efforts to reach their goals, they may also not recognize their own limitations. Obstacles and unfortunate circumstances occur in financial matters, and success or recognition for hard work may not be realized until later in life. Natural disasters such as hurricanes, floods, earthquakes, tornadoes, and drought seem to follow this person.

Jupiter Square Uranus

This person seeks change but is often in too much of a hurry to be effective. This is an enthusiastic but restless person who is independent, outspoken and opinionated. Some are impulsive adventure seekers who desire to travel and experience new surroundings. Unlucky in speculation.

Jupiter Square Neptune

A persuasive speaker who is creative and talented, but who may be unable to produce effective results. This is an emotional person who tends to exaggerate and can be extravagant and pleasure seeking. Ideals are formulated in a dream world that can result in harsh realities and unfortunate circumstances.

Jupiter Square Pluto

This person has difficulty conforming to what is perceived as unfair societal practices and seeks change and improvement, but often meets with disappointment and harsh consequences. A daring risk taker who seeks adventure through travel, gambling or speculation. This person vacillates between enthusiastic self-confidence and an emotional self doubt.

Jupiter Trines

Jupiter Trine Saturn

Realistic and attainable goals are set by a determined and constructive person who possesses financial, business and management abilities. A traditional person who accepts family responsibilities with enthusiasm. This is a dignified person known for his or her honesty and integrity.

Jupiter Trine Uranus

This aspect can produce a creative genius who jumps over obstacles or limitations of any kind. This is an unconventional person who seizes opportunities in an effort to accomplish goals. Sudden luck and protection are found.

Jupiter Trine Neptune

Spiritualism and psychic abilities are enhanced as are intuition and sensitivities allowing the person to experience heightened awareness. A creative person who uses their talents in a productive and useful manner.

Jupiter Trine Pluto

This aspect confers great powers of concentration which may be focused on creative endeavors. An enthusiastic determination and will power are applied to ambitions and endeavors. opportunities are realized.

Jupiter Oppositions

Jupiter Opposition Saturn

This person may lack ambition and drive or may be overwhelmed by problems and obstacles. There is an indication for legal difficulties, lawsuits, and trouble through foreign travel with this aspect. There is a tendency to be discontent and unhappy.

Jupiter Opposition Uranus

Good intentions are distracted by a restless, impulsive nature prone to sudden and abrupt changes. An enthusiastic and independent person who can also be extravagant and wasteful. This person looks for the easy and fast track to success. A poor aspect for speculation and gambling.

Jupiter Opposition Neptune

An impractical dreamer who fails to focus creativity and energies in a productive manner. Exaggerated promises cannot be fulfilled. There is an indication for legal difficulties and financial problems through speculation or carelessness. Emotional and sensitive, this person's over active imagination leads to misgivings.

Jupiter Opposition Pluto

A lack of humility and a determination to impose their philosophical views on others results in conflict. This is a person prone to exaggeration and arrogant, dogmatic ways. Idealistic beliefs and self-doubt confuse this individual who fails to learn from experience.

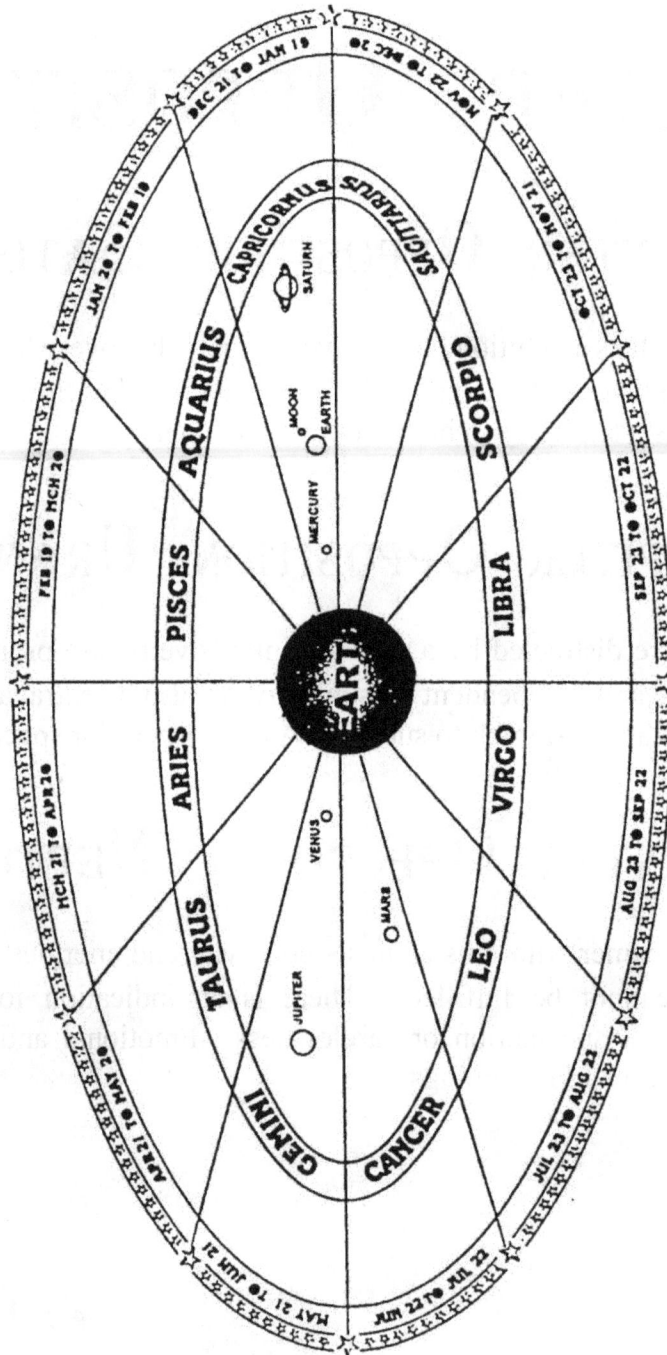

Saturn

Saturn teaches the necessary lessons in life through restrictions and limitations. Saturn enhances the ability to develop self-discipline, responsibility, and determination. Hard work and diligent efforts produce results against difficulties and obstacles.

Saturn Conjuncts

Saturn Conjunct Uranus

This aspect brings self-discipline and direction to the creativity of Uranus producing original and innovative thinking and the possibility for genius. This person is a serious and practical student with a powerful intellect who learns to apply diligent efforts and persistence to problems and difficulties. This person knows how to appear conventional and traditional but quite often is an unconventional thinker who produces innovative ideas and initiates new concepts.

Saturn Conjunct Neptune

Organizational skill and a methodical approach are applied to creative talents. A sensible, practical person who possesses an aptitude for business and finance. The intuition and perceptions are focused on direction and ambitions. This aspect can produce an inspiring leader or a person involved in secretive endeavors. The magical quality of Neptune is enhanced in a person who is not easily fooled by others. This person may be distrustful and questioning and when poorly afflicted by other aspects this can lead to depression and moodiness.

SATURN CONJUNCT PLUTO

A capable person ambitious for power, prestige and position. This aspect endows a person with strong powers of self-control and self-discipline. A person who is persevering and able to endure hardships in order to achieve goals. Highly intuitive, this individual perceives the weaknesses of others and can use that knowledge for his or her own purposes, either for good or bad, and can be secretive about intentions.

SATURN SEXTILES

SATURN SEXTILE URANUS

A responsible, dependable and dutiful person who is loyal to friends. This aspect endows the initiative to use original and innovative ideas in practical endeavors. This is an independent, determined person who is willing to work hard for what he or she believes in. Personal freedom is very important to this person who seeks to acquire knowledge and put it to use in a sensible manner.

SATURN SEXTILE NEPTUNE

In this aspect discipline enhances the insights and intuitions of the individual allowing the person to apply a methodical and well organized approach to inspirational and imaginative endeavors. This ambitious person likes to succeed at endeavors while being responsible and accepting obligations and duties.

SATURN SEXTILE PLUTO

This aspect brings wisdom to power, the influence and strength of which is dependent on the other indicators in the birth chart. The ability to concentrate and to focus on realistic goals is indicated along with the ability to inspire others.

Saturn Squares

Saturn Square Uranus

An individualistic approach is enhanced by this aspect in a person who may suddenly take drastic actions. This person can be indecisive on some matters but is strongly opinionated on others, to the point of being radical. There is a resentment of others who don't agree with him or her.

Saturn Square Neptune

This aspect brings a fear of failure and rejection in a person who is capable and talented. Abilities are either put to use in an ambitious manner who not developed by a person who is pleasure seeking. Fears and phobias have a tendency to unbalance the emotions. This person is susceptible to being taken advantage of.

Saturn Square Pluto

Insecurities can lead this person to be overly controlling and competitive. Impulsive and at times unpredictable, this person's fear of failure drive him or her to strive for attainment, sometimes seeking the fast road to success. This aspect brings burdens, obstacles and disappointments to a person who must learn patience and endurance.

Saturn Trines

Saturn Trine Uranus

A practical person with a strong will power and determination. Personal freedoms are important to this person who applies discipline to furthering ideals. A sensible leader who desires knowledge and applies innovative ideas in a methodical manner.

Saturn Trine Neptune

Insight and intuition are focused on a strong perceptive foresight and ability to plan and organize. This person may keep his or her plans to themselves, in a secretive manner, until they're ready to divulge their intentions. This aspect may produce a person adept at analyzing or one capable of working with confidential information.

Saturn Trine Pluto

This person has few fears and strives to succeed as if on a mission in life. Energetic and tireless, this person endures and focuses an intense concentration on realistic pursuits. The ability to inspire enthusiasm in others is found in this aspect.

Saturn Oppositions

Saturn Opposition Uranus

A unique nature that draws the attentions of others. A hard working but unpredictable person who may plan one thing but do another in the face of sudden obstacles. May be indecisive or strongly opinionated on some subjects. Radical decisions bring unexpected results.

Saturn Opposition Neptune

This person may feel the world is against them producing a person who is distrustful of others. This aspect may lead a person to decide against seeking a partner or spouse. A talented person who may strive to succeed or may not be at all ambitious. Emotions bring self-made obstacles that can lead to psychological or nervous disorders.

Saturn Opposition Pluto

An emotionally intense person whose ambitions can lead them to be manipulating, controlling and seeking. Their fears lead them to be ambitious of fast-producing results. Highly competitive, this person lacks patience and tact in their relationships with others.

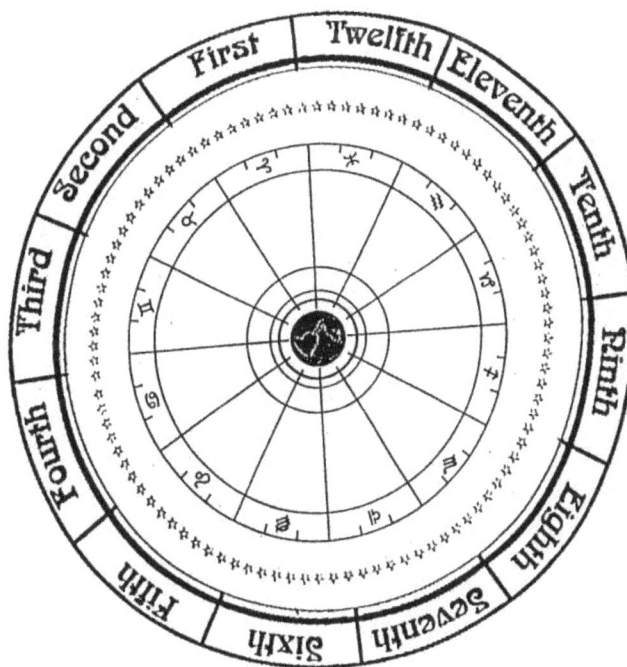

URANUS

Uranus is the planet of originality, freedom, individualism, and sudden or unexpected events and opportunities. Inventive capabilities and genius are enhanced by this aspect.

URANUS CONJUNCTS

URANUS CONJUNCT NEPTUNE

This aspect occurs every 171 years and is considered to have more of a generational influence on people. It enhances creative, original, and innovative ideas and influences the abilities to be visionary and prophetic. Freedoms and personal liberties are important.

URANUS CONJUNCT PLUTO

This conjunction occurs every 115 years and is considered to have a generational effect. Humanitarism is enhanced, however, power can be focused on either positive or negative pursuits.

Uranus

Sextiles/Trines

Uranus Sextile/Trine Neptune

This is usually not an individually prominent aspect. The influences are generational in nature. Artistic abilities, emotions, and ideals may be enhanced.

Uranus Sextile/Trine Pluto

Sudden intuitions or inspirations in an idealistic nature are combined with intellectual abilities. Psychic abilities may be enhanced.

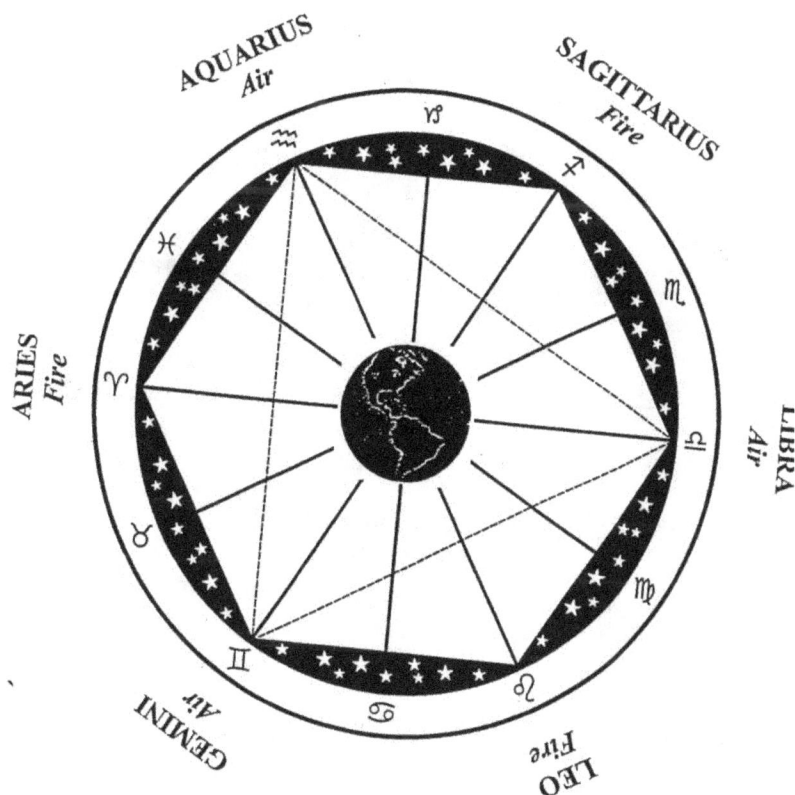

Uranus

Squares/Oppositions

Uranus Square/Opposition Neptune

This aspect has a generational affect and is generally not of importance in an individual chart unless these two planets are strong in the chart. This person can be highstrung, nervous, sensitive and opinionated and either an idealist or indecisive.

Uranus Square/Opposition Pluto

This aspect is influential in an individuals chart only if Uranus and Pluto are prominent. This makes for a discontented person who can be thoughtless and argumentative.

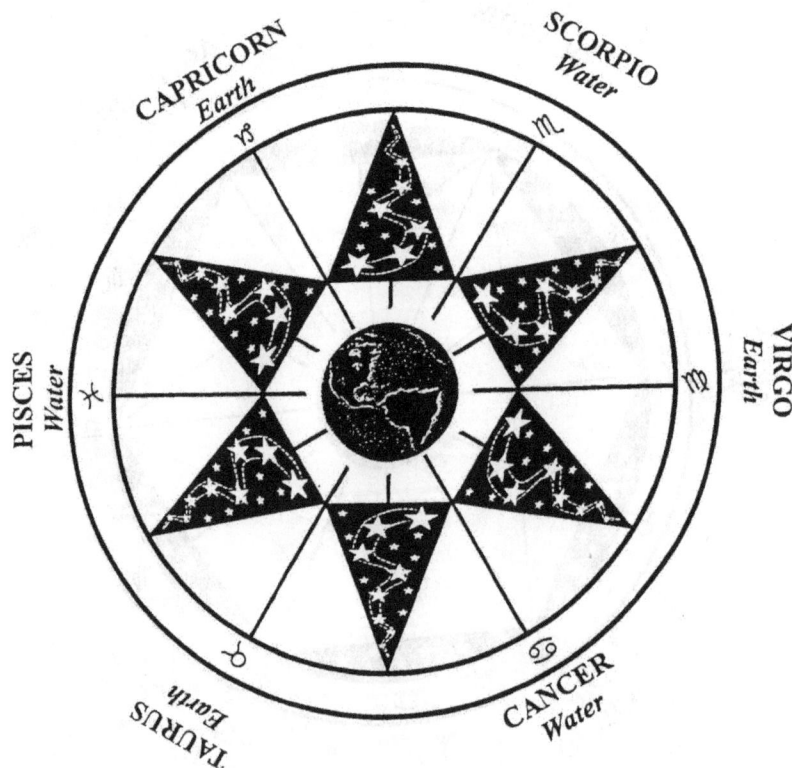

NEPTUNE

Neptune's influence is emotional, sensitive, imaginative and intuitive. Psychic abilities are enhanced.

NEPTUNE CONJUNCT

NEPTUNE CONJUNCT PLUTO

This aspect has a generational influence that appearsin a subtle manner that enhances idealism and sensitivities.

NEPTUNE SEXTILE

NEPTUNE SEXTILE PLUTO

A generational influence that only affects individual charts if Neptune or Pluto are strong in the chart. It enhances psychic abilities.

Neptune Trine

Neptune Trine Pluto

This aspect is not prominent in individual charts.

Neptune Square/Opposition

Neptune Square/Opposition Pluto

This aspect produces a generational influence.

Pluto

Pluto is a powerful planet influencing powers of both enlightenment and destruction. It brings regenerative forces of transformation and elimination.

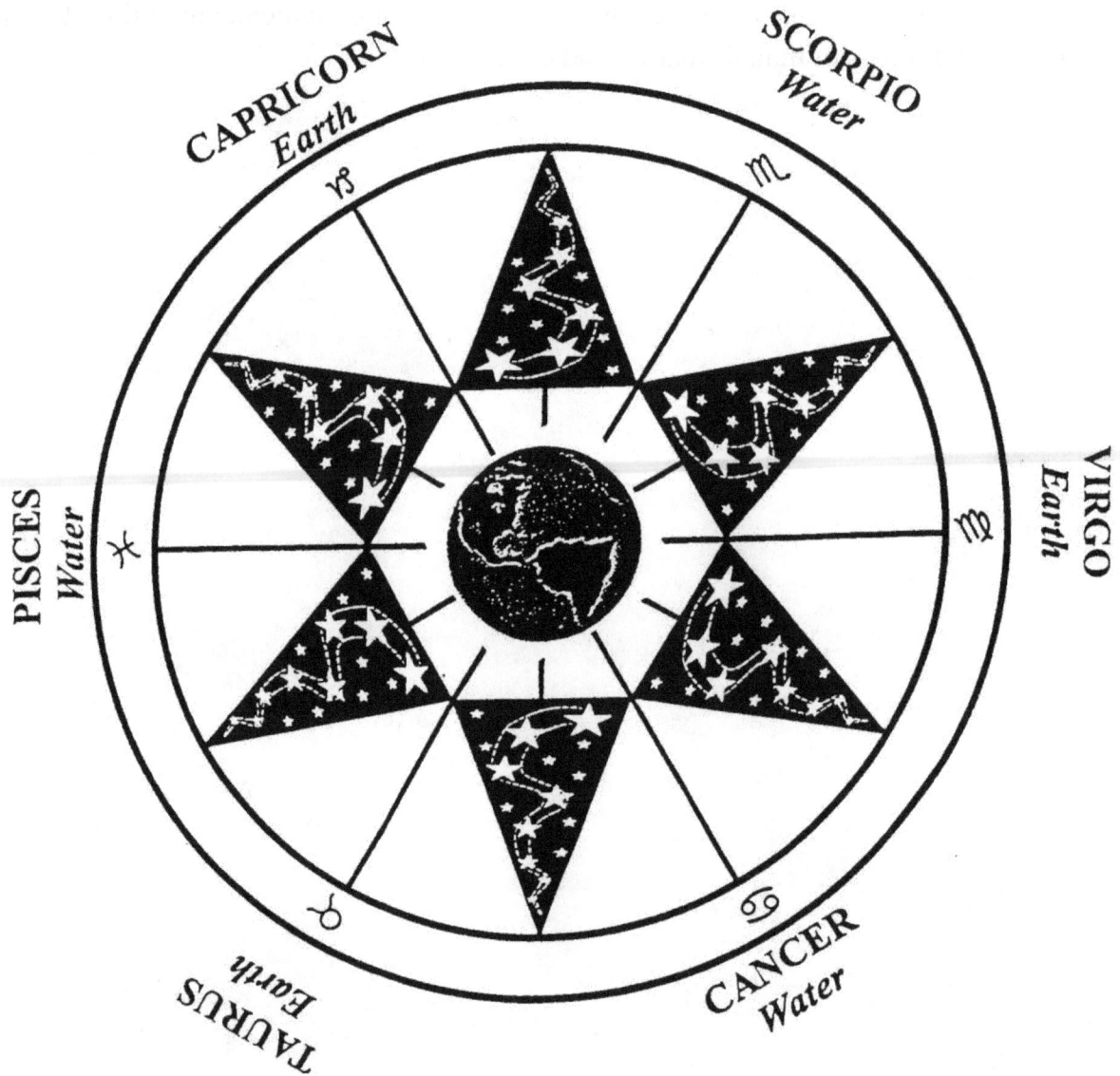

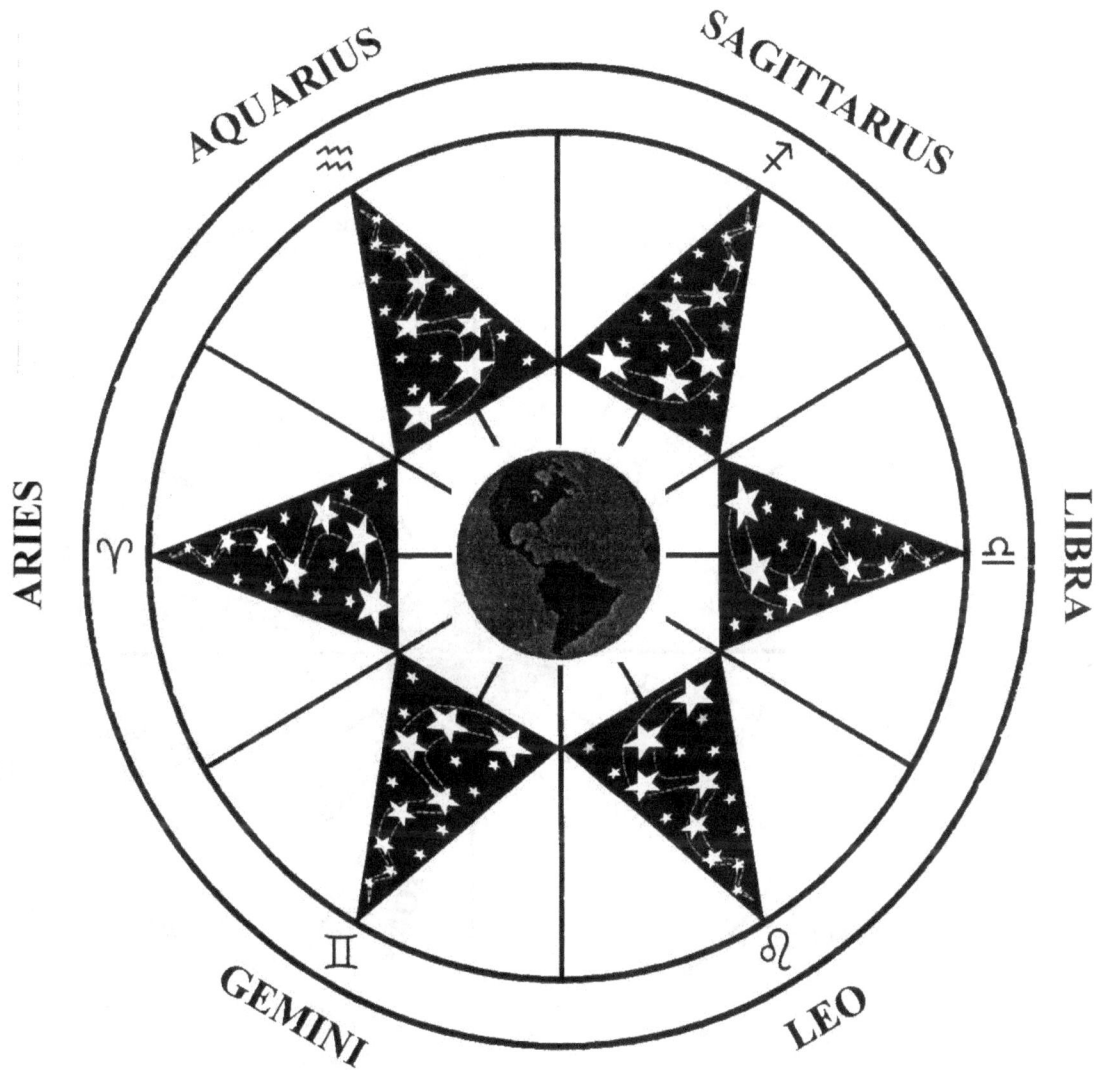

House Rulership

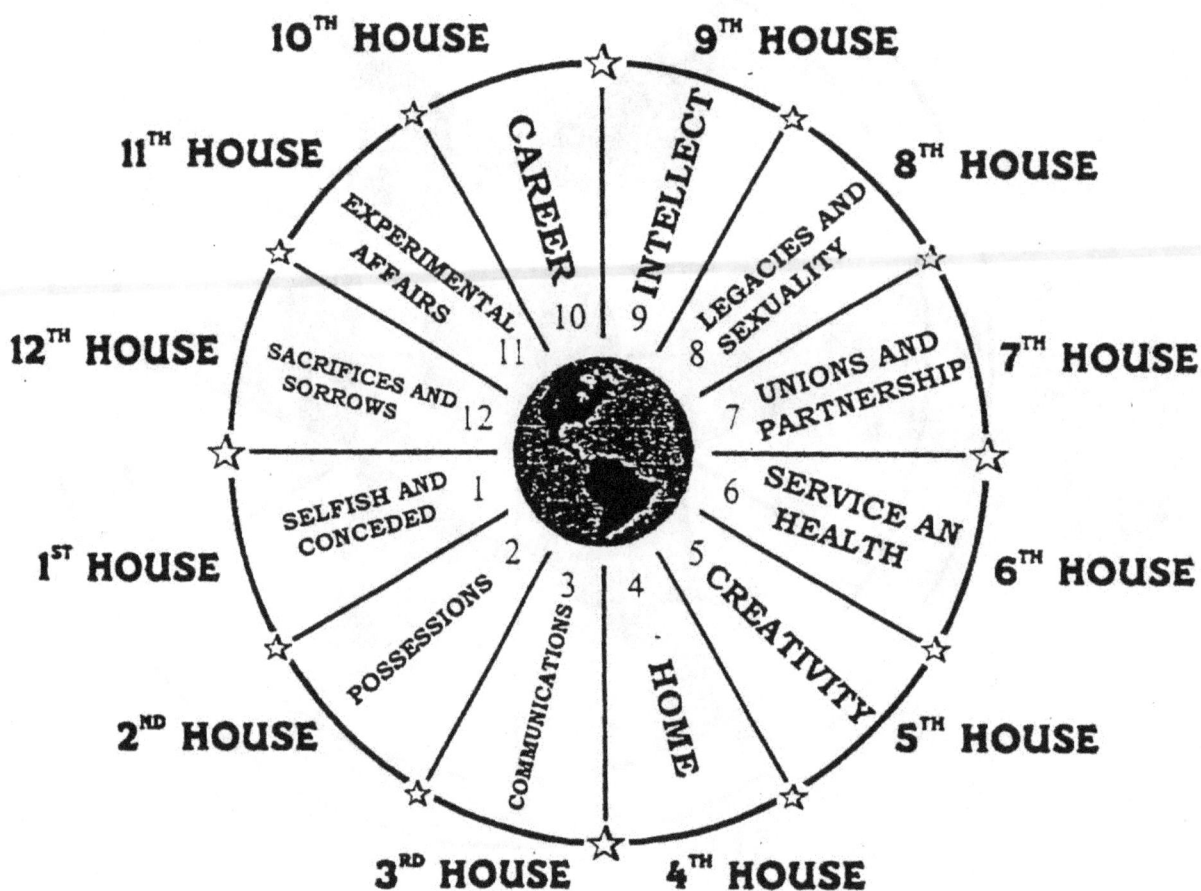

10TH HOUSE
9TH HOUSE
11TH HOUSE
8TH HOUSE
12TH HOUSE
7TH HOUSE
1ST HOUSE
6TH HOUSE
2ND HOUSE
5TH HOUSE
3RD HOUSE
4TH HOUSE

CAREER
INTELLECT
EXPERIMENTAL AFFAIRS
LEGACIES AND SEXUALITY
SACRIFICES AND SORROWS
UNIONS AND PARTNERSHIP
SELFISH AND CONCEDED
SERVICE AN HEALTH
POSSESSIONS
CREATIVITY
COMMUNICATIONS
HOME

10 9 8 11 7 12 1 6 2 5 3 4

The Ascendant to Your Rising Sign

The Ascendant, or Rising sign, is the sign or specific degree of the zodiac that was rising on the eastern horizon of the birthplace at the time of your birth. It is considered the sign located on the cusp of the First House of your birth chart, or the eastern point of intersection of the horizon and the ecliptic which is at the nine o'clock position on the birth chart. In order to interpret a birth chart accurately, the Ascendant must first be determined.

Only the Sun and the Moon are more significant in your chart than the Ascendant. It indicates a strong influence on your personality and must be considered along with your Sun and Moon signs when determining your astrological makeup. The zodiacal sign that the Sun was moving through at the time of your birth is your Sun sign and influences your individual characteristics and personality, that is your personal style as well as how you pursue your goals in life. The Moon, on the other hand, represents your emotions, your subjective feelings and how you react to life. Compared to the Sun and Moon signs, the Ascendant is that image which you project to others. This image is the first impression that others have of you, and it may be a subtle combination of the Ascendant and the Sun sign, with your Moon influences, more private or personal, being concealed and protected from the general public. Some astrologers describe the Ascendant as a mask which is used to face the world while others see it as a positive goal that each of us is working toward. Your Ascendant most influences your self-awareness, self-sufficiency, and self-interest while indicating how you determine your goals in life and how you put to best use your creativity.

To determine the Ascendant it is necessary to know the time and place of birth. That is because one sign of the zodiac may take thirty minutes to more than three hours to cross the horizon, but all twelve signs will have risen every twenty-four hours with the average time it takes for a sign to rise being approximately two hours. This is considered the most important point on a birth chart, being that other calculations of the horoscope are based on the sign of the Ascendant. The descriptions of the Ascendant in the following pages are taken alone, without any planets in the First House. The planets that fall within each House can possess either negative or positive influences.

ARIES ASCENDANT

Those individuals born at the time when Aries was rising are best recognized by their fiery energy and driving ambition. You want most to lead and to be in control or to dominate the situation. You apply all of your courage and enterprising nature into directions of your own making, and then you strive to see those plans carried out. In order to do that, you want to direct, to rearrange the various facets influencing your plan, to influence others to participate and to carry out your ideas, and then to complete your objective. In your social circle, you may be the person who most usually initiates activities. You may ask for input from others, but your influence most determines the time and place. Your strong will power, self-assurance, and confident nature grants you the ability to guide, control and lead others who in many instances are only too glad to follow along. Your active mind may process an idea a minute, and your daring, adventurous nature is only too receptive to impulsively grasping one of these innovative ideas. And once you have a definite idea in mind, it becomes tantamount to you, and you use all your personal influence to not only present it to others, but to insure that is adopted and that your plans are put into action. Your independent nature is forceful, and you are not above running rough shod over any criticism. Your ingenuity and belief in your own abilities combines with your combativeness to push aside any and all obstacles. And you are at your best when devising plans and determining the best manner in which to proceed.

You most love your independence, and you make every possible effort to maintain it. In business, social, or personal affairs, in order to feel secure, you must also feel that you are in control of your situation. You resent any form of limitation, restriction, or inconvenience placed on your time and efforts. And you can become adamantly determined, quick-tempered, or resentful of any impositions. More than likely, you don't hold a grudge for long, preferring to remain on friendly terms with as many people as possible. In love you are responsible, caring, generous and attentive with an aggressive sex drive, but unless things go your way, you are not always faithful.

Aries, ruled by Mars, is a very strong Rising sign indicating a penetrating, determined, and forceful personality. The negative influences of this sign include nervous stress, strain, tension, and selfishness with a quick temper and a combative nature that is prone to conflict. You are a person who puts to use all the resources at your disposal as well as all of your effort, strength, and will power in order to gain success in your endeavors, achievement, and recognition. You possess a strong intellect, philosophical nature, and an interest in science and discoveries. You excel in positions requiring action, decisiveness, and executive or leadership abilities.

Taurus Ascendant

A Taurus Ascendant makes for a reliable, persevering, self-reliant, and hard working individual. You are determined, proud and ambitious, but with Venus ruling Taurus, you are also graceful, charming, sociable, affectionate, kind, caring, and loving. Basically a calm, easy going person, your persistent nature leads you to pursue your goals and dreams in a steadfast manner. Taurus is the sign of wealth and material possessions, making these aspects of life important to you. You most aspire to feeling secure, and this is realized only after you have attained a steady source of income, possessions, and a home. You also strive to create a sense of beauty and harmony in your life, and this may be reflected in your home decorating skills and in the possessions which you collect for yourself, your family, and your home. Then too, music, art, writing, or some form of creativity is important in your life, and you may find yourself drawn to natural settings which expands your appreciation of beauty.

Personal strength, vitality, energy, and endurance enhances your life, and you prefer to use your personal resources in a serious, conservative, and conscientious manner. You may take your time cautiously pondering and arriving at well-thought out decisions, but once you have made a decision, you stick to it obstinately and stubbornly. Your gentle nature leads you to patiently listen to or wait out any opposition or criticisms, but once cornered, you can become as angry as a bull choosing to charge through any obstacles in your path. At the same time, you also possess a tendency to become set in your ways, preferring a daily routine that best suits your schedule, plans, and organization. You have a strong preference for being practical (except when it comes to shopping), and you are known for being reliable and trustworthy. You are also a compassionate listener who expresses sincere concern for the problems of others, and your friends and family may seek you out for a strong shoulder to lean on. In love, you are earthy, passionate, and devoted. The negative influences of Taurus include a tendency to be stubborn, lazy, unreasonable, prejudiced, indolent, or jealous and possessive with a love of sensual pleasures, food and drink which can lead to weight problems.

You are well suited to responsible positions requiring executive ability and may find that you are blessed with the ability to successfully produce or earn an income which allows you to support and care for your family. You may find success in areas dealing with the earth, nature, and its products.

Gemini Ascendant

Gemini, ruled by Mercury the communicator, is delightfully energetic and engaging. You may appear to be in constant motion, and the truth be known, you are capable of engaging in more than one activity at a time. You are ambitious and aspiring of success in your endeavors, but at the same time you are curious with an investigative nature that leads you in many directions. Somewhat restless, you resent restrictions to your abilities, thinking, and freedom of movement. You love to meet new people and to be exposed to their ideas and viewpoints. You also possess a desire to travel in order to be exposed to new experiences, people, ideas, and lifestyles. Adventuresome and ever ready for anything new and different, you can be daring and bold. You are highly imaginative, perceptive, and intuitive with a desire to learn and to expand your intellect. You collect information, facts, and people with ease. Sociable and entertaining, you love to move from one activity to another remaining actively engaged in all that is going on around you. You most desire change, variety, and diversity. Ingenious, quick-witted, and clever, you have a way with words that others find entertaining. Never at a lost for social invitations, you find your list of phone numbers growing daily. In romance, you are drawn to variety and new experiences, and in love you most admire a person who challenges you intellectually.

In addition to an active intellect, you are dexterous and do well in skills requiring the use of your hands. You may develop an interest in the field of communication such as writing, editing, publishing, journalism, radio or TV, or you may find yourself more interested in an occupation requiring mechanical ability. Then too, as much as you appreciate an audience, you may be drawn to acting or entertaining. You possess an appreciation for literature, science, art, music, languages, dancing, travel, inventions, and gadgets.

The negative traits of Gemini includes an excitable nature which is overly restless, anxious, worrisome, sarcastic, fretful, and easily upset with a tendency to be discontent, irritable, impatient, arrogant, critical, and snobbish.

You are drawn to encounters, change, and new ideas and places. You may travel, meet new people daily, change residences or occupations, or find some form of instilling variety into your life.

Cancer Ascendant

Cancer is ruled by the Moon which influences an individual to be changeable, sensitive, and reflective. You may find that you are often ruled by your emotions, perceiving life and the actions of others based on your moods and temperament. While you most desire friendships, admiration and recognition, you may be somewhat reserved, at times even shy, withdrawn, and preoccupied with your family, home, security, and possessions. You are highly creative and imaginative, and possess the ability to retain facts, figures, and information. You are also intuitive with perhaps psychic tendencies as well. Highly reflective, you are capable of emulating the personalities of others or adapting easily to changing situations, people, and places. There is an indication for change to be a part of your life whether this is in your home, social, or professional life. There are times when your sensitivities and imagination gets the better of you, and you see slights and insults where none were intended. This more than likely leads back to your strong wish to be accepted, loved and appreciated. To this end, you can be sociable, pleasant, charming, and caring, but you have difficulty allowing others to really get to know you. You have a very private and personal world of your own making which you withdraw to in order to enjoy the fantasies of your creative imagination. Your flair for languages, impersonations, and drama enhances your highly creative nature as well. Then too, when it comes to business and finance, you can be economical or if necessary frugal and shrewd, knowing instinctually where to find the best buy, the best investment, or the most advantageous savings program. Your personal and financial security is a primary concern for you, and you save and plan for the future with an intense desire to be prepared. Your love of your family and home can lead you to develop an interest in your ancestry, heritage, genealogy, and perhaps history in general. You possess an active and receptive mind with a preference toward developing your intellect. Once you set your goals, you can be industrious and tenacious, pursuing your interests in life despite any obstacles. You possess a strong love of nature, animals, wildlife, and beautiful settings. You also love to travel in order to see the sights and to be exposed to change and different surroundings. In love, you can be faithful and devoted as long as you feel appreciated.

The negative influences of Cancer include a tendency to be selfish, possessive, lazy, restless, untidy, timid, overly sensitive, gossipy and with a feeling of being unappreciated and picked on.

You are deeply emotional with a sympathetic and caring nature which leads you to be generous and concerned about the welfare of others. You are most remembered for your highly creative and imaginative abilities.

Leo Ascendant

Leo, ruled by the Sun, is magnanimous and magnetic with a noble and powerful strength of character and high ideals. Your pride and self-assured manner leads you to set high standards for yourself and others. You are best known for being good-natured, warm hearted, kind, and generous. With a tendency to be philosophical, you are frankly outspoken with a strong feeling of independence and forcefulness. Ambitious and confident, you strive to lead others in a responsible manner. And with your regal bearing, it is more than likely that others take notice of you the moment you walk in a room. You can be energetic, dignified, and fun loving all at the same time. But most importantly, your perseverance, determination, and conscientious efforts inspire others to follow your lead. You appear to be at your best when in a position of power, prestige, and distinction. Your expansive nature also exhibits a flair for the dramatic, and you may love to be center stage, effortlessly drawing to you the attention and admiration of others. At the same time, you display a remarkable courage and fortitude which at times can lead you to be impulsive. You can be quick tempered, but you prefer to forgive and forget. In love, you are ardent, passionate, and loyal as long as you are admired. There is an indication that you find favor from others, gain through luck and contacts, and find success in money, career, and friendship. Success may come to you when you least expect it.

A poorly developed Leo might be faulted for getting through life more on charm and pleasantries than by hard work, but the chances are you are efficient at organizing, planning, motivating, and leading others. You possess executive abilities in whatever field of endeavor you select as a career.

The negative traits of Leo influences an individual to be snobbish, arrogant, critical, lazy, judgmental, quick tempered, egotistical, conceited, condescending, and temperamental.

You are an enthusiastic and energetic person who aspires to ambitious pursuits and goals. You are creative and expressive with an inclination that leads you to inspire others.

Virgo Ascendant

Virgo, ruled by Mercury, influences an individual to be thoughtful, reserved, cautious, analytical, and a prodigious thinker and ponderer. Contemplative, productive, and industrious, you use your intellect to collect and sort data, facts, and necessary resources in order to obtain your goals and security in life. You size up situations, people, places, and events based on your methodical and well organized approach to life. As far as you are concerned, life would be easier if everyone simply followed the same rules and made sensible decisions. Your friends admire and appreciate your responsible attitude and caring and kind nature knowing they can depend on you for reliable advice. And you are probably one of the most dependable and loyal friends a person could hope to have. While you possess a deep concern for your friends and family, you view the world through a non-emotional and intellectually perceptive viewpoint. Perhaps this is because you realize that your gains in life will be realized through perseverance, hard work, and tenacious effort. You apply yourself diligently to your tasks using your energies, strength, and vitality to promote your efforts. You have a tendency to set high standards for yourself and others, and, being analytical, you also have a remarkable ability to be critical, pinpointing in great detail the pros and cons of any situation. You are more than likely an active person who is not easily contented to sit still and wait for what life may bring. You aspire to security and with this in mind, you seek position and financial well being, training yourself to be an economical, smart shopper and investor. But then in finance, you apply a shrewdness and the same attention to detail as you do in the other areas of your life.

Diplomatic and tactful, you generally get along well with others, however, you can be most cautious and discriminating in your choice of friends and associates. When it comes to love, you are caring, ardent and passionate, but you apply a great deal of caution to this area of your life as well. In fact, you may let many an opportunity slip away while you ponder and consider the outcome of each and every introduction. But then, you are prudent and practical in most every aspect of your life. Prone to certain health problems, you develop a diet and exercise program to improve your health and, if smart, you learn to watch your weight. The negative traits of Virgo leads an individual to be overly critical, sarcastic, cautious, and prone to worry resulting in a nervous disposition. You prefer to gain through the merits of your own effort, and with Virgo this can often be the case. Considering that you possess strong intellectual abilities, a logical thinking process, good intuition and perception, and the ability to observe and collect information, you are well endowed to succeed at your endeavors.

Libra Ascendant

Libra is ruled by Venus, the planet of romantic grace, charm, and poise. You are sociable, diplomatic, and sincerely caring and concerned with the welfare of others. You aspire to balance and seek harmonious, pleasant atmospheres and congenial companions. While you may have a bit of a temper, you prefer to use your charm and diplomacy to appease other people by instilling good will and fairness into situations. And fairness is very important to you. All forms of injustices and cruelty are appalling to you, and you find yourself drawn to upholding the standards which you strive to establish in your life. You may appear to stick to the middle ground in any disagreements, but the truth be known, you are capable of perceiving both sides to an argument, and being judicious, you consider all sides before making up your mind. At times, this ability to understand the viewpoints of others leaves you too open to accepting other people which results in you being lead into situations or decisions that you wouldn't have otherwise considered. But you learn from your errors, becoming more discriminating with time and developing a rather aloof demeanor which you hope will hold people at somewhat of a distance until you have time to discern their intentions. You are fond of beauty in nature, art, music, literature, and poetry, preferring refined and cultured activities. Sociable and fun loving, you are drawn to other people who share your interests, and with your pleasing nature, you make others feel at ease and comfortable. The negative traits of this sign include a tendency to be too pleasure seeking, superficial, jealous, lazy, easily influenced by others, dependent, or a tendency to aspire to success but to avoid hard work.

You are intellectually capable, imaginative, goodnatured, and optimistic with a creative flair. You can be thrown off your natural balance by obstacles, problems or the interference of other people, but you regain your optimism, preferring to believe that the future holds good things. There is indication for luck through partners, associates, the spouse, or endeavors related to women. Your refined attitude and good taste lead you to creating pleasant surroundings for yourself and family. You develop an interest in home decorating, fashion, style, jewelry, art and all else that adds beauty and refinement to your life. You possess a strong preference to be cheerfully on the go and actively engaged in life. Kind, good-natured, generous, charming and humane, you are well liked and accepted by others wherever you go.

Scorpio Ascendant

With a Scorpio Ascendant, you are a person who focuses your determination, will power, and forceful nature on every aspect of life. At times it appears that you succeed on the strength of your will power alone. You can appear outwardly reserved, but that is a mask for a magnetic, intriguing and subliminally mysterious nature that you prefer not to exhibit too overtly. You can even be a bit secretive preferring to protect your private life and personal thoughts from the curiosity of others. Your mind is quick, and this is reflected in your actions, speech, and that witty humor which can be dry and sarcastic at times. You are also highly observant with a strong preference for investigation or research. You may prefer to work alone, or you may find that you are quite capable of influencing and leading other people in group projects. You can be direct and outspoken, even blunt, when you want to get your point across, but at other times you prefer to keep your motives to yourself and to use persuasion (some say manipulation) to sway others to your viewpoint. Self-reliant and confident, you don't back away from a debate, argument, contest of wills, or obstacles. You can be a daring and formidable opponent and will patiently wait until you perceive your advantage in any situation. With friends, you are faithful, loyal, and kind, but you don't hesitate to back away from any relationships that hint at restrictions or limitations. You want your own space, mobility, and privacy of thoughts.

You are tenaciously enterprising and find that you succeed on the strength of your efforts. You are fond of travel, pleasurable activities, companionship of others who share your interests, and you are strongly attracted to the opposite sex. In love you are ardently passionate, and there are indications for secret love affairs. You may aspire to the good life and luxurious surroundings, or you may be quite capable of economically managing on very little. You can be shrewd when necessary with a critical and penetrating concentration and good judgment. You most generally make your own decisions, are not easily influenced, and develop bold, fixed opinions. When left to your own devices, you are adventuresome, daring, impulsive, energetic, and courageous, but when restricted, preoccupied, or otherwise limited in your actions, you can become restless, disinterested and indolent. The negative traits of Scorpio influence an individual to be too sarcastic, critical, jealous, possessive, revengeful, and cruel.

Ruled by Pluto, Scorpio is bestowed with strong emotions, will power, determination, forcefulness, and the ability to face obstacles and problems. You overcome reversals and limitations in your life by being unafraid of change and a fresh start.

Sagittarius Ascendant

A Sagittarius Ascendant and you are drawn to travel, to explore, and to seek new adventures, people, and experiences. The lure of what's around the next corner or what people might be doing in a totally different location is too much for you to resist. And with your friendly, cheerful, congenial, and optimistic nature, you meet people easily and are swept into their lives and adventures. You are open, accepting and tolerant of other people, making a faithful, kind, loyal and devoted friend. In fact, sympathetic, caring, and sincbre, the lengths you'll go to for a friend are without compare. That is, you are a wonderful friend for as long as you decide to stick around. But chances are you keep in touch with a number of favorite people while you move on down the road. You possess the strongest love for freedom of thought and movement, and you resist any restrictions to your free wheeling, independent ways. You are not possessive or jealous, preferring to allow others the same freedoms that you enjoy. At the same time, you possess a strong humanitarian outlook, are philosophical, and have an active and quick mind.

Injustice, unfairness, cruelty, and selfish motives offend your fair-minded thinking, and you strive to overcome these types of obstacles not only for yourself but for your friends as well. Your quick mind can lead you to be frank and outspoken at times, and then too, you can be impulsive in speech, actions, and decision making. Your curious and exploring intellect leads you to seek out knowledge and information from a variety of sources in an effort to be well informed of events, discoveries, and new ideas. You may be somewhat unconventional, preferring the latest concepts to outdated and traditional methods, thoughts, and practices. Neither are you very receptive to rules, traditions, laws and regulations that apply restrictions to your ideas and lifestyle. Your personal observations and knowledge combined with your experiences and intuitive insights may lead you to be prophetic and farseeing, often accurately deducing the outcome of a situation. You love sports, the outdoors, nature, and friendly competitions, and you may well be athletic with a casual appearance and outgoing nature. More settled types of people may advice you that you'd be better off if you'd just settle down and stick to one thing, but chances are you aren't particularly attached to possessions, people and places, preferring new experiences to obligations. And while you may prefer the simpler, less complicated life, you aspire to financial security not out of love of money but for the freedom it provides. That walk on a faraway beach with a pleasant companion is a memory dear to your heart. You are an idealist and a romantic in affairs of the heart, but you may find yourself breaking off the relationship and moving on. Indications are that this position will impulsively marry and separate at least once. You can be happy in marriage as long as your partner understands your need for personal freedom. The negative traits of this sign lead an individual to be too restless, impulsive, reckless, conceited, impatient, self-indulgent, and speculative. You may find luck and success in a foreign country or in association with a person from a foreign country.

Capricorn Ascendant

The Capricorn Ascendant is a person honed to perfection by the limitations and restrictions of the planet Saturn. You are self-disciplined, reserved and patient, waiting tenaciously for whatever obstacles life presents to you. Hard working and diligent, you aspire to financial success and an influential or leadership position. You accept ,responsibilities seriously and make concise and carefully thought out decisions. Serious, quiet, and thoughtful, you possess a strong cognitive ability and a contemplative nature. Although dignified and reserved, you can be warm, friendly and witty once you get to know people, but your alert mind is actively engaged in thought, gathering information and processing it for future use. You are goaloriented, economical, and conservative. You value society's safe and sure traditions and conventions, feeling secure within a system that allows for predictability. Careful and practical, you can be cautious and prudent but you apply your strong determination to achieving your goals and ambition. You are self-assured and display remarkable selfesteem, but you may have a tendency to worry excessively, being concerned about the future and what it will bring. With your strong powers of concentration, you develop well organized systems and apply well thought out procedures to your endeavors and plans. You are a natural at business enterprises, capable of making decisions and plans and leading others. Chances are you are more intellectual than emotional, and you think out your attachments, associations, and friendships. You can be discerning and discriminating, but your actions are based on sensible decisions. There is indication that you marry for either social or business reasons as well, but you are a devoted, responsible partner. The negative traits of Capricorn lead an individual to be stubborn, worrisome, pessimistic, selfish, melancholy, snobbish, and aloof. You are an ambitious person who faces obstacles with determination and endurance. You may have a tendency to want to control your personal situation, but this is out of concern for what is best for yourself and others. on a personal level, you are industrious with an interest in philosophy, theology, spiritualism, law, medicine, or world affairs. If you travel, it may be more for business or personal reasons than for pleasure. In later years, once you feel financially secure, the chances are you become more generous and relaxed, taking the time to enjoy life, home and family.

Aquarius Ascendant

The Aquarius Ascendant leads a person to value friendships and humanitarian causes. You are friendly, sincere, kind, and sociable. At the same time, you have a tendency to be quietly determined, patient, faithful and unobtrusive. In a group of people, you are the quieter person who doesntt necessarily seek the spotlight, but who possesses a quality that draws other people to you anyway. You can be quite philosophical, concerning yourself with the well being of people, cultures, and mores. And while you surround yourself with friends, many times your discussions will turn to causes and concerns. However, you most value your freedom of time, thought, and movement as well, and there are times when you withdraw from the group to seek isolation and privacy. Relationships are fine with you as long as they don't become restrictive or limiting. And if another person does attempt to control you, you withdraw from the situation. But you make every effort to remain friends, and usually that is exactly what happens. Other people simply like you and want to share your companionship. You are intelligent and charming, with an easy going manner and a gift for story telling. Aquarius are known for being futuristic, modern thinkers who prefer innovation and who seek new ideas. You are a person not tied to convention, traditions, and the accepted mode of operation. So progressive is your thinking that there are times when your thoughts may be light years ahead of other people. Then too, your foresight may lead you to be a bit impatient with less progressive thinkers. You also possess strong likes and dislikes and are more than a bit stubborn. Once you arrive at a decision, you aren't easily influenced or swayed. You are futuristic in your thinking, and there are times when you don't feel that the rules and regulations apply to you. You would much prefer to re-envision society, inventing the rules as you go along. That's not to say that you aren't practical, economical, and ambitious. You make an adroit business person, applying your observations and insights into decision making and productive methods and plans. The negative traits of Aquarius incline an individual to be radical, rebellious, eccentric, tactless, unpredictable, and contrary. There are indications for gains and reversals in life with unexpected experiences, changes, encounters, and obstacles. Your adventuresome spirit, however, leads you to be independent, original, reforming, and versatile. Romance is fun, but when you find love you become happy in marriage if it allows for your freedom of thought and expression. Friends and companions are helpful to you in your personal and professional life. This is the sign of inventive genius, and if you don't qualify for genius, then you remain inventive and a profound thinker.

Pisces Ascendant

The Pisces Ascendant leads you to be creative with a strong artistic flair. You are also caring, nurturing, kind, sympathetic and loving. Others are drawn to your appealing charms, and you display a courteous, friendly, and hospitable nature. Highly idealistic, imaginative, and romantic, you can be emotional and impressionable. Chances are you are talented, aspiring and ambitious, but you find yourself drawn to being dreamy, restless and pleasure seeking. You are sociable and fond of entertaining, fun activities, and interacting with companionable people. You much prefer the good life and being surrounded by beautiful, even luxurious possessions. You can be changeable and sentimental, easily moved by emotions to moodiness and melancholy and a sad story distresses you. But your preference is for happy times and pleasant experiences. You are successful when working on creative projects in a group situation, or can just as easily produce successfully by yourself. You may be drawn to the arts, music, theater, acting, entertaining or any field which requires an active and creative nature. or you may develop an interest in a field which involves caring for or counseling other people. In love, you are romantically sensual, devoted, and nurturing, and can become self-sacrificing when necessary. You are drawn to travel, new experiences, and meeting people. Beauty inspires you, and you possess a love of nature, wildlife, animals and the out of doors. You are capable of perceiving all of the colors of the rainbow, and the sunset over a lake inspires your creative talents. You like family, home and children, and strive to create a pleasant environment for your family. Your strong intuition and perceptive ability may lead you to develop your psychic abilities or an interest in spiritualism, the occult, mysticism, or unexplained phenomena. While you are most generally outgoing, you also cherish your privacy and there is a great deal about you which you prefer to remain personal and secret. The negative traits of Pisces lead an individual to be indecisive, vague, pleasure seeking, lazy, easily influenced, susceptible to drugs and alcohol, and dependent on other people. Neptune, the ruling planet of Pisces, grants you a strong intuition, imagination, compassion, and creativity. Easy going and good natured, you possess a versatility that others admire. You can be insightfully inspiring to others. You most create your own self-esteem by developing your talents and utilizing your strengths and abilities to find success in your endeavors.

THE PLANETS ARE PROJECTING THE ASTRAL INFLUX UPON THE EARTH.

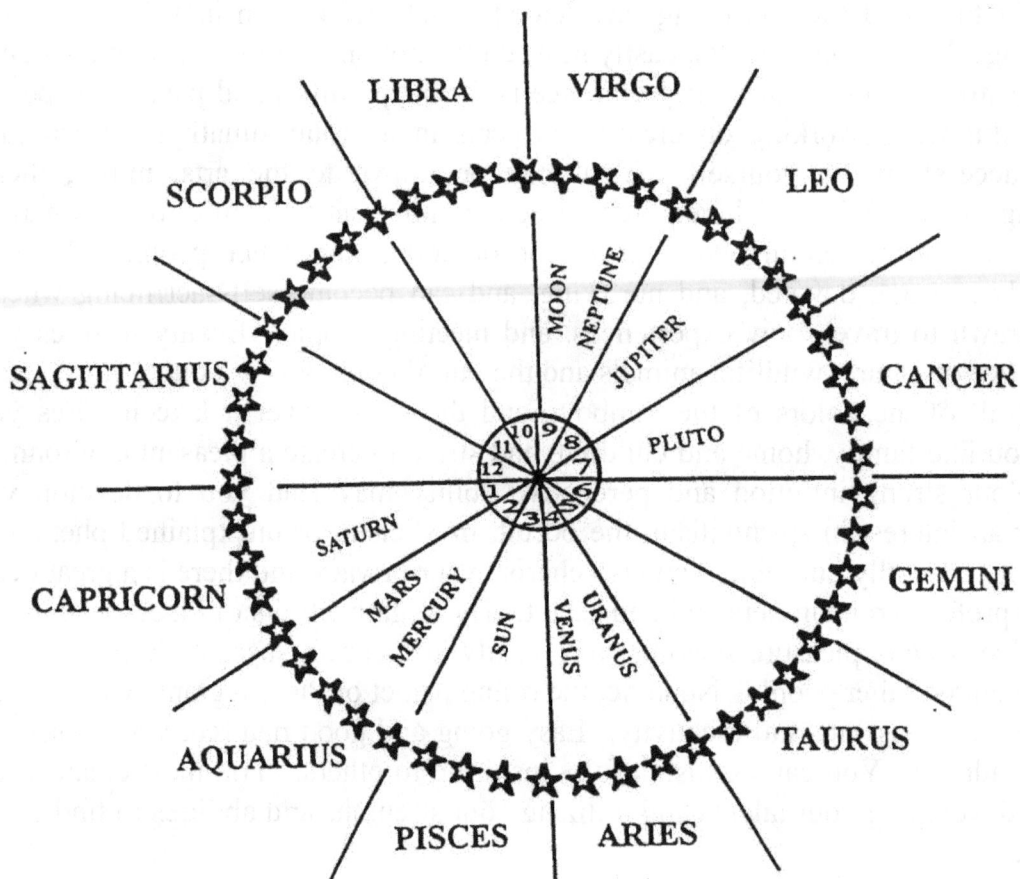

The Planets are The Daughters of The Sun

1. **The Sun** represents the vital principle, the health, the wealth, the social situation, the authority, the father, and it is the planet which, if it is favorable, is situated into a birthing sky, can improve the whole horoscope.

2. **The Moon** represents sensibility, receptivity, all the fantasy, reverie, the female principle, the crowd, the cash, the popularity, the secret, the mother. A strong moon in the horoscope shows adventure.

3. **Mercury** shows writing, the word, trade, and living intelligence.

4. **Venus** shows a modest chance, money, love, kindness. In the theme, it represents a providential chance.

5. **Mars** shows violence, enemy, and death.

6. **Jupiter** shows equilibrium, justice, order, high authority, abundance, and health. In the theme, if it has a good aspect, indicates a hppy life.

7. **Saturn** shows the impediment, the misery, the social fall. In the theme, if it has a bad aspect, indicates endless unhappiness.

8. **Neptune** represents the unsolved in all the domains. Can be a good medium, with prophetic spirit, the mysticism, the hallucination, the illusion, the obsession.

9. **URANUS** shows the originate, under all its aspects, the inventive spirit, the revolution, the human loving, the eccentricity. A strong Uranus indicates a reformative spirit, an inventive, and unselfish one.

10. **PLUTO** shows intensity, obsession, domination, control, and sexuality. A strong Pluto indicates empowerment and commitment.

BIRTH CHART

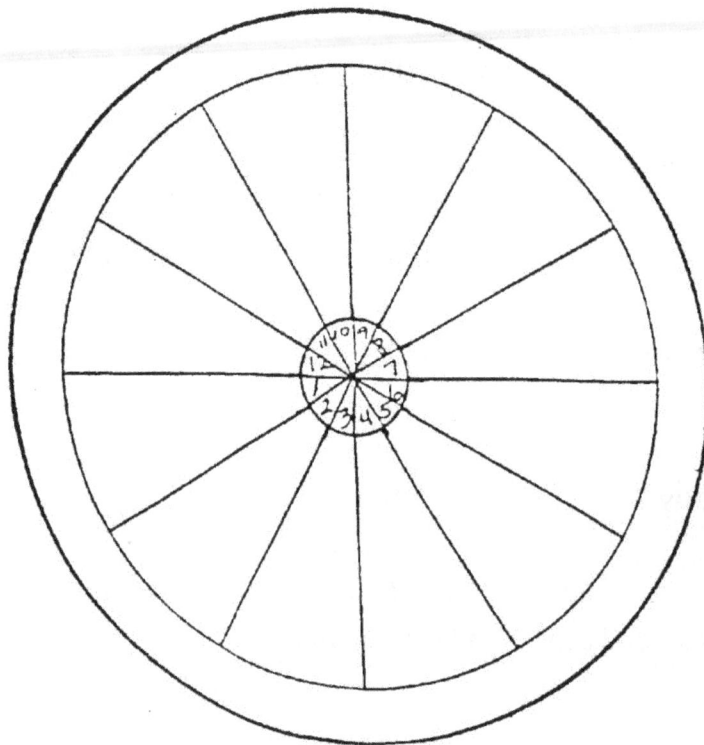

CONSTRUCTING THE BIRTH CHART

Constructing the Birth Chart can be easy if you follow rhese directions step by step. Use the following Tables to find your Ascendant and the location of your Planets. Then, write this information on the blank Birth Chart provided for your use. Next, make notes to interpret the meaning of the location of the Planets: the signs where the Planets are located and the Houses where the Planets are located. Finally, make notes on the interpretation of the Aspects of your Planets. The inner circle of the Birth Circle provided is divided into twelve sections and numbered to represent each of the Houses.

STEP 1:

Turn to the Ascendant Tables to find your Ascendant or Rising Sign. This is the sign rising at the hour of your birth and is your First House. Moving counter-clockwise around the Chart, write down the name of each of the remaining signs. For example:

Aries, Taurus, Gemini, Cancer, Leo Virgo, Libra, Scorpio, Sagittarius, Capricorn, Aquarius, Pisces

Taurus, Gemini, Cancer, Leo Virgo, Libra, Scorpio, Sagittarius, Capricorn, Aquarius, Pisces, Aries

Gemini, Cancer, Leo Virgo, Libra, Scorpio, Sagittarius, Capricorn, Aquarius, Pisces, Aries, Taurus

Cancer, Leo Virgo, Libra, Scorpio, Sagittarius, Capricorn, Aquarius, Pisces, Aries, Taurus, Gemini

Leo, Virgo, Libra, Scorpio, Sagittarius, Capricorn, Aquarius, Pisces, Aries, Taurus, Gemini, Cancer

Virgo, Libra, Scorpio, Sagittarius, Capricorn, Aquarius, Pisces, Aries, Taurus, Gemini, Cancer, Leo

Libra, Scorpio, Sagittarius, Capricorn, Aquarius, Pisces, Aries, Taurus, Gemini, Cancer, Leo, Virgo

Scorpio, Sagittarius, Capricorn, Aquarius, Pisces, Aries, Taurus, Gemini, Cancer, Leo, Virgo, Libra

Sagittarius, Capricorn, Aquarius, Pisces, Aries, Taurus, Gemini, Cancer, Leo, Virgo, Libra, Scorpio

Capricorn, Aquarius, Pisces, Aries, Taurus, Gemini, Cancer, Leo, Virgo, Libra, Scorpio, Sagittarius

Aquarius, Pisces, Aries, Taurus, Gemini, Cancer, Leo, Virgo, Libra, Scorpio, Sagittarius, Capricorn

Pisces, Aries, Taurus, Gemini, Cancer, Leo, Virgo, Libra, Scorpio, Sagittarius, Capricorn, Aquarius

In each of these lists, the first sign is the Ascendant and the signs which follow are in a House. The Ascendant is in the First House, and the other signs are in the subsequent Houses from Two through Twelve. Each sign of the Zodiac represents or is interpreted to mean something different in each of the Houses.

STEP 2:

The second step is to refer to the Tables for the Sun, Moon, and Planets to find which sign of the Zodiac they were in at the time of your birth. Now, on your Birth Chart, write down or place the Sun, the Moon, and each of the Planets in the House where its Zodiac sign is located. Again, the Sun, the Moon, and the Planets have different influences depending up on which House they are located.

STEP 3:

Divide a sheet of paper into three sections and title these sections: (1) Planet, (2) Sign, (3) House. In each of the sections, make notes on the meanings of these locations.

STEP 4:

On the same or a different sheet of paper, list the different Aspects. Then read the section of this book explaining Aspects, and again take notes on which ones apply to your Birth Chart.

When you have finished, you will have not only a completed Birth Chart, but an interpretation of how these influences effect your life, your decisions, and your destiny. Ater you become more familiar with the terms and symbols of Astrology, you may prefer to use th e glyphs below like a professional astrologer rather than the names of the signs and planets. This makes the Chart much neater and easier to read.

BIRTH CHART

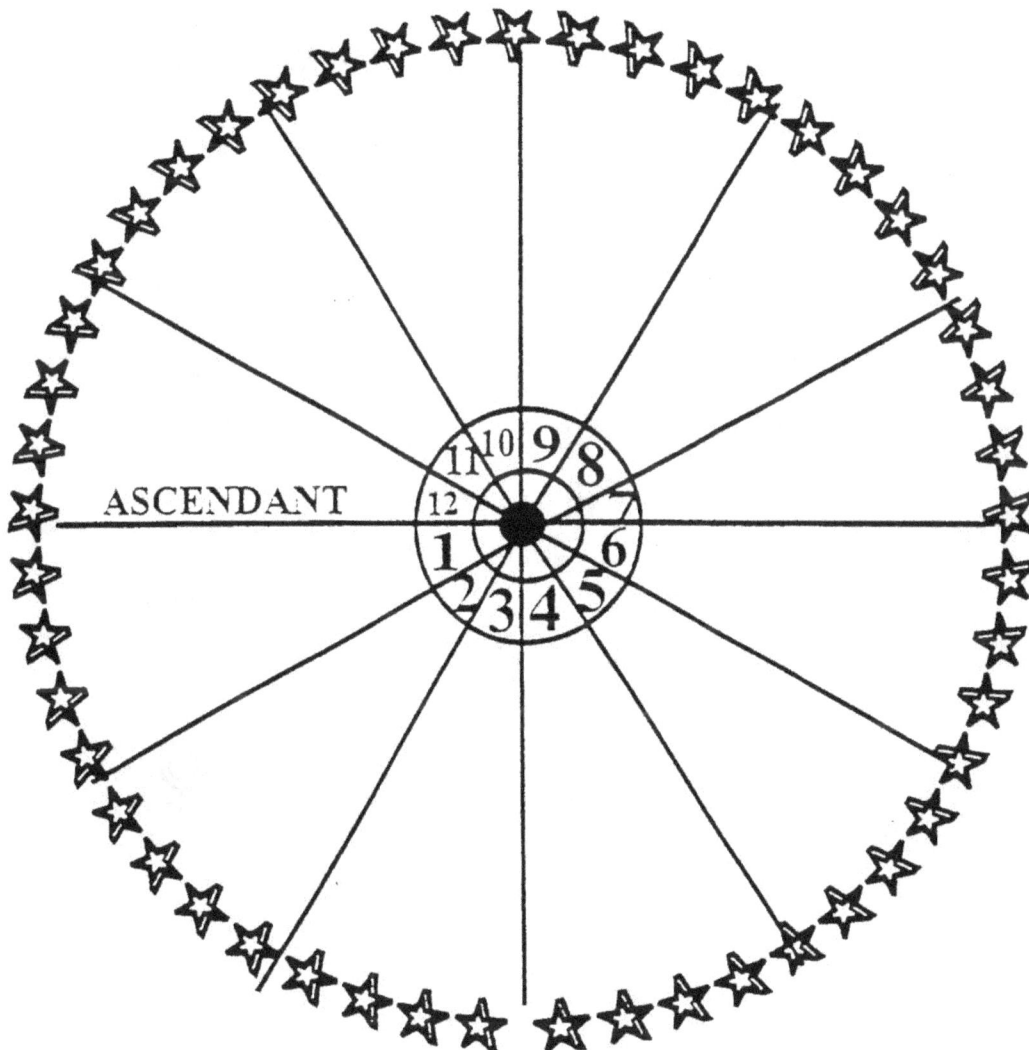

ZODIAC PLANETS

ARIES	♈		SUN	☉
TAURUS	♉		MOON	☽
GEMINI	♊		MERCURY	☿
CANCER	♋		VENUS	♀
LEO	♌		MARS	♂
VIRGO	♍		JUPITER	♃
LIBRA	♎		SATURN	♄
SCORPIO	♏		URANUS	♅
SAGITTARIUS	♐		NEPTUNE	♆
CAPRICORN	♑		PLUTO	P or ♇
AQUARIUS	♒			
PISCES	♓			

The Hour of Your Birth Determines The Orientation of Your Destiny

From "the time of birth" point of view, people are divided into two categories. In the first category are those who are born in the daytime who are called extroverts. In the second category are those people born at night, and they are called introverts. The first category is influenced by the solar magnetism that charges them like electric batteries which gives to them dynamism and the ability to influence others and to attract others like a magnet. Those people born in the second category or at night don't receive the Sun's magnetism. Rather, they receive it only through the Earth's crust. This influence is weaker, and, as a consequence, the people are weaker, more internalized, and more to themselves. They have a tendency to stay more inside their homes or offices.

The people born into these two categories who do not choose to follow their path or their destiny suffer social failures, obstacles, problems, and losses. For example, there are leaders who were born during the night and who later became tyrants and instigated catastrophies. The greatest leaders of the world, it has been noted, were born during the middle of the day. The solar magnetism of those born during the day is reflected in the individual's look which can be fascinating and lively. If people are born during the night, the eyes are more rigid and have a fixed look.

(1) Those people born two hours before sunrise have good chances to better themselves through there their aspirations and dreams. They are born with much promise in the presence of the Sun. They possess a demonstrative and strong character. They are less concerned about other people's opinions. They are strongly individualistic. The women born at this time may have a difficult time marrying or never get married at all.

(2) Those born between from two hours to four hours before sunrise are the people who throughout life want to make money. They possess a strong economical aptitude. The wealthy people who have the Sun and another planet or maybe two in the 2nd House are very sensitive to the acquisition of money. This person desires an abundance of money, but they obtain it only through hard work. Whether or not they have money, depends on how well they manage their money. Those who spend as much as they make will not balance their money well, and the end of their life will be poor.

(3) Those who are born between midnight and 2 a.m. are persons with writing skills and also inclinations for short journeys which will be very useful for them. Sometimes this person will have a scientific spirit that leads them to research and lab work. These individuals during their life have close relationships with their brothers and sisters.

(4) Those who are born from 10 p.m. to midnight will have an attraction for impossible dreams and aspirations they cannot reach. They possess a strong love of family and their country. Then the Sun is in the 4th House and its position represents long-term love, feelings, sentiments, and relationships with their family, home, and Earth. Also, there are strong feelings for heritage and ancestors because this House shows the origins of the native.

(5) Those who are born from 8 p.m. and 10 p.m. possess strong feelings and instincts such as for sexual desires, gambling, clubs, drinking, and/or teaching and acting. They possess a lot of sex appeal.

(6) Those who are born from 6 p.m. and 8 p.m. very often possess a negative, pessimistic, or sad faith. They are born to live with physical illnesses or moral crisis. Also, they are forced to work hard. Usually they have a weaker vitality because the Earth's magnetism is weaker at this time. This is the reason that these persons who are exposed to lots of sickness will try to take care if other sick people. Many times they become famous doctors.

(7)   Those wo are born in the time period two hours before sunset possess the tendency to have a feeling of togetherness. Their motto is, "me and you" because they cannot live alone. The men and women want associations, contracts, to process, and they look hard to find a partner in life. If there are found more than two Planets in this House with the Sun, there is the possibility of multiple marriages.

(8)   Those who are born from 2 p.m. to 4 p.m. are born in the 8th House. The natives are fascinated by the afterlife and the occult sciences. They have a strong feeling of sacrifice. Because the Sun is in the 8th House, throughout their lives they will receive gifts, legacies, and they enjoy a good retirement, but their lives may be shorter than others.

(9)   Those who are born from midday to 2 p.m., when the Sun is in 9th House, love to travel and have a tendency to become well educated. There is a connection between man and reality and a connection between man and God through religion, between man and philosophy through noble thinking, and between men and the world through long journeys. Those born at this time of the day are highly spiritual and intellectual. There is a feeling for speculation and travel. They achieve prominence in their social circles and earn honors.

(10)   Those who are born from 10 a.m. and midday are very social and reach a high position in society, earning honors and prestige. They have a strong feeling to dominate.

(11) Those who area born from 8 a.m. and 10 a.m. have a life orientated to friendship. They will have a lot of chances for protection, and they will reach a high level in life because of this.

(12) Those who area born from 6 a.m. and 8 a.m. have a very turbulent destiny. The Sun is weak and you can look directly at it because its rays are weaker. The magnetism of the Sun does not cover all of the Earth yet. The native will be almost all the time without vitality and will have a disposition to be isolated, either willingly or forcefully. They are opposed by restrictions and limitations. The women born during this time will become prisoners, with or without their will, in their own home.

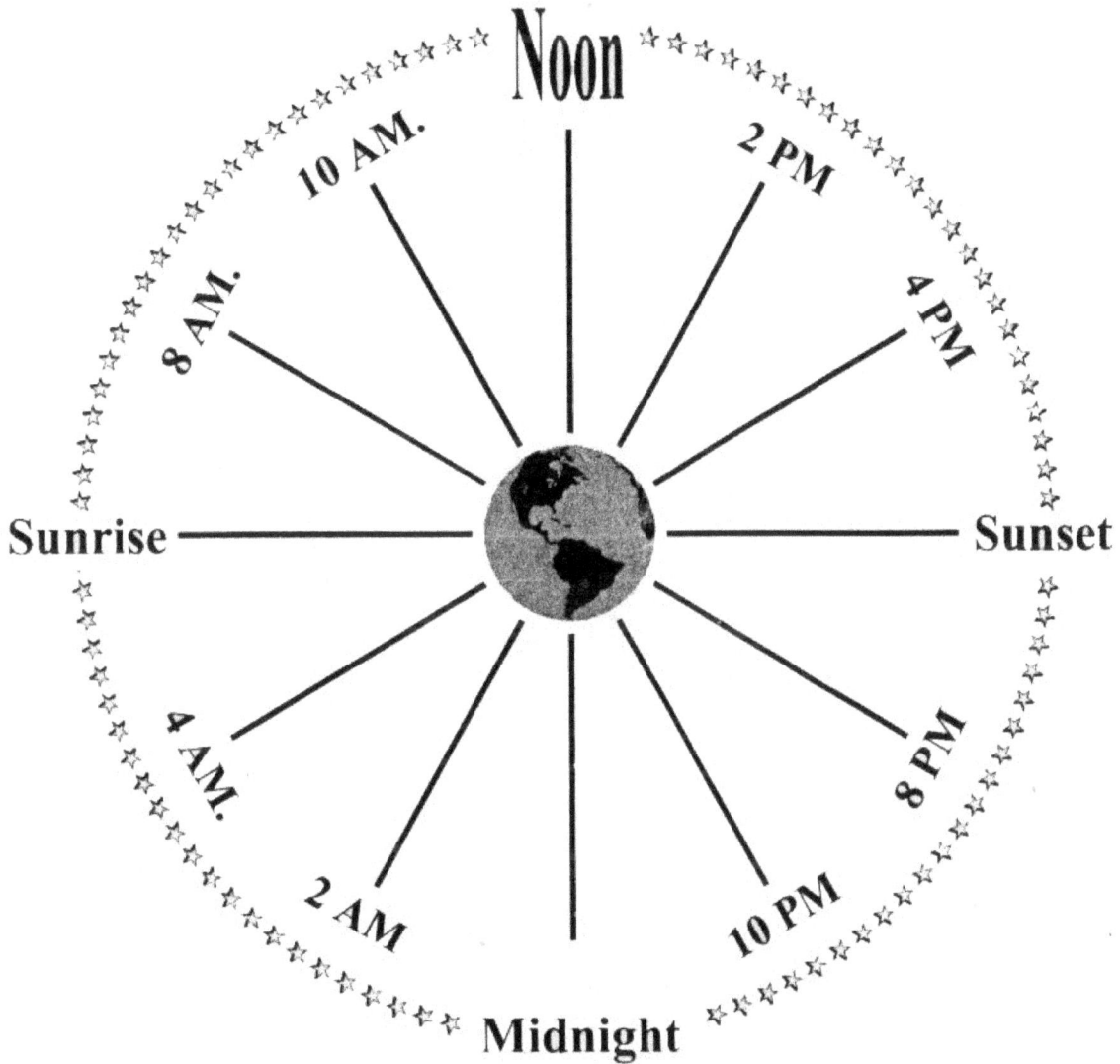

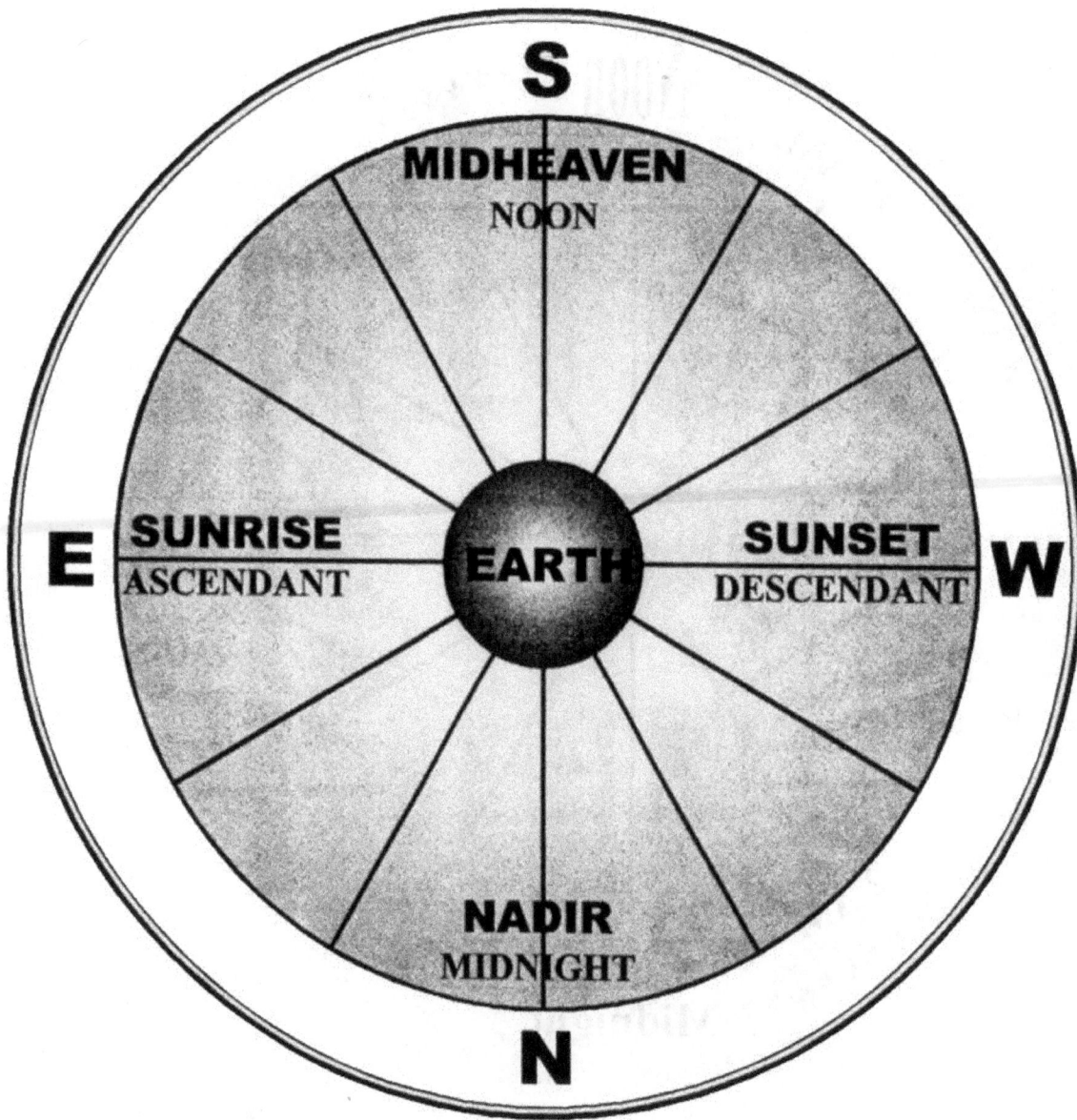

Mercury Tables

	1900	1901	1902	1903	1904	1905	1906	1907
JAN	8 CAP 28 AQU	2 CAP 21 AQU	13 AQU	6 AQU	2 AQU 13 CAP	CAP	12 CAP	6 CAP 25 AQU
FEB	14 PIS	7 PIS	1 PIS 18 AQU	AQU	15AQU	9 AQU 27 PIS	2 AQU 19 PIS	12 PIS
MAR	3 ARI 29 PIS	PIS	18 PIS	14 PIS	7 PIS 23 ARI	15 ARI	7 ARI	3 ARI 13 PIS
APR	16 ARI	15 ARI	9 ARI 25 TAU	1 ARI 16 TAU	7 TAU	1 TAU 28 ARI	ARI	18 ARI
MAY	10 TAU 26GEM	3 TAU 17GEM	9 GEM 29 CAN	2 GEM	TAU	15 TAU	14 TAU 31GEM	8 TAU 23GEM
JUN	9 CAN 27 LEO	1 CAN	26GEM	GEM	14GEM	8 GEM 23CAN	14 CAN 30LEO	6 CAN 27LEO
JUL	LEO	CAN	13 CAN	10 CAN 25 LEO	1 CAN 15 LEO	7 LEO 27 VIR	LEO	26 CAN
AUG	LEO	9 LEO 25 VIR	2 LEO 17 VIR	9 VIR 29 LIB	1 VIR 28 LIB	VIR	LEO	12 LEO 31 VIR
SEP	2 VIR 18 LIB	11 LIB 30 SCO	3 LIB 28 SCO	LIB	7 VIR	VIR	7 VIR 23 LIB	16 LIB
OCT	7 SCO 30 SAG	SCO	15 LIB	LIB	8 LIB 26 SCO	1 LIB 19 SCO	11 SCO	4 SCO
NOV	18 SCO	SCO	10 SCO 29 SAG	3 SCO 22 SAG	14 SAG	7 SAG	1 SAG	SCO
DEC	12 SAG	6 SAG 26 CAP	18 CAP	11 CAP	4 CAP	1 CAP 9 SAG	6 SCO 12 SAG	10 SAG 30 CAP

MERCURY TABLES

	1908	1909	1910	1911	1912	1913	1914	1915
JAN	18 AQU	10 AQU	3 AQU 30 CAP	CAP	15 CAP	10 CAP 29 AQU	3 CAP 22 AQU	14 AQU
FEB	4 PIS	AQU	15 AQU	12 AQU	6 AQU 25 PIS	16 PIS	8 PIS	2 PIS 23 AQU
MAR	PIS	17 PIS	11 PIS 29 ARI	4 PIS 20 ARI	11 ARI	4 ARI	PIS	19 PIS
APR	12 ARI 29 TAU	5 ARI 21 TAU	12 TAU 30GEM	5 TAU	ARI	7 PIS 13 ARI	16 ARI	10 ARI 26 TAU
MAY	13GEM 29 CAN	5 GEM	GEM	TAU	16 TAU	12 TAU 26GEM	4 TAU 19GEM	10GEM 29 CAN
JUN	CAN	GEM	1 TAU 11GEM	12GEM 28 CAN	5 GEM 19 CAN	10 CAN 28 LEO	3 CAN	CAN
JUL	CAN	13 CAN 29 LEO	6 CAN 21 LEO	12 LEO 30 VIR	4 LEO 26 VIR	LEO	CAN	CAN
AUG	6 LEO 21 VIR	13 VIR 31 LIB	5 VIR 27 LIB	VIR	20 LEO	LEO	11 LEO 27 VIR	4 LEO 18 VIR
SEP	7 LIB 28 SCO	LIB	28 VIR	VIR	10 VIR 28 LIB	4 VIR 28 LIB	12 LIB	5 LIB 28 SCO
OCT	SCO	LIB	11 LIB 31 SCO	6 LIB 24 SCO	15 SCO	8 SCO 30 SAG	2 SCO	20 LIB
NOV	1 LIB 11 SCO	7 SCO 26 SAG	19 SAG	11 SAG	4 SAG	23 SCO	SCO	11 SCO
DEC	3 SAG 22 CAP	15 CAP	8 CAP	2 CAP 27 SAG	SAG	13 SAG	7 SAG 27 CAP	1 SAG 20 CAP

Mercury Tables

	1916	1917	1918	1919	1920	1921	1922	1923
JAN	7 AQU	1 AQU 17 CAP	CAP	13 CAP	8 CAP 27 AQU	18 AQU	11 AQU	4 AQU
FEB	AQU	14 AQR	10 AQU	3 AQU 21 PIS	13 PIS	5 PIS	1 PIS 8 AQU	6 CAP 13 AQU
MAR	14 PIS	8 PIS 25 ARI	1 PIS 17 ARI	9 ARI	2 ARI 19 PIS	PIS	18 PIS	12 PIS 30 ARI
APR	2 ARI 17 TAU	9 TAU	2 TAU	ARI	17 ARI	13 ARI	7 ARI 22 TAU	14 TAU
MAY	2 GEM	TAU	TAU	15 TAU	8 TAU 23 GEM	1 TAU 15 GEM 31 CAN	7 GEM 31 CAN	1 GEM
JUN	GEM	14 GEM	9 GEM 24 CAN	2 GEM 16 CAN	6 CAN 26 LEO	CAN	10 GEM	GEM
JUL	10 CAN 25 LEO	3 CAN 17 LEO	9 LEO 27 VIR	1 LEO	LEO	CAN	13 CAN 31 LEO	8 CAN 22 LEO
AUG	9 VIR 28 LIB	2 VIR 26 LIB	VIR	LEO	2 CAN 10 LEO 31 VIR	8 LEO 23 VIR	15 VIR	7 VIR 27 LIB
SEP	LIB	14 VIR	VIR	8 VIR 25 LIB	16 LIB	8 LIB 29 SCO	1 LIB	LIB
OCT	LIB	9 LIB 28 SCO	3 LIB 20 SCO	13 SCO	5 SCO 30 SAG	SCO	1 SCO 4 LIB	4 VIR 11 LIB
NOV	4 SCO 22 SAG	15 SAG	8 SAG	2 SAG	10 SAG	SCO	8 SCO 27 SAG	1 SCO 20 SAG
DEC	14 CAP	5 CAP	1 CAP 15 SAG	SAG	10 SAG 31 CAP	4 SAG 24 CAP	16 CAP	9 CAP

Mercury Tables

	1924	1925	1926	1927	1928	1929	1930	1931
JAN	CAP	14 CAP	11 CAP 31 AQU	4 CAP 23 AQU	16 AQU	8 AQU	2 AQU 22 CAP	CAP
FEB	13 AQU	7 AQU 25 PIS	17 PIS	9 PIS	3 PIS 29 AQU	AQU	15 AQU	11 AQU
MAR	5 PIS 21 ARI	13 ARI	5 ARI	PIS	17 PIS	15 PIS	9 PIS 26 ARI	2 PIS 18 ARI
APR	5 TAU	1 TAU 15 ARI	ARI	17 ARI	10 ARI 27 TAU	3 ARI 18 TAU	10 TAU	3 TAU
MAY	TAU	16 TAU	13 TAU 29GEM	6 TAU 20GEM	11GEM 28 CAN	3 GEM	1 GEM 17 TAU	TAU
JUN	12GEM 29 CAN	6 GEM 20 CAN	12 CAN 29 LEO	4 CAN 28 LEO	CAN	GEM	14GEM	11GEM 26 CAN
JUL	13 LEO 30 VIR	5 LEO 26 VIR	LEO	13 CAN	CAN	11 CAN 27 LEO	4 CAN 18 LEO	10 LEO 28 VIR
AUG	VIR	27 LEO	LEO	11 LEO 28 VIR	4 LEO 19 VIR	11 VIR 30 LIB	3 VIR 26 LIB	VIR
SEP	VIR	10 VIR 29 LIB	5 VIR 21 LIB	13 LIB	5 LIB 27 SCO	LIB	19 VIR	VIR
OCT	6 LIB 24 SCO	16 SCO	9 SCO 31 SAG	3 SCO	24 LIB	LIB	10 LIB 29 SCO	4 LIB 21 SCO
NOV	12 SAG	5 SAG	28 SCO	SCO	11 SCO	5 SCO 24 SAG	17 SAG	9 SAG
DEC	2 CAP 31 SAG	SAG	13 SAG	9 SAG 28 CAP	1 SAG 20 CAP	13 CAP	6 CAP	1 CAP 20 SAG

Mercury Tables

	1932	1933	1934	1935	1936	1937	1938	1939
JAN	14 CAP	8 CAP 27 AQU	1 CAP 20 AQU	12 AQU	5 AQU	1 AQU 9 CAP	6 SAG 12 CAP	12 CAP
FEB	4 AQU 22 PIS	13 PIS	6 PIS	1 PIS 14 AQU	AQU	13 AQU	8 AQU 26 PIS	1 AQU 19 PIS
MAR	9 ARI	3 ARI 25 PIS	PIS	18 PIS	13 PIS 31 ARI	6 PIS 22 ARI	14 ARI	7 ARI
APR	ARI	17 ARI	14 ARI	8 ARI 24 TAU	14 TAU 30 GEM	6 TAU	1 TAU 23 ARI	ARI
MAY	15 TAU	10 TAU 25 GEM	2 TAU 16 GEM	8 GEM 29 CAN	GEM	TAU	16 TAU	14 TAU 30 GEM
JUN	2 GEM 16 CAN	8 CAN 26 LEO	1 CAN	20 GEM	GEM	13 GEM 30 CAN	7 GEM 22 CAN	13 CAN 30 LEO
JUL	2 LEO 27 VIR	LEO	CAN	13 CAN	8 CAN 23 LEO	14 LEO 31 VIR	6 LEO 26 VIR	LEO
AUG	10 LEO	LEO	9 LEO 24 VIR	1 LEO 16 VIR	7 VIR 27 LIB	VIR	VIR	LEO
SEP	9 VIR 27 LIB	2 VIR 17 LIB	10 LIB 30 SCO	3 LIB 28 SCO	LIB	VIR	2 LEO 10 VIR 30 LIB	6 VIR 23 LIB
OCT	13 SCO	6 SCO 29 SAG	SCO	12 LIB	LIB	8 LIB 25 SCO	18 SCO	11 SCO
NOV	2 SAG	15 SCO	SCO	9 SCO 29 SAG	2 SCO 20 SAG	13 SAG	6 SAG	1 SAG
DEC	SAG	11 SAG	6 SAG 25 CAP	18 CAP	10 CAP	3 CAP	SAG	3 SCO 13 SAG

Mercury Tables

	1940	1941	1942	1943	1944	1945	1946	1947
JAN	6 CAP 25 AQU	16 AQU	9 AQU	3 AQU 27 CAP	CAP	13 CAP	9 CAP 29 AQU	2 CAP 21 AQU
FEB	11 PIS	3 PIS	AQU	15 AQU	12 AQU	5 AQU 23 PIS	15 PIS	7 PIS
MAR	4 ARI 7 PIS	6 AQU 16 PIS	16 PIS	10 PIS 28 ARI	2 PIS 19 ARI	11 ARI	4 ARI	PIS
APR	16 ARI	12 ARI 28 TAU	5 ARI 20 TAU	11 TAU 30GEM	3 TAU	ARI	1 PIS 16 ARI	15 ARI
MAY	6 TAU 21GEM	12GEM 29 CAN	4 GEM	26 TAU	TAU	16 TAU	11 TAU 26GEM	4 TAU 18GEM
JUN	4 CAN 26 LEO	CAN	GEM	13GEM	11GEM 26 CAN	4 GEM 18 CAN	9 CAN 27 LEO	2 CAN
JUL	20 CAN	CAN	12 CAN 28 LEO	6 CAN 20 LEO	11 LEO 28 VIR	3 LEO 26 VIR	LEO	CAN
AUG	11 LEO 29 VIR	6 LEO 21 VIR	12 VIR 31 LIB	5 VIR 26 LIB	VIR	17 LEO	LEO	10 LEO 26 VIR
SEP	14 LIB	6 LIB 28 SCO	LIB	25 VIR	VIR	10 VIR 27 LIB	3 VIR 19 LIB	11 LIB
OCT	3 SCO	29 LIB	LIB	11 LIB 30 SCO	4 LIB 22 SCO	14 SCO	7 SCO 30 SAG	1 SCO
NOV	SCO	11 SCO	6 SCO 25 SAG	18 SAG	10 SAG	3 SAG	20 SCO	SCO
DEC	9 SAG 29 CAP	2 SAG 21 CAP	14 CAP	7 CAP	1 CAP 23 SAG	SAG	12 SAG	7 SAG 26 CAP

Mercury Tables

	1948	1949	1950	1951	1952	1953	1954	1955
JAN	14 AQU	6 AQU	1 AQU 15 AQU	CAP	13 CAP	6 CAP 25 AQU	18 AQU	10 AQU
FEB	1 PIS 20 AQU	AQU	14 AQU	9 AQU 28 PIS	2 AQU 28 PIS	11 PIS	4 PIS	AQU
MAR	18 PIS	14 PIS	7 PIS 24 ARI	16 ARI	7 ARI	2 ARI 15 PIS	PIS	17 PIS
APR	8 ARI 24 TAU	1 ARI 16 TAU	8 TAU	1 TAU	ARI	17 ARI	13 ARI 30 TAU	6 ARI 21 TAU
MAY	8 GEM 28 CAN	1 GEM	TAU	1 ARI 14 TAU	14 TAU 31 GEM	8 TAU 22 GEM	14 GEM 30 CAN	6 GEM
JUN	28 GEM	GEM	14 GEM	9 GEM 23 CAN	14 CAN 30 LEO	6 CAN 26 LEO	CAN	GEM
JUL	11 CAN	9 CAN 25 LEO	2 CAN 16 LEO	8 LEO 27 VIR	LEO	28 CAN	CAN	13 CAN 30 LEO
AUG	2 LEO 17 VIR	9 VIR 28 LIB	1 VIR 27 LIB	VIR	LEO	11 LEO 30 VIR	7 LEO 22 VIR	14 VIR
SEP	3 LIB 27 SCO	LIB	10 VIR	LIB	7 VIR 23 LIB	15 LIB	8 LIB 28 SCO	1 LIB
OCT	16 LIB	LIB	9 LIB 27 SCO	2 LIB 19 SCO	11 SCO	4 SCO 31 SAG	SCO	LIB
NOV	9 SCO 29 SAG	3 SCO 22 SAG	14 SAG	7 SAG	1 SAG	6 SCO	4 LIB 11 SCO	8 SCO 26 SAG
DEC	18 CAP	11 CAP	4 CAP	1 CAP 12 SAG	SAG	10 SAG 30 CAP	4 SAG 23 CAP	16 CAP

Mercury Tables

	1956	1957	1958	1959	1960	1961	1962	1963
JAN	4 AQU	CAP	14 CAP	10 CAP 30 AQU	4 CAP 23 AQU	14 AQU	7 AQU	1 AQU 20 CAP
FEB	2 CAP 15 AQU	12 AQU	6 AQU 24 PIS	16 PIS	9 PIS	1 PIS 24 AQU	AQU	15 AQU
MAR	11 PIS 28 ARI	4 PIS 20 ARI	12 ARI	5 ARI	PIS	18 PIS	15 PIS	9 PIS 25 ARI
APR	12 TAU 29 GEM	4 TAU	2 TAU 10 ARI	ARI	15 ARI	10 ARI 26 TAU	2 ARI 17 TAU	9 TAU
MAY	GEM	TAU	16 TAU	12 TAU 28 GEM	4 TAU 18 GEM	10 GEM 28 CAN	3 GEM	2 GEM 10 TAU

	1956	1957	1958	1959	1960	1961	1962	1963
JUN	GEM	12 GEM 28 CAN	5 GEM 19 CAN	11 CAN 28 LEO	2 CAN 30 LEO	CAN	GEM	14 GEM
JUL	6 CAN 21 LEO	12 LEO 29 VIR	4 LEO 26 VIR	LEO	5 CAN	CAN	11 CAN 26 LEO	3 CAN 18 LEO
AUG	5 VIR 26 LIB	VIR	23 LEO	LEO	10 LEO 26 VIR	3 LEO 18 VIR	10 VIR 29 LIB	3 VIR 26 LIB
SEP	29 VIR	VIR	10 VIR 28 LIB	4 VIR 20 LIB	12 LIB	4 LIB 27 SCO	LIB	16 VIR
OCT	11 LIB 31 SCO	6 LIB 23 SCO	16 SCO	8 SCO 30 SAG	1 SCO	21 LIB	LIB	10 LIB 28 SCO
NOV	18 SAG	11 SAG	4 SAG	25 SCO	SCO	10 SCO 30 SAG	4 SCO 23 SAG	16 SAG
DEC	8 CAP	2 CAP 28 SAG	SAG	13 SAG	7 SAG 27 CAP	19 CAP	12 CAP	5 CAP

Mercury Tables

	1964	1965	1966	1967	1968	1969	1970	1971
JAN	CAP	12 CAP	7 CAP 26 AQU	19 AQU	12 AQU	4 AQU	3 AQU 4 CAP	2 SAG 13 CAP
FEB	10 AQU 29 PIS	3 AQU 21 PIS	13 PIS	5 PIS	1 PIS 11 AQU	AQU	13 AQU	7 AQU 26 PIS
MAR	16 ARI	8 ARI	2 ARI 21 PIS	PIS	17 PIS	12 PIS 30 ARI	5 PIS 22 ARI	13 ARI
APR	1 TAU	ARI	17 ARI	14 ARI	6 ARI 22 TAU	14 TAU 31 GEM	6 TAU	1 TAU 18 ARI
MAY	TAU	15 TAU	9 TAU 24 GEM	1 TAU 15 GEM 31 CAN	6 GEM 29 CAN	GEM	TAU	16 TAU
JUN	9 GEM 24 CAN	1 GEM 15 CAN	7 CAN 26 LEO	CAN	13 GEM	GEM	13 GEM 30 CAN	7 GEM 21 CAN
JUL	8 LEO 27 VIR	1 LEO 31 VIR	LEO	CAN	12 CAN 31 LEO	7 CAN 22 LEO	14 LEO 31 VIR	6 LEO 26 VIR
AUG	VIR	3 LEO	LEO	8 LEO 24 VIR	14 VIR	6 VIR 27 LIB	VIR	29 LEO
SEP	VIR	8 VIR 25 LIB	1 VIR 17 LIB	9 LIB 29 SCO	1 LIB 28 SCO	LIB	VIR	11 VIR 30 LIB
OCT	2 LIB 20 SCO	12 SCO	5 SCO 30 SAG	SCO	7 LIB	6 VIR 9 LIB	7 LIB 25 SCO	17 SCO
NOV	8 SAG 30 CAP	2 SAG	12 SCO	SCO	8 SCO 27 SAG	1 SCO 20 SAG	12 SAG	6 SAG
DEC	16 SAG	SAG	11 SAG 31 CAP	5 SAG 24 CAP	16 CAP	9 CAP	3 CAP	SAG

Mercury Tables

	1972	1973	1974	1975	1976	1977	1978	1979
JAN	11 CAP 31 AQU	4 CAP 23 AQU	15 AQU	8 AQU	2 AQU 24 CAP	CAP	13 CAP	8 CAP 28 AQU
FEB	18 PIS	9 PIS	2 PIS	AQU	15 AQU	10 AQU	4 AQU 22 PIS	14 PIS
MAR	5 ARI	PIS	2 AQU 17 PIS	16 PIS	9 PIS 26 ARI	2 PIS 18 ARI	10 ARI	3 ARI 28 PIS
APR	ARI	16 ARI	11 ARI 27 TAU	4 ARI 19 TAU	10 TAU 29 GEM	2 TAU	ARI	17 ARI
MAY	12 TAU 29 GEM	5 TAU 20 GEM	11 GEM 29 CAN	4 GEM	19 TAU	TAU	16 TAU	10 TAU 26 GEM
JUN	11 CAN 28 LEO	3 CAN 27 LEO	CAN	GEM	13 GEM	10 GEM 26 CAN	3 GEM 17 CAN	9 CAN 27 LEO
JUL	LEO	16 CAN	CAN	12 CAN 28 LEO	4 CAN 18 LEO	10 LEO 28 VIR	2 LEO 27 VIR	LEO
AUG	LEO	11 LEO 28 VIR	5 LEO 20 VIR	12 VIR 30 LIB	3 VIR 25 LIB	VIR	13 LEO	LEO
SEP	5 VIR 21 LIB	13 LIB	6 LIB 27 SCO	LIB	21 VIR	VIR	9 VIR 26 LIB	2 VIR 18 LIB
OCT	9 SCO 30 SAG	2 SCO	26 LIB	LIB	10 LIB 28 SCO	4 LIB 21 SCO	14 SCO	6 SCO 30 SAG
NOV	29 SCO	SCO	11 SCO	6 SCO 24 SAG	16 SCO	9 SAG	3 SAG	17 SCO
DEC	12 SAG	8 SAG 28 CAP	2 SAG 21 CAP	13 CAP	6 CAP	1 CAP 21 SAG	SAG	12 SAG

Mercury Tables

	1980	1981	1982	1983	1984	1985	1986	1987
JAN	2 CAP 20 AQU	12 AQU 31 PIS	5 AQU	1 AQU 12 CAP	CAP	11 CAP	5 CAP 24 AQU	17 AQU
FEB	7 PIS	16 AQU	AQU	14 AQU	8 AQU 27 PIS	1 AQU 18 PIS	11 PIS	3 PIS
MAR	PIS	17 PIS	13 PIS 31 ARI	6 PIS 23 ARI	14 ARI 31 TAU	6 ARI	3 ARI 11 PIS	11 AQU 13 PIS
APR	14 ARI	8 ARI 24 TAU	15 TAU	7 TAU	25 RAI	ARI	17 ARI	12 ARI 29 TAU
MAY	2 TAU 16GEM 31 CAN	8 GEM 28 CAN	1 GEM	TAU	15 TAU	13 TAU 30GEM	7 TAU 22GEM	13GEM 29 CAN

	1980	1981	1982	1983	1984	1985	1986	1987
JUN	CAN	22GEM	GEM	14GEM	7 GEM 22 CAN	13 CAN 29 LEO	5 CAN 26 LEO	CAN
JUL	CAN	12 CAN	9 CAN 24 LEO	1 CAN 15 LEO	6 LEO 26 VIR	LEO	23 CAN	CAN
AUG	8 LEO 24 VIR	1 LEO 16 VIR	8 VIR 27 LIB	1 VIR 29 LIB	VIR	LEO	11 LEO 29 VIR	6 LEO 21 VIR
SEP	9 LIB 29 SCO	2 LIB 27 SCO	LIB	5 VIR	30 LIB	6 VIR 22 LIB	14 LIB	7 LIB 28 SCO
OCT	SCO	13 LIB	LIB	8 LIB 26 SCO	17 SCO	10 SCO 31 SAG	3 SCO	31 LIB
NOV	SCO	9 SCO 28 SAG	2 SCO 21 SAG	14 SAG	6 SAG	SAG	SCO	11 SCO
DEC	5 SAG 24 CAP	17 CAP	10 CAP	4 CAP	1 CAP 7 SAG	4 SCO 12 SAG	9 SAG 29 CAP	3 SAG 22 CAP

361

MERCURY TABLES

	1988	1989	1990	1991	1992	1993	1994	1995
JAN	10 AQU	2 AQU 28 CAP	CAP	14 CAP	9 CAP 29 AQU	2 CAP 21 AQU	13 AQU	6 AQU
FEB	AQU	14 AQU	11 AQU	5 AQU 23 PIS	16 PIS	7 PIS	1 PIS 21 AQU	AQU
MAR	16 PIS	10 PIS 27 ARI	3 PIS 19 ARI	11 ARI	3 ARI	PIS	18 PIS	14 PIS
APR	4 ARI 20 TAU	11 TAU 29GEM	4 TAU	ARI	3 PIS 14 ARI	15 ARI	9 ARI 25 TAU	2 ARI 17 TAU
MAY	4 GEM	28 TAU	TAU	16 TAU	10 TAU 26GEM	3 TAU 18GEM	9 GEM 28 CAN	2 GEM
JUN	GEM	12GEM	11GEM 27 CAN	4 GEM 19 CAN	9 CAN 27 LEO	1 CAN	CAN	GEM
JUL	12 CAN 28 LEO	5 CAN 20 LEO	11 LEO 29 VIR	4 LEO 29 VIR	LEO	CAN	2 GEM 10 CAN	10 CAN 25 LEO
AUG	12 VIR 30 LIB	4 VIR 26 LIB	VIR	19 LEO	LEO	10 LEO 26 VIR	3 LEO 17 VIR	9 VIR 28 LIB
SEP	LIB	26 VIR	VIR	10 VIR 27 LIB	3 VIR 19 LIB	11 LIB 30 SCO	3 LIB 27 SCO	LIB
OCT	LIB	11 LIB 30 SCO	5 LIB 22 SCO	15 SCO	7 SCO 29 SAG	SCO	19 LIB	LIB
NOV	6 SCO 25 SAG	17 SAG	10 SAG	4 SAG	21 SCO	SCO	10 SCO 29 SAG	4 SCO 22 SAG
DEC	14 CAP	7 CAP	1 CAP	SAG	12 SAG	6 SAG	19 CAP	11 CAP

362

Mercury Tables

	1996	1997	1998	1999	2000	2001	2002	2003
JAN	1 AQU 17 CAP	CAP	12 CAP	6 CAP 26 AQU	18 AQU	10 AQU	3 AQU	CAP
FEB	14 AQU	9 AQU 27 PIS	2 AQU 20 PIS	12 PIS	5 PIS	1 PIS 6 AQU	3 CAP 13 AQU	12 AQU
MAR	7 PIS 24 ARI	15 ARI	8 ARI	2 ARI 18 PIS	PIS	17 PIS	11 PIS 29 ARI	4 PIS 21 ARI
APR	7 TAU	1 TAU	ARI	17 ARI	12 ARI 29 TAU	6 ARI 21 TAU	13 TAU 30 GEM	5 TAU
MAY	TAU	4 ARI 12 TAU	14 TAU	8 TAU 23 GEM	14 GEM 29 CAN	5 GEM	GEM	TAU
JUN	13 GEM	8 GEM 23 CAN	1 GEM 15 CAN	6 CAN 26 LEO	CAN	GEM	GEM	12 GEM 29 CAN
JUL	2 CAN 16 LEO	8 LEO 26 VIR	LEO	31 CAN	CAN	12 CAN 30 LEO	7 CAN 21 LEO	13 LEO 30 VIR
AUG	1 VIR 26 LIB	VIR	LEO	10 LEO 31 VIR	7 LEO 22 VIR	13 VIR 31 LIB	6 VIR 26 LIB	VIR
SEP	12 VIR	VIR	7 VIR 24 LIB	16 LIB	7 LIB 28 SCO	LIB	LIB	VIR
OCT	8 LIB 26 SCO	2 LIB 19 SCO	11 SCO	4 SCO 30 SAG	SCO	LIB	2 VIR 11 LIB 31 SCO	6 LIB 24 SCO
NOV	14 SAG	7 SAG 30 CAP	1 SAG	9 SCO	SCO	7 SCO 26 SAG	19 SAG	12 SAG
DEC	4 CAP	13 SAG	SAG	10 SAG	3 SAG	15 CAP	8 CAP	2 CAP 30 SAG

Mercury Tables

	2004	2005	2006	2007	2008	2009	2010	2011
JAN	14 CAP	9 CAP 30 AQU	3 CAP 22 AQU	15 AQU	7 AQU	1 AQU 21 CAP	CAP	13 CAP
FEB	6 AQU 25 PIS	16 PIS	8 PIS	2 PIS 26 AQU	AQU	14 AQU	10 AQU	3 AQU 21 PIS
MAR	12 ARI 31 TAU	4 ARI	PIS	18 PIS	14 PIS	8 PIS 25 ARI	1 PIS 17 ARI	9 ARI
APR	12 ARI	ARI	16 ARI	10 ARI 27 TAU	2 ARI 17 TAU	9 TAU 30 GEM	2 TAU	ARI
MAY	16 TAU	12 TAU 28 GEM	5 TAU 19 GEM	11 GEM 28 CAN	2 GEM	13 TAU	TAU	15 TAU

	2004	2005	2006	2007	2008	2009	2010	2011
JUN	5 GEM 19 CAN	11 CAN 27 LEO	3 CAN 28 LEO	CAN	GEM	13 GEM	10 GEM 25 CAN	2 GEM 16 CAN
JUL	4 LEO 25 VIR	LEO	10 CAN	CAN	10 CAN 26 LEO	3 CAN 17 LEO	9 LEO 27 VIR	2 LEO 28 VIR
AUG	24 LEO	LEO	10 LEO 27 VIR	4 LEO 19 VIR	10 VIR 28 LIB	2 VIR 25 LIB	VIR	8 LEO
SEP	10 VIR 28 LIB	4 VIR 20 LIB	12 LIB	5 LIB 27 SCO	LIB	17 VIR	VIR	9 VIR 25 LIB
OCT	15 SCO	8 SCO 30 SAG	1 SCO	23 LIB	LIB	9 LIB 28 SCO	3 LIB 20 SCO	13 SCO
NOV	4 SAG	26 SCO	SCO	11 SCO	4 SCO 23 SAG	15 SAG	8 SAG 30 CAP	2 SAG
DEC	SAG	12 SAG	8 SAG 27 CAP	1 SAG 20 CAP	12 CAP	5 CAP	18 SAG	SAG

MERCUR TABLES

	2012	2013	2014	2015	2016	2017	2018	2019
JAN	8 CAP 27 AQU	19 AQU	11 AQU 31 PIS	4 AQU	1 AQU 8 CAP	4 SAG 12 CAP	10 CAP 31 AQU	4 CAP 24 AQU
FEB	13 PIS	5 PIS	12 AQU	AQU	13 AQU	7 AQU 25 PIS	17 PIS	10 PIS
MAR	2 ARI 23 PIS	PIS	17 PIS	12 PIS 30 ARI	5 PIS 21 ARI	13 ARI 31 TAU	6 ARI	PIS
APR	16 ARI	13 ARI	7 ARI 23 TAU	14 TAU 30GEM	5 TAU	20 ARI	ARI	17 ARI
MAY	9 TAU 24GEM	1 TAU 15GEM	7 GEM 29 CAN	GEM	TAU	15 TAU	13 TAU 29GEM	6 TAU 21GEM
JUN	7 CAN 25 LEO	CAN	17GEM	8 CAN	29 CAN 13 LEO	21 CAN 5 LEO	28 LEO LEO	26 LEO 19 CAN
JUL	LEO	CAN	12 CAN 31 LEO	23 LEO 7 VIR	30 VIR VIR	25 VIR 31 LEO	LEO	11 LEO
AUG	31 VIR	8 LEO 23 VIR	15 VIR	27 LIB LIB	VIR	9 VIR	5 VIR	29 VIR 14 LIB
SEP	16 LIB	9 LIB 29 SCO	2 LIB 27 SCO	LIB	7 LIB	29 LIB 17 SCO	21 LIB 9 SCO	3 SCO
OCT	5 SCO 29 SAG	SCO	10 LIB		24 SCO		30 SAG	
NOV	14 SCO	SCO	8 SCO 27 SAG	2 SCO 20 SAG	12 SAG	5 SAG	SAG	SCO
DEC	10 SAG 31 CAP	4 SAG 24 CAP	16 CAP	9 CAP	2 CAP	SAG	1 SAG 12 SAG	9 SAG 28 CAP

MERCURY TABLES

	2020	2021	2022	2023	2024	2025
JAN	16 AQU	8 AQU	2 AQU 25 CAO	CAP	13 CAP	8 CAP 27 AQU
FEB	3 PIS	AQU	14 AQU	11 AQU	4 AQU 23 PIS	14 PIS
MAR	4 AQU 16 PIS	15 PIS	9 PIS 27 ARI	2 PIS 18 ARI	9 ARI	3 ARI 29 PIS
APR	10 ARI 27 TAU	3 ARI 19 TAU	10 TAU 29GEM	3 TAU	ARI	16 ARI
MAY	11GEM 28 CAN CAN	3 GEM GEM	22 TAU	TAU	15 TAU	10 TAU 25GEM
JUN	 CAN	 11 CAN	13GEM	11GEM 26 CAN	3 GEM 17 CAN	8 CAN 26 LEO
JUL	 4 LEO	27 LEO 11 VIR	5 CAN 19 LEO	10 LEO 28 VIR	2 LEO 25 VIR	LEO
AUG	19 VIR 5 LIB	29 LIB LIB	4 VIR 25 LIB	VIR	14 LEO	LEO
SEP	27 SCO 27 LIB	LIB	23 VIR	VIR	9 VIR 26 LIB	2 LIB 18 LIB
OCT		5 SCO 24 SAG	10 LIB 29 SCO	4 LIB 22 SCO	13 SCO	6 SCO 29 SAG
NOV	10 SCO	13 CAP	17 SAG	10 SAG	2 SAG	18 SCO
DEC	1 SAG 20 CAP	4 SAG 24 CAP	6 CAP	1 CAP 23 SAG	SAG	11 SAG

Venus Tables

	1900	1901	1902	1903	1904	1905	1906	1907
JAN	19 PIS	16 CAP	11 PIS	10 AQU	4 SAG 30 CAP	7 PIS	1 CAP 25 AQU	SAG
FEB	13 ARI	9 AQU	6 AQU	3 PIS 27 ARI	23 AQU	2 ARI	18 PIS	6 CAP
MAR	10 TAU	5 PIS 29 ARI	AQU	24 TAU	19 PIS	6 TAU	14 ARI	6 AQU
APR	5 GEM	22 TAU	4 PIS	18GEM	12 ARI	TAU	7 TAU	1 PIS 27 ARI
MAY	5 CAN	17GEM	7 ARI	13 CAN	7 TAU 31GEM	9 ARI 28 TAU	1 GEM 26 CAN	22 TAU
JUN	CAN	10 CAN	3 TAU 30GEM	8 LEO	25 CAN	TAU	20 LEO	16GEM
JUL	CAN	5 LEO 29 VIR	25 CAN	7 VIR	19 LEO	8 GEM	15 VIR	11 CAN
AUG	CAN	23 LIB	19 LEO	17 LIB	13 VIR	6 CAN	10 LIB	4 LEO 28 VIR
SEP	8 LEO	17 SCO	13 VIR	5 VIR	6 LIB 30 SCO	1 LEO 26 VIR	7 SCO	22 LIB
OCT	8 VIR	12 SAG	7 LIB 31 SCO	VIR	25 SAG	21 LIB	9 SAG	16 SCO
NOV	3 LIB 28 SCO	7 CAP	24 SAG	8 LIB	18 CAP	14 SCO	SAG	9 SAG
DEC	23 SAG	5 AQU	18 CAP	9 SCO	13 AQU	8 SAG	15 SCO 25 SAG	3 CAP 27 AQU

Venus Tables

	1908	1909	1910	1911	1912	1913	1914	1915
JAN	20 PIS	15 CAP	15 PIS 29 AQU	10 AQU	4 SAG 29 CAP	7 PIS	1 CAP 24 AQU	SAG
FEB	13 ARI	8 AQU	AQU	3 PIS 27 ARI	23 AQU	2 ARI	17 PIS	6 CAP
MAR	10 TAU	4 PIS 28 ARI	AQU	23 TAU	18 PIS	6 TAU	13 ARI	6 AQU
APR	5 GEM	22 TAU	5 PIS	17GEM	12 ARI	TAU	6 TAU	1 PIS 26 ARI
MAY	5 CAN	16GEM	6 ARI	13 CAN	6 TAU 31GEM	1 ARI 31 TAU	1 GEM 26 CAN	21 TAU

	1908	1909	1910	1911	1912	1913	1914	1915
JUN	CAN	9 CAN	3 TAU 29GEM	8 LEO	24 CAN	TAU	19 LEO	15GEM
JUL	CAN	4 LEO 29 VIR	25 CAN	7 VIR	19 LEO	8 GEM	15 VIR	10 CAN
AUG	CAN	22 LIB	19 LEO	VIR	12 VIR	5 CAN	10 LIB	4 LEO 28 VIR
SEP	8 LEO	16 SCO	12 VIR	VIR	5 LIB 30 SCO	1 LEO 26 VIR	7 SCO	21 LIB
OCT	8 VIR	12 SAG	6 LIB 30 SCO	VIR	24 SAG	21 LIB	9 SAG	15 SCO
NOV	3 LIB 28 SCO	7 CAP	23 SAG	8 LIB	17 CAP	14 SCO	SAG	8 SAG
DEC	22 SAG	5 AQU	17 CAP	9 SCO	12 AQU	8 SAG	5 SCO 30 SAG	2 CAP 26 AQU

368

Venus Tables

	1916	1917	1918	1919	1920	1921	1922	1923
JAN	19 PIS	15 CAP	AQU	9 AQU	4 SAG 29 CAP	6 PIS	24 AQU	2 SAG
FEB	13 ARI	8 AQU	AQU	2 PIS 26 ARI	22 ARI	2 ARI	17 PIS	6 CAP
MAR	9 TAU	4 PIS 28 ARI	AQU	23 TAU	18 PIS	7 TAU	13 ARI	6 AQU
APR	5 GEM	21 TAU	5 PIS	17GEM	11 ARI	25 ARI	6 TAU 30GEM	1 PIS 26 ARI
MAY	5 CAN	15GEM	6 ARI	12 CAN	6 TAU 30GEM	ARI	25 CAN	21 TAU
JUN	CAN	9 CAN	3 TAU 29GEM	8 LEO	24 CAN	1 TAU	19 LEO	15GEM
JUL	CAN	3 LEO 28 VIR	24 CAN	7 VIR	18 LEO	8 GEM	14 VIR	9 CAN
AUG	CAN	22 LIB	18 LEO	VIR	11 VIR	5 CAN 31 LEO	10 LIB	3 LEO 27 VIR
SEP	8 LEO	16 SCO	12 VIR	VIR	5 LIB 29 SCO	25 VIR	7 SCO	20 LIB
OCT	7 VIR	11 SAG	6 LIB 30 SCO	VIR	24 SAG	20 LIB	10 SAG	14 SCO
NOV	2 LIB 27 SCO	6 CAP	23 SAG	9 LIB	17 CAP	13 SCO	28 SCO	8 SAG
DEC	22 SAG	5 AQU	16 CAP	8 SCO	12 AQU	7 SAG 31 CAP	SCO	2 CAP 26 AQU

Venus Tables

	1924	1925	1926	1927	1928	1929	1930	1931
JAN	19 PIS	14 CAP	AQU	9 AQU	3 AQU 28 CAP	6 PIS	23 AQU	3 SAG
FEB	12 ARI	7 AQU	AQU	2 PIS 26 ARI	22AQU	2 ARI	12 ARI	5 AQU 31 PIS
MAR	9 TAU	3 PIS 27 ARI	AQU	22 TAU	17 PIS	8 TAU	12 ARI	5 AQU 31 PIS
APR	5 GEM	21 TAU	5 PIS	16GEM	11 ARI	19 ARI	5 TAU 30GEM	25 ARI
MAY	5 CAN	15GEM	6 ARI	12 CAN	5 TAU 30GEM	ARI	24 CAN	20 TAU
JUN	CAN	8 CAN	2 TAU 28GEM	7 LEO	23 CAN	3 TAU	19 LEO	14GEM
JUL	CAN	3 LEO 28 VIR	24 CAN	7 VIR	18 LEO	7 GEM	14 VIR	9 CAN
AUG	CAN	21 LIB	17 LEO	VIR	11 VIR	5 CAN 31 LEO	9 LIB	2 LEO 27 VIR
SEP	8 LEO	15 SCO	11 VIR	VIR	4 LIB 29 SCO	25 VIR	6 SCO	20 LIB
OCT	7 VIR	11 SAG	5 LIB 29 SCO	VIR	23 SAG	19 LIB	11 SAG	14 SCO
NOV	2 LIB 27 SCO	6 CAP	22 SAG	9 LIB	17 CAP	13 SCO	22 SCO	7 SAG
DEC	21 SAG	5 AQU	16 CAP	8 SCO	11 AQU	7 SAG 30 CAP	SCO	1 CAP 25 AQU

Venus Tables

	1932	1933	1934	1935	1936	1937	1938	1939
JAN	18 PIS	14 CAP	AQU	8 AQU	3 SAG 28 CAP	5 PIS	23 AQU	4 SAG
FEB	12 ARI	7 AQU	AQU	1 PIS 25 ARI	21 AQU	2 ARI	16 PIS	6 CAP
MAR	8 TAU	3 PIS 27 ARI	AQU	22 TAU	17 PIS	9 TAU	12 ARI	5 AQU 31 PIS
APR	4 GEM	20 TAU	6 PIS	16GEM	10 ARI	13 ARI	5 TAU 29GEM	25 ARI
MAY	6 CAN	14GEM	6 ARI	11 CAN	5 TAU 29GEM	ARI	24 CAN	20 TAU
JUN	CAN	8 CAN	2 TAU 28GEM	7 LEO	23 CAN	4 TAU	18 LEO	14GEM

	1932	1933	1934	1935	1936	1937	1938	1939
JUL	13GEM 28 CAN	2 LEO 27 VIR	23 CAN	7 VIR	17 LEO	7 GEM	14 VIR	8 CAN
AUG	CAN	21 LIB	17 LOE	VIR	10 VIR	4 CAN 30 LEO	9 LIB	2 LEO 26 VIR
SEP	8 LEO	15 SCO	10 VIR	VIR	4 LIB 28 SCO	24 VIR	6 SCO	19 LIB
OCT	7 VIR	10 SAG	5 LIB 29 SCO	VIR	22 SAG	19 LIB	13 SAG	13 SCO
NOV	1 LIB 26 SCO	6 CAP	21 SAG	9 LIB	16 CAP	12 SCO	15 SCO	6 SAG 30 CAP
DEC	21 SAG	5 AQU	15 CAP	8 SCO	11 AQU	6 SAG 30 CAP	SCO	25 AQU

Venus Tables

	1940	1941	1942	1943	1944	1945	1946	1947
JAN	18 PIS	13 CAP	AQU	8 AQU	2 SAG 27 CAP	5 PIS	22 AQU	5 SAG
FEB	12 ARI	6 AQU	AQU	1 PIS 25 ARI	21 AQU	2 ARI	15 PIS	6 CAP
MAR	8 TAU	2 PIS 26 ARI	AQU	21 TAU	16 PIS	11 TAU	11 ARI	4 AQU 30 PIS
APR	4 GEM	20 TAU	6 PIS	15GEM	10 ARI	7 ARI	4 TAU 28GEM	24 ARI
MAY	6 CAN	14GEM	5 ARI	11 CAN	4 TAU 29GEM	ARI	23 CAN	19 TAU
JUN	CAN	7 CAN	1 TAU 27GEM	7 LEO	22 CAN	4 TAU	17 LEO	13GEM
JUL	5 GEM 31 CAN	2 LEO 26 VIR	23 CAN	7 VIR	16 LEO	7 GEM	13 VIR	8 CAN
AUG	CAN	20 LIB	16 LEO	VIR	10 VIR	4 CAN 30 LEO	9 LIB	1 LEO 26 VIR
SEP	8 LEO	14 SCO	10 VIR	VIR	3 LIB 28 SCO	24 VIR	6 SCO	19 LIB
OCT	6 VIR	10 SAG	4 LIB 28 SCO	VIR	22 SAG	18 LIB	16 SAG	13 SCO
NOV	1 LIB 26 SCO	6 CAP	21 SAG	9 LIB	16 CAP	12 SCO	8 SCO	6 SAG 30 CAP
DEC	20 SAG	5 AQU	15 CAP	8 SCO	10 AQU	6 SAG 29 CAP	SCO	24 AQU

VENUS TABLES

	1948	1949	1950	1951	1952	1953	1954	1955
JAN	17 PIS	13 CAP	AQU	7 AQU 31 PIS	2 SAG 27 CAP	5 PIS	22 AQU	6 SAG
FEB	11 ARI	6 AQU	AQU	24 ARI	20 AQU	2 ARI	15 PIS	5 CAP
MAR	8 TAU	2 PIS 26 ARI	AQU	21 TAU	16 PIS	14 TAU 31 ARI	11 ARI	4 AQU 30 PIS
APR	4 GEM	19 TAU	6 PIS	15GEM	9 ARI	ARI	4 TAU 28GEM	24 ARI
MAY	7 CAN	13GEM	5 ARI	10 CAN	4 TAU 28GEM	ARI	23 CAN	19 TAU
JUN	29GEM	7 CAN	1 TAU 27GEM	7 LEO	22 CAN	5 TAU	17 LEO	13GEM
JUL	GEM	1 LEO 26 VIR	22 CAN	7 VIR	16 LEO	7 GEM	13 VIR	7 CAN
AUG	2 CAN	20 LIB	16 LEO	VIR	9 VIR	3 CAN 29 LEO	8 LIB	1 LEO
25 VIR	8 LEO	14 SCO	9 VIR	VIR	3 LIB 27 SCO	23 VIR	6 SCO	18 LIB
OCT	6 VIR	10 SAG	4 LIB 28 SCO	VIR	21 SAG	18 LIB	23 SAG 27 SCO	12 SCO
NOV	1 LIB 25 SCO	5 CAP	20 SAG	9 LIB	15 CAP	11 SCO	SCO	5 SAG 29 CAP
DEC	20 SAG	6 AQU	14 CAP	7 SCO	10 AQU	5 SAG 29 CAP	SCO	24 AQU

Venus Tables

	1956	1957	1958	1959	1960	1961	1962	1963
JAN	17 PIS	12 CAP	AQU	7 AQU 31 PIS	2 SAG 26 CAP	4 PIS	21 AQU	6 SAG
FEB	11 ARI	5 AQU	AQU	24 ARI	20 AQU	1 ARI	14 PIS	5 CAP
MAR	7 TAU	1 PIS 25 ARI	AQU	20 TAU	15 PIS	ARI	10 ARI	4 AQU 29 PIS
APR	4 GEM	18 TAU	6 PIS	14 GEM	9 ARI	ARI	3 TAU 28 GEM	23 ARI
MAY	7 CAN	13 GEM	5 ARI 31 TAU	10 CAN	3 TAU 28 GEM	ARI	22 CAN	18 TAU
JUN	23 GEM	6 CAN	26 GEM	6 LEO	21 CAN	5 TAU	17 LEO	12 GEM
JUL	GEM	1 LEO 25 VIR	22 CAN	8 VIR	15 LEO	6 GEM	12 VIR	7 CAN 31 LEO

	1956	1957	1958	1959	1960	1961	1962	1963
AUG	4 CAN	19 LIB	15 LEO	VIR	9 VIR	3 CAN 29 LEO	8 LIB	25 VIR
SEP	8 LOE	14 SCO	9 VIR	19 LEO 25 VIR	2 LIB 26 SCO	23 VIR	6 SCO	18 LIB
OCT	5 VIR 31 LIB	9 SAG	3 LIB 27 SCO	VIR	21 SAG	17 LIB	SCO	12 SCO
NOV	25 SCO	5 CAP	20 SAG	9 LIB	15 CAP	11 SCO	SCO	5 SAG 29 CAP
DEC	19 SAG	6 AQU	14 CAP	7 SCO	10 AQU	4 SAG 28 CAP	SCO	23 AQU

Venus Tables

	1964	1965	1966	1967	1968	1969	1970	1971
JAN	16 PIS	12 CAP	AQU	6 AQU 30 PIS	1 SAG 26 CAP	4 PIS	21 AQU	6 SAG
FEB	10 ARI	5 AQU	6 CAP 25 AQU	23 ARI	19 AQU	1 ARI	13 PIS	5 CAP
MAR	7 TAU	1 PIS 25 ARI	AQU	20 TAU	15 PIS	ARI	10 ARI	3 AQU 29 PIS
APR	3 GEM	18 TAU	6 PIS	14 GEM	8 ARI	ARI	3 TAU 27 GEM	23 ARI
MAY	8 CAN	12 GEM	4 ARI 31 TAU	10 CAN	3 TAU 27 GEM	ARI	22 CAN	18 TAU
JUN	17 GEM	6 CAN 30 LEO	26 GEM	6 LEO	20 CAN	5 TAU	16 LEO	12 GEM
JUL	GEM	25 VIR	21 CAN	8 VIR	15 LEO	6 GEM	12 VIR	6 CAN 31 LEO
AUG	5 CAN	19 LIB	15 LEO	VIR	8 VIR	3 CAN 28 LEO	8 LIB	24 VIR
SEP	7 LEO	13 SCO	8 VIR	9 LEO	2 LIB 26 SCO	22 VIR	6 SCO	17 LIB
OCT	5 VIR 31 LIB	9 SAG	2 LIB 26 SCO	1 VIR	21 SCO	17 LIB	SCO	11 SCO
NOV	24 SCO	5 CAP	19 SAG	9 LIB	14 CAP	10 SCO	SCO	4 SAG 28 CAP
DEC	19 SAG	6 AQU	13 CAP	7 SCO	9 AQU	4 SAG 28 CAP	SCO	23 AQU

Venus Tables

	1972	1973	1974	1975	1976	1977	1978	1979
JAN	16 PIS	11 CAP	29 CAP	6 AQU 30 PIS	1 SAG 26 CAP	4 PIS	20 AQU	7 SAG
FEB	10 ARI	4 AQU 28 PIS	28 AQU	23 ARI	19 AQU	2 ARI	13 PIS	5 CAP
MAR	6 TAU	24 ARI	AQU	19 TAU	14 PIS	ARI	9 ARI	3 AQU 28 PIS
APR	3 GEM	17 TAU	6 PIS	13GEM	8 ARI	ARI	2 TAU 27GEM	22 ARI
MAY	10 CAN	12GEM	4 ARI 31 TAU	9 CAN	2 TAU 26GEM	ARI	21 CAN	17 TAU
JUN	11GEM	5 CAN 30 LEO	25GEM	6 LEO	20 CAN	6 TAU	16 LEO	11GEM
JUL	GEM	24 VIR	20 CAN	9 VIR	14 LEO	6 GEM	11 VIR	6 CAN 30 LEO
AUG	5 CAN	18 LIB	14 LEO	VIR	8 VIR	2 CAN 28 LEO	7 LIB	23 VIR
SEP	7 LEO	13 SCO	8 VIR	2 LEO	1 LIB 25 SCO	22 VIR	7 SCO	17 LIB
OCT	5 VIR 30 LIB	9 SAG	2 LIB 26 SCO	4 VIR	20 SAG	16 LIB	SCO	11 SCO
NOV	4 SCO	5 CAP	19 SAG	9 LIB	14 CAP	9 SCO	SCO	4 SAG 28 CAP
DEC	18 SAG	7 AQU	13 CAP	6 SCO	9 AQU	3 SAG 27 CAP	SCO	22 AQU

Venus Tables

	1980	1981	1982	1983	1984	1985	1986	1987
JAN	15 PIS	11 CAP	22 CAP	5 AQU 29 PIS	25 CAP	4 PIS	20 AQU	7 SAG
FEB	9 ARI	4 AQU 28 PIS	CAP	22 ARI	18 AQU	2 ARI	12 PIS	4 CAP
MAR	6 TAU	24 ARI	2 AQU	19 TAU	14 PIS	ARI	8 ARI	3 AQU 28 PIS
APR	3 GEM	17 TAU	6 PIS	13GEM	7 ARI	ARI	2 TAU 26GEM	22 ARI
MAY	12 CAN	11GEM	4 ARI 30 TAU	9 CAN	1 TAU 26GEM	ARI	21 CAN	17 TAU

	1980	1981	1982	1983	1984	1985	1986	1987
JUN	5 GEM	5 CAN 29 LEO	25GEM	6 LEO	19 CAN	6 TAU	15 LEO	11GEM
JUL	GEM	24 VIR	20 CAN	10 VIR	14 LEO	6 GEM	11 VIR	5 CAN 30 LEO
AUG	6 CAN	18 LIB	14 LEO	27 LEO	7 VIR 31 LIB	2 CAP 27 LEO	7 LIB	23 VIR
SEP	7 LEO	12 SCO	7 VIR	LEO	25 SCO	21 VIR	7 SCO	16 LIB
OCT	4 VIR 30 LIB	8 SAG	1 LIB 25 SCO	5 VIR	20 SAG	16 LIB	SCO	10 SCO
NOV	23 SCO	5 CAP	18 SAG	9 LIB	13 CAP	9 SCO	SCO	3 SAG 27 CAP
DEC	18 SAG	8 AQU	12 CAP	6 SCO 31 SAG	8 AQU	3 SAG 27 CAP	SCO	22 AQU

Venus Tables

	1988	1989	1990	1991	1992	1993	1994	1995
JAN	15 PIS	10 CAP	16 CAP	4 AQU 28 PIS	25 CAP	3 PIS	19 AQU	7 SAG
FEB	9 ARI	3 AQU 27 PIS	CAP	22 ARI	18 AQU	2 ARI	12 PIS	4 CAP
MAR	6 TAU	23 ARI	3 AQU	18 TAU	13 PIS	ARI	8 ARI	2 AQU 28 PIS
APR	3 GEM	16 TAU	6 PIS	12GEM	7 ARI	ARI	1 TAU 26GEM	21 ARI
MAY	17 CAN 27GEM	11GEM	3 ARI 30 TAU	8 CAN	1 TAU 25GEM	ARI	20 CAN	16 TAU
JUN	GEM	4 CAN 29 LEO	24GEM	5 LEO	19 CAN	6 TAU	15 LEO	10GEM
JUL	GEM	23 VIR	19 CAN	11 VIR	13 LEO	5 GEM	11 VIR	5 CAN 29 LEO
AUG	6 CAN	17 LIB	13 LEO	21 LOE	7 VIR 31 LIB	1 CAN 27 LEO	7 LIB	22 VIR
SEP	7 LEO	12 SCO	7 VIR	LEO	24 SCO	21 VIR	7 SCO	15 LIB
OCT	4 VIR 29 LIB	8 SAG	1 LIB 25 SCO	6 VIR	19 SAG	15 LIB	SCO	10 SCO
NOV	23 SCO	5 CAP	18 SAG	9 LIB	13 CAP	8 SCO	SCO	3 SAG 27 CAP
DEC	17 SAG	9 AQU	12 CAP	6 SCO 31 SAG	8 AQU	2 SAG 26 CAP	SCO	21 AQU

Venus Tables

	1996	1997	1998	1999	2000	2001	2002	2003
JAN	14 PIS	10 CAP	9 CAP	4 AQU 28 PIS	24 CAP	3 PIS	18 AQU	7 SAG
FEB	8 ARI	2 AQU 26 PIS	CAP	21 ARI	17 AQU	2 ARI	11 PIS	4 CAP
MAR	5 TAU	23 ARI	4 AQU	18 TAU	13 PIS	ARI	7 ARI	2 AQU 27 PIS
APR	3 GEM	16 TAU	6 PIS	12GEM	6 ARI 30 TAU	ARI	1 TAU 25GEM	21 ARI
MAY	GEM	10GEM	3 ARI 29 TAU	8 CAN	25GEM	ARI	20 CAN	16 TAU
JUN	GEM	3 CAN 28 LEO	24GEM	5 LEO	18 CAN	6 TAU	14 LEO	9 GEM
JUL	GEM	23 VIR	19 CAN	12 VIR	13 LEO	5 GEM	10 VIR	4 CAN 29 LEO
AUG	7 CAN	17 LIB	13 LEO	15 LEO	6 VIR 30 LIB	1 CAN 26 LEO	7 LIB	22 VIR
SEP	6 LEO	11 SCO	6 VIR 30 LIB	LEO	24 SCO	20 VIR	7 SCO	15 LIB
OCT	3 VIR 29 LIB	8 SAG	24 SCO	7 VIR	19 SAG	15 LIB	SCO	9 SCO
NOV	22 SCO	5 CAP	17 SAG	8 LIB	12 CAP	8 SCO	SCO	2 SAG 26 CAP
DEC	17 SAG	11 AQU	11 CAP	5 SCO 30 SAG	8 AQU	2 SAG 26 CAP	SCO	21 AQU

Venus Tables

	2004	2005	2006	2007	2008	2009	2010	2011
JAN	14 PIS	9 CAP	1 CAP	3 AQU 27 PIS	24 CAP	3 PIS	18 AQU	7 SAG
FEB	8 ARI	2 AQU 26 PIS	CAP	21 ARI	17 AQU	2 ARI	11 PIS	4 CAP
MAR	5 TAU	22 ARI	5 AQU	17 TAU	12 PIS	ARI	7 ARI 31 TAU	1 AQU 27 PIS
APR	3 GEM	15 TAU	5 PIS	11GEM	6 ARI 30 TAU	11 PIS 24 ARI	24GEM	20 ARI
MAY	GEM	9 GEM	3 ARI 29 TAU	8 CAN	24GEM	ARI	19 CAN	15 TAU
JUN	GEM	3 CAN 28 LEO	23GEM	5 LEO	18 CAN	6 TAU	14 LEO	9 GEM

	2004	2005	2006	2007	2008	2009	2010	2011
JUL	GEM	22 VIR	18 CAN	14 VIR	12 LEO	5 GEM 31 CAN	10 VIR	3 CAN 28 LEO
AUG	7 CAN	16 LIB	12 LEO	8 LEO	5 VIR 30 LIB	26 LEO	6 LIB	21 VIR
SEP	6 LEO	11 SCO	6 VIR 30 LIB	LEO	23 SCO	20 VIR	8 SCO	14 LIB
OCT	3 VIR 28 LIB	7 SAG	24 SCO	8 VIR	18 SAG	14 LIB	SCO	9 SCO
NOV	22 SCO	5 CAP	17 SAG	8 LIB	12 CAP	7 SCO	7 LIB 29 SCO	2 SAG 26 CAP
DEC	16 SAG	15 AQU	11 CAP	5 SCO 30 SAG	7 AQU	1 SAG 25 CAP	SCO	20 AQU

Venus Tables

	2012	2013	2014	2015	2016	2017	2018	2019
JAN	14 PIS	8 CAP	CAP	3 AQU 27 PIS	23 CAP	3 PIS	17 AQU	7 SAG
FEB	8 ARI	1 AQU 25 PIS	CAP	20 ARI	16 AQU	3 ARI	10 PIS	3 CAP
MAR	5 TAU	21 ARI	5 AQU	17 TAU	12 PIS	ARI	6 ARI 30 TAU	1 AQU 26 PIS
APR	3 GEM	15 TAU	5 PIS	11GEM	5 ARI 29 TAU	2 PIS 28 ARI	24GEM	20 ARI
MAY	GEM	9 GEM	2 ARI 28 TAU	7 CAN	24 GEM	ARI	19 CAN	15 TAU
JUN	GEM	2 CAN 27 LEO	23 GEM	5 LEO	17 CAN	6 TAU	13 LEO	8 GEM
JUL	GEM	22 VIR	18 CAN	18 VIR 31 LEO	12 LEO	4 GEM 31 CAN	9 VIR	3 CAN 27 LEO
AUG	7 CAN	16 LIB	12 LEO	LEO	5 VIR 29 LIB	25 LEO	6 LIB	21 VIR
SEP	6 LEO	11 SCO	5 VIR 29 LIB	LEO	23 SCO	19 VIR	9 SCO	14 LIB
OCT	3 VIR 28 LIB	7 SAG	23 SCO	8 LIB	18 SCO	14 LIB	31 LIB	8 SCO
NOV	21 SCO	5 CAP	16 SAG	8 LIB	11 CAP	7 SCO	LIB	1 SAG 25 CAP
DEC	15 SAG	CAP	10 CAP	4 SCO 30 SAG	7 AQU	1 SAG 25 CAP	2 SCO	20 AQU

VENUS TABLES

	2020	2021	2022	2023	2024	2025
JAN	13 PIS	8 CAP	CAP	2 AQU 26 PIS	23 CAP	2 PIS
FEB	7 ARI	1 AQU 26 PIS	CAP	20 ARI	16 AQU	4 ARI
MAR	4 TAU	21 ARI	6 AQU	16 TAU	11 PIS	27 PIS
APR	3 GEM	14 TAU	5 PIS	10 GEM	4 ARI 29 TAU	30 ARI
MAY	GEM	8 GEM	2 ARI 28 TAU	7 CAN	23 GEM	ARI
JUN	GEM	2 CAN 26 LEO	22 GEM	5 LEO	17 CAN	5 TAU
JUL	GEM	21 VIR	17 CAN	LEO	11 LEO	4 GEM 30 CAN
AUG	7 CAN	15 LIB	11 LEO	LEO	4 VIR 29 LIB	25 LEO
SEP	6 LEO	10 SCO	4 VIR 29 LIB	LEO	22 SCO	19 VIR
OCT	2 VIR 27 LIB	7 SAG	23 SCO	8 VIR	17 SAG	13 LIB
NOV	11 SCO	5 CAP	16 SAG	8 LIB	11 CAP	6 SCO 30 SAG
DEC	15 SAG	CAP	9 CAP	4 SCO 29 SAG	7 AQU	4 CAP

Mars Tables

	1900	1901	1902	1903	1904	1905	1906	1907
JAN	21 AQU	VIR	1 AQU	LIB	19 PIS	13 SCO	PIS	SCO
FEB	28 PIS	VIR	8 PIS	LIB	26 ARI	SCO	4 ARI	5 SAG
MAR	PIS	1 LEO	18 ARI	LIB	ARI	SCO	17 TAU	SAG
APR	7 ARI	LEO	27 TAU	19 VIR	6 TAU	SCO	28 GEM	1 CAP
MAY	17 TAU	11 VIR	TAU	30 LIB	17 GEM	SCO	GEM	CAP
JUN	27 GEM	VIR	7 GEM	LIB	30 CAN	SCO	11 CAN	CAP
JUL	GEM	13 LIB	20 CAN	LIB	CAN	SCO	27 LEO	CAP
AUG	9 CAN	31 SCO	CAN	6 SCO	14 LEO	21 SAG	LEO	CAP
SEP	26 LEO	SCO	4 LEO	22 SAG	LEO	SAG	12 VIR	CAP
OCT	LEO	14 SAG	23 VIR	SAG	1 VIR	7 CAP	29 LIB	13 AQU
NOV	23 VIR	23 CAP	VIR	3 CAP	20 LIB	17 AQU	LIB	27 PIS
DEC	VIR	CAP	19 LIB	12 AQU	LIB	27 PIS	17 SCO	PIS

Mars Tables

	1908	1909	1910	1911	1912	1913	1914	1915
JAN	10 ARI	9 SAG	22 TAU	31 CAP	30 GEM	10 CAP	CAN	30 AQU
FEB	22 TAU	23 CAP	TAU	CAP	GEM	19 AQU	CAN	AQU
MAR	TAU	CAP	14 GEM	13 AQU	GEM	30 PIS	CAN	9 PIS
APR	6 GEM	9 AQU	GEM	23 PIS	5 CAN	PIS	CAN	16 ARI
MAY	22 CAN	25 PIS	1 CAN	PIS	28 LEO	7 ARI	1 LEO	25 TAU
JUN	CAN	PIS	18 LEO	2 ARI	LEO	16 TAU	25 VIR	TAU
JUL	7 LEO	21 ARI	LEO	15 TAU	16 VIR	29 GEM	VIR	6 GEM
AUG	24 VIR	ARI	5 VIR	TAU	VIR	GEM	14 LIB	19 CAN
SEP	VIR	26 PIS	21 LIB	5 GEM	2 LIB	15 CAN	29 SCO	CAN
OCT	10 LIB	PIS	LIB	GEM	17 SCO	CAN	SCO	7 LEO
NOV	25 SCO	20 ARI	6 SCO	30 TAU	30 SAG	CAN	11 SAG	LEO
DEC	SCO	ARI	20 SAG	TAU	SAG	CAN	21 CAP	LEO

Mars Tables

	1916	1917	1918	1919	1920	1921	1922	1923
JAN	LEO	9 AQU	11 LIB	27 PIS	31 SCO	5 PIS	SCO	21 ARI
FEB	LEO	16 PIS	25 VIR	PIS	SCO	13 ARI	18 SAG	ARI
MAR	LEO	26 ARI	VIR	6 ARI	SCO	25 TAU	SAG	3 TAU
APR	LEO	ARI	VIR	14 TAU	23 LIB	TAU	SAG	15GEM

	1916	1917	1918	1919	1920	1921	1922	1923
MAY	28 VIR	4 TAU	VIR	26GEM	LIB	5 GEM	SAG	30 CAN
JUN	VIR	14GEM	23 LIB	GEM	LIB	18 CAN	SAG	CAN
JUL	23 LIB	27 CAN	LIB	8 CAN	10 SCO	CAN	SAG	15 LEO
AUG	LIB	CAN	16 SCO	23 LEO	SCO	3 LEO	SAG	31 VIR
SEP	8 SCO	12 LEO	SCO	LEO	4 SAG	19 VIR	13 CAP	VIR
OCT	21 SAG	LEO	1 SAG	9 VIR	18 CAP	VIR	30 AQU	17 LIB
NOV	SAG	2 VIR	11 CAP	30 LIB	27 AQU	6 LIB	AQU	LIB
DEC	1 CAP	VIR	20 AQU	LIB	AQU	26 SCO	11 PIS	3 SCO

Mars Tables

	1924	1925	1926	1927	1928	1929	1930	1931
JAN	19 SAG	ARI	SAG	TAU	18 CAP	GEM	CAP	LEO
FEB	SAG	5 TAU	8 CAP	21GEM	28 AQU	GEM	6 AQU	16 CAN
MAR	6 CAP	23GEM	22 AQU	GEM	AQU	10 CAN	17 PIS	29 LEO
APR	24 AQU	GEM	AQU	16 CAN	7 PIS	CAN	24 ARI	LEO
MAY	AQU	9 CAN	3 PIS	CAN	16 ARI	12 LEO	ARI	LEO
JUN	24 PIS	26 LEO	14 ARI	6 LEO	26 TAU	LEO	2 TAU	10 VIR
JUL	PIS	LEO	ARI	25 VIR	TAU	4 VIR	14GEM	VIR
AUG	24 AQU	12 VIR	1 TAU	VIR	8 GEM	21 LIB	28 CAN	1 LIB
SEP	AQU	28 LIB	TAU	10 LIB	GEM	LIB	CAN	17 SCO
OCT	19 PIS	LIB	TAU	25 SCO	2 CAN	6 SCO	20 LEO	30 SAG
NOV	PIS	13 SCO	TAU	SCO	CAN	18 SAG	LEO	SAG
DEC	19 ARI	27 SAG	TAU	8 SAG	20GEM	29 CAP	LEO	9 CAP

Mars Tables

	1932	1933	1934	1935	1936	1937	1938	1939
JAN	17 AQU	VIR	AQU	LIB	14 PIS	5 SCO	30 ARI	29 SAG
FEB	24 PIS	VIR	3 PIS	LIB	24 ARI	SCO	ARI	SAG
MAR	PIS	VIR	14 ARI	LIB	ARI	12 SAG	12 TAU	21 CAP
APR	3 ARI	VIR	22 TAU	LIB	1 TAU	SAG	23GEM	CAP
MAY	12 TAU	VIR	TAU	LIB	13GEM	14 SCO	GEM	24 AQU
JUN	22GEM	VIR	2 GEM	LIB	25 CAN	SCO	6 CAN	AQU
JUL	GEM	6 LIB	15 CAN	29 SCO	CAN	SCO	22 LEO	21 CAP
AUG	4 CAN	26 SCO	30 LEO	SCO	10 LEO	8 SAG	LEO	CAP
SEP	20 LEO	SCO	LEO	16 SAG	26 VIR	30 CAP	7 VIR	23 AQU
OCT	LEO	9 SAG	18 VIR	28 CAP	VIR	CAP	25 LIB	AQU
NOV	13 VIR	19 CAP	VIR	CAP	14 LIB	11 AQU	LIB	19 PIS
DEC	VIR	27 AQU	11 LIB	6 AQU	LIB	21 PIS	11 SCO	PIS

Mars Tables

	1940	1941	1942	1943	1944	1945	1946	1947
JAN	3 ARI	4 SAG	11 TAU	26 CAP	GEM	5 CAP	CAN	25 AQU
FEB	16 TAU	17 CAP	TAU	CAP	GEM	14 AQU	CAN	AQU
MAR	TAU	CAP	7 GEM	8 AQU	28 CAN	24 PIS	CAN	4 PIS
APR	1 GEM	2 AQU	26 CAN	17 PIS	CAN	PIS	22 LEO	11 ARI
MAY	17 CAN	16 PIS	CAN	27 ARI	22 LEO	2 ARI	LEO	20 TAU
JUN	CAN	PIS	13 LEO	ARI	LEO	11 TAU	20 VIR	30GEM
JUL	3 LEO	2 ARI	LEO	7 TAU	11 VIR	23GEM	VIR	GEM
AUG	19 VIR	ARI	1 VIR	23GEM	28 LIB	GEM	9 LIB	13 CAN
SEP	VIR	ARI	17 LIB	GEM	LIB	7 CAN	24 SCO	30 LEO
OCT	5 LIB	ARI	LIB	GEM	13 SCO	CAN	SCO	LEO
NOV	20 SCO	ARI	1 SCO	GEM	25 SAG	11 LEO	6 SAG	LEO
DEC	SCO	ARI	15 SAG	GEM	SAG	26 CAN	17 CAP	1 VIR

385

Mars Tables

	1948	1949	1950	1951	1952	1953	1954	1955
JAN	VIR	4 AQU	LIB	22 PIS	19 SCO	PIS	SCO	15 ARI
FEB	12 LEO	11 PIS	LIB	PIS	SCO	7 ARI	9 SAG	26 TAU
MAR	LEO	21 ARI	28 VIR	1 ARI	SCO	20 TAU	SAG	TAU
APR	LEO	29 TAU	VIR	10 TAU	SCO	TAU	12 CAP	10GEM
MAY	18 VIR	TAU	VIR	21GEM	SCO	1 GEM	CAP	25 CAN
JUN	VIR	9GEM	11 LIB	GEM	SCO	13 CAN	CAP	CAN
JUL	17 LIB	23 CAN	LIB	3 CAN	SCO	29 LEO	2 SAG	11 LEO
AUG	LIB	CAN	10 SCO	18 LEO	27 SAG	LEO	24 CAP	27 VIR
SEP	3 SCO	6 LEO	25 SAG	LEO	SAG	14 VIR	CAP	VIR
OCT	17 SAG	26 VIR	SAG	4 VIR	12 CAP	VIR	21 AQU	13 LIB
NOV	24 CAP	VIR	6 CAP	24 LIB	21 AQU	1 LIB	AQU	28 SCO
DEC	CAP	26 LIB	15 AQU	LIB	30 PIS	20 SCO	4 PIS	SCO

Mars Tables

	1956	1957	1958	1959	1960	1961	1962	1963
JAN	13 SAG	28 TAU	SAG	TAU	13 CAP	CAN	CAP	LEO
FEB	28 CAP	TAU	3 CAP	10GEM	22 AQU	CAP	1 AQU	LEO
MAR	CAP	17GEM	17 AQU	GEM	AQU	CAN	12 PIS	LEO
APR	14 AQU	GEM	26 PIS	10 CAN	2 PIS	CAN	19 ARI	LEO
MAY	AQU	4 CAN	PIS	31 LEO	11 ARI	5 LEO	28 TAU	LEO
JUN	3 PIS	21 LEO	7 ARI	LEO	20 TAU	28 VIR	TAU	3 VIR
JUL	PIS	LEO	21 TAU	20 VIR	TAU	VIR	8 GEM	26 LIB
AUG	PIS	8 VIR	TAU	VIR	1 GEM	16 LIB	22 CAN	LIB
SEP	PIS	23 LIB	20GEM	5 LIB	20 CAN	LIB	CAN`	12 SCO
OCT	PIS	LIB	28 TAU	21 SCO	CAN	1 SCO	11 LEO	25 SAG
NOV	PIS	8 SCO	TAU	SCO	CAN	13 SAG	LEO	SAG
DEC	6 ARI	22 SAG	TAU	3 SAG	CAN	24 CAP	LEO	5 CAP

MARS TABLES

	1964	1965	1966	1967	1968	1969	1970	1971
JAN	13 AQU	VIR	30 PIS	LIB	9 PIS	SCO	24 ARI	22 SAG
FEB	20 PIS	VIR	PIS	12 SCO	16 ARI	25 SAG	ARI	SAG
MAR	29 ARI	VIR	9 ARI	31 LIB	27 TAU	SAG	6 TAU	12 CAP
APR	ARI	VIR	17 TAU	LIB	TAU	SAG	18GEM	CAP
MAY	7 TAU	VIR	28GEM	LIB	8 GEM	SAG	GEM	3 AQU
JUN	17GEM	28 LIB	GEM	LIB	20 CAN	SAG	2 CAN	AQU
JUL	30 CAN	LIB	10 CAN	19 SCO	CAN	SAG	18 LEO	AQU
AUG	CAN	20 SCO	25 LEO	SCO	25 LEO	SAG	LEO	AQU
SEP	14 LEO	SCO	LEO	9 SAG	21 VIR	21 CAP	2 VIR	AQU
OCT	LEO	4 SAG	12 VIR	22 CAP	VIR	CAP	20 LIB	AQU
NOV	5 VIR	14 CAP	VIR	CAP	9 LIB	4 AQU	LIB	6 PIS
DEC	VIR	23 AQU	3 LIB	1 AQU	29 SCO	15 PIS	6 SCO	26 ARI

MARS TABLES

	1972	1973	1974	1975	1976	1977	1978	1979
JAN	ARI	SAG	TAU	21 CAP	GEM	CAP	25 CAN	20 AQU
FEB	10 TAU	12 CAP	27GEM	CAP	GEM	9 AQU	CAN	27 PIS
MAR	26GEM	26 AQU	GEM	3 AQU	18 CAN	19 [PIS	CAN	PIS
APR	GEM	AQU	20 CAN	11 PIS	CAN	27 ARI	10 LEO	6 ARI
MAY	12 CAN	7 PIS	CAN	21 ARI	16 LEO	ARI	LEO	15 TAU
JUN	28 LEO	20 ARI	8 LEO	30 TAU	LEO	5 TAU	13 VIR	25GEM
JUL	LEO	ARI	27 VIR	TAU	6 VIR	17GEM	VIR	GEM
AUG	14 VIR	12 TAU	VIR	14GEM	24 LIB	31 CAN	4 LIB	8 CAN
SEP	30 LIB	TAU	12 LIB	GEM	LIB	CAN	19 SCO	24 LEO
OCT	LIB	29 ARI	28 SCO	17 CAN	8 SCO	26 LEO	SCO	LEO
NOV	15 SCO	ARI	SCO	25GEM	20 SAG	LEO	1 SAG	19 VIR
DEC	30 SAG	24 TAU	10 SAG	GEM	31 CAP	LEO	12 CAP	VIR

Mars Tables

	1980	1981	1982	1983	1984	1985	1986	1987
JAN	VIR	AQU	LIB	17 PIS	10 SCO	PIS	SCO	8 ARI
FEB	VIR	6 PIS	LIB	24 ARI	SCO	2 ARI	2 SAG	20 TAU
MAR	11 LEO	16 ARI	LIB	ARI	SCO	14 TAU	27 CAP	TAU
APR	LEO	25 TAU	LIB	5 TAU	SCO	26GEM	CAP	5 GEM
MAY	3 VIR	TAU	LIB	16GEM	SCO	GEM	CAP	20 CAN
JUN	VIR	5 GEM	LIB	29 CAN	SCO	9 CAN	CAP	CAN
JUL	10 LIB	18 CAN	LIB	CAN	SCO	24 LEO	CAP	6 LEO
AUG	29 SCO	CAN	3 SCO	13 LEO	17 SAG	LEO	CAP	22 VIR
SEP	SCO	1 LEO	19 SAG	29 VIR	SAG	9 VIR	CAP	VIR
OCT	12 SAG	20 VIR	31 CAP	VIR	5 CAP	27 LIB	8 AQU	8 LIB
NOV	21 CAP	VIR	CAP	18 LIB	15 AQU	LIB	25 PIS	23 SCO
DEC	30 AQU	15 LIB	10 AQU	LIB	25 PIS	14 SCO	PIS	SCO

Mars Tables

	1988	1989	1990	1991	1992	1993	1994	1995
JAN	8 SAG	19 TAU	29 CAP	20GEM	9 CAP	CAN	27 AQU	22 LEO
FEB	22 CAP	TAU	CAP	GEM	17 AQU	CAN	AQU	LEO
MAR	CAP	11GEM	11 AQU	GEM	27 PIS	CAN	7 PIS	LEO
APR	6 AQU	28 CAN	20 PIS	2 CAN	PIS	27 LEO	14 ARI	LEO
MAY	22 PIS	CAN	31 ARI	26 LEO	5 ARI	LEO	23 TAU	25 VIR
JUN	PIS	16 LEO	ARI	LEO	14 TAU	23 VIR	TAU	VIR
JUL	13 ARI	LEO	12 TAU	15 VIR	26GEM	VIR	3 GEM	21 LIB
AUG	ARI	3 VIR	31GEM	VIR	GEM	11 LIB	16 CAN	LIB
SEP	ARI	19 LIB	GEM	1 LIB	12 CAN	26 SCO	CAN	7 SCO
OCT	24 PIS	LIB	GEM	16 SCO	CAN	SCO	4 LEO	20 SAG
NOV	1 ARI	3 SCO	GEM	28 SAG	CAN	9 SAG	LEO	30 CAP
DEC	ARI	17 SAG	14 TAU	SAG	CAN	19 CAP	12 VIR	CAP

388

Mars Tables

	1996	1997	1998	1999	2000	2001	2002	2003
JAN	8 AQU	3 LIB	25 PIS	26 SCO	3 PIS	SCO	18 ARI	16 SAG
FEB	15 PIS	LIB	PIS	SCO	11 ARI	14 SAG	ARI	SAG
MAR	24 ARI	8 VIR	4 ARI	SCO	22 TAU	SAG	1 TAU	4 CAP
APR	ARI	VIR	12 TAU	SCO	TAU	SAG	13GEM	21 AQU
MAY	2 TAU	VIR	23GEM	5 LIB	3 GEM	SAG	28 CAN	AQU
JUN	12GEM	19 LIB	GEM	LIB	16 CAN	SAG	CAN	16 PIS
JUL	25 CAN	LIB	6 CAN	4 SCO	31 LEO	SAG	13 LEO	PIS
AUG	CAN	14 SCO	20 LEO	SCO	LEO	SAG	29 VIR	PIS
SEP	9 LEO	28 SAG	LEO	2 SAG	16 VIR	8 CAP	VIR	PIS
OCT	30 VIR	SAG	7 VIR	16 CAP	VIR	28 AQU	15 LIB	PIS
NOV	VIR	9 CAP	27 LIB	26 AQU	3 LIB	AQU	LIB	PIS
DEC	VIR	18 AQU	LIB	AQU	23 SCO	8 PIS	1 SCO	16 ARI

Mars Tables

	2004	2005	2006	2007	2008	2009	2010	2011
JAN	ARI	SAG	TAU	16 CAP	GEM	CAP	LEO	15 AQU
FEB	3 TAU	6 CAP	17GEM	25 AQU	GEM	4 AQU	LEO	22 PIS
MAR	21GEM	20 AQU	GEM	AQU	4 CAN	14 PIS	LEO	PIS
APR	GEM	30 PIS	13 CAN	6 PIS	CAN	22 ARI	LEO	1 ARI
MAY	7 CAN	PIS	CAN	15 ARI	9 LEO	31 TAU	LEO	11 TAU
JUN	23 LEO	11 ARI	3 LEO	24 TAU	LEO	TAU	7 VIR	20GEM
JUL	LEO	27 TAU	22 VIR	TAU	1 VIR	11GEM	29 LIB	GEM
AUG	10 VIR	TAU	VIR	7 GEM	19 LIB	25 CAN	LIB	3 CAN
SEP	26 LIB	TAU	7 LIB	28 CAN	LIB	CAN	14 SCO	18 LEO
OCT	LIB	TAU	23 SCO	CAN	3 SCO	16 LEO	28 SAG	LEO
NOV	10 SCO	TAU	SCO	CAN	16 SAG	LEO	SAG	10 VIR
DEC	25 SAG	TAU	5 SAG	31GEM	27 CAP	LEO	7 CAP	VIR

Mars Tables

	2012	2013	2014	2015	2016	2017	2018	2019
JAN	VIR	AQU	LIB	12 PIS	3 SCO	28 ARI	26 SAG	ARI
FEB	VIR	1 PIS	LIB	19 ARI	SCO	ARI	SAG	14 TAU
MAR	VIR	12 ARI	LIB	31 TAU	5 SAG	9 TAU	17 CAP	31 GEM
APR	VIR	20 TAU	LIB	TAU	SAG	21 GEM	CAP	GEM
MAY	VIR	31 GEM	LIB	11 GEM	27 SCO	GEM	15 AQU	15 CAN
JUN	VIR	GEM	LIB	24 CAN	SCO	4 CAN	AQU	CAN
JUL	3 LIB	13 CAN	25 SCO	CAN	SCO	20 LEO	AQU	1 LEO
AUG	23 SCO	27 LEO	SCO	8 LEO	2 SAG	LEO	12 CAP	17 VIR
SEP	SCO	LEO	13 SAG	24 VIR	27 CAP	5 VIR	10 AQU	VIR
OCT	6 SAG	15 VIR	26 CAP	VIR	CAP	22 LIB	AQU	3 LIB
NOV	16 CAP	VIR	CAP	12 LIB	9 AQU	LIB	15 PIS	19 SCO
DEC	25 AQU	7 LIB	4 AQU	LIB	19 PIS	9 SCO	31 ARI	SCO

Mars Tables

	2020	2021	2022	2023	2024	2025
JAN	3 SAG	6 TAU	24 CAP	GEM	4 CAP	6 CAN
FEB	16 CAP	TAU	CAP	GEM	13 AQU	CAN
MAR	30 AQU	3 GEM	6 AQU	25 CAN	22 PIS	CAN
APR	AQU	23 CAN	14 PIS	CAN	30 ARI	17 LEO
MAY	12 PIS	CAN	24 ARI	20 LEO	ARI	LEO
JUN	27 ARI	11 LEO	ARI	LEO	8 TAU	17 VIR
JUL	ARI	29 VIR	5 TAU	10 VIR	20 GEM	VIR
AUG	ARI	VIR	20 GEM	27 LIB	GEM	6 LIB
SEP	ARI	14 LIB	GEM	LIB	4 CAN	22 SCO
OCT	ARI	30 SCO	GEM	11 SCO	CAN	SCO
NOV	ARI	SCO	GEM	24 SAG	3 LEO	4 SCO
DEC	ARI	13 SAG	GEM	SAG	LEO	15 CAP

JUPITER TABLES

1900	SAG	1901	JAN 1 - JAN 18 SAG JAN 19 - DEC 31 CAP	1902	JAN 1 - FEB 6 CAP FEB 7 - DEC 31 AQU
1903	JAN 1 - FEB 19 AQU FEB 20 - DEC 31 PIS	1904	JAN 1 - FEB 29 PIS MAR 1 - AUG 8 ARI AUG 9 - AUG 31 TAU SEP 1 - DEC 31 ARI	1905	JAN 1 - MAR 7 ARI MAR 8 - JUN 20 TAU JUL 21 - DEC 4 GEM DEC 5 - DEC 31 TAU
1906	JAN 1 - MAR 9 TAU MAR 10 - JUN 30 GEM JUL 31 - DEC 31 CAN	1907	JAN 1 - AUG 18 CAN AUG 19 - DEC 31 LEO	1908	JAN 1 - SEP 11 LEO SEP 12 - DEC 31 VIR
1909	JAN 1 - OCT 11 VIR OCT 12 - DEC 31 LIB	1910	JAN 1 - OCT 11 VIR OCT 12 - DEC 31 LIB	1911	JAN 1 - DEC 2 SCO DEC 10 - DEC 31 SAG
1912	SAG	1913	JAN 1 - JAN 2 SAG JAN 3 - DEC 31 CAP	1914	JAN 1 - JAN 21 CAP JAN 22 - DEC 31 AQU
1915	JAN 1 - FEB 3 AQU FEB 4 - DEC 31 PIS	1916	JAN 1 - FEB 11 PIS FEB 12 - JUN 25 ARI JUN 26 - OCT 26 TAU OCT 27 - DEC 31 ARI	1917	JAN 1 - FEB 12 ARI FEB 13 - JUN 29 TAU JUN 30 - DEC 31 GEM
1918	JAN 1 - JUN 12 GEM JUL 13 - DEC 31 CAN	1919	JAN 1 - AUG 1 CAN AUG 2 - DEC 31 LEO	1920	JAN 1 - AUG 26 LEO AUG 27 - DEC 31 VIR
1921	JAN 1 - SEP 25 VIR SEP 26 - DEC 31 LIB	1922	JAN 1 - OCT 26 LIB OCT 27 - DEC 31 SCO	1923	JAN 1 - NOV 24 SCO NOV 26 - DEC 31 SAG
1923	JAN 1 - NOV 24 SCO NOV 25 - DEC 31 SAG	1924	JAN 1 - DEC 17 SAG DEC 18 - DEC 31 CAP	1925	CAP
1926	JAN 1 - JAN 5 CAP JAN 6 - DEC 31 AQU	1927	JAN 1 - JAN 17 AQU JAN 18 - JUN 5 PIS JUN 6 - SEP 10 ARI SEP 11 - DEC 31 PIS	1928	JAN 1 - JAN 22 PIS JAN 23 - JUN 3 ARI JUN 4 - DEC 31 TAU
1929	JAN 1 - JUN 11 TAU JUN 12 - DEC 12 GEM	1930	JAN 1 - JUN 26 GEM JUN 27 - DEC 31 CAN	1931	JAN 1 - JUL 16 CAN JUL 17 - DEC 31 LEO
1932	JAN 1 - AUG 10 LEO AUG 11 - DEC 31 VIR	1933	JAN 1 - SEP 9 VIR SEP 10 - DEC 31 LIB	1934	JAN 1 - OCT 10 LIB OCT 11 - DEC 31 SCO
1935	JAN 1 NOV 8 SCO NOV 9 - DEC 31 SAG	1936	JAN 1 - DEC 1 SAG DEC 2 - DEC 31 CAP	1937	JAN 1 - MAY 13 AQU MAY 14 - JUL 29 PIS JUL 30 - DEC 29 AQU DEC 30 - DEC 31 PIS
1939	JAN 1 - MAY 11 PIS MAY 12 - OCT 29 ARI OCT 30 - DEC 20 PIS DEC 21 - DEC 31 ARI	1940	JAN 1 - MAY 15 ARI MAY 16 - DEC 31 TAU	1941	JAN 1 - MAY 25 TAU MAY 26 - DEC 31 GEM
1942	JAN 1 - JUN 9 GEM JUN 10 - DEC 31 CAN	1943	JAN 1 - JUN 30 CAN JUL 1 - DEC 31 LEO	1944	JAN 1 - JUL 25 LEO JUL 26 - DEC 31 VIR
1945	JAN 1 - AUG 24 VIR AUG 25 - DEC 31 LIB	1946	JAN 1 - SEP 24 LIB SEP 25 - DEC 31 SCO	1947	JAN 1 - OCT 23 SCO OCT 24 - DEC 31 SAG
1949	JAN 1 - APR 12 CAP APR 13 - JUN 27 AQU JUN 28 - NOV 30 CAP DEC 1 - DEC 31 AQU	1950	JAN 1 - APR 14 AQU APR 15 - SEP 14 PIS SEP 15 - DEC 1 AQU DEC 2 - DEC 31 PIS	1951	JAN 1 - APR 21 PIS APR 22 - DEC 31 ARI

JUPITER TABLES

1952	JAN 1 - APR 28 ARI APR 29 - DEC 31 TAU	1953	JAN 1 - MAY 9 TAU MAY 10 - DEC 31 GEM	1954	JAN 1 - MAY 23 GEM MAY 24 - DEC 31 CAN
1955	JAN 1 - JUN 12 CAN JUN 13 - NOV 16 LEO NOV 17 - DEC 31 VIR	1956	JAN 1 - JAN 17 VIR JAN 18 - JUL 7 LEO JUL 8 - DEC 12 VIR	1957	JAN 1 - FEB 19 LIB FEB 20 - AUG 6 VIR AUG 7 - DEC 31 LIB
1958	JAN 1 - JAN 13 LIB JAN 14 - MAR 20 SCO MAR 21 - SEP 6 LIB SEP 7 - DEC 31 SCO	1959	JAN 1 - FEB 10 SCO FEB 11 - APR 24 SAG APR 25 - OCT 5 SCO OCT 6 - DEC 31 SAG	1960	JAN 1 - MAR 1 SAG MAR 2 - JUN 9 CAP JUN 10 - OCT 25 SAG OCT 26 - DEC 31 CAP
1961	JAN 1 - MAR 14 CAP MAR 15 - AUG11 AQU AUG 12 - NOV 3 CAP NOV 4 - DEC 31 AQU	1062	JAN 1 - MAR 25 AQU MAR 26 - DEC 31 PIS	1963	JAN 1 - APR 3 PIS APR 4 - DEC 31 ARI
1964	JAN 1 - APR 11 ARI APR 12 - DEC 31 TAU	1965	JAN 1 - APR 22 TAU APR 23 - SEP 20 GEM SEP 21 - NOV 16 CAN NOV 17 - DEC 31 GEM	1966	JAN 1 - MAY 5 GEM MAY 6 - SEP 27 CAN SEP 28 - DEC 31 LEO
1967	JAN 1 - JAN 15 LEO JAN 16 - MAY 22 CAN MAY 23 - OCT 18 LEO OCT 19 - DEC 31 VIR	1968	JAN 1 - FEB 26 VIR FEB 27 - JUN 15 LEO JUN 16 - NOV 15 VIR NOV 16 - DEC 31 LIB	1969	JAN 1 - MAR 30 LIB MAR 31 - JUL 15 VIR JUL 16 - DEC 16 LIB DEC 17 - DEC 31 SCO
1970	JAN 1 - APR 29 SCO APR 30 - AUG 15 LIB AUG 16 - DEC 31 SCO	1971	JAN 1 - JAN 13 SCO JAN 14 - JUN 4 SAG JUN 5 - SEP 11 SCO SEP 12 - DEC 31 SAG	1972	JAN 1 - FEB 6 SAG FEB 7 - JUL 24 CAP JUL 25 - SEP 25 SAG SEP 26 - DEC 31 CAP
1073	JAN 1 - FEB 22 CAP FEB 23 - DEC 31 AQU	1974	JAN 1 - MAR 7 AQU MAR 8 - DEC 31 AQU	1075	JAN 1 - MAR 18 PIS MAR 19 - DEC 31 ARI
1976	JAN 1 - MAR 25 ARI MAR 26 - AUG 22 TAU AUG 23 - OCT 16 GEM OCT 17 - DEC 31 TAU	1977	JAN 1 - APR 3 TAU APR 4 - AUG 20 GEM AUG 21 - DEC 30 CAN DEC 31 GEM	1978	JAN 1 - APR 11 GEM APR 12 - SEP 4 CAN SEP 5 - DEC 31 LEO
1979	JAN 1 - FEB 28 LEO MAR 1 - APR 19 CAN APR 20 - SEP 28 LEO SEP 29 - DEC 31 VIR	1980	JAN 1 - OCT 26 VIR OCT 27 - DEC 31 LIB	1981	JAN 1 - NOV 26 LIB NOV 27 - DEC 31 SCO
1982	JAN 1 - DEC 25 SCO DEC 26 - DEC 31 SAG	1983	SAG	1984	JAN 1 - JAN 19 SAG JAN 20 - DEC 31 CAP
1985	JAN 1 - FEB 6 CAP FEB 7 - DEC 31 AQU	1986	JAN 1 - FEB 20 AQU FEB 21 - DEC 31 PIS	1987	JAN 1 - MAR 2 PIS MAR 3 - DEC 31 ARI
1988	JAN 1 - MAR 8 ARI MAR 9 - JUL 21 TAU JUL 22 NOV 30 GEM DEC 1 - DEC 31 TAU	1989	JAN 1 - MAR 10 TAU MAR 11 - JUL 30 GEM JUL 31 - DEC 31 CAN	1990	JAN 1 - AUG 17 CAN AUG 18 - DEC 31 LEO
1991	JAN 1 - SEP 11 LEO SEP 12 - DEC 31 VIR	1992	JAN 1 - OCT 10 VIR OCT 11 - DEC 31 LIB	1993	JAN 1 - NOV 9 LIB NOV 10 - DEC 31 SCO
1994	JAN 1 - DEC 8 SCO DEC 9 - DEC 31 SAG	1995	SAG	1996	JAN 1 - JAN 2 SAG JAN 3 - DEC 31 CAP

JUPITER TABLES

1997	JAN 1 - JAN 21 CAP JAN 22 - DEC 31 AQU	1998	JAN 1 - FEB 3 AQU FEB 4 - DEC 31 PIS	1999	JAN 1 - FEB 12 PIS FEB 13 - JUN 27 ARI JUN 28 - OCT 22 TAU OCT 23 - DEC 31 ARI
2000	JAN 1 - FEB 14 ARI FEB 15 - JUN 30 TAU JUL 1 - DEC 31 GEM	2001	JAN 1 - JUL 12 GEM JUL 13 - DEC 31 CAN	2002	JAN 1 - AUG 1 CAN AUG 2 - DEC 31 LEO
2003	JAN 1 - AUG 27 LEO AUG 28 - DEC 31 VIR	2004	JAN 1 - SEP 25 VIR SEP 26 - DEC 31 LIB	2005	JAN 1 - OCT 25 LIB OCT 26 - DEC 31 SCO
2006	JAN 1 - NOV 23 SCO NOV 24 - DEC 31 SAG	2007	JAN 1 - DEC 18 SAG DEC 19 - DEC 31 CAP	2008	CAP
2009	JAN 1 - JAN 5 CAP JAN 6 - DEC 31 AQU	2010	JAN 1 - JAN 17 AQU JAN 18 - JUN 5 PIS JUN 6 - SEP 8 ARI SEP 8 - DEC 31 PIS	2011	JAN 1 - JAN 22 PIS JAN 23 - JUN 4 ARI JUN 5 - DEC 31 TAU
2012	JAN 1 - JUN 11 TAU JUN 12 - DEC 31 GEM	2013	JAN 1 - JUN 25 GEM JUN 26 - DEC 31 CAN	2014	JAN 1 - JUL 16 CAN JUL 17 - DEC31 LEO
2015	JAN 1 - AUG 11 LEO AUG 12 - DEC 31 VIR	2016	JAN 1 - SEP 9 VIR SEP 10 - DEC 31 LIB	2017	JAN 1 - OCT 10 LIB OCT 11 - DEC 31 SCO
2018	JAN 1 - NOV 8 SCO NOV 9 - DEC 31 SAG	2019	JAN 1 - DEC 2 SAG DEC 3 - DEC 31 CAP	2020	JAN 1 - DEC 19 CAP DEC 20 - DEC 31 AQU
2021	JAN 1 - MAY 13 AQU MAY 15 - JUL 27 PIS JUL 28 - DEC 28 AQU DEC 29 - DEC 31 PIS	2022	JAN 1 - MAY 10 PIS MAY 11 - OCT 27 ARI OCT 28 - DEC 20 PIS DEC 21 - DEC 31 ARI	2023	JAN 1 - MAY 16 ARI MAY 17 - DEC 31 TAU
2024	JAN 1- MAY 25 TAU MAY 26 - DEC31 GEM	2025	JAN 1 - JUN 9 GEM JUN 10 - DEC 31 CAN		

Saturn Tables

1900	JAN 1 - JAN 20 SAG JAN 21 - JUL 18 CAP JUL 19 - OCT 16 SAG OCT 17 - DEC 31 CAP	1901	CAP	1902	CAP
1903	JAN 1 - JAN 19 CAP JAN 20 - DEC 31 AQU	1904	AQU	1905	JAN 1 - APR 12 AQU APR 12 - AUG 16 PIS AUG 17 - DEC 31 AQU
1906	JAN 1 - JAN 7 AQU JAN 8 - DEC 31 PIS	1907	PIS	1908	JAN 1 - MAR 19 PIS MAR 20 - DEC 31 ARI
1909	ARI	1910	JAN 1 - MAY 16 ARI MAY 17 - DEC 14 TAU DEC 15 - DEC 31 ARI	1911	JAN 1 - JAN 19 ARI JAN 20 - DEC 31 TAU
1912	JAN 1 - JUL 6 TAU JUL 7 - NOV 30 GEM DEC 1 - DEC 31 TAU	1913	JAN 1 - MAR 25 TAU MAR 26 DEC 31 GEM	1914	JAN 1 - AUG 24 GEM AUG 25 - DEC 6 CAN DEC 7 - DEC 31 GEM
1915	JAN 1 - MAY 11 GEM MAY 12- DEC 31 CAN	1916	JAN 1 - OCT 17 CAN OCT 18 - DEC 7 LEO DEC 8 - DEC 31 CAN	1917	JAN 1 - JUN 24 CAN JUN 25 - DEC 31 LEO
1918	LEO	1919	JAN 1 - AUG 12 LEO AUG 13 - DEC 31 VIR	1920	VIR
1921	JAN 1 - OCT 7 VIR OCT 8 - DEC 31 LIB	1922	LIB	1923	JAN 1 - DEC 19 LIB DEC 20 - DEC 31 SCO
1924	JAN 1 - APR 5 SCO APR 6 - SEP 13 LIB SEP 14 - DEC 31 SCO	1925	SCO	1926	JAN 1 - DEC 2 SCO DEC 3 - DEC 31 SAG
1927	SAG	1928	SAG	1929	JAN 1 - MAR 15 SAG MAR 16 - MAY 5 CAP MAY 6 - NOV 29 SAG NOV 30 - DEC 31 CAP
1930	CAP	1931	CAP	1932	JAN 1 - FEB 23 CAP FEB 24 - AUG 12 AQU AUG 13 - NOV 19 CAP NOV 20 - DEC 31 AQU
1933	AQU	1934	AQU	1935	JAN 1 - FEB 14 AQU FEB 15- DEC 31 PIS
1936	PIS	1937	JAN 1 - APR 24 PIS ARP 25 - OCT 17 ARI OCT 18 - DEC31 PIS	1938	JAN 1- JAN 13 PIS JAN 14 - DEC 31 ARI
1939	JAN 1 - JUL 5 ARI JUL 6 - SEP 21 TAU SEP 22 - DEC 31 ARI	1940	JAN 1 - MAR 19 ARI MAR 20 - DEC 31 TAU	1941	TAU
1942	JAN 1 - MAY 8 TAU MAY 9 - DEC 31 GEM	1943	GEM	1944	JAN 1 - JUN 19 GEM JUN 20 - DEC 31 CAN
1945	CAN	1946	JAN 1 - AUG 2 CAN AUG 3 - DEC 31 LEO	1947	LEO
1948	JAN 1 - SEP 18 LEO SEP 19 - DEC 31 VIR	1949	JAN 1 - APR 2 VIR APR 3 - MAY 28 LEO MAY 29 - DEC 31 VIR	1950	JAN 1 - NOV 20 VIR NOV 21 - DEC 31 LIB

Saturn Tables

1951	JAN 1 - MAR 6 LIB MAR 7 - AUG 13 VIR AUG 14 - DEC 31 LIB	1952	LIB	1953	JAN 1 - OCT 22 LIB OCT 23 - DEC 31 SCO
1954	SCO	1955	SCO	1956	JAN 1 - JAN 12 SCO JAN 13 - MAY 13 SAG MAY 14 - OCT 10 SCO OCT 11 - DEC 31 SAG
1957	SAG	1958	SAG	1959	JAN 1 - JAN 5 SAG JAN 6 - DEC 31 CAP
1960	CAP	1061	CAP	1962	JAN 1 - JAN 3 CAP JAN 4 - DEC 31 AQU
1963	AQU	1964	JAN 1 - MAR 23 AQU MAR 24 - SEP 16 PIS SEP 17 - DEC 15 AQU DEC 16 - DEC 31 PIS	1965	PIS
1966	PIS	1967	JAN 1 - MAR 3 PIS MAR 4 - DEC 31 ARI	1968	ARI
1969	JAN 1 - APR 29 ARI APR 30 - DEC 31 TAU	1970	TAU	1971	JAN 1 - JUN 18 TAU JUN 19 - DEC 31 GEM
1072	JAN 1 - JAN 9 GEM JAN 10 - FEB 21 TAU FEB 22 - DEC 31 GEM	1973	JAN 1 - AUG 1 GEM AUG 2 - DEC 31 CAN	1074	JAN 1 - JAN 7 CAN JAN 8 - APR 18 GEM ARI 19 - -DEC 31 CAN
1975	JAN 1 - SEP 16 CAN SEP 17 - DEC 31 LEO	1976	JAN 1 - JAN 14 LEO JAN 15 - JUN 4 CAN JUN 5 - DEC 31 LEO	1977	JAN 1 - NOV 16 LEO NOV 17 - DEC 31 VIR
1978	JAN 1 - JAN 4 VIR JAN 5 - JUL 25 LEO JUL 26 - DEC 31 VIR SEP 29 - DEC 31 VIR	1979	VIR	1980	JAN 1 - SEP 20 VIR SEP 21 - DEC 31 LIB
1981	LIB	1982	JAN 1 - NOV 28 LIB NOV 29 - DEC 31 SCO	1983	JAN 1 - MAY 5 SCO MAY 7 - AUG 23 LIB AUG 24 - DEC 31 SCO
1984	SCO	1985	JAN 1 - NOV 16 SCO NOV 17 - DEC 31 SAG	1986	SAG
1987	SAG	1988	JAN 1 - FEB 13 SAG FEB 14 - JUN 9 CAP JUN 10 - NOV 11 SAG NOV 12 - DEC 31 CAP	1989	CAP
1990	CAP	1991	JAN 1 - FEB 6 CAP FEB 7 - DEC 31 AQU	1992	AQU
1993	JAN 1 - MAY 20 AQU MAY 21- JUN 29 PIS JUN 30 - DEC 31 AQU	1994	JAN 1 - JAN 28 AQU JAN 29 - DEC 31 PIS	1995	PIS
1996	JAN 1 - APR 6 PIS ARP 7 - DEC 31 ARI	1997	ARI	1998	JAN 1 - JUN 8 ARI JUN 9 - OCT 25 TAU OCT 26 - DEC 31 ARI
1999	JAN 1 - FEB 28 ARI MAR 1 - DEC 31 TAU	2000	TAU	2001	JAN 1 - APR 20 TAU APR 21 - DEC 31 GEM

Saturn Tables

2002	GEM	2003	JAN 1 - JUN 4 GEM JUN 5 - DEC 31 CAN	2004	CAN
2005	JAN 1 - JUL 16 CAN JUL 17 - DEC 31 LEO	2006	LEO	2007	JAN 1 - SEP 2 LEO SEP 3 - DEC 31 VIR
2008	VIR	2009	JAN 1 - OCT 29 VIR OCT 30 - DEC 31 LIB	2010	JAN 1 - APR 7 LIB APR 8 - JUL 21 VIR JUL 22- DEC 31 LIB
2011	LIB	2012	JAN 1 - OCT 5 LIB OCT 6 - DEC 31 SCO	2013	SCO
2014	JAN 1 - DEC 23 SCO DEC 24 - DEC 31 SAG	2015	JAN 1 - JUN 14 SAG JUN 15 - SEP 18 SCO SEP 19 - DEC 31 SAG	2016	SAG
2017	JAN 1 - DEC 19 SAG DEC 20 - DEC 31 CAP	2018	CAP	2019	CAP
2020	JAN 1 - MAR 21 CAP MAR 22 - JUL 1 AQU JUL 2 - DEC 17 CAP DEC 18 - DEC 31 AQU	2021	AQU	2022	AQU
2023	JAN 1- MAR 7 AQU MAR 8 - DEC 31 PIS	2024	PIS	2025	JAN 1 - MAY 24 PIS MAY 25 - AUG 31 ARI SEP 1 - DEC 31 PIS

Uranus Tables

1900	SAG	1901	SAG	1902	SAG
1903	SAG	1904	JAN 1 - DEC 19 SAG DEC 20 - DEC 31 CAP	1905	CAP
1906	CAP	1907	CAP	1908	CAP
1909	CAP	1910	CAP	1911	CAP
1912	JAN 1 - JAN 30 CAP JAN 31 - SEP 4 AQU SAP 5 - NOV 11 CAP NOV 12 - DEC 31 AQU	1913	AQU	1914	AQU
1915	AQU	1916	AQU	1917	AQU
1918	AQU	1919	JAN 1 - MAR 31 AQU APR 1 - AUG 16 PIS AUG 17 - DEC 31 AQU	1920	JAN 1 - JAN 22 AQU JAN 23 - DEC 31 PIS
1921	PIS	1922	PIS	1923	PIS
1924	PIS	1925	PIS	1926	PIS
1927	JAN 1 - MAR 31 PIS APR 1 - NOV 4 ARI NOV 5 - DEC 31 PIS	1928	JAN 1 - JAN 12 PIS JAN 13 - DEC 31 ARI	1929	ARI
1930	ARI	1931	ARI	1932	ARI
1933	ARI	1934	JAN 1 - JUN 6 ARI JUN 7 - OCT 9 TAU OCT 10 - DEC 31 ARI	1935	JAN 1 - MAR 27 ARI MAR 28- DEC 31 TAU
1936	TAU	1937	TAU	1938	TAU
1939	TAU	1940	TAU	1941	JAN 1 - AUG 7 TAU AUG 8 - OCT 4 GEM OCT 5 - DEC 31 TAU
1942	JAN 1 - MAY 14 TAU MAY 15 - DEC 31 GEM	1943	GEM	1944	GEM
1945	GEM	1946	GEM	1947	GEM
1948	JAN 1 - AUG 30 GEM AUG 31 - NOV 11 CAN NOV 12 - DEC 31 GEM	1949	JAN 1 - JUN 9 GEM JUN 10 - DEC 31 CAP	1950	CAN
1951	CAN	1952	CAN	1953	CAN
1954	CAN	1955	JAN 1 - AUG 24 CAN AUG 25 - DEC 31 LEO	1956	JAN 1 - JAN 27 LEO JAN 28 - JUN 9 CAN JUN 10 - DEC 31 LEO
1957	LEO	1958	LEO	1959	LEO
1960	LEO	1061	JAN 1 - NOV 1 LEO NOV 2 - DEC 31 VIR	1962	JAN 1 - JAN 9 VIR JAN 10 - AUG 9 LEO AUG 10 - DEC 31 VIR
1963	VIR	1964	VIR	1965	VIR
1966	VIR	1967	VIR	1968	JAN 1 - SEP 28 VIR SEP 29 - DEC 31 LIB
1969	JAN 1 - MAY 20 LIB MAY 21 - JUN 23 VIR JUN 24 - DEC 31 LIB	1970	LIB	1971	LIB

Uranus Tables

1072	LIB	1973	LIB	1974	JAN 1 - NOV 20 LIB NOV 21 - DEC 31 SCO
1975	JAN 1 - MAY 1 SCO MAY 2 - SEP 7 LIB SEP 8 - DEC 31 SCO	1976	SCO	1977	SCO
1978	SCO	1979	SCO	1980	SCO
1981	JAN 1 - FEB 16 SCO FEB 17 - MAR 20 SAG MAR 21 - NOV 15 SCO NOV 16 - DEC 31 SAG	1982	SAG	1983	SAG
1984	SAG	1985	SAG	1986	SAG
1987	SAG	1988	JAN 1 - FEB 13 SAG FEB 14 - MAY 26 CAP MAY 27 - DEC 2 SAG DEC 3 - DEC 31 CAP	1989	CAP
1990	CAP	1991	CAP	1992	CAP
1993	CAP	1994	CAP	1995	JAN 1 - MAR 31 CAP APR 1 - JUN 8 AQU JUN 9 - DEC 31 CAP
1996	JAN 1 - JAN 11 CAP JAN 12 - DEC 31 AQU	1997	AQU	1998	AQU
1999	AQU	2000	AQU	2001	AQU
2002	AQU	2003	JAN 1 - MAR 10 AQU MAR 11 - SEP 14 PIS SEP 15 - DEC 30 AQU DEC 31 PIS	2004	PIS
2005	PIS	2006	PIS	2007	PIS
2008	PIS	2009	PIS	2010	JAN 1 - MAY 28 PIS MAY 29 - AUG 13 ARI AUG 14 - DEC 31 PIS
2011	JAN 1 - MAR 11 PIS MAR 12 - DEC 31 ARI	2012	ARI	2013	ARI
2014	ARI	2015	ARI	2016	ARI
2017	ARI	2018	JAN 1 - MAY 15 ARI MAY 16 - NOV 5 TAU NOV 6 - DEC 31 ARI	2019	JAN 1 - MAR 5 ARI MAR 6 - DEC 31 TAU
2020	TAU	2021	TAU	2022	TAU
2023	TAU	2024	TAU	2025	JAN 1 - JUL 6 TAU JUL 7 - NOV 7 GEM NOV 8 - DEC 31 TAU

Neptune Tables

1900	GEM	1901	JAN 1 - JUL 19 GEM JUL 20 - DEC 24 CAN DEC 25 - DEC 31 GEM	1902	JAN 1 - MAY 20 GEM MAY 21 - DEC 31 CAN
1903	CAN	1904	CAN	1905	CAN
1906	CAN	1907	CAN	1908	CAN
1909	CAN	1910	CAN	1911	CAN
1912	CAN	1913	CAN	1914	JAN 1 - SEP 23 CAN SEP 24 - DEC 14 LEO DEC 15 - DEC 31 CAN
1915	JAN 1 - JUL 18 CAN JUL 19 - DEC 31 LEO	1916	JAN 1 - MAR 18 LEO MAR 19 - MAY 1 CAN MAY 2 - DEC 31 LEO	1917	LEO
1918	LEO	1919	LEO	1920	LEO
1921	LEO	1922	LEO	1923	LEO
1924	LEO	1925	LEO	1926	LEO
1927	LEO	1928	JAN 1 - SEP 20 LEO SEP 21 - DEC 31 VIR	1929	JAN 1 - FEB 18 VIR FEB 19 - JUL 23 LEO JUL 24 - DEC 31 VIR
1930	VIR	1931	VIR	1932	VIR
1933	VIR	1934	VIR	1935	VIR
1936	VIR	1937	VIR	1938	VIR
1939	VIR	1940	VIR	1941	VIR
1942	JAN 1 - OCT 3 VIR OCT 4 - DEC 31 LIB	1943	JAN 1 - APR 16 LIB APR 17 - AUG 2 VIR AUG 3 - DEC 31 LIB	1944	LIB
1945	LIB	1946	LIB	1947	LIB
1948	LIB	1949	LIB	1950	LIB
1951	LIB	1952	LIB	1953	LIB
1954	LIB	1955	JAN 1 - DEC 23 LIB DEC 24 - DEC 31 SCO	1956	JAN 1 - MAR 11 SCO MAR 12 - OCT 18 LIB OCT 19 - DEC 31 SCO
1957	JAN 1 - JUN 15 SCO JUN 16 - AUG 5 LIB AUG 6 - DEC 31 SCO	1958	SCO	1959	SCO
1960	SCO	1061	SCO	1962	SCO
1963	SCO	1964	SCO	1965	SCO
1966	SCO	1967	SCO	1968	SCO
1969	SCO	1970	JAN 1 - JAN 4 SCO JAN 5 - MAY 2 SAG MAY 3 - NOV 6 SCO NOV 7 - DEC 31 SAG	1971	SAG
1072	SAG	1973	SAG	1974	SAG
1975	SAG	1976	SAG	1977	SAG
1978	SAG	1979	SAG	1980	SAG
1981	SAG	1982	SAG	1983	SAG

Neptune Tables

1984	JAN 1 - JAN 18 SAG JAN 19 - JUN 22 CAP JUN 23 - NOV 20 SAG NOV 21 - DEC 31 CAP	1985	CAP	1986	CAP
1987	CAP	1988	CAP	1989	CAP
1990	CAP	1991	CAP	1992	CAP
1993	CAP	1994	CAP	1995	CAP
1996	CAP	1997	CAP	1998	JAN 1 - JAN 28 CAP JAN 29 - AUG 22 AQU AUG 23 - NOV 27 CAP NOV 28 - DEC 31 AQU
1999	AQU	2000	AQU	2001	AQU
2002	AQU	2003	AQU	2004	AQU
2005	AQU	2006	AQU	2007	AQU
2008	AQU	2009	AQU	2010	AQU
2011	JAN 1 - APR 4 AQU APR 5 - AUG 5 PIS AUG 6 - DEC 31 AQU	2012	JAN 1 - FEB 3 AQU FEB 4 - DEC 31 PIS	2013	PIS
2014	PIS	2015	PIS	2016	PIS
2017	PIS	2018	PIS	2019	PIS
2020	PIS	2021	PISA	2022	PIS
2023	PIS	2024	PIS	2025	JAN 1 - MAR 29 PIS MAR 30 - OCT 21ARI OCT 22 - DEC 31 PIS

PLUTO TABLES

1900	GEM	1901	GEM	1902	GEM
1903	GEM	1904	GEM	1905	GEM
1906	GEM	1907	GEM	1908	GEM
1909	GEM	1910	GEM	1911	GEM
1912	JAN 1 SEP 9 GEM SAP 10 - OCT 19 CAN OCT 20 - DEC 31 GEM	1913	JAN 1 - JUL 9 GEM JUL 10 - DEC 27 CAN DEC 28 - DEC 31 CAN	1914	JAN 1 - MAY 26 GEM MAY 27 - DEC 31 CAN
1915	CAN	1916	CAN	1917	CAN
1918	CAN	1919	CAN	1920	CAN
1921	CAN	1922	CAN	1923	CAN
1924	CAN	1925	CAN	1926	CAN
1927	CAN	1928	CAN	1929	CAN
1930	CAN	1931	CAN	1932	CAN
1933	CAN	1934	CAN	1935	CAN
1936	CAN	1937	JAN 1 - OCT 6 CAN OCT 7 - NOV 24 LEO NOV 25 - DEC 31 CAN	1938	JAN 1 - AUG 3 CAN AUG 4 - DEC 31 LEO
1939	JAN 1 - FEB 6 LEO FEB 7 - JUN 13 CAN JUN 14 - DEC 31 LEO	1940	LEO	1941	LEO
1942	LEO	1943	LEO	1944	LEO
1945	LEO	1946	LEO	1947	LEO
1948	LEO	1949	LEO	1950	LEO
1951	LEO	1952	LEO	1953	LEO
1954	LEO	1955	LEO	1956	JAN 1 - OCT 19 LEO OCT 20 - DEC 31 VIR
1957	JAN 1 - JAN 14 VIR JAN 15 - AUG 18 LEO AUG 19 - DEC 31 VIR	1958	JAN 1 - APR 10 VIR APR 11 - JUN 10 LEO JUN 11 - DEC 31 VIR	1959	VIR
1960	VIR	1061	VIR	1962	VIR
1963	VIR	1964	VIR	1965	VIR
1966	VIR	1967	VIR	1968	VIR
1969	VIR	1970	VIR	1971	JAN 1 - OCT 4 VIR OCT 5 - DEC 31 LIB
1072	JAN 1 - NOV 5 LIB NOV 6 - DEC 31 SCO	1973	LIB	1074	LIB
1975	LIB	1976	LIB	1977	LIB
1978	LIB	1979	LIB	1980	LIB
1981	LIB	1982	LIB	1983	JAN 1 - MAY 17 SCO MAY 18 - AUG 27 LIB AUG 28 - DEC 31 SCO
1984	JAN 1 - MAY 17 SCO MAY 18 - AUG 27 LIB AUG 28 - DEC 31 SCO	1985	SCO	1986	SCO
1987	SCO	1988	SCO	1989	SCO
1990	SCO	1991	SCO	1992	SCO

Pluto Tables

1993	SCO	1994	SCO	1995	JAN 1 - JAN 16 SCO JAN 17 - APR 21 SAG APR 22 - NOV 9 SCO NOV 10 - DEC 31 SAG
1996	SAG	1997	SAG	1998	SAG
1999	SAG	2000	SAG	2001	SAG
2002	SAG	2003	SAG	2004	SAG
2005	SAG	2006	SAG	2007	SAG
2008	JAN 1 - JAN 26 SAG JAN 27 - JUN 14 CAP JUN 15 - NOV 27 SAG NOV 28 - DEC 31 CAP	2009	CAP	2010	CAP
2011	CAP	2012	CAP	2013	CAP
2014	CAP	2015	CAP	2016	CAP
2017	CAP	2018	CAP	2019	CAP
2020	CAP	2021	CAP	2022	CAP
2023	JAN 1 - MAR 23 CAP MAR 24 - JUN 11 AQU JUN 12 - DEC 31 CAP	2024	JAN 1 - JAN 20 CAP JAN 21 - AUG 31 AQU SEP 1 - NOV 19 CAP NOV 20 - DEC 31 AQU	2025	AQU

Sun Tables

	1900	1901	1902	1903	1904	1905	1906	1907
JAN	20 AQU	20AQU	20AQU	21AQU	21AQU	20AQU	20AQU	20AQU
FEB	18 PIS	19 PIS	19 PIS	19 PIS	19 PIS	19 PIS	19 PIS	19 PIS
MAR	20 ARI	21 ARI	21 ARI	21 ARI	20 ARI	21 ARI	21 ARI	21 ARI
APR	20 TAU	20 TAU	20 TAU	21 TAU	20 TAU	20 TAU	20 TAU	21 TAU
MAY	21GEM	21GEM	21GEM	22GEM	21GEM	21GEM	21GEM	22GEM
JUN	21 CAN	21 CAN	22 CAN	22 CAN	21 CAN	21 CAN	22 CAN	22 CAN
JUL	23 LEO	23 LEO	23 LEO	23 LEO	23 LEO	23 LEO	23 LEO	23 LEO
AUG	23 VIR	23 VIR	23 VIR	24 VIR	23 VIR	23 VIR	23 VIR	24 VIR
SEP	23 LIB	23 LIB	23 LIB	24 LIB	23 LIB	23 LIB	23 LIB	23 LIB
OCT	23 SCO	23 SCO	24 SCO	24 SCO	23 SCO	23 SCO	24 SCO	24 SCO
NOV	22 SAG	22 SAG	23 SAG	23 SAG	22 SAG	22 SAG	22 SAG	23 SAG
DEC	22 CAP	22 CAP	22 CAP	22 CAP	22 CAP	22 CAP	22 CAP	22 CAP

Sun Tables

	1908	1909	1910	1911	1912	1913	1914	1915
JAN	21 AQU	20 AQU	20 AQU	20 AQU	21 AQU	20 AQU	20 AQU	20 AQU
FEB	19 PIS	19 PIS	19 PIS	19 PIS	19 PIS	19 PIS	19 PIS	19 PIS
MAR	20 ARI	21 ARI	21 ARI	21 ARI	20 ARI	21 ARI	21 ARI	21 ARI
APR	20 TAU	20 TAU	20 TAU	21 TAU	20 TAU	20 TAU	20 TAU	20 TAU
MAY	21GEM	21GEM	21GEM	22GEM	21GEM	21GEM	21GEM	21GEM
JUN	21 CAN	21 CAN	22 CAN	22 CAN	21 CAN	21 CAN	22 CAN	22 CAN
JUL	23 LEO	23 LEO	23 LEO	23 LEO	23 LEO	23 LEO	23 LEO	23 LEO
AUG	23 VIR	23 VIR	23 VIR	24 VIR	23 VIR	23 VIR	23 VIR	24 VIR
SEP	23 LIB	23 LIB	23 LIB	23 LIB	23 LIB	23 LIB	23 LIB	23 LIB
OCT	23 SCO	23 SCO	24 SCO	24 SCO	23 SCO	23 SCO	24 SCO	24 SCO
NOV	22 SAG	22 SAG	22 SAG	23 SAG	22 SAG	22 SAG	22 SAG	23 SAG
DEC	22 CAP	22 CAP	22 CAP	22 CAP	21 CAP	22 CAP	22 CAP	22 CAP

Sun Tables

	1916	1917	1918	1919	1920	1921	1922	1923
JAN	21 AQU	20 AQU	20 AQU	20 AQU	21 AQU	20 AQU	20 AQU	20 AQU
FEB	19 PIS	19 PIS	19 PIS	19 PIS	19 PIS	18 PIS	19 PIS	19 PIS
MAR	20 ARI	20 ARI	21 ARI	21 ARI	20 ARI	20 ARI	21 ARI	21 ARI
APR	20 TAU	20 TAU	20 TAU	21 TAU	20 TAU	20 TAU	20 TAU	20 TAU
MAY	21GEM	21GEM	21GEM	22GEM	21GEM	21GEM	21GEM	21GEM
JUN	21 CAN	21 CAN	22 CAN	22 CAN	21 CAN	21 CAN	22 CAN	22 CAN
JUL	23 LEO	23 LEO	23 LEO	23 LEO	22LEO	23 LEO	23 LEO	23 LEO
AUG	23 VIR	23 VIR	23 VIR	24 VIR	23 VIR	23 VIR	23 VIR	23 VIR
SEP	23 LIB	23 LIB	23 LIB	23 LIB	23 LIB	23 LIB	23 LIB	23 LIB
OCT	23 SCO	23 SCO	24 SCO	24 SCO	23 SCO	23 SCO	23 SCO	24 SCO
NOV	22 SAG	22 SAG	22 SAG	23 SAG	22 SAG	22 SAG	22 SAG	23 SAG
DEC	21 CAP	22 CAP	22 CAP	22 CAP	21 CAP	22 CAP	22 CAP	22 CAP

Sun Tables

	1924	1925	1926	1927	1928	1929	1930	1931
JAN	21 AQU	20 AQU	20 AQU	20 AQU	21 AQU	20 AQU	20 AQU	20 AQU
FEB	19 PIS	18 PIS	19 PIS	19 PIS	19 PIS	18 PIS	19 PIS	19 PIS
MAR	20 ARI	20 ARI	21 ARI	21 ARI	20 ARI	20 ARI	21 ARI	21 ARI
APR	20 TAU	20 TAU	20 TAU	20 TAU	20 TAU	20 TAU	20 TAU	20 TAU
MAY	21GEM	21GEM	21GEM	21GEM	21GEM	21GEM	21GEM	21GEM
JUN	21 CAN	21 CAN	21 CAN	22 CAN	21 CAN	21 CAN	22 CAN	22 CAN
JUL	22 LEO	23 LEO	23 LEO	23 LEO	22 LEO	23 LEO	23 LEO	23 LEO
AUG	23 VIR	23 VIR	23 VIR	23 VIR	23 VIR	23 VIR	23 VIR	23 VIR
SEP	23 LIB	23 LIB	23 LIB	23 LIB	23 LIB	23 LIB	23 LIB	23 LIB
OCT	23 SCO	23 SCO	23 SCO	24 SCO	23 SCO	23 SCO	23 SCO	24 SCO
NOV	22 SAG	22 SAG	22 SAG	23 SAG	22 SAG	22 SAG	22 SAG	23 SAG
DEC	21 CAP	22 CAP	22 CAP	22 CAP	21 CAP	22 CAP	22 CAP	22 CAP

Sun Tables

	1932	1933	1934	1935	1936	1937	1938	1939
JAN	21 AQU	20 AQU	20 AQU	20 AQU	21 AQU	20 AQU	20 AQU	20 AQU
FEB	19 PIS	18 PIS	19 PIS	19 PIS	19 PIS	18 PIS	19 PIS	19 PIS
MAR	20 ARI	20 ARI	21 ARI	21 ARI	20 ARI	20 ARI	21 ARI	21 ARI
APR	20 TAU	20 TAU	20 TAU	20 TAU	20 TAU	20 TAU	20 TAU	20 TAU
MAY	21GEM	21GEM	21GEM	21GEM	21GEM	21GEM	21GEM	21GEM
JUN	21 CAN	21 CAN	21 CAN	22 CAN	21 CAN	21 CAN	21 CAN	22 CAN
JUL	22 LEO	23 LEO	23 LEO	23 LEO	22 LEO	23 LEO	23 LEO	23 LEO
AUG	23 VIR	23 VIR	23 VIR	23 VIR	23 VIR	23 VIR	23 VIR	23 VIR
SEP	23 LIB	23 LIB	23 LIB	23 LIB	23 LIB	23 LIB	23 LIB	23 LIB
OCT	23 SCO	23 SCO	23 SCO	24 SCO	23 SCO	23 SCO	23 SCO	24 SCO
NOV	22 SAG	22 SAG	22 SAG	23 SAG	22 SAG	22 SAG	22 SAG	22 SAG
DEC	21 CAP	22 CAP	22 CAP	22 CAP	21 CAP	22 CAP	22 CAP	22 CAP

Sun Tables

	1940	1941	1942	1943	1944	1945	1946	1947
JAN	20 AQU	20 AQU	20 AQU	20 AQU	20 AQU	20 AQU	20 AQU	20 AQU
FEB	19 PIS	18 PIS	19 PIS	19 PIS	19 PIS	18 PIS	19 PIS	19 PIS
MAR	20 ARI	20 ARI	21 ARI	21 ARI	20 ARI	20 ARI	21 ARI	21 ARI
APR	20 TAU	20 TAU	20 TAU	20 TAU	20 TAU	20 TAU	20 TAU	20 TAU
MAY	21GEM	21GEM	21GEM	21GEM	20GEM	21GEM	21GEM	21GEM
JUN	21 CAN	21 CAN	21 CAN	22 CAN	21 CAN	21 CAN	21 CAN	22 CAN
JUL	22 LEO	23 LEO	23 LEO	23 LEO	22 LEO	23 LEO	23 LEO	23 LEO
AUG	23 VIR	23 VIR	23 VIR	23 VIR	23 VIR	23 VIR	23 VIR	23 VIR
SEP	22 LIB	23 LIB	23 LIB	23 LIB	22 LIB	23 LIB	23 LIB	23 LIB
OCT	23 SCO	23 SCO	23 SCO	24 SCO	23 SCO	23 SCO	23 SCO	24 SCO
NOV	22 SAG	22 SAG	22 SAG	22 SAG	22 SAG	22 SAG	22 SAG	22 SAG
DEC	21 CAP	22 CAP	22 CAP	22 CAP	21 CAP	21 CAP	22 CAP	22 CAP

Sun Tables

	1948	1949	1950	1951	1952	1953	1954	1955
JAN	20 AQU	20 AQU	20 AQU	20 AQU	20 AQU	20 AQU	20 AQU	20 AQU
FEB	19 PIS	18 PIS	19 PIS	19 PIS	19 PIS	18 PIS	18 PIS	19 PIS
MAR	20 ARI	20 ARI	20 ARI	21 ARI	20 ARI	20 ARI	20 ARI	21 ARI
APR	19 TAU	20 TAU	20 TAU	20 TAU	19 TAU	20 TAU	20 TAU	20 TAU
MAY	20GEM	21GEM	21GEM	21GEM	20GEM	21GEM	21GEM	21GEM
JUN	21 CAN	21 CAN	21 CAN	22 CAN	21 CAN	21 CAN	21 CAN	21 CAN
JUL	22 LEO	22 LEO	23 LEO	23 LEO	22 LEO	22 LEO	23 LEO	23 LEO
AUG	23 VIR	23 VIR	23 VIR	23 VIR	22 VIR	23 VIR	23 VIR	23 VIR
SEP	22 LIB	23 LIB	23 LIB	23 LIB	22 LIB	23 LIB	23 LIB	23 LIB
OCT	23 SCO	23 SCO	23 SCO	24 SCO	23 SCO	23 SCO	23 SCO	23 SCO
NOV	22 SAG	22 SAG	22 SAG	22 SAG	22 SAG	22 SAG	22 SAG	22 SAG
DEC	21 CAP	21 CAP	22 CAP	22 CAP	21 CAP	21 CAP	22 CAP	22 CAP

Sun Tables

	1956	1957	1958	1959	1960	1961	1962	1963
JAN	20 AQU	20 AQU	20 AQU	20 AQU	20 AQU	20 AQU	20 AQU	20 AQU
FEB	19 PIS	18 PIS	18 PIS	19 PIS	19 PIS	18 PIS	18 PIS	19 PIS
MAR	20 ARI	20 ARI	20 ARI	21 ARI	20 ARI	20 ARI	20 ARI	21 ARI
APR	19 TAU	20 TAU	20 TAU	20 TAU	19 TAU	20 TAU	20 TAU	20 TAU
MAY	20GEM	21GEM	21GEM	21GEM	20GEM	21GEM	21GEM	21GEM
JUN	21 CAN	21 CAN	21 CAN	21 CAN	21 CAN	21 CAN	21 CAN	21 CAN
JUL	22 LEO	22 LEO	23 LEO	23 LEO	22 LEO	22 LEO	23 LEO	23 LEO
AUG	22 VIR	23 VIR	23 VIR	23 VIR	22 VIR	23 VIR	23 VIR	23 VIR
SEP	22 LIB	23 LIB	23 LIB	23 LIB	22 LIB	23 LIB	23 LIB	23 LIB
OCT	23 SCO	23 SCO	23 SCO	23 SCO	23 SCO	23 SCO	23 SCO	23 SCO
NOV	22 SAG	22 SAG	22 SAG	22 SAG	22 SAG	22 SAG	22 SAG	22 SAG
DEC	21 CAP	21 CAP	22 CAP	22 CAP	21 CAP	21 CAP	22 CAP	22 CAP

Sun Tables

	1964	1965	1966	1967	1968	1969	1970	1971
JAN	20 AQU	20 AQU	20 AQU	20 AQU	20 AQU	20 AQU	20 AQU	20 AQU
FEB	19 PIS	18 PIS	18 PIS	19 PIS	19 PIS	18 PIS	18 PIS	19 PIS
MAR	20 ARI	20 ARI	20 ARI	21 ARI	20 ARI	20 ARI	20 ARI	21 ARI
APR	19 TAU	20 TAU	20 TAU	20 TAU	19 TAU	20 TAU	20 TAU	20 TAU
MAY	20GEM	21GEM	21GEM	21GEM	20GEM	21GEM	21GEM	21GEM
JUN	21 CAN	21 CAN	21 CAN	21 CAN	21 CAN	21 CAN	21 CAN	21 CAN
JUL	22 LEO	22 LEO	23 LEO	23 LEO	22 LEO	22 LEO	23 LEO	23 LEO
AUG	22 VIR	23 VIR	23 VIR	23 VIR	22 VIR	23 VIR	23 VIR	23 VIR
SEP	22 LIB	23 LIB	23 LIB	23 LIB	22 LIB	23 LIB	23 LIB	23 LIB
OCT	23 SCO	23 SCO	23 SCO	23 SCO	23 SCO	23 SCO	23 SCO	23 SCO
NOV	22 SAG	22 SAG	22 SAG	22 SAG	22 SAG	22 SAG	22 SAG	22 SAG
DEC	21 CAP	21 CAP	22 CAP	22 CAP	21 CAP	21 CAP	22 CAP	22 CAP

Sun Tables

	1972	1973	1974	1975	1976	1977	1978	1979
JAN	20 AQU	19 AQU	20 AQU	20 AQU	20 AQU	19 AQU	20 AQU	20 AQU
FEB	19 PIS	18 PIS	18 PIS	19 PIS	19 PIS	18 PIS	18 PIS	19 PIS
MAR	20 ARI	20 ARI	20 ARI	21 ARI	20 ARI	20 ARI	20 ARI	21 ARI
APR	19 TAU	20 TAU	20 TAU	20 TAU	19 TAU	19 TAU	20 TAU	20 TAU
MAY	20GEM	20GEM	21GEM	21GEM	20GEM	20GEM	21GEM	21GEM
JUN	21 CAN	21 CAN	21 CAN	21 CAN	21 CAN	21 CAN	21 CAN	21 CAN
JUL	22 LEO	22 LEO	23 LEO	23 LEO	22 LEO	22 LEO	22 LEO	23 LEO
AUG	22 VIR	23 VIR	23 VIR	23 VIR	22 VIR	23 VIR	23 VIR	23 VIR
SEP	22 LIB	23 LIB	23 LIB	23 LIB	22 LIB	22 LIB	23 LIB	23 LIB
OCT	23 SCO	23 SCO	23 SCO	23 SCO	23 SCO	23 SCO	23 SCO	23 SCO
NOV	21 SAG	22 SAG	22 SAG	22 SAG	21 SAG	22 SAG	22 SAG	22 SAG
DEC	21 CAP	21 CAP	22 CAP	22 CAP	21 CAP	21 CAP	22 CAP	22 CAP

Sun Tables

	1980	1981	1982	1983	1984	1985	1986	1987
JAN	20 AQU	19 AQU	20 AQU	20 AQU	20 AQU	19 AQU	20 AQU	20 AQU
FEB	19 PIS	18 PIS	18 PIS	19 PIS	19 PIS	18 PIS	18 PIS	18 PIS
MAR	20 ARI	20 ARI	20 ARI	20 ARI	20 ARI	20 ARI	20 ARI	20 ARI
APR	19 TAU	19 TAU	20 TAU	20 TAU	19 TAU	19 TAU	20 TAU	20 TAU
MAY	20GEM	20GEM	21GEM	21GEM	20GEM	20GEM	21GEM	21GEM
JUN	21 CAN	21 CAN	21 CAN	21 CAN	20 CAN	21 CAN	21 CAN	21 CAN
JUL	22 LEO	22 LEO	22 LEO	23 LEO	22 LEO	22 LEO	22 LEO	23 LEO
AUG	22 VIR	23 VIR	23 VIR	23 VIR	22 VIR	22 VIR	23 VIR	23 VIR
SEP	22 LIB	22 LIB	23 LIB	23 LIB	22 LIB	22 LIB	23 LIB	23 LIB
OCT	23 SCO	23 SCO	23 SCO	23 SCO	23 SCO	23 SCO	23 SCO	23 SCO
NOV	21 SAG	22 SAG	22 SAG	22 SAG	21 SAG	22 SAG	22 SAG	22 SAG
DEC	21 CAP	21 CAP	21 CAP	22 CAP	21 CAP	21 CAP	21 CAP	22 CAP

Sun Tables

	1988	1989	1990	1991	1992	1993	1994	1995
JAN	20 AQU	19 AQU	20 AQU	20 AQU	20 AQU	19 AQU	20 AQU	20 AQU
FEB	19 PIS	18 PIS	18 PIS	18 PIS	19 PIS	18 PIS	18 PIS	18 PIS
MAR	20 ARI	20 ARI	20 ARI	20 ARI	20 ARI	20 ARI	20 ARI	20 ARI
APR	19 TAU	19 TAU	20 TAU	20 TAU	19 TAU	19 TAU	20 TAU	20 TAU
MAY	20GEM	20GEM	21GEM	21GEM	20GEM	20GEM	21GEM	21GEM
JUN	20 CAN	21 CAN	21 CAN	21 CAN	20 CAN	21 CAN	21 CAN	21 CAN
JUL	22 LEO	22 LEO	22 LEO	23 LEO	22 LEO	22 LEO	22 LEO	23 LEO
AUG	22 VIR	22 VIR	23 VIR	23 VIR	22 VIR	22 VIR	23 VIR	23 VIR
SEP	22 LIB	22 LIB	23 LIB	23 LIB	22 LIB	22 LIB	23 LIB	23 LIB
OCT	22 SCO	23 SCO	23 SCO	23 SCO	23 SCO	23 SCO	23 SCO	23 SCO
NOV	21 SAG	22 SAG	22 SAG	22 SAG	21 SAG	22 SAG	22 SAG	22 SAG
DEC	21 CAP	21 CAP	21 CAP	22 CAP	21 CAP	21 CAP	21 CAP	22 CAP

Sun Tables

	1996	1997	1998	1999	2000	2001	2002	2003
JAN	20 AQU	19 AQU	20 AQU	20 AQU	20 AQU	19 AQU	20 AQU	20 AQU
FEB	19 PIS	18 PIS	18 PIS	18 PIS	19 PIS	18 PIS	18 PIS	18 PIS
MAR	20 ARI	20 ARI	20 ARI	20 ARI	20 ARI	20 ARI	20 ARI	20 ARI
APR	19 TAU	19 TAU	20 TAU	20 TAU	19 TAU	19 TAU	20 TAU	20 TAU
MAY	20GEM	20GEM	21GEM	21GEM	20GEM	20GEM	21GEM	21GEM
JUN	20 CAN	21 CAN	21 CAN	21 CAN	20 CAN	21 CAN	21 CAN	21 CAN
JUL	22 LEO	22 LEO	22 LEO	23 LEO	22 LEO	22 LEO	22 LEO	23 LEO
AUG	22 VIR	22 VIR	23 VIR	23 VIR	22 VIR	22 VIR	23 VIR	23 VIR
SEP	22 LIB	22 LIB	23 LIB	23 LIB	22 LIB	22 LIB	22 LIB	23 LIB
OCT	22 SCO	23 SCO	23 SCO	23 SCO	22 SCO	23 SCO	23 SCO	23 SCO
NOV	21 SAG	22 SAG	22 SAG	22 SAG	21 SAG	22 SAG	22 SAG	22 SAG
DEC	21 CAP	21 CAP	21 CAP	22 CAP	21 CAP	21 CAP	21 CAP	22 CAP

Sun Tables

	2004	2005	2006	2007	2008	2009	2010	2011
JAN	20 AQU	19 AQU	19 AQU	20 AQU	20 AQU	19 AQU	19 AQU	20 AQU
FEB	19 PIS	18 PIS	18 PIS	18 PIS	19 PIS	18 PIS	18 PIS	18 PIS
MAR	20 ARI	20 ARI	20 ARI	20 ARI	20 ARI	20 ARI	20 ARI	20 ARI
APR	19 TAU	19 TAU	20 TAU	20 TAU	19 TAU	19 TAU	19 TAU	20 TAU
MAY	20GEM	20GEM	20GEM	21GEM	20GEM	20GEM	20GEM	21GEM
JUN	20 CAN	21 CAN	21 CAN	21 CAN	20 CAN	21 CAN	21 CAN	21 CAN
JUL	22 LEO	22 LEO	22 LEO	22 LEO	22 LEO	22 LEO	22 LEO	22 LEO
AUG	22 VIR	22 VIR	23 VIR	23 VIR	22 VIR	22 VIR	23 VIR	23 VIR
SEP	22 LIB	22 LIB	22 LIB	23 LIB	22 LIB	22 LIB	22 LIB	23 LIB
OCT	22 SCO	23 SCO	23 SCO	23 SCO	22 SCO	23 SCO	23 SCO	23 SCO
NOV	21 SAG	22 SAG	22 SAG	22 SAG	21 SAG	21 SAG	22 SAG	22 SAG
DEC	21 CAP	21 CAP	21 CAP	22 CAP	21 CAP	21 CAP	21 CAP	22 CAP

Sun Tables

	2012	2013	2014	2015	2016	2017	2018	2019
JAN	20 AQU	19 AQU	19 AQU	20 AQU	20 AQU	19 AQU	19 AQU	20 AQU
FEB	19 PIS	18 PIS	18 PIS	18 PIS	19 PIS	18 PIS	18 PIS	18 PIS
MAR	19 ARI	20 ARI	20 ARI	20 ARI	19 ARI	20 ARI	20 ARI	20 ARI
APR	19 TAU	19 TAU	19 TAU	20 TAU	19 TAU	19 TAU	19 TAU	20 TAU
MAY	20GEM	20GEM	20GEM	21GEM	20GEM	20GEM	20GEM	21GEM
JUN	20 CAN	20 CAN	21 CAN	21 CAN	20 CAN	20 CAN	21 CAN	21 CAN
JUL	22 LEO	22 LEO	22 LEO	22 LEO	22 LEO	22 LEO	22 LEO	22 LEO
AUG	22 VIR	22 VIR	22 VIR	23 VIR	22 VIR	22 VIR	23 VIR	23 VIR
SEP	22 LIB	22 LIB	22 LIB	23 LIB	22 LIB	22 LIB	22 LIB	23 LIB
OCT	22 SCO	23 SCO	23 SCO	23 SCO	22 SCO	23 SCO	23 SCO	23 SCO
NOV	21 SAG	21 SAG	22 SAG	22 SAG	21 SAG	21 SAG	22 SAG	22 SAG
DEC	21 CAP	21 CAP	21 CAP	21 CAP	21 CAP	21 CAP	21 CAP	21 CAP

Sun Tables

	2020	2021	2022	2023	2024	2025
JAN	20 AQU	19 AQU	19 AQU	20 AQU	20 AQU	19 AQU
FEB	18 PIS	18 PIS	18 PIS	18 PIS	18 PIS	18 PIS
MAR	19 ARI	20 ARI	20 ARI	20 ARI	19 ARI	20 ARI
APR	19 TAU	19 TAU	19 TAU	20 TAU	19 TAU	19 TAU
MAY	20GEM	20GEM	20GEM	21GEM	20GEM	20GEM
JUN	20 CAN	20 CAN	21 CAN	21 CAN	20 CAN	20 CAN
JUL	22 LEO	22 LEO	22 LEO	22 LEO	22 LEO	22 LEO
AUG	22 VIR	22 VIR	22 VIR	23 VIR	22 VIR	22 VIR
SEP	22 LIB	22 LIB	22 LIB	23 LIB	22 LIB	22 LIB
OCT	22 SCO	22 SCO	23 SCO	23 SCO	22 SCO	22 SCO
NOV	21 SAG	21 SAG	22 SAG	22 SAG	21 SAG	21 SAG
DEC	21 CAP	21 CAP	21 CAP	21 CAP	21 CAP	21 CAP

Rising Signs A.M. Births

	1 AM	2 AM	3 AM	4 AM	5 AM	6 AM
JAN 1	LIB	SCO	SCO	SCO	SAG	SAG
JAN 9	LIB	SCO	SCO	SAG	SAG	SAG
JAN 17	SCO	SCO	SCO	SAG	SAG	CAP
JAN 25	SCO	SCO	SAG	SAG	SAG	CAP
FEB 2	SCO	SCO	SAG	SAG	CAP	CAP
FEB 10	SCO	SAG	SAG	SAG	CAP	CAP
FEB 18	SCO	SAG	SAG	CAP	CAP	AQU
FEB 26	SAG	SAG	SAG	CAP	AQU	AQU
MAR 6	SAG	SAG	CAP	CAP	AQU	PIS
MAR 14	SAG	CAP	CAP	AQU	AQU	PIS
MAR 22	SAG	CAP	CAP	AQU	PIS	ARI
MAR 30	CAP	CAP	AQU	PIS	PIS	ARI
APR 7	CAP	CAP	AQU	PIS	ARI	ARI
APR 14	CAP	AQU	AQU	PIS	ARI	TAU
APR 22	CAP	AQU	PIS	ARI	ARI	TAU
APR 30	AQU	AQU	PIS	ARI	TAU	TAU
MAY 8	AQU	PIS	ARI	ARI	TAU	GEM
MAY 16	AQU	PIS	ARI	TAU	GEM	GEM
MAY 24	PIS	ARI	ARI	TAU	GEM	GEM
JUN 1	PIS	ARI	TAU	GEM	GEM	CAN
JUN 9	ARI	ARI	TAU	GEM	GEM	CAN
JUN 17	ARI	TAU	GEM	GEM	CAN	CAN
JUN 25	TAU	TAU	GEM	GEM	CAN	CAN
JULY 3	TAU	GEM	GEM	CAN	CAN	CAN
JULY 11	TAU	GEM	GEM	CAN	CAN	LEO
JULY 18	GEM	GEM	CAN	CAN	CAN	LEO
JULY 26	GEM	GEM	CAN	CAN	LEO	LEO
AUG 3	GEM	CAN	CAN	CAN	LEO	LEO
AUG 11	GEM	CAN	CAN	LEO	LEO	LEO
AUG 18	CAN	CAN	CAN	LEO	LEO	VIR
AUG 17	CAN	CAN	LEO	LEO	LEO	VIR
SEP 4	CAN	CAN	LEO	LEO	LEO	VIR
SEP 12	CAN	LEO	LEO	LEO	VIR	VIR
SEP 20	LEO	LEO	LEO	VIR	VIR	VIR
SEP 28	LEO	LEO	LEO	VIR	VIR	LIB
OCT 6	LEO	LEO	VIR	VIR	VIR	LIB
OCT 14	LEO	VIR	VIR	VIR	LIB	LIB
OCT 22	LEO	VIR	VIR	LIB	LIB	LIB

Rising Signs A.M. Births

	1 AM	2 AM	3 AM	4 AM	5 AM	6 AM
OCT 30	VIR	VIR	VIR	LIB	LIB	SCO
NOV 7	VIR	VIR	LIB	LIB	LIB	SCO
NOV 15	VIR	VIR	LIB	LIB	SCO	SCO
NOV 23	VIR	LIB	LIB	LIB	SCO	SCO
DEC 1	VIR	LIB	LIB	SCO	SCO	SCO
DEC 9	LIB	LIB	LIB	SCO	SCO	SAG
DEC 18	LIB	LIB	SCO	SCO	SCO	SAG
DEC 28	LIB	LIB	SCO	SCO	SAG	SAG

Rising Signs A.M. Births

	7 AM	8 AM	9 AM	10 AM	11 AM	12 NOON
JAN 1	CAP	CAP	AQU	AQU	PIS	ARI
JAN 9	CAP	CAP	AQU	PIS	ARI	TAU
JAN 17	CAP	AQU	AQU	PIS	ARI	TAU
JAN 25	CAP	AQU	PIS	ARI	TAU	TAU
FEB 2	AQU	PIS	PIS	ARI	TAU	GEM
FEB 10	AQU	PIS	ARI	TAU	TAU	GEM
FEB 18	PIS	PIS	ARI	TAU	GEM	GEM
FEB 26	PIS	ARI	TAU	TAU	GEM	GEM
MAR 6	PIS	ARI	TAU	GEM	GEM	CAP
MAR 14	ARI	TAU	TAU	GEM	GEM	CAN
MAR 22	ARI	TAU	GEM	GEM	CAN	CAN
MAR 30	TAU	TAU	GEM	CAN	CAN	CAN
APR 7	TAU	GEM	GEM	CAN	CAN	LEO
APR 14	TAU	GEM	GEM	CAN	CAN	LEO
APR 22	GEM	GEM	GEM	CAN	LEO	LEO
APR 30	GEM	CAN	CAN	CAN	LEO	LEO
MAY 8	GEM	CAN	CAN	LEO	LEO	LEO
MAY 16	CAN	CAN	CAN	LEO	LEO	VIR
MAY 24	CAN	CAN	LEO	LEO	LEO	VRI
JUN 1	CAN	CAN	LEO	LEO	VIR	VIR
JUN 9	CAN	LEO	LEO	LEO	VIR	VIR
JUN 17	CAN	LEO	LEO	VIR	VIR	VIR
JUN 25	LEO	LEO	LEO	VIR	VIR	LIB
JULY 3	LEO	LEO	VIR	VIR	VIR	LIB
JULY 11	LEO	LEO	VIR	VIR	LIB	LIB
JULY 18	LEO	VIR	VIR	VIR	LIB	LIB
JULY 26	VIR	VIR	VIR	LIB	LIB	LIB
AUG 3	VIR	VIR	VIR	LIB	LIB	SCO

Rising Signs A.M. Births

	7 AM	8 AM	9 AM	10 AM	11 AM	12 NOON
AUG 18	VIR	VIR	LIB	LIB	SCO	SCO
AUG 27	VIR	LIB	LIB	LIB	SCO	SCO
SEP 4	VIR	VIR	LIB	LIB	SCO	SCO
SEP 12	LIB	LIB	LIB	SCO	SCO	SAG
SEP 20	LIB	LIB	SCO	SCO	SCO	SAG
SEP 28	LIB	LIB	SCO	SCO	SAG	SAG
OCT 6	LIB	SCO	SCO	SCO	SAG	SAG
OCT 14	LIB	SCO	SCO	SAG	SAG	CAP
OCT 22	SCO	SCO	SCO	SAG	SAG	CAP
OCT 30	SCO	SCO	SAG	SAG	CAP	CAP
NOV 7	SCO	SCO	SAG	SAG	CAP	CAP
NOV 15	SCO	SAG	SAG	CAP	CAP	AQU
NOV 23	SAG	SAG	SAG	CAP	CAP	AQU
DEC 1	SAG	SAG	CAP	CAP	AQU	AQU
DEC 9	SAG	SAG	CAP	CAP	AQU	PIS
DEC 18	SAG	CAP	CAP	AQU	AQU	PIS
DEC 28	SAG	CAP	AQU	AQU	PIS	ARI

Rising Signs P.M. Births

	1 PM	2 PM	3 AM	4 PM	5 PM	6 PM
JAN 1	TAU	GEM	GEM	CAN	CAN	CAN
JAN 9	TAU	GEM	GEM	CAN	CAN	LEO
JAN 17	GEM	GEM	CAN	CAN	CAN	LEO
JAN 25	GEM	GEM	CAN	CAN	LEO	LEO
FEB 2	GEM	CAN	CAN	CAN	LEO	LEO
FEB 10	GEM	CAN	CAN	LEO	LEO	LEO
FEB 18	CAN	CAN	CAN	LEO	LEO	VIR
FEB 26	CAN	CAN	LEO	LEO	LEO	VIR
MAR 6	CAN	LEO	LEO	LEO	VIR	VIR
MAR 14	CAN	LEO	LEO	VIR	VIR	VIR
MAR 22	LEO	LEO	LEO	VIR	VIR	LIB
MAR 30	LEO	LEO	VIR	VIR	VIR	LIB
APR 7	LEO	LEO	VIR	VIR	LIB	LIB
APR 14	LEO	VIR	VIR	VIR	LIB	LIB
APR 22	LEO	VIR	VIR	LIB	LIB	LIB
APR 30	VIR	VIR	VIR	LIB	LIB	SCO
MAY 8	VIR	VIR	LIB	LIB	LIB	SCO
MAY 16	VIR	VIR	LIB	LIB	SCO	SCO
MAY 24	VIR	LIB	LIB	LIB	SCO	SCO
JUN 1	VIR	LIB	LIB	SCO	SCO	SCO
JUN 9	LIB	LIB	LIB	SCO	SCO	SAG
JUN 17	LIB	LIB	SCO	SCO	SCO	SAG
JUN 25	LIB	LIB	SCO	SCO	SAG	SAG
JULY 3	LIB	SCO	SCO	SCO	SAG	SAG
JULY 11	LIB	SCO	SCO	SAG	SAG	SAG
JULY 18	SCO	SCO	SCO	SAG	SAG	CAP
JULY 26	SCO	SCO	SAG	SAG	SAG	CAP
AUG 3	SCO	SCO	SAG	SAG	CAP	CAP
AUG 11	SCO	SAG	SAG	SAG	CAP	CAP
AUG 18	SCO	SAG	SAG	CAP	CAP	AQU
AUG 27	SAG	SAG	SAG	CAP	CAP	AQU
SEP 4	SAG	SAG	CAP	CAP	AQU	PIS
SEP 12	SAG	SAG	CAP	AQU	AQU	PIS
SEP 20	SAG	CAP	CAP	AQU	PIS	PIS
SEP 28	CAP	CAP	AQU	AQU	PIS	ARI
OCT 6	CAP	CAP	AQU	PIS	ARI	ARI
OCT 14	CAP	AQU	AQU	PIS	ARI	TAU
OCT 22	CAP	AQU	PIS	ARI	ARI	TAU

413

Rising Signs P.M. Births

	1 PM	2 PM	3 PM	4 PM	5 PM	6 PM
NOV 7	AQU	AQU	PIS	ARI	TAU	TAU
NOV 15	AQU	PIS	ARI	TAU	GEM	GEM
NOV 23	PIS	ARI	ATI	TAU	GEM	GEM
DEC 1	PIS	ARI	TAU	GEM	GEM	CAN
DEC 9	ARI	TAU	TAU	GEM	GEM	CAN
DEC 18	ARI	TAU	GEM	GEM	CAN	CAN
DEC 28	TAU	TAU	GEM	GEM	CAN	CAN

RISING SIGNS P.M. BIRTHS

	7 PM	8 PM	9 PM	10 PM	11 PM	12 MID-NIGHT
JAN 1	LEO	LEO	VIR	VIR	VIR	LIB
JAN 9	LEO	LEO	VIR	VIR	VIR	LIB
JAN 17	LEO	VIR	VIR	VIR	LIB	LIB
JAN 25	LEO	VIR	VIR	LIB	LIB	LIB
FEB 2	VIR	VIR	VIR	LIB	LIB	SCO
FEB 10	VIR	VIR	LIB	LIB	LIB	SCO
FEB 18	VIR	VIR	LIB	LIB	SCO	SCO
FEB 26	VIR	LIB	LIB	LIB	SCO	SCO
MAR 6	VIR	LIB	LIB	SCO	SCO	SCO
MAR 14	LIB	LIB	LIB	SCO	SCO	SAG
MAR 22	LIB	LIB	CSO	SCO	SCO	SAG
MAR 30	LIB	SCO	SCO	SCO	SAG	SAG
APR 7	LIB	SCO	SCO	SCO	SAG	SAG
APR 14	SCO	SCO	SCO	SAG	SAG	CAP
APR 22	SCO	SCO	SCO	SAG	SAG	CAP
APR 30	SCO	SCO	SAG	SAG	CAP	CAP
MAY 8	SCO	SAG	SAG	SAG	CAP	CAP
MAY 16	SCO	SAG	SAG	CAP	CAP	AQU
MAY 24	SAG	SAG	SAG	CAP	CAP	AQU
JUN 1	SAG	SAG	CAP	CAP	AQU	AQU
JUN 9	SAG	SAG	CAP	CAP	AQU	PIS
JUN 17	SAG	CAP	CAP	AQU	AQU	PIS
JUN 25	SAG	CAP	CAP	AQU	PIS	ARI
JULY 3	CAP	CAP	AQU	AQU	PIS	ARI
JULY 11	CAP	CAP	AQU	PIS	ARI	TAU
JULY 18	CAP	AQU	AQU	PIS	ARI	TAU
JULY 26	CAP	AQU	PIS	ARI	TAU	TAU

Rising Signs P.M. Births

	7 PM	8 PM	9 PM	10 PM	11 PM	12 MID-NIGHT
AUG 3	AQU	AQU	PIS	ARI	TAU	GEM
AUG 11	AQU	PIS	ARI	TAU	TAU	GEM
AUG 18	PIS	PIS	ARI	TAU	GEM	GEM
AUG 27	PIS	ARI	TAU	TAU	GEM	GEM
SEP 4	PIS	ARI	TAU	GEM	GEM	CAN
SEP 12	ARI	TAU	TAU	GEM	GRM	CAN
SEP 20	ARI	TAU	GEM	GEM	CAN	CAN
SEP 28	TAU	TAU	GEM	GEM	CAN	CAN
OCT 6	TAU	GEM	GEM	CAN	CAN	LEO
OCT 14	TAU	GEM	GEM	CAN	CAN	LEO
OCT 22	GEM	GEM	CAN	CAN	LEO	LEO
OCT 30	GEM	CAN	CAN	CAN	LEO	LEO
NOV 7	GEM	CAN	CAN	CAN	LEO	LEO
NOV 15	CAN	CAN	CAN	LEO	LEO	VIR
NOV 23	CAN	CAN	LEO	LEO	LEO	VIR
DEC 1	CAN	CAN	LEO	LEO	VIR	VIR
DEC 9	CAN	LEO	LEO	LEO	VIR	VIR
DEC 18	CAN	LEO	LEO	VIR	VIR	VIR
DEC 28	LEO	LEO	VIR	VIR	VIR	LIB

Date With Destiny

Share With Me Your Fantasy

DATE WITH DESTINY

SHARE WITH ME YOUR FANTASY

Numerology - What Do Numbers Tell About You?

There are certain vibrations in the letters of your name and in the date of birth, which could influence you life and destiny. First, let's find the' numerical sign'. The science of numbers is the oldest magic science in the world. Next, you have a table with the numerical values from the mathematician and philosopher Pythagoras. Here is how you can find your number!.

1	2	3	4	5	6	7	8	9
A	B	C	D	E	F	G	H	I
J	K	L	M	N	O	P	Q	R
S	T	U	V	W	X	Y	Z	

Let's suppose that your name is Loredana Elena Balu and that you were born in **11/27/1972**. Write the corresponding numbers under your name, that is: **36954151 53551 2133**. Add up all these numbers. The result is **62**. Then you add up **6 + 2 = 8**. If this sum is also a two digit number, add these two digits together. You were born under this number. If you are married you can use that name also. Use the name that you feel closer to.

Let's see where your destiny leads you! Using your date of birth you can determine your destiny or, in other words, your life partner. Using the example above (**11/27/1972**), add up all these numbers - **1 + 1 + 2 + 7 + 7 + 2 = 30**. Since **3 + 0 = 3**, your destiny is number **3**.

ONE - THE LEADER

This number has the following symbols: the Sun and the Sword. Number one means first. You are a leader, or you consider yourself a leader. You think you are in the center of the Universe. You are strong, and you can influence everybody around you. You have the talent to lead. You are ambitious, authoritarian, and always in a hurry. Sometimes you make decisions too fast, and as a result, you might make mistakes. If you make mistakes, you push things to their limits just to show that you are right. You are stubborn, and you never give in. You have to win. You are aggressive; you are the warrior type. It's good for you to be busy in a sport. You have to select a career or position in life that brings you satisfaction with the responsibility and authority that you have. Your destiny is to be a leader, but if you are obliged to be a subordinate you will become bitter and rude, and you will not collaborate with the others. You may also win a great deal of money, but you will spend it fast. You try to do your best in understanding other people socially and romantically. You can get along with number 3, who is charming; with number 6, who is obedient; and with number 9, who loves you. Be careful, don't be too bossy!

YOUR ROAD leads you toward the top of the mountain; once there, you should stay there. Never give up! This road requires ambition, strong will, and determination!

TWO - GOOD AND QUIET

This number has the following symbols: The Book and the Moon. This number is characterized by good balance and good health. It can also be a sign of poverty and death. You are generous and a diplomat. Your mind is very clear and bright. You are such an optimist that you forget your own interest and so generous that you might lose everything. But this doesn't bother you much. You are attracted to people who have problems. You love to give an good advice. Tolerant and understanding, you forgive a lot. Some people will take advantage of you. You are very friendly, full of imagination, and you love peace. Sometimes you are lazy; you accept undesirable conditions without complaining. You adapt to any situation. You don't seek power. In a difficult situation you can become a dreamer, looking for shelter in the misty world of imagination. You never aim very high. Two is an emotional number. Your partners should not hurt your feelings. You will get along well with the numbers 4, 7, and 8. Be careful about the numbers 6 and 9, who are also emotional. If you run into these numbers, the journey is not going to be very pleasant.

YOUR ROAD winds smoothly along the green meadows without ups and downs. This number is well balanced and quiet, never pushing himself ahead of others.

Three - Smart

This number has the following symbols: the Triangle and Mars. The past, present, and future play an important role in your life. You are sociable, attractive, convincing, charming, and have influence upon the people around you. You adapt easily to any situation because you have a quick mind, you are intuitive, and you are a native psychologist. In any situation, you choose the best way to solve the problem. You have talent, and you are gifted in any field of work you choose. You struggle to satisfy everybody around you; you do your best to entertain others, and it's hard for you to break relationships. Number 3 lives in the present, never regretting the past and never thinking about the future. You have a tendency to buy on credit and to spend beyond your means. You are an optimist; you know how to take advantage of any opportunity. You are inpatient, and you cannot stand dealing with the every day troubles of life. You cannot be careless about your duties. You are very independent, and you have a lot of ideas. It is hard for you to be obedient. You can become too proud and without scruples. The number 3 has to make some efforts to understand the emotional signs. You get along well with numbers 1, 5, 6, and 9.

YOUR ROAD is well built running here and there, going round obstacles, and adapting to certain situations. You are clever and flexible.

Four - Solidarity

This number has the following symbols: the Square and Mercury, and represents the four seasons and the four elements. This is the number of solidarity. You are honest and loyal. Anybody can count on you, but you are not very happy. Entertainment and leisure is not for you. You take life very seriously, and you are attracted by convention. You are conservative, and you enjoy family life, being faithful to its obligations and duties. You don't really enjoy new ideas without studying them for awhile. You are methodical, spending a lot of time before deciding what to do, and your plans are very detailed. You appreciate receiving high honors, and they make you very happy. You appreciate stability and security in life. Sometimes you underestimate yourself. Number 4 can also become the symbol of poverty and defeat. You always fear the future, and you hate gambling and taking risks - unless you are well rewarded. You have no one particular talent, and that's why you have to work hard to succeed. You are good in any activity that requires precession, method, and observation. Number 4 is a rational number, and he gets along well with numbers 2, 7, and 8.

YOUR ROAD is the main road. It is straight; it has many traffic signs and speed restrictions. It is the road of hard work and duty. It does not lead to danger and misfortune.

Five - Adventure

All risks are for number 5, regardless of the game. You are an adventurer; you love any unexpected change in your life; you love new experiences and the unknown. You are right most of the time, and you are annoyingly lucky. You are a revolutionary under the sign of Jupiter. You take advantage of all opportunities. You feel at home everywhere, and you have a real talent for foreign languages. You hate to be bored, so you look for obstacles and dangers. You are independent, very active, and practical. You are not very educated, but you have a lot of life experiences. You love to live your life fully with your friends, and you can do a lot for them. However, you do not accept their advice; it is too boring for you. Nothing is monotonous around you; you always look for excitement in a variety of directions. Nothing scares you. You can count on your luck no matter what happens. Sometimes you get very angry, and you might even become dangerous to the people around you. You succeed in any job that implies risk. Banks fascinate you. You are intuitive and will yield to the emotional signs in order to get along with them. Number 5 will associate with the following numbers: 1, 3, 6, and 9, but they should be careful because you will quickly get tired of them. To maintain the relationship, they will have to surprise you frequently.

Your Road is abrupt and has ups and downs. It is full of surprises, and of course, it is not boring at all. You need a lot of good luck to pass all dangers and barriers.

Six - Good Heart

This number has the following symbols: the 6 colors of the rainbow and Venus. This sign means attraction, harmony, and trust. You like nice things because you are sensitive. You are also a little shy and naive. You love children, animals, nature, flowers, and birds. You are always happy, nice, open, and always changing your ideas. You love to be appreciated, and you also love receiving honors - even if they are minor. You hate to argue when your friend's opinion is different from yours. You are easily influenced by someone else, and you have a hard time making up your own mind. Only the people closest to you might be right, everybody else is wrong. People might think you are hypocritical. Sometimes you are; you agree in order to avoid arguing. This diplomacy might be due to fear. You are very sensitive; you hurt many times; and you might cry. You will succeed in life because you are an artist, but if you aren't lucky, it means that you lack ambition. You may be lucky in politics because you are very conventional. Your closest friends are numbers: 1, 3, 5, and 9.

Your Road is a smooth, winding path along a quiet river with plenty of resting places. You are happy and patient. This is the path of duty and altruism.

Seven - Mystery

There are 7 Cardinal points, 7 planets and 7 days of the week. The number 7 is a magical number. It is the number of poetry and mystery. It is under the influence of Saturn, and it means knowledge and meditation. You have a universal intelligence, an exceptional intuition, and an active imagination that can lead you to success. You are not open, you don't say much. You like to be by yourself. You hate mediocrity and take offense to your partners' vulgarity and bad actions. You want to be above everybody else. Material issues are not a problem for you. You will be good as a monk or philosopher. If you reach the top, you will be isolated. You have good intuition and clairvoyance, but you lack tact. This might bring you trouble. You should have no connection with 2, 6, or 9, but you can get along fine with 1, 3, 4, and 8.

Your Road is scattered with crosses and cannibalistic signs. It is up to you to interpret all these. The road goes through thick forests, and there are many places good for meditation.

Eight - Money

This number is under the influence of Uranus. You never change direction, you are straight in your thoughts and actions, and nobody can stop you. If you come to a wall that you cannot climb, you break it down. You are strong, and you know how to make people love you. You have an excellent memory, and you are highly resilient. You don't have patience for vague ideas. You have a tendency of taking advantage of your friends. Sometimes they take advantage of you as well. You have magic in yourself in making money. You have to be careful, though, or you might go to jail. Your best companion is number 2 because he or she brings you balance. You also feel secure with number 4 and number 7.

Your Road is wide and functional. Walking on it, you should be careful. You are efficient and willing to keep on going. Your fate will pay you for your effort.

Nine - Genius

This number is under the influence of Neptune. The symbol is the vast sea. You are strong, and you have a strong personality. You have genius, intelligence, and a great energy for creation. You will get anything you wish for, and you will always want more because you are ambitious. You are honest, good, generous, and optimistic. You are also wise, and this will keep you away from troubles and extremes. You might become very proud, arrogant, and preoccupied with yourself. Stop admiring and thinking of yourself so highly while ignoring others. You can do well in art, industry, or science. People tend to admire you and appreciate you, attributing to you qualities that you might not in reality have. You are like a magnet; you are controlled by the stars. Avoid number 2, who is very emotional, and number 6, who likes your spirit and flexibility. Number 9s will get along very well with number 1, will love number 3's charm, and will become number 5's accomplice.

YOUR ROAD is privileged and protected and has lots of priorities. It leads you everywhere - though fields and oceans, seas and mountains - and nothing stops it. This is truly the successful road.

LIGIA BALU
www.ligiabalu.com

ASTROLOGICAL BOOKS:

A COMPLETE COMPILATION of ancient and modern reflections
on the individual zodiac signs. Learn how to chart your **PERSONAL DESTINY**, understand your **SEXUALITY**, discover your **COMPATIBILITY WITH YOUR SOUL-MATE**, determine the **DESTINY OF YOUR SOUL-MATE**, find your **STARS,** and gain other information about your **FUTURE. Complete astrological information on Love Signs, Sun Signs, Moon Signs, Planets, Houses, Numerology**, as well as the complete **Astrological Tables for the Years from 1900 to 2025 and instruction on how to cast your very own Chart**.

Order the book for your specific Zodiac sign or the COMPLETE ASTROLOGY book series which contains information on *ALL TWELVE SIGNS* - from **ARIES** to **PISCES**. Each book contain over 550 pages.

INDIVIDUAL ZODIAC BOOKS

Individual Love Signs, Sun Signs, Moon Signs, Planets, Houses, Astrological Tables, Numerology, Relationships With Other Signs and a lot more in each book.

ASTROLOGY - ARIES
HOW TO FIND YOUR SOUL-MATE, STARS AND DESTINY
ISBN: 0-9651186-2-2 Price: $49.95

ASTROLOGY - TAURUS
HOW TO FIND YOUR SOUL-MATE, STARS AND DESTINY
ISBN: 0-9651186-3-0 Price: $49.95

Astrology - Gemini

HOW TO FIND YOUR SOUL-MATE, STARS AND DESTINY
ISBN: 0-9651186-4-9   Price: $49.95

Astrology - Cancer

HOW TO FIND YOUR SOUL-MATE, STARS AND DESTINY
ISBN: 0-9651186-5-7   Price: $49.95

Astrology - Leo

HOW TO FIND YOUR SOUL-MATE, STARS AND DESTINY
ISBN: 0-9651186-6-5   Price: $49.95

Astrology - Virgo

HOW TO FIND YOUR SOUL-MATE, STARS AND DESTINY
ISBN: 0-9651186-7-3   Price: $49.95

Astrology - Libra

HOW TO FIND YOUR SOUL-MATE, STARS AND DESTINY
ISBN: 0-9651186-8-1   Price: $49.95

Astrology - Scorpio

HOW TO FIND YOUR SOUL-MATE, STARS AND DESTINY
ISBN: 0-9651186-9-X   Price: $49.95

Astrology - Sagittarius

HOW TO FIND YOUR SOUL-MATE, STARS AND DESTINY
ISBN: 1-892530-00-7   Price: $49.95

Astrology - Capricorn

HOW TO FIND YOUR SOUL-MATE, STARS AND DESTINY
ISBN: 1-892530-02-3 Price: $49.95

Astrology - Aquarius

HOW TO FIND YOUR SOUL-MATE, STARS AND DESTINY
ISBN: 1-892530-01-5 Price: $49.95

Astrology - Pisces

HOW TO FIND YOUR SOUL-MATE, STARS AND DESTINY
ISBN: 1-892530-03-1 Price: $49.95

True - Stories:

Believe In Your Dreams, Not In Your Fears

Is the inspiring and unforgettable story of a young girl who possessed an **INDOMITABLE SPIRIT** that r5efused to give up no matter what life gave her. I*n her own brutally honest words, **Ligia Balu** paints a haunting picture of her life. She tells how she survived not only the **ABUSE** of her family, but also a man who took **ADVANTAGE** of her, and finally a government that wanted to **IMPRISION** her. **RISKING HER LIFE**, she **ESCAPED** two communist countries and finally found freedom for herself and her daughters in America. "Running from the **WOLVES** and finding the **BIG BAD BEARS**"
ISBN: 0-9651186-0-6

American Dream Made Me Cry and Scream

Is Ligia Balu's INCREDIBLE TRUE STORY of arriving in America and finding the AMERICAN DREAM THAT MADE HER CRY and SCREAM? Once in America, Ligia became ENTRAPPED and VICTIMIZED by the POWERFULL BUREAUCRACY that RAPED, BETRAYED. EXTORTED, and ROBBED.. She was abused by the greed fro money of the BUREAUCRACTIC JUNGLE. She was DISCRIMINATED against, and her CHARACTER was ASSASSINATED. The PAIN, SUFFERING, MANIPULATION, DISAPPOINTMENT, and ISOLATION inflicted upon her by the justice system and her legal advisors were unbelievable. She found that only those who have the power have the rights and the freedom. What kind of system would robe her DREAM and make her CRY and SCREAM? You have to read the book to believe it.\
ISBN: 1-892530-04-X

Adult Romantic Fiction:

Talk Dirty To Me is a collection of light-hearted adult SATRICAL

FANTASIES of MODER N SEXUALITY. **Each story is filled with WILD, UNIHIBITED, and INTIMATE SEXUAL DISIRES and PLEAURES.**
ISBN: 1-892530-05-8   Price: $39.99

Order Online

To order or learn more about Ligia Balu's books, visit www.ligiabalu.com

www.ingramcontent.com/pod-product-compliance
Lightning Source LLC
Chambersburg PA
CBHW062024210326
41519CB00060B/6980